Handbook of
Practical Electrical
Design

Handbook of Practical Electrical Design

J. F. McPARTLAND
Editorial Director
McGraw-Hill Electrical Construction Communications Center
New York, N.Y.

Conforms to Prevailing National Electrical Code^R Requirements

assistant editors

Brian J. McPartland
Engineer
Leviton Mfg. Co., Inc.
Little Neck, N.Y.

James L. McPartland
Sales Engineer
Klockner-Moeller Corp.
Fairfield, N.J.

Steven P. McPartland
Applications Engineer
SDM, Electronic Marketing Reps, Inc.
Lodi, N.J.

Brendan A. McPartland
Foreman
Tenafly Electric Service
Tenafly, N.J.

McGRAW-HILL BOOK COMPANY
New York   St. Louis   San Francisco   Auckland
Bogotá   Hamburg   Johannesburg   London   Madrid
Mexico   Montreal   New Delhi   Panama   Paris
São Paulo   Singapore   Sydney   Tokyo   Toronto

Library of Congress Cataloging in Publication Data

McPartland, Joseph F.
    Handbook of practical electrical design.

    Includes index.
    1.Electric wiring—Handbooks, manuals, etc.
I.Title.
TK3271.M35   1983   621.31   82-20798
ISBN 0-07-045695-X

 234567890   KGP/KGP   89210987654

ISBN 0-07-045695-X

The editors for this book were Harold B. Crawford, Edward M. Millman, Peggy
Lamb, and Earle Resnick, the designer was Mark E. Safran, and the production
supervisor was Teresa F. Leaden. It was set in Baskerville by University Graph-
ics, Inc.

Printed and bound by The Kingsport Press.

Contents

ABOUT THE AUTHOR

Joseph F. McPartland is the Editorial Director of McGraw-Hill's Electrical Construction Communications Center, which produces *Electrical Construction and Maintenance, Electrical Construction and Maintenance Products Yearbook, Electrical Wholesaling,* and *Electrical Marketing Newsletter.* He has authored and coauthored a wide variety of books on electrical design, electrical construction methods, and electrical equipment. His most notable work is *McGraw-Hill's NATIONAL ELECTRICAL CODE® HANDBOOK* published every three years in conjunction with CODE® release. Throughout his career he has been teaching classes for electrical contractors, consulting engineers, plant electrical people, and electrical inspectors. In addition he has produced several audio and video programs of electrical study and presents courses many times a year throughout the United States. Mr. McPartland received a Bachelor of Science degree in Electrical Engineering from the Thayer School of Engineering, Dartmouth.

Preface

"How to design electrical systems" could be the subtitle for this handbook. It is directed to electrical personnel and also to persons in mechanical disciplines—heating, ventilating, air conditioning, manufacturing, processing—who are involved with related electrical applications. This handbook has been prepared as a guide and reference for all designers of electrical systems. It addresses the everyday needs of electrical designers—consulting engineers, electrical and mechanical contractors, plant personnel, architects, electricians, and maintenance personnel. The book explains, in simple terms, the practical procedures involved in the design of electrical circuits and systems to supply lighting motors, appliances, machines, fans, air conditioning, heating, signals and controls in industrial, commercial, and residential applications.

Various phases of electrical design are covered in books, booklets, and magazine articles, but this handbook is a unified and complete treatment of the art of electrical design. It treats all facets of the design of electrical systems, from specific electrical loads and the circuits that serve them to connection to main power-supply lines, and relates the many parts to the whole. To satisfy the ever-increasing need for modern electrical design data, accepted practices are covered thoroughly, and design trends which have particular merit are emphasized. The presentation is based on sound engineering principles and intelligent conformity to the safety provisions of the **National Electrical Code,** and due recognition is given to substantial spare capacity as an essential element of modern electrical design.

The material is presented in the order in which design normally proceeds. First, general considerations relative to the contemplated system and its design are identified. Then, the detailed design of the system begins with provision of the branch circuits to supply the loads directly. From the nature and extent of the branch circuiting, the required system of feeders and subfeeders is developed. Next, the overall distribution scheme is planned. And, finally, the details of the energy supply to the entire system are developed.

Step-by-step details are presented on the sizing, selection, and application of conductors, raceways, switches, fuses, circuit breakers, motor starters, panel-boards, switchboards, motor control centers, transformers, contactors, relays, wiring devices, lighting fixtures, and other related system components. In addition, a very wide variety of practical pointers, illustrations, and examples are provided. All the design information is presented so as to be readily translated into a set of electrical diagrams and sketches—called "the plans"— which constitute the designer's instructions to the electrical installer. To complement these plans, a set of electrical specifications should be prepared. Together, the plans and specifications must fully and clearly instruct the installer in constructing the overall electrical system to achieve the design intent and implement the engineering concepts.

J. F. McPartland

1

Planning for Electrical Design

Every electrical design project should begin with careful planning based on four vital requirements. The designer must:

1. Ensure conformity to *applicable codes and standards*.
2. Study and establish the *electrical needs of the building*.
3. Determine the *characteristics of the energy supply* to the overall system.
4. Scale details of the overall electrical system to meet the *limitations of budgeted funds*.

The design of an electrical system for any building or outdoor area is basically a matter of providing an arrangement of conductors and equipment to safely and effectively transfer electric energy from a source of power to lamps, motors, and other functional devices which operate on electricity. This simple task is readily reduced to three basic steps which set the outline for the detailed design of any electrical system:

1. Select basic wiring concepts and configurations which will supply electric power of the required characteristics at each point of utilization.
2. Implement the electrical circuiting concepts with actual conductors, apparatus, and hardware, selecting types, sizes, models, characteristics, appearance, ratings, and other specifics of the required equipment.
3. Account for the installation of the overall electrical system, as determined in the first two steps, within the physical dimensions and structural makeup of a building, showing, as clearly as possible, the locations and details of equipment mountings, raceway runs, connections to main power-supply lines, and other elements that require special attention.

Of course, these three steps are interrelated, and particular decisions made within any one step will affect corresponding elements within either or both of the other steps.

As can be seen from the three steps, the design of an electrical system is expressed in the form of electrical plans. All phases of a design, including the design of subsystems within the major system, should be reduced to a set of blueprints which present schematic wiring diagrams, single-line diagrams, full-wiring hookups, riser diagrams, isometric and other sketches, detail drawings, and equipment schedules—all as necessary to convey a clear picture of the system to the installer.

But before the actual design work begins, many factors must be considered and understood in relation to the contemplated design of an electrical system.

Approaching the Design Task

Successful electrical design for all types of modern buildings depends first of all upon the right approach. Electrical designers must be thoroughly familiar with all the background factors of design. They must, of course, have depth of engineering ability. But over and above this, they must fully understand the relation between the pure technology of an electrical system and such considerations as safe application, capacity for load growth, flexibility in the use of the system, and effective layout. Based on such understanding, the right approach to electrical design is inevitable.

Then the right approach must be correlated with modern standards. Because electrical design is a dynamic thing—the continually evolving product of years of accumulating technology—the electrical designer must learn new techniques and follow promising trends. Electrical design is not an automatic procedure that merely requires filling in formulas and adding parts together. It involves clear understanding of old, accepted techniques and the reasons why those techniques have survived the test of years of application. It requires combinations of old and new wiring techniques and the ability to devise original circuits and layouts for new or special equipment applications. In short, it demands that designers be as dynamic as the art and science they practice. Only with an all-around grasp of the subject can designers exercise judgment in the selection of circuit and feeder arrangements, equipment types and ratings, and installation methods.

Every electrical system must provide power and light without hazard to life or property, with sufficient extra capacity to meet foreseeable load growth, with ready adaptability to load modifications and revised layouts, and with provisions for necessary accessibility in the distribution arrangement.

Safety Is Basic

Compliance with the provisions of the **National Electrical Code (NEC)** can effectively minimize fire and accident hazards in any electrical design. The **NEC** sets forth requirements, recommendations, and suggestions and constitutes a minimum standard for the framework of electrical design. As is stated in its own introduction, the **NEC** is concerned with the "practical safeguarding of persons and property from hazards arising from the use of electricity" for light, heat, power, signaling, and other purposes.

Although the **NEC** assures minimum safety provisions, actual design work must include the constant consideration of safety as required by special types or conditions of electrical application. For example, the effective provision of automatic protective devices and the selection of control equipment for particular applications involve engineering skills beyond routine adherence to **NEC** requirements. Then, too, designers must know the physical characteristics—the application advantages and limitations—of the many materials they use for enclosing, supporting, insulating, isolating, and, in general, protecting electrical equipment. The task of safe application based on skill and experience is particularly important in designing for hazardous locations. Safety is not automatically made a characteristic of a system by simply observing codes. Safety must be designed into a system.

The **NEC** is recognized as a legal criterion of safe electrical design and installation. It is used in court litigation and by insurance companies as a basis for insuring buildings. Because the **NEC** is such an important instrument of safe design, it must be thoroughly understood by electrical designers. They must be familiar with all its sections and should know the accepted interpretations which have been given to many of its specific rulings. They should keep abreast of the official interpretations issued by the **NEC** committee. They should know the intent of **NEC** requirements—i.e., the spirit as well as the letter of each provision. They should keep informed on interim amendments to the **NEC**. And, most important, they should keep the **NEC** text handy and study it often.

In addition to the **NEC** itself, the electrical designer should have on hand and be familiar with other standards and recommended practices made available in pamphlet form by the National Fire Protection Association. Designers may write to the Association at Batterymarch Park, Quincy, MA 02269 for any of the following:

70	National Electrical Code
70B	Recommended Practices for Electrical Equipment Maintenance
70A	Electrical Code for One- and Two-Family Dwellings
70L	Model State Electrical Law, Inspection of Electrical Installations
76A	Essential Electrical Systems for Health Care Facilities
79	Electrical Metalworking Machine Tools
71	Central Station Protective Signaling Systems
72A	Local Protective Signaling Systems
72B	Auxiliary Protective Signaling Systems
72C	Remote Station Protective Signaling Systems
72D	Proprietary Protective Signaling Systems
73	Public Fire Service Communications
74	Household Fire Warning Equipment
78	Lightning Protection Code
496	Purged Enclosures for Electrical Equipment
493	Standard on Intrinsically Safe Process Control Equipment
30	Flammable Combustible Liquids Code
33	Spray Application
101	Life Safety Code
501	Mobile Homes

Electrical designers must also be familiar with insurance company regulations and all local codes which affect particular installations. They should know in detail where and how local codes depart from the **NEC**. Again, they must understand the basis for special provisions to assure most effective conformity throughout a system. Often, a local condition which requires special code rulings has a significant effect upon the entire design.

OSHA and Third-Party Certification

All modern electrical work must also be related to Occupational Safety and Health Administration (OSHA) regulations and to the standards and application instructions of nationally recognized product testing and certifying organizations, such as Underwriters Laboratories Inc. (UL),

Factory Mutual Engineering Corp., and Electrical Testing Laboratories, Inc. OSHA's relation to electrical work is clear. Subpart S, "Electrical," of the Occupational Safety and Health Standards sets forth the electrical regulations that OSHA makes mandatory, especially with respect to design safety standards for electrical systems.

OSHA regulations on electrical installations require third-party certification of the essential safety of the equipment and component products used to assemble an electrical installation and set forth the following:

Approval. The conductors and equipment required or permitted by this subpart shall be acceptable only if approved.

Approved. Acceptable to the authority enforcing this subpart. The authority enforcing this subpart is the Assistant Secretary of Labor for Occupational Safety and Health. The definition of "Acceptable" indicates what is acceptable to the Assistant Secretary of Labor, and therefore approved within the meaning of this subpart.

Acceptable. An installation or equipment is acceptable to the Assistant Secretary of Labor, and approved within the meaning of this Subpart S: (i) if it is accepted, or certified, or listed, or labeled, or otherwise determined to be safe by a nationally recognized testing laboratory, such as, but not limited to, Underwriters Laboratories Inc. and Factory Mutual Engineering Corp.; or (ii) with respect to an installation or equipment of a kind which no nationally recognized testing laboratory accepts, certifies, lists, labels, or determines to be safe, if it is inspected or tested by another federal agency, or by a state, municipal, or other local authority responsible for enforcing occupational safety provisions of the **National Electrical Code**, and found in compliance with the provisions of the **National Electrical Code** as applied in this subpart; or (iii) with respect to custom-made equipment or related installations which are designed, fabricated for, and intended for use by, a particular customer, if it is determined to be safe for its intended use by its manufacturer on the basis of test data which the employer keeps and makes available for inspection to the Assistant Secretary and his authorized representatives.

Other definitions contained in OSHA's electrical design standard are:

Listed. Equipment is "listed" if it is of a kind mentioned in a list which (a) is published by a nationally recognized laboratory which makes periodic inspection of the production of such equipment and (b) states such equipment meets nationally recognized standards or has been tested and found safe for use in a specified manner;

Labeled. Equipment is "labeled" if there is attached to it a label, symbol, or other identifying mark of a nationally recognized testing laboratory which (a) makes periodic inspections of the production of such equipment and (b) whose labeling indicates compliance with nationally recognized standards or tests to determine safe use in a specified manner:

Accepted. An installation is "accepted" if it has been inspected and found by a nationally recognized testing laboratory to conform to specified plans or to procedures of applicable codes;

Certified. Equipment is "certified" if it (a) has been tested and found by a nationally recognized testing laboratory to meet nationally recognized standards or to be safe for use in a specified manner, or (b) is of a kind whose production is periodically inspected by a nationally recognized testing laboratory, and (c) it bears a label, tag, or other record of certification;

Utilization equipment. Utilization equipment means equipment which utilizes electric energy for mechanical, chemical, heating, lighting, or similar useful purpose.

The use of custom-made equipment is covered in (iii) of the above-quoted OSHA rule. Every manufacturer of custom equipment must provide documentary safety-test data to the owner on whose work premises the custom equipment is installed. And it seems to be a reasonable conclusion from the whole rule itself that custom equipment assemblies must make maximum use of "listed," "labeled," or "certified" components.

Section 110-2 of the **NEC** regulates the use of electrical products and equipment in a rule similar to the OSHA regulation on acceptability of equipment, as described above. It states:

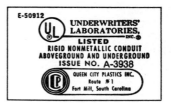

Third-party certification, such as listing by Underwriters Laboratories Inc., is an absolute must for products utilized in modern electrical design work. These typical labels, which are affixed to the products themselves, are visible evidence for the installer and electrical inspector that the products are certified to satisfy Section 110-2 of the **NEC**.

Approval. The conductors and equipment required or permitted by this Code shall be acceptable only if approved.

The **NEC** definition of "approved" is "acceptable to the authority having jurisdiction." This means that the electrical inspector having jurisdiction at any specific installation is the person who will decide what conductors and/or equipment are "approved." Although an inspector is not required to use "listing" or "labeling" by a national testing lab as the deciding factor in the approval of products, inspectors invariably base their acceptance of products on listings by testing labs. Certainly, the **NEC** almost exhibits the same insistence as OSHA that, whenever possible, acceptability must be based on some kind of listing or certification by a national lab. But on this matter, the OSHA law takes precedence—a "listed," "labeled," or otherwise "certified" product must always be used in preference to the same "kind" of product that is not recognized by a national testing lab.

NEC Section 110-3 gives general rules on the "examination, identification, installation and use of equipment." Part (b) of that section says:

> Listed or labeled equipment shall be used or installed in accordance with any instructions included in the listing or labeling.

That rule clearly and certainly says, for instance, that any and every product listed in the UL *Electrical Construction Materials Directory (Green Book)* must be used exactly as described in the application data given with the listing in the book. Because the *Electrical Construction Materials Directory* and the other UL books of product listings, such as the *Hazardous Location Equipment Directory (Red Book)* and the *Electrical Appliance and Utilization Equipment Directory (Orange Book)*, contain massive amounts of installation and application instructions, all those specific bits of application data become mandatory **NEC** regulations as a result of the rule in Section 110-3(b). The data given in the UL listings books supplement and expand upon rules given in the **NEC**. In fact, effective compliance with **NEC** regulations can only be assured by careful study and observance of the limitations and conditions spelled out in the application instructions given in the UL listings books or similar instructions provided by other national testing labs.

Capacity for Present-plus

In general, every electrical system should have sufficient capacity to serve the loads for which it is designed, plus spare capacity to meet anticipated growth in the load on the system. In particular, this means that conductors and raceways must be sized liberally for computed loads; substations, transformers, and switching and protective devices must have the needed capacity and ratings. And spare capacity throughout the branch circuiting should be reflected back through the entire electrical system to the point of power supply.

Allowance for load growth is probably the most neglected consideration in electrical design today. Although the **NEC** contains an almost obscure recommendation that electrical plans and specifications include allowances "for future increases in the use of electricity," experience with current electrical modernization practice indicates that lack of spare capacity plagues existing systems. In all types of buildings, conduit risers are filled to capacity, and the conductors in them are either loaded fully or overloaded. In most commercial and institutional buildings, the entire electrical system is at or near saturation—in branch circuits, feeders, and service. And in the majority of these buildings, modernization of the existing electrical system to handle increased load demands is impeded by an absence of space in which new risers and circuits may be run. It is obvious that the overall design of these electrical systems did not account sufficiently for future load growth.

Modern electrical design, therefore, must include careful planning for future increases in electrical utilization. Depending upon the particular conditions in any installation, the mains, switchgear, transformers, feeders, panelboards, and circuits should be sized to handle considerable load growth. Conductors should be selected on the basis of carrying capacity, voltage drop, and estimated future requirements. Conduit, wireways, troughs, and other raceways should be sized to allow future increases in occupancy. And the space used to house electrical equipment— electric closets, switchgear rooms, substation cages, riser and pipe shafts, etc.—should also be able to accommodate more equipment at a later date.

Every design must include:

Flexibility: Depending upon the type of building—industrial, commercial, or institutional—the electrical system must be designed to provide the required flexibility in distribution and circuiting. The layout and type of equipment should readily accommodate changes in the locations of motors and other utilization devices. Feeders, distribution panelboards, and circuits should be suited to a wide range of utilization patterns, allowing full and efficient use of power capacity for activities in the building's various areas.

Accessibility: Every electrical system should rate high in accessibility. In its final form, the design of the system must provide ease of access to equipment for maintenance and repair and for any possible extensions, modifications, or alterations to the system. The system of conductors, raceways, and equipment must allow for full use of its power-handling ability.

Reliability: Depending upon the nature of the activities within a building, the continuity of electrical supply and overall reliability of the wiring system itself can be a more or less important consideration. Where electric utility companies have a good record of supply continuity, and temporary loss of power would not be a direct life or property hazard, special provisions for a separate emergency power supply are usually not necessary. But in many industrial plants, hospitals, and buildings with essential electrically powered equipment, standby power plants or multiple services must be provided for absolute reliability of supply.

ANALYZING A SPECIFIC SYSTEM

Based on full recognition and appreciation of the foregoing system characteristics, the electrical design for a specific building begins with an analysis of the building type, its loads, and the source of supply. This analysis involves the careful determination of all usual and special electrical requirements for the type of building. The activities to be performed in the building and the nature of electrical usage by the occupants must be considered. Whether the building is a school, an office building, an industrial plant, or a hospital, the designer must know the history of electrical applications in that type of building and must be well informed on current trends and practices.

What Kind of Building?

The general design approach to any contemplated electrical system takes into account the characteristics of the building: Is it small or large; single-story or multilevel; an industrial plant, office building, apartment house, school, or hospital; and so on. These characteristics give insight to the types of electrical utilization, the need for flexibility, the accessibility of the system, and the duty cycles of various load devices. The many and varied requirements for hazardous locations also serve to give direction to the design approach. In such a general approach, the designer should select many design possibilities which suit the particular building and reject all methods and techniques which are immediately seen to be inapplicable.

In general, planning for the design of an electrical system should begin with determination and study of the size and nature of the total load to be served. This means approximation of lighting loads on a watts per square foot basis, analysis of the number and sizes of motors to be served in various areas of the building, and determination of the amounts of other utilization loads and their concentrations through the building. A full understanding of all loads—their sizes and points of application—is essential to the selection of the best type of distribution system.

Standardizing Equipment

Maximum standardization in equipment type and ratings should always be a design objective. The selection of standard supply voltage and standard values at all voltage levels results in significant economy since standard-rated transformers, switchgear, motors, and other equipment cost less than special equipment for nonstandard voltages.

Lack of standardization in an electrical system complicates maintenance in that replacement parts are not easy to obtain, a large inventory of parts and equipment must be maintained, and the efficiency of maintenance personnel is reduced. And the use of special, nonstandard equipment and voltages may seriously impair expansion or alteration of the electrical system at a later date. However, where special, nonstandard equipment is necessary for the particular function of a building, such equipment must be carefully selected and integrated into the system.

Providing the System Power Supply

Another preliminary consideration that fundamentally affects the design procedure is the nature of the power supply which will serve the building's electrical system. The power may be supplied by either the distribution system of an electric utility company or a private generating plant.

Utility company distribution is the most common type of supply to buildings and plants. The purchase of energy usually represents decided economy over the private generation of electric energy within the building or plant. Of course, in certain cases such as paper and pulp mills, the need for large amounts of process steam makes possible the use of excess steam to generate energy economically. In such cases, the generating plant may be operated independently of a utility company supply to the building or in parallel with it.

If a private generating plant operates independently, a part of the total electrical load may be connected to it and part to the utility system. The use of a generating plant in parallel with a utility company supply must be checked and coordinated with utility company engineers.

If power is obtained from a utility company line, the characteristics of the supply must be matched to the requirements of the building. Depending upon the voltage and capacity of the supply, one particular type of distribution is often best suited to carrying the electrical loads in the building.

The purchase of power at utilization voltages indicates certain types of distribution within the building; higher-voltage services and primary supplies also indicate particular types of distribution systems. When several different supply voltages are available, each should be appraised in relation to the various distribution methods which might be used. Consultation with the utility company concerning the relation between the different services and interior distribution systems should precede any decisions.

CHECKLISTS OF MODERN DESIGN TECHNIQUES

Industrial Projects

Electrical load density varies with the particular type of industrial operation and can run well over 20 W/ft². Typical elements of modern industrial electrical systems include the following:

- Primary distribution to loadcenter substations
- Armored-cable feeders rated for 5-kV or 15-kV circuits
- Extensive use of 480-V power and 277-V fluorescent and/or metal-halide lighting

- Busway and/or armored cable for secondary voltage feeders
- Cable troughs and/or racks carrying suitably protected cables, for maximum system flexibility and accessibility
- Rigid nonmetallic conduit in corrosive locations
- Underfloor power ducts for flexible circuiting to machine loads throughout widespread plant areas
- Plug-in busway for ready tapping to power and light loads
- Carefully designed hazardous-location installations
- Thoroughly engineered protection against system faults
- Adequate interrupting capacity for all devices that interrupt current flow—both control switching and protective equipment
- Coordinated use of short-circuit and overload protective devices to effectively isolate faulted sections of the system without interrupting service to other sections
- Modern, totally enclosed, dead-front switchgear for maximum safety, designed for easy maintenance
- High-power-factor operation with excellent regulation of voltage at all levels in the plant

Office Buildings

Typical load densities for modern office buildings can run over 15 W/ft² where electric usage is at a maximum for light and power facilities. General area lighting may account for up to 8W/ft²; power loads, 4 W/ft² or more; and small machines and incidental loads, up to 2 W/ft². Typical elements include:

- Primary supply to two or more unit substations strategically placed within the building
- Adequate interrupting capacities and coordinated protection
- Effective protection against ground-fault currents of low values
- Three-phase, 4-wire, 480/277-V distribution for power and general area fluorescent lighting
- Busway risers
- Air-handling luminaires to provide lighting and air conditioning from single units for module layouts
- Step-down local transformers for 120/208-V, 3-phase, 4-wire supply to incandescent lighting and convenience-receptacle circuits
- Underfloor raceway, through-floor fittings, or under-carpet flat-conductor cable systems for power, light, telephones, and signals in large office areas with shifting loads or layouts
- Effective alarms, signals, and communication facilities—fire alarms, burglar alarms, security-officer systems
- Remote-control switching
- Carrier-frequency relay switching without control wires
- Raised floors for underfloor circuiting to computers in data-processing centers

Shopping Centers

Lighting-load densities vary widely among the different areas of shopping centers—from 3 to more than 10 W/ft^2—depending on the sizes and types of stores, the relative amounts of fluorescent and incandescent lighting, the light used for decoration, and the amount of outside lighting. Power-load densities depend upon the facilities selected. Typical elements include:

- Primary distribution to loadcenter subs
- Use of 480/277 V among and within buildings
- Contactors and split-bus panelboards for time-clock control of large blocks of sign, display, and outdoor lighting
- Modern store lighting
- Music-distribution, intercom, and paging systems
- Modern parking-lot lighting: fluorescent, mercury-vapor, metal-halide, high-pressure sodium, and/or low-pressure sodium luminaires

Schools

Lighting loads for modern schools vary with the type and layout of the building and the extent to which daylighting is utilized in classrooms. Typical load densities range from 3 to 7 W/ft^2, depending upon the type of lighting (incandescent or fluorescent) and luminaires. Power loads may include air conditioning, electric heating, ventilation, compressors, and elevators. Typical elements are:

- Distribution at 4160 V (or higher, where direct utility supply requires) to two or more load-centers in large-area, multibuilding school layouts
- On-site electric generation for air-conditioned schools
- Use of 480/277-V distribution for power and lighting loads
- Electric heating throughout school buildings for maximum comfort and effective climate control, minimizing maintenance and eliminating the hazards of combustible fuels
- Modern, high-level lighting with maximum visual comfort
- "Showcase" wiring and equipment application in laboratories and craft shops, providing integration in the teaching of electricity and its use
- Modern gymnasium lighting, electric scoreboards, and public-address equipment
- Engineered lighting and power facilities for auditoriums, including programmed dimmer control of stage lighting and house lights
- Intercom and paging systems
- Closed-circuit television systems for instructional purposes
- Clock-and-bell program system
- Modern fire-alarm system
- Vandal-alarm system to protect against vandals and other intruders during closed hours

Hospitals

Typical electrical demand for modern hospital facilities can range up to 3000 W per bed where maximum use is made of electrical equipment. Typical elements include:

- Loadcenter distribution where possible
- Careful provision for electric supply continuity in case of failure of normal supply: two separate services, a special emergency service, an emergency standby power generator, or a standby supply for essential lighting and other essential facilities
- Modern lighting throughout
- Engineered application of modern signal and communication equipment: call-and-register systems, paging systems, intercom, closed-circuit television, radio-program distribution
- Modern wiring of laboratory facilities and x-ray equipment
- Careful compliance with all **NEC** requirements on wiring in operating rooms (ungrounded circuits and ground-detector-and-alarm system) and in hazardous (anesthetizing) areas

Lighting and Appliance Branch Circuits

A *branch circuit* is any segment of a wiring system extending beyond the final automatic overcurrent protective device that is approved for use as branch-circuit protection and designated by the **NEC** as the branch-circuit protective device (Fig. 1). Branch circuits generally originate in panelboards, but individual branch circuits to motors commonly originate at individual fused switches or circuit breakers tapped from busways or nippled out of auxiliary gutters.

In an electrical system, the branch circuits are the circuits of lowest capacity and current rating. They are the circuits to which load devices—lights, motors, etc.—are connected. Of course, it is possible (and not at all uncommon) to have branch circuits of very high rating—for instance, a branch circuit with conductors rated at 450 A to supply a 150-hp, 230-V, 3-phase motor. But the vast majority of branch circuits are rated not over 30 A.

Thermal cutouts or motor overload devices are not branch-circuit protection. Neither are fuses in luminaires or in plug connections, where they are used for ballast protection or individual fixture protection. Such "supplementary overcurrent protection" is the type of individual fixture or appliance protection that is connected on the load side of branch-circuit protection. **NEC** Sections 424-22(c) to (e) cover the use of the supplementary overcurrent protection required for electric-resistance duct heaters.

Figure 2 shows a common situation for bridge and highway circuits, in which a number of large lamp loads are fed by a feeder with individual fusing for each fixture; this does result in the tap circuits becoming the branch circuits of the system.

In its simplest form, a branch circuit consists of two wires which carry current at a particular voltage from protective device to utilization device. Although the branch circuit represents the last step in the transfer of power from the service or source of

FIG. 1   Definition of a branch circuit.

FIG. 2   Common variation from standard branch-circuit layout.

energy to the utilization devices, it is the starting point for modern design procedures. First, the loads are circuited. Then, the circuits are lumped on the feeders. Finally, the distribution system is connected to one or more sources of power.

Each and every branch circuit—whether for a power or lighting load in a commercial, industrial, or residential building—should be sized for its load, with spare capacity added where possible load growth is indicated, where necessary to reduce heating in continuously operating circuits, and/or for voltage stability. The design should also provide for the economical addition of circuits to handle future loads. Each circuit must provide the required power capacity at full utilization voltage at every outlet.

Accessibility and flexibility are two important characteristics of effective branch circuiting. In commercial and industrial buildings, the shifting of load devices is common, as a result of tenant changes, rearrangement of office layouts, changes in production schedules, or relocation of production lines. Circuits in such areas must readily accommodate these shifts and must allow extensions.

Selecting Branch-Circuit Voltage

In general, modern branch-circuit design has developed from the trend toward high distribution voltages, incorporating the favorable economic and operating characteristics of loadcenter layout. As a result, branch circuits are short runs from strategically located panelboards or transformer loadcenters, instead of long circuits from a central distribution panelboard. The high voltage is delivered to the loadcenter, where it is reduced to utilization levels. Or, in a one-voltage system, the feeder is carried to the approximate physical center of a load concentration area before it is subdivided into branch circuits. The use of loadcenters and the resulting short branch circuits provide better-regulated voltage supplies to utilization devices and minimum circuit disturbances as a result of load changes.

Modern electrical systems use four basic voltage configurations (or systems) for branch circuits supplying utilization voltages at light and power outlets. These are derived from the common secondary distribution systems, as follows.

Single-phase, 3-wire system

This is a commonly used system in individual residences and small apartment and commercial buildings. Both lighting and single-phase motor loads can be served. Two-wire, 240-V branch circuits can be used for power loads; 3-wire, 120/240-V circuits for lighting outlets, split-wired duplex receptacles, and some power devices such as electric ranges; and 2-wire, 120-V circuits for lighting and receptacle outlets.

Three-phase, 4-wire wye system

This is the most widely used 3-phase secondary distribution system. The most common use of this configuration is the 120/208-V, 3-phase, 4-wire system with the neutral grounded. With this system, a variety of circuits is available: 4-wire, 120/208-V circuits; 3-wire, 120/208-V circuits; 3-wire, 208-V circuits; 2-wire 208-V circuits; and 2-wire, 120-V circuits. Such a system can serve a combination of power and lighting loads; it offers flexibility for circuit layout and the application of required utilization equipment.

Another 3-phase, 4-wire, wye-connected system which has gained rapidly in acceptance and application is the 480/277-V wye-connected system (also called the 460/265-V system, owing to the voltage spread). Under certain conditions, this system offers more advantages and economy in commercial building applications than the 120/208-V system. The system makes available three types of branch circuits: 480-V, 3-phase circuits for motor loads; 277-V, single-phase circuits for fluorescent or mercury-vapor lighting; and 120-V, 240-V or 120/208-V circuits from step-down transformers for receptacle circuits and miscellaneous loads.

Three-phase, 3-wire system

This is the common delta-connected secondary system with phase-to-phase voltage of 240, 480, or 600 V between each pair of phase conductors. This system is used where the motor load represents a large part of the total load. In some such systems, 3-phase, 3-wire circuits at 480 V feed individual motor loads, with single- or 3-phase step-down transformers, stepping from 480 to 120 V used to supply lighting and receptacle circuits.

Three-phase, 4-wire delta system

This is a variation on the 3-phase, 3-wire delta system above; it is often called a "red-leg" delta system because the phase leg with higher voltage to ground than the other two phase legs is commonly painted red to differentiate it. One of the transformer secondary windings supplying the system is center-tapped to derive a grounded neutral conductor to the two phase legs between which it is connected. Motors are supplied at 240 V, either 3-phase or single-phase, with single-phase 120-V branch circuits taken from the grounded center-tapped conductor and its associated phases.

In some areas, electrical systems are supplied from two coexisting services—a 3-phase, 3-wire system for power loads and a separate 120/240-V, single-phase or 120/208-V, 3-phase grounded system for lighting and appliance loads.

Other system configurations are commonly used to supply branch circuits, but they are derived from these basic systems. Particular local conditions and/or requirements for load balance throughout an overall system frequently dictate the use of less common derivations of branch circuits. Such arrangements include 120/208-V, 3-wire feeds to branch-circuit panelboards, open-delta 3-phase supply to light and power branch circuits, and other special transformer hookups supplying branch-circuit loads. Motor branch circuits may be served directly from high-voltage supplies.

Basic Circuit Concepts

The design of a lighting layout for any commercial, institutional, or industrial area includes determination of the required intensity of lighting and selection of the general type of lighting (incandescent, fluorescent, mercury-vapor, metal-halide, high-pressure sodium, low-pressure

FIG. 3 Types of branch circuits.

sodium, or some combination of these), number of lamps per luminaire, number of luminaires, types of luminaires, mounting details, wiring methods, and other specifics. The lighting design for a particular area provides a value for the required circuit power (wattage) capacity. This value is used in determining the number and type of lighting circuits needed.

As shown in Fig. 3, **NEC** Article 100 gives definitions covering three types of branch circuits. The fundamentals of safety in lighting and appliance branch-circuit design are given in **NEC** Articles 200, 210, and 220. As set forth in Section 210-1, Article 210 covers branch circuits supplying lighting or appliance loads or combinations of such loads. Where motors or motor-operated appliances are connected to any branch circuit that also supplies lighting or other appliance loads, the provisions of both Article 210 and Article 430 apply. Article 430 applies where branch circuits supply motor loads only.

A "branch circuit" as covered by Article 210 may be a 2-wire circuit or a "multiwire" branch circuit. A "multiwire" branch circuit consists of two or more ungrounded conductors having a potential difference between them and an identified grounded conductor connected to the neutral conductor of the system and having equal potential difference between it and each of the ungrounded conductors. Thus, a 3-wire circuit consisting of two opposite-polarity ungrounded conductors and a neutral derived from a 3-wire, single-phase system or a 4-wire circuit consisting of three different phase conductors and a neutral of a 3-phase, 4-wire system is a *single multiwire branch circuit,* not three circuits (Fig. 4). It is only one circuit even though it may be protected by two or three single-pole protective devices or a multipole protective device in the panelboard. This is important, because other sections of the **NEC** refer to conditions involving "one branch circuit" or "a two-wire or multiwire branch circuit." (See Sections 250-24 and 410-31.)

A branch circuit is rated according to the setting or rating of the overcurrent device used to protect the circuit. Any branch circuit with more than one outlet, other than an "individual

FIG. 4 "Multiwire" branch circuit must include a neutral conductor.

FIG. 5 Each point at which a plug cap may be inserted is a receptacle.

branch circuit" (see definition in Fig. 3), must be rated at 15, 20, 30, 40, or 50 A. That is, the protective device must have one of those ratings for multioutlet circuits, and the conductors must meet the other size requirements of Article 210. However, in industrial plants a circuit rated over 50 A is permitted to have two or more receptacle outlets to provide for easy relocation and connection of a single load such as a welder.

Under the definition for "receptacle" in **NEC** Article 100, it is indicated that a duplex convenience outlet (a duplex receptacle) is considered to be two receptacles and not one—even though there is only one box (Fig. 5). Thus, a circuit that supplies only one duplex receptacle is not an "individual branch circuit" as defined in Article 100 because two loads (appliances, tools, etc.) could be connected to it. Even though there is only one box and, technically, only one "receptacle outlet," the circuit must conform to the rules for circuits with "more than one outlet."

It is very important to remember that the size of the overcurrent device determines the rating of any circuit covered by Article 210, even when the conductors used for the branch circuit have a current rating higher than that of the protective device. In a typical case, for example, a 20-A circuit breaker in a panelboard might be used to protect a branch circuit in which No. 10 conductors are used as the circuit wires. Although the load on the circuit does not exceed 20 A, and No. 12 conductors would have sufficient current-carrying capacity to be used in the circuit, the No. 10 conductors with their rating of 30 A were probably selected to reduce the voltage drop in a long homerun. But the rating of the circuit is 20 A because that is the size of the overcurrent device. The current rating of the wire must generally be not lower than the rating of the protective device, but it does not enter into the current rating of the circuit.

Although multioutlet branch circuits are limited in rating to 15, 20, 30, 40, and 50 A, a branch circuit to a single-load outlet (for instance, a branch circuit to one machine or to one receptacle outlet) may have any current rating. For instance, there could be a 200-A branch circuit to a special receptacle outlet, or a 1000-A branch circuit to a single machine.

Number of Circuits

The number of branch circuits required to handle the general lighting load depends on the total load to be served, the layout of the lighting system and outlets, the amount of load to be placed on each circuit, and the capacity of the circuits to be used. According to the **NEC**, the number of circuits shall be not less than that determined from "the total computed load" and the capacity of the circuits to be used. The "total computed load" for general lighting is the load determined from **NEC** Section 220-2(b) and represents the minimum load for which branch-circuit capacity must be provided—even if the actual connected load for general lighting is lower in kilowatts or kilovoltamperes.

Throughout any building or any given area of a building, the "total computed load" for general lighting must be taken as not less than the value obtained by multiplying the total number of square feet of floor area covered by the general lighting by the minimum unit load in watts per square foot as specified in **NEC** Table 220-2(b) for the particular type of occupancy served by that general lighting. Note that the minimum-unit load values given in the table apply to the type of area that the general lighting serves. In a particular building, the entire area may be of the same type—all store areas or all office areas—or a number of different areas may have to be considered. For instance, for any and every area of a building that is office space, the unit load of $3\frac{1}{2}$ W/ft^2 must be multiplied by the number of square feet of such area to determine the minimum "branch-circuit load" or "total computed load." For an area of the building that is a restaurant, a load of 2 W/ft^2 must be used. A courtroom in the same building has a minimum branch-circuit load ("total computed load") of 2 W/ft^2. Each store area must be taken at 3 W/ft^2.

In all cases, calculation of the "total computed load" establishes the minimum general lighting load for which branch circuits must be provided—regardless of the fact that very efficient modern light sources and fixtures can readily provide a high-quality high-footcandle lighting environment at unit loads of much less than those shown in the table and would require less branch-circuit capacity than that needed for the "total computed load." However, in any case where the actual connected load of general lighting equipment exceeds the watts-per-square-foot load shown in the table, the required number of branch circuits (and the capacity of the branch circuits) must be increased to serve the connected loads adequately in accordance with all applicable **NEC** rules. Section 220-3(a) says, "In all installations, the number of circuits shall be sufficient to supply the load served."

WARNING: Although Section 220-3(a) does say that "the minimum number of branch circuits shall be determined from the total computed load and the size or rating of the circuits used," Section 220-3(d) makes an important exception to that requirement. Although Section 220-3 is headed "Branch Circuits Required," part (d) goes beyond branch circuits and makes clear that a feeder to a branch-circuit panelboard and the main busbars in the panelboard must have a minimum ampacity sufficient to serve the calculated total load of lighting, appliances, motors, and other loads supplied. And the amount of feeder and panel ampacity required for the general lighting load must not be less than the ampere value determined from the circuit voltage and the total power (watts), where the latter is obtained by multiplying the minimum unit load from Table 220-2(b) (watts per square foot) by the area of the occupancy supplied by the feeder—even if the actual connected load is less than the calculated load determined on the watts-per-square-foot basis. (Of course, as noted above, if the connected load is greater than that calculated on the watts-per-square-foot basis, the greater load value must be used in determining the number of branch circuits, the panelboard capacity, and the feeder capacity.)

It should be carefully noted that the second sentence of Section 220-3(d) states, "Where the load is computed on a watts-per-square-foot basis, the wiring system up to and including the branch-circuit panelboard(s) shall be provided to serve not less than the calculated load." The phrase "wiring system up to and including" requires that a feeder must have sufficient capacity for the total minimum branch-circuit load, determined as the area (square feet) times the minimum unit load [watts per square foot, from Table 220-2(b)]. And the phrase clearly requires that amount of capacity in every part of the distribution system supplying the load. The required capacity would, for instance, be required in a subfeeder to the panel, in the main feeder from which the subfeeder is tapped, and in the service conductors supplying the whole system.

Actually, reference to "wiring system" in the wording of Section 220-3(d) presents a requirement that goes beyond the heading "Branch Circuits Required" of Section 220-3 and, in fact, constitutes a requirement on feeder capacity that supplements the rule of the second sentence of Section 220-10(a). This requires a feeder to be sized to have enough capacity for the "computed load" as determined by Part **A** of this article (which means, as computed in accordance with Section 220-2).

A second part of Section 220-3(d) affects the required minimum number of branch circuits. Although the feeder and panelboard must have a minimum ampacity equivalent to the "computed load," it is only necessary to install the number of branch-circuit overcurrent devices and circuits required to handle a connected load that is less than the calculated load. The last sentence of Section 220-3(d) is clearly an exception to the basic rule of the first sentence of Section 220-3(a), which says that the minimum number of branch circuits "shall be determined from the total computed load." Instead of supplying that minimum number of branch circuits, it is necessary to have only the number of branch circuits required for the actual total "connected load."

For general lighting in an office area, Fig. 6 shows a typical comparison between the actual load and the minimum "total computed load." The layout of 265-V lighting fixtures is the result of a lighting design that provides a high-quality illumination level using efficient fluorescent lamps. The total load of all the lighting fixtures comes to only 92,000 W. However, the rule of

Fixture layout for 265-V lighting system in an office area is a 92,000-W actual load . . .

200 ft

200 ft

Floor area calculated from outside dimensions:
200 ft × 200 ft
= 40,000 sq ft

. . . but calculating minimum required branch-circuit load as required by NEC Sec. 220-2 calls for branch-circuit capacity at least equal to
40,000 sq ft × 3.5 W/sq ft
= 140,000 W, computed load.

FIG. 6 Minimum required branch-circuit capacity that must be provided for general lighting is based on actual load and "computed load."

Section 220-3 says that branch circuits for lighting "shall be provided to supply the loads computed in accordance with Section 220-2." Because the "total computed load" must be taken as 3.5 W/ft^2 × 40,000 ft^2 of floor area, or 140,000 W, the first sentence of Section 220-3 and the first sentence of part (a) of that section would appear to require a total branch-circuit capacity of 140,000 W. If 20-A, 265-V circuits are used, and the load on each circuit must be limited to 16 A because the lighting is "continuous" (operates for 3 hours or more), that basic rule would seem to call for 140,000 ÷ (16 A × 265 V) or 33 branch-circuit protective devices in the panel serving the office area. But, the last sentence of Section 220-3(d) says, "Branch-circuit overcurrent devices and circuits need only be installed to serve the connected load." As a result, the minimum required number of branch-circuit poles then becomes 92,000 W ÷ (16 A × 265 V), or 22 branch-circuit protective devices. However, as discussed later, the panelboard busbars, the feeder to the panel, and the service-entrance conductors for the building must have their load-current ratings based on the "total computed load" of 140,000 W and not on the 92,000-W actual load.

Load Limitation for Continuous Operation

Where a branch-circuit load will be in "continuous" operation—such as for general illumination in large office areas, private offices, schools, hospitals, industrial plants, and the like—the total load on each branch circuit may not exceed 80 percent of the rating of the circuit, and the "rating of the circuit" is established as the ampere rating of the branch-circuit protective device. The **NEC** defines a continuous load as a load "where the maximum current is expected to continue for three hours or more."

This 80 percent limitation on circuit loading is set forth in three different sections of the **NEC**: Sections 210-22(c), 220-2(a), and 384-16(c) (Fig. 7).

Section 210-22(c) says that, for loads other than motor loads, the total load on a branch circuit must not exceed 80 percent of the circuit rating when the load will constitute a continuous load—such as store lighting.

FIG. 7 Unless "listed" for continuous operation at 100 percent of its rating, a branch-circuit overcurrent device must not be loaded over 80 percent of its rating.

Because branch circuits for multioutlet circuits, such as lighting circuits, are rated in accordance with the rating or setting of the overcurrent device, the maximum permitted load is 80 percent of the rating of the protective device. Although this limitation applies only to loads other than motor loads, Section 384-16(c) says that "the total load on any overcurrent device located in a panelboard shall not exceed 80 percent of its rating where in normal operation the load will continue for 3 hours or more."

NOTE: In both the above cases, the 80 percent load limitation does not apply "where the assembly including the overcurrent device is approved for continuous duty at 100 percent of its rating," which means where it is listed by Underwriters Laboratories Inc. (UL) or another qualified electrical testing laboratory. But Section 384-16(c) requires any circuit breaker (CB) or fuse in a panelboard to have its load limited to 80 percent; and only one exception is made for such protective devices in a panel: A continuous load of 100 percent is permitted only when the protective-device assembly (fuse in switch or CB) is approved for 100 percent continuous duty. (And there are no such devices rated less than 600 A. UL has a hard-and-fast rule that any breaker not marked for continuous load must have its load limited to 80 percent of its rating.)

For panelboard applications, the use of conductors that have had their ampacity derated because more than three are located in a raceway (Note 8 to **NEC** Tables 310-16 through 310-19) does not eliminate the 80 percent limit on the continuous load on the circuit protective device. An example would be six No. 12 TW copper wires in a conduit, with the ampacity of each derated from the ampacity of 25 A shown in **NEC** Table 310-16 to 20 A (0.80 × 25 A) and then protected at 20 A. Such circuits may not be continuously loaded up to the 20-A value. Section 384-16(c) would require that a continuous load be not over 80 percent of the 20-A rating of the protective device, or 16 A in this example (Fig. 8).

For supplying continuous load, 6 No. 12 TW wires have an ampacity of 25 A each, but each must have its load current limited to 80% × 25, or 20 A, as required by Note 8 of Table 310-16, and must be protected at not more than 20 A, as required by footnote of table.

20-A CBs in panel

Even though wires are "derated" in accordance with Note 8, Sec. 384-16(c) limits circuit load to 16 A on CB or fuse protection.

FIG. 8 Overcurrent-device load limitation of 80 percent is required even for circuit conductors derated in accordance with Note 8 of **NEC** Table 310-16.

Circuit Voltage Limits

Voltage limitations for branch circuits are presented in Section 210-6. In general, branch circuits serving lampholders, fixtures, or receptacles of the standard 15-A or lower rating are limited to operation on 240/120-V, single-phase or 208/120-V, 3-phase, 4-wire systems—or, as the **NEC** puts it, at a maximum voltage rating of 150 V to ground. But exceptions to this rule are necessary to allow more efficient application of the higher-voltage lamps and appliances available today, as follows.

1. Industrial locations: In both indoor and outdoor industrial locations, where the conditions of maintenance and supervision assure that only qualified individuals will service lighting fixtures with mogul-base, screw-shell lampholders or other approved types of lampholders, circuits may be operated above 150 V to ground, but not above 300 V to ground. This means that fluorescent, mercury-vapor, metal-halide, high-pressure-sodium, low-pressure-sodium, and/or incandescent fixtures may be supplied by 480/277-V, grounded-wye circuits, with loads connected phase to neutral and/or phase to phase. Such circuits operate at 277 V to ground—that is, not over 300 V to ground—even, say, when 480-V ballasts are connected phase to phase on such circuits. Or, lighting could be supplied by 240-V delta systems, either ungrounded or with one of the phase legs grounded, because such systems operate at not more than 300 V to ground.

The use of incandescent lighting at over 150 V to ground is accepted by the **NEC** for industrial plants, but incandescent fixtures must be mounted at least 8 ft above the floor—if sufficient mounting height is available. If an 8-ft mounting height is not available, as in pipe tunnels or other low-overhead industrial passages or areas, then 277-V incandescent units may be mounted lower than 8 ft (Fig. 9).

FIG. 9 277-V incandescent lighting is limited to use only in "industrial establishments."

It should be noted that Exception No. 1 to Section 210-6(a) applies only to "industrial establishments." And, although this rule requires a minimum 8-ft mounting height for incandescent mogul-base lampholders, there is no requirement that electric-discharge fixtures with mogul-base, screw-shell lampholders must be mounted 8 ft above the floor.

Although 277-V incandescent lamps are available with medium screw bases, and fixtures for them are available with medium screw-shell lampholders, the use of such equipment violates Exception No. 1*b* to **NEC** Section 210-6(a). And Section 210-6(c)(2) limits the use of medium-base, screw-shell lampholders only to where they are connected to circuit wires with not more than 150 V between the wires, as in 2-wire, 120-V circuits.

2. Commercial and institutional occupancies: In commercial and institutional occupancies, branch circuits serving only ballasts for electric-discharge lighting may be rated over 150 V to ground, but not above 300 V to ground, if the ballasts are in permanently installed fixtures. Note that incandescent fixtures are not recognized for supply by 277-V circuits in these nonindustrial-type occupancies. Only fluorescent, mercury-vapor, metal-halide, high-pressure-sodium, and low-pressure-sodium, of the common lamps, are permitted on 480/277-V circuits in stores, health-care facilities, office buildings, schools, public and commercial areas of hotels, transportation terminals, etc. And when any of those electric-discharge lamps are used with screw-shell lampholders, the fixtures must be installed not less than 8 ft above the floor.

Both the above exceptions permit installations on 480/277-V, 3-phase, 4-wire wye systems, with equipment connected phase to phase (480-V circuits) or phase to neutral (277-V circuits). In either case, the voltage to ground is only 277 V and therefore does not exceed the 300-V maximum. Figure 10 covers the use of electric-discharge lighting on such circuits in the various types of occupancies. In any such application, it is important that the neutral point of the 480/277-V wye be grounded to limit the voltage above ground to 277 V. If the neutral were not grounded and the system operated ungrounded, the voltage to ground, according to the **NEC**, would be 480 V (see the definition of "voltage to ground" in Article 100), and lighting equipment could be used on such circuits only for outdoor applications as specified in Section 210-6(b) (discussed later).

On a neutral-grounded 480/277-V system, incandescent, fluorescent, mercury-vapor, metal-halide, high-pressure-sodium, and low-pressure-sodium equipment can be connected from phase to neutral on the 277-V circuits. If fluorescent or mercury-vapor fixtures are to be connected phase to phase, some **NEC** authorities say that autotransformer-type ballasts cannot be used when they raise the voltage to more than 300 V because, they contend, the **NEC** calls for connection to a circuit made up of a grounded wire and a hot wire (see **NEC** Section 410-78.) On phase-to-phase connection they would require the use of 2-winding, electrically isolating ballast transformers. The wording of Section 410-78 does, however, lend itself to the interpretation that only the supply system to the ballast need be grounded—thus permitting the two hot legs of a 480-V circuit to supply an autotransformer because the hot legs are derived from a neutral-grounded "system." But Section 210-9 contains the basic rule that "branch circuits shall not be supplied by autotransformers," and that can become a complicating factor. Use of a two-winding (isolating) ballast is clearly acceptable and avoids all confusion.

The rules noted above would also permit the use of lighting equipment on 2-wire or 3-wire circuits derived from 240 V, 3-phase, 3-wire delta-connected systems of either the corner-grounded or the ungrounded type, inasmuch as the voltage to ground would be only 240 V (under the 300-V maximum) in either case (Fig. 11). Lamps rated at 240 V and 240-V ballasts would be used. Again, however, there might be objection to the use of autotransformer ballasts between the ungrounded phase wires of the grounded delta system; and, in some cases, there would be definite objection to their use on the ungrounded system because that would be a clear violation of Section 410-78, which requires that autotransformer ballasts be supplied from a grounded system when they raise the voltage to more than 300 V.

Section 210-6 permits the use of an integral switch in any 277- or 480-V lighting fixture, provided the switch is not readily accessible. That means that a manual or relay switch (for on-off control of the fixture or for setting the light output of lamps fed by ballasts with two-level or three-level light control) is acceptable in the fixture if the fixture is mounted up out of the reach of persons not qualified in electrical work, and access to the fixture by qualified maintenance personnel would require the use of a ladder or some type of lift or platform. Remote-control switching is also accepted, with the relay as a switch in the fixture, and control exercised at low voltage by another switch (manual or automatic) away from the fixture. The use of multilevel ballasts to obtain various light outputs from lamps can be facilitated by having integral switches in the fixture or the box supplying the fixture.

Approved for use in:

1. Industrial establishments
2. Office buildings
3. Schools
4. Stores
5. Hospitals
6. Hotels
7. Transportation terminals

480Y/277 V 4 wire

Nominal 277 V

Nominal 480 V

A
B
C
N

Voltage to ground above 150 V but not more than 300 V

Circuit connection may be phase-to-neutral or phase-to-phase

Switch control may be an integral part of fixture, if not readily accessible. Remote-control relay switching may be used.

Fluorescent lighting units are limited to use on 15-A or 20-A circuits

Electric discharge lamp ballast

Fixtures must be permanently installed in other than industrial establishments.

Fluorescent, mercury, metal-halide, and/or sodium fixtures

No restriction on mounting height for fixtures with other than screw-shell lampholders. Screw-shell electric-discharge lampholders must be 8 ft above floor in commercial and institutional occupancies.

FIG. 10 Electric-discharge lighting on 480/227-V grounded-wye circuits must satisfy varying rules.

FIG. 11 Fluorescent, mercury-vapor, metal-halide, sodium-vapor or incandescent lighting on circuits not over 300 V to ground.

3. Other locations: NEC Section 210-6(b) permits fluorescent and/or other electric discharge units to be installed on circuits rated up to 600 V between conductors—but only where the lamps are mounted in permanently installed fixtures on poles or similar structures for the illumination of areas such as highways, bridges, athletic fields, and parking lots at a height of not less than 22 ft, or on other structures such as tunnels at a height of not less than 18 ft (Fig. 12).

This permission for the use of any of the electric discharge units under the conditions described is based on phase-to-phase voltage rather than phase-to-ground voltage. This rule has the effect of permitting the use of 240- or 480-V ungrounded circuits for the lighting applications described. But, as described above, autotransformer-type ballasts are not permitted on an ungrounded system if they raise the voltage to more than 300 V (Section 410-78). In such cases, ballasts with two-winding transformation would have to be used.

Section 210-6(c)(1) applies to "dwelling unit(s) and guest rooms in hotels, motels, and similar occupancies." Note that the wording differentiates between "dwelling unit(s)" and rooms in hotels, motels, and other living or sleeping units that do not satisfy the definition of "dwelling unit" in Article 100. This is an important clarification that arises in a number of other **NEC** rules. Note that a hotel or motel room or a room in a dormitory is not a "dwelling unit" if it does not contain "permanent provisions for cooking" (a kitchen) and is therefore not subject to those rules that apply only to "dwelling unit(s)."

The rule of Section 210-6(c)(1) prohibits the use of circuits with over 150 V between conductors to supply screw-shell lampholders, receptacles, or appliances in dwelling units and the other specified rooms, except for permanently connected appliances and cord-and-plug–connected appliances rated over 1380 W or ¼ hp or more. The rule limits incandescent lighting to the common nominal 120-V rating at all lighting outlets in those places. But this rule has caused considerable discussion and controversy in the past when applied to split-wired receptacles and duplex receptacles of two voltage levels.

In past years, some inspectors have argued that split-wired general-purpose duplex receptacles are not acceptable in dwelling units and in hotel and motel guest rooms because they are supplied by conductors with more than 150 V between them—i.e., 240 V on the 3-wire, single-phase, 240/120-V circuit so commonly used in residences. The two hot legs connect to the brass-colored terminals on the receptacle, with the shorting tab broken off, and the voltage between those conductors does exceed 150 V. The same condition applies when a 240/120-V duplex receptacle is used: The 240-V receptacle is fed by conductors with more than 150 V between them. But the **Code**-making panel ruled that the use of duplex receptacles connected on 240/120-V, 3-wire, single-phase circuits in dwelling units and guest rooms in hotels, motels, dormitories, etc. is not prohibited (Fig. 13).

The objection to split-wired receptacles in dwelling units has a sound basis. Two appliances connected by 2-wire cords to a split-wired receptacle in a kitchen do present a real potential hazard. With, say, a coffee maker and a toaster plugged into a split-wired duplex, if the hot wire in each appliance should contact the metal enclosure of the appliance, there would be 240 V between the two appliance enclosures. The user would be exposed to the extremely dangerous chance of touching the two appliances with different hands, which would put 240 V across the use from hand to hand, on a path through the heart (Fig. 14). The use of non-split-wired receptacles at the usual spacing of up to 12 ft does tend to separate appliances on different hot legs.

The use of split-wired receptacles and other receptacles with more than 150 V between terminals is not ruled out by Section 210-6(c)(1) in any commercial, institutional, or industrial location. Of course, it is perfectly acceptable at any time to use a split-wired receptacle for switch control of one plug-in point, with the other left hot all the time, where both the switched and unswitched wires that are run to the duplex receptacle are tapped from the same hot leg of the 2-wire circuit feeding the switch box (Fig. 15).

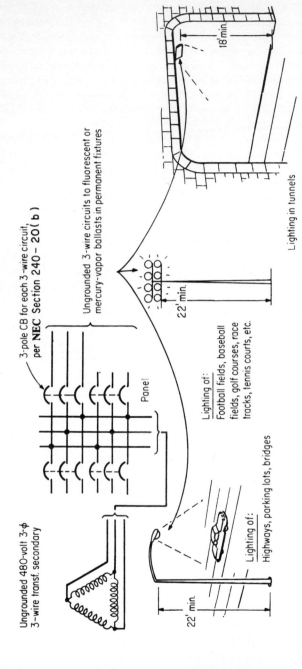

Ungrounded 480-volt 3-φ
3-wire transf. secondary

3-pole CB for each 3-wire circuit,
per **NEC** Section 240 – 20(**b**)

Ungrounded 3-wire circuits to fluorescent or
mercury-vapor ballasts in permanent fixtures

Panel

Lighting of:
Football fields, baseball
fields, golf courses, race
tracks, tennis courts, etc.

Lighting of:
Highways, parking lots, bridges

22' min.

22' min.

18' min.

Lighting in tunnels

FIG. 12 Fixtures at over 300 V to ground (up to 600 V between conductors) may supply lighting
as shown.

General purpose
split-wired receptacles
(top receptacle on one
circuit, bottom receptacle
on other circuit)

FIG. 13 Split-wired receptacles are permitted in residential occupancies (dwelling units) and in all other types of occupancies (commercial, institutional, industrial, etc.).

FIG. 14 This is a potential hazard with split-wiring of duplex receptacles.

FIG. 15 This is not *really* a split-wired receptacle.

Exception No. 1 of Section 210-6(c)(1) permits "permanently connected appliances" to be supplied by a circuit with voltage between conductors in excess of 150 V; and permission is intended for the use of 277-V heaters in dwelling units, as in high-rise apartment buildings and similar large buildings that may be served at 480/277 V.

Section 210-6(c)(2) is worded in such a way as to make the use of 277-V lamps acceptable only when such lamps have mogul bases. A medium-base lamp may be used only on a circuit that does not have more than 150 V between conductors. The use of medium-base 277-V incandescent lamps (which are available) is clearly prohibited by that rule as well as Exception No. 1 of Section 210-6(a), as noted in Fig. 9.

Acceptable Circuit Loads

A single branch circuit to one outlet (for a single machine, appliance, or other individual utilization device) may serve any load and is unrestricted as to ampere rating. Circuits with more than one outlet (even a circuit that supplies a single duplex receptacle—which is two receptacles and may supply two load devices) are subject to **NEC** load limitations. The **NEC** limitations on the use of branch circuits with two or more outlets are as follows:

1. Branch circuits rated 15 and 20 A may serve lighting units and/or appliances. The rating of any one-cord and plug–connected appliance must not exceed 80 percent of the branch-circuit rating. Appliances fastened in place may be connected to a circuit serving lighting units and/or plug-connected appliances, provided the total rating of the fixed appliances fastened in place does not exceed 50 percent of the circuit rating (Fig. 16). However, modern design provides separate circuits for individual fixed appliances. In commercial and industrial buildings, separate circuits should be provided for lighting and separate circuits for receptacles.

FIG. 16   General-purpose branch circuits.

2. Branch circuits rated 30 A may serve fixed lighting units (with heavy-duty-type lampholders) in other than dwelling units or appliances in any occupancy. An individual cord-and-plug–connected appliance that draws more than 24 A may not be connected to this type of circuit (Fig. 17).

3. Branch circuits rated 40 and 50 A may serve fixed lighting units (with heavy-duty lampholders) or infrared heating devices in other than dwelling units, or cooking appliances in any occupancy (Fig. 18).

Note: Lighting or appliances, not both

FIG. 17 Multioutlet 30-A circuits.

Application of the above rules—and other **NEC** rules that refer to dwelling units—must take into consideration the **NEC** definition of that phrase: A dwelling unit is defined as "one or more rooms" used "as a housekeeping unit" and must contain space or areas specifically dedicated to "eating, living, and sleeping" and must have "permanent provisions for cooking and sanitation" (Fig. 19). A one-family house is a "dwelling unit." So is an apartment in an apartment house or a condominium unit. But a guest room in a hotel or motel or a dormitory room or unit is not a "dwelling unit" if it does not contain "permanent provisions for cooking"—which must mean a built-in range or counter-mounted cooking unit (with or without an oven). Nevertheless, the rule of Section 210-8, which requires ground-fault circuit interrupter (GFCI) protection of receptacles in bathrooms of "dwelling units," applies to bathroom receptacles in standard hotel or motel units, even though such units do not contain "permanent provisions for cooking."

It should be noted that the requirement for heavy-duty-type lampholders in lighting units on 30-, 40-, and 50-A multioutlet branch circuits excludes the use of fluorescent lighting on these circuits because fluorescent lampholders are not rated "heavy duty" in accordance with Section 210-21(a). Other electric-discharge lighting fixtures (mercury-vapor, metal-halide, sodium) with mogul lampholders may be used on these circuits provided tap-conductor requirements are satisfied.

As indicated, multioutlet branch circuits for lighting are limited to a maximum loading of 50 A. Individual branch circuits may supply any loads. Excepting motors, this means that an individual piece of equipment may be supplied by a branch circuit that has sufficient carrying capacity in its conductors, is protected against current in excess of the capacity of the conductors, and supplies only the single outlet for the load device.

Fixed outdoor electric snow-melting and deicing installations may be supplied by any of the above-described branch circuits or by individual branch circuits of higher rating, provided the circuit conductors and overcurrent protection are rated not less than 125 percent of the total heater load current. (See Section 426-4 in Article 426, "Fixed Outdoor Electric De-Icing and Snow Melting Equipment.")

Lighting-Circuit Details

A number of specific **NEC** rules affect the design and installation of lighting equipment:

1. Branch-circuit wires that come within 3 in of a ballast within a luminaire (that is, wires running through end-to-end–connected fluorescent fixtures) must be rated at 90°C, as are RHH, THHN, XHHW, and THW wires. THW wire is normally rated at 75°C, but it has a 90°C rating for use in fluorescent fixtures (Fig. 20). As explained later, these 75 and

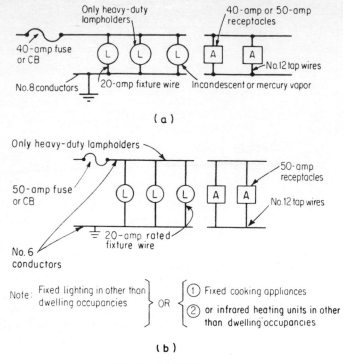

(a)

Only heavy-duty lampholders

40-amp or 50-amp receptacles

40-amp fuse or CB

No.8 conductors

20-amp fixture wire

No.12 tap wires

Incandescent or mercury vapor

Only heavy-duty lampholders

50-amp fuse or CB

50-amp receptacles

No.12 tap wires

No. 6 conductors

20-amp rated fixture wire

Note: Fixed lighting in other than dwelling occupancies } OR { ① Fixed cooking appliances
② or infrared heating units in other than dwelling occupancies

(b)

TYPICAL EXAMPLE

50-amp, 3-phase, 4-wire circuits to supply incandescent floodlights on pole for lighting of a baseball field.

Panel with 400-amp bus and all 50-amp, 1-pole CB's

It is interesting to note that this is not a "lighting and appliance" panelboard as defined in Sec. 384-14 and does not require main protection per Sec 384-16(a)

37.5-amp load per phase

Three 1500-watt floodlights per phase leg

Neutral

Four, single-conductor No.6, type-UF cables underground from panel to pole

Typical 9-light pole

(c)

FIG. 18 (*a*) Multioutlet 40-A circuits. (*b*) Multioutlet 50-A circuits. (*c*) Typical example.

"LIVING" space

"SLEEPING" space (bedrooms)

"SANITATION" (bathroom)

"COOKING" (range or built-in cooktop and oven)

"EATING" space

FIG. 19 Specific conditions identify a "dwelling unit." Eating, living, and sleeping space could be one individual area, as in an efficiency apartment. But the unit must contain a "bathroom," defined in Section 210-8(a) as "an area including a basin with one or more of the following: a toilet, a tub, or a shower." And the unit must contain permanent cooking equipment.

Home run ⟶

Type THW wire now has a 90°C rating for use in wiring through fixtures. Previously, Types RHH, THHN or other 90°C wire had to be used.

Continuous-row fluorescent fixtures

FIG. 20 Type THW wire has a 90°C rating for through-fixture use, within 3 in of a ballast.

90°C wires must be used at the ampacities of 60°C wires (such as TW) of corresponding sizes for circuits up to 100 A, unless equipment is marked to permit the use of wires rated higher than 60°C.

2. For supplying recessed incandescent luminaires, ⅜-in flexible metal conduit, liquid-tight flexible metal conduit, or flexible metallic tubing is acceptable for the 4- to 6-ft length that carries the 150°C fixture wires from the branch-circuit junction box to the luminaire to satisfy Section 410-67(c) (Fig. 21). **NEC** Table 350-3 shows that ⅜-in flexible metal conduit may contain up to four No. 14 AF wires and up to seven No. 18 wires. For a typical recessed incandescent lighting fixture, two AF wires may be used, without a grounding conductor, in a 4- to 6-ft

Typical use of flexible metallic tubing, flexible metal conduit or liquidtight flex : 4-to-6-ft for length fixture whip in ceiling, containing two No. 18 Type AF wires (for 6-amp fixture load, see Section 402-5) or two No. 16 Type AF wires (for 8-amp fixture load). Section 240-4 permits No. 16 and No. 18 fixture wire to be protected at 20 amps.

Lighting supplied by branch circuit protected at not over 20 A

Outlet box

Conduit or cable circuit

May be 3/8-in. size

Suspended ceiling

Recessed incandescent fixture

IMPORTANT!   Flexible metal conduit, liquidtight flex, or flex tubing is equipment grounding conductor because AF wires in flex are tapped from circuit protected at not over 20 amps, as permitted in Section 250-91(b), Exception No 1 or Exception No. 2.

FIG. 21   Flexible raceways and fixture wires must be carefully applied to recessed fixtures.

length of ⅜-in flex in accordance with Exception No. 1*b* of Section 210-19(c). Two No. 16 AF wires (rated at 8 A) or two No. 18 wires (rated at 6 A) may be used on circuits rated not over 20 A, where the current rating of the fixture does not exceed the AF wire rating. Fixture wire used in the flex from the box to the fixture may be protected and sized according to Exception No. 2 of Section 210-19(c) and Section 240-4, which specifically permits the use of fixture wires for taps as covered in Exception No. 1 of Section 210-19(c) and states that No. 16 and No. 18 fixture wires are considered protected by a 20-A CB or fuse.

3. Although flexible metal conduit, flex tubing, and liquid-tight flex are not approved as equipment grounding means in long lengths, Exceptions No. 1 and No. 2 of Section 250-91(b) do recognize the three types of flex as a grounding means if the total length of flex is not over 6 ft. Standard metal flex (Greenfield) may be used as an equipment ground only where the wires within the flex are protected at not over 20 A (Fig. 22). However, an equipment grounding conductor *must* be used (either inside or outside the flex) in every case where metal flex supplies equipment that is not fixed in one place or location. In such cases, where the flex is subject to moving and flexing, an equipment grounding conductor must be used, regardless of length or conductor protection.

4. An electric-discharge lighting fixture, when supported independently of the outlet box, may be supplied by cord-and-plug connection to a receptacle installed above the fixture, provided it is an open installation so that the cord is continuously visible for its entire length and is

FLEX NOT OVER 6 FT LONG is suitable as a grounding means (without a separate ground wire) if the conductors in it are protected by OC devices rated not more than 20 amps.

FIG. 22   Flex not over 6 ft long may serve as grounding means.

not subject to strain or physical damage. Figure 23 shows the hookups covered by **NEC** Section 410-30.

5. Lampholders must not have a rating lower than the load to be served; and lampholders connected to circuits rated over 20 A must be of the heavy-duty type (that is, rated at least 660 W for an "admedium" type, and at least 750 W for other types). Because fluorescent lampholders are not of the heavy-duty type, this excludes the use of fluorescent luminaires on 30-, 40-, and 50-A circuits. The intent is to limit the rating of lighting branch circuits supplying fluorescent fixtures to 20 A. The ballast (not the lamp) is directly connected to the branch circuit, but by controlling the lampholder rating, a 20-A limit is established for the ballast circuit.

Most lampholders manufactured and intended for use with electric-discharge lighting for illumination purposes are rated less than 750 W and are not classified as heavy-duty lampholders. If the luminaires are individually protected, such as by a fuse in the cord plug of a luminaire cord connected to, say, a 50-A trolley or plug-in busway, some inspectors have permitted use of fluorescent luminaires on 30-, 40-, and 50-A circuits. But such protection in the cord plug or in the luminaire is supplementary (Section 240-10), and branch-circuit protection rated at 30, 40, and 50 A would still exclude the use of fluorescent fixtures according to Section 210-21(a).

Modular Wiring

So-called "modular wiring systems" are highly engineered, ultraflexible, plug-in branch-circuit systems for supplying and controlling lighting fixtures in suspended ceilings (Fig. 24). These systems are UL-listed and made of UL-listed raceway, cable, and connector components.

NEC Article 604 recognizes the various types of modular wiring systems that provide plug-in connections to lighting fixtures, switches, and receptacles in all kinds of commercial and institutional interiors that use suspended ceilings made of lift-out ceiling tiles. Except for those fixtures that are UL-listed for cord-and-plug connection as part of listed modular wiring systems for use in suspended-ceiling spaces, cord connection may not be used for fixtures installed in lift-out ceilings.

These manufactured wiring systems are a logical solution to a variety of needs in electrical systems for commercial and institutional occupancies. In the interest of giving the public a better way at a better price, a number of manufacturers have developed basic wiring systems to provide plug-and-receptacle–type interconnection of branch-circuit wires to lighting fixtures in suspended-ceiling spaces. Such systems afford ready connection between the hard-wired circuit

Area light
or floodlight
mounted on
building, pole,
pedestal, or
other structure
or support

Adjustable
mounting
bracket

Cord <u>must</u> be of
the hard-usage or
extra-hard-usage type
and must be as short
as possible

Box secured
in position
and fed by
conduit

Fixed cord connection
from outlet to fixture
(no cord-cap or connector
body used) is permitted
for lighting fixtures
that require adjusting
or aiming after
installation.

This technique may now be used . . .

Ceiling

Fixture hung
from chain or
rod hangers
or otherwise
supported
independently
of ceiling box

Box on ceiling
with hole-cover
to support
cord pendant

Connector body on
end of cord provides
separable connection
to flanged inlet (with
recessed blades) set
in fixture housing.

Fluorescent or
other electric
discharge
fixture

. . . As well as this technique—

Electric discharge
lighting fixture
equipped with
grounding-type
cord and plug

Plug on upper
end of cord
connects to
receptacle

Box must be
"directly" above
the fixture

FIG. 23 Cord connection of lighting fixtures.

GROUND WIRE REQUIRED
In prewired flex assemblies
and in BX or MC cable assemblies

Ceiling
distribution
box

Box connector

Pipe
and
wire

Different prewired cable
assemblies, each with
number of wires and
type of connector
to provide for various
circuit functions

Switch

Lighting
panel

Incandescent, fluorescent, mercury-vapor,
HPS, or metal-halide fixtures—or
feed can be made to receptacle outlets

FIG. 24 Modular, plug-in branch-circuit prewired cable assemblies are
called "manufactured wiring systems."

homerun and the cables and/or ducts that form a grid- or tree-like layout of circuiting to supply
incandescent, fluorescent, or high-intensity discharge (HID) luminaires in the ceiling.

The acknowledged advantages of modular wiring systems are numerous and significant:

- Factory-prewired raceways and cables provide highly flexible and accessible plug-in connection to multicircuit runs of 120- and/or 227-V conductors.
- Drastic reductions can be made in requirements for conventional pipe-and-wire hookups of individual circuits, which are costly and inflexible.
- Plug receptacles afford a multiplicity of connection points for fixtures to satisfy needs for specific types and locations of lighting units. They permit the location of fixtures to serve any initial layout of desks or work stations and still offer unlimited, easy, and extremely economical modification of layouts and the addition of fixtures for any future rearrangements of office landscaping or activities.
- These systems may include switches and/or convenience receptacles in walls, partitions, or floor-to-ceiling poles, with readily altered switching provisions to provide energy conservation through effective on-off control of any revised lighting layout.
- Work on the systems has been covered by agreement between the IBEW and associated trades.
- Such systems have the potential for tax advantage via accelerated depreciation as office equipment rather than real estate.

Modular systems may be used in air-handling ceilings. The equipment may be used in the specific applications and environments for which it is listed by UL.

Residential Circuits

The design of general-purpose lighting and appliance branch circuits for dwelling units (single-family houses and apartments) should be based on a minimum provision of one 20-A, 2-wire,

120-V circuit for each 800 ft² of floor area or one 15-A, 2-wire, 120-V circuit for each 600 ft² of floor area. This is an **NEC** minimum that works out to be a capacity of 3 W/ft². For hotel or motel rooms, at least 2 W/ft² should be used. The number of circuits determined in this way will handle general illumination and convenience receptacles spaced (so no point is more than 6 ft from a receptacle) along the floor line in any wall space of 2 ft width or more in kitchen, family room, dining room, breakfast room, living room, bedrooms, parlor, library, den, sun room, and recreation room (Fig. 25). The wall space occupied by sliding panels in exterior walls must be counted as wall space, and, since room is not available on the wall for mounting wall outlets, floor outlets may have to be installed to ensure that no point along the floor line is more than 6 ft from an outlet.

Location of the receptacle as shown will permit the plugging in of a lamp or appliance located 6 ft on either side of the receptacle [Sec. 210-52(a)].

FIG. 25 Spacing of residential receptacles (**NEC** Section 210-52).

Care should be exercised to provide receptacle outlets in smaller sections of wall space (of width 2 ft or more) segregated by doors, fireplaces, bookcases, or windows. Although receptacles are not required by the **NEC** in hallways, good design practice dictates that convenience receptacles should be spaced to provide at least one for each 20 ft of wall length.

In determining the location of a receptacle outlet, the measurement is to be made along the floor line of the wall and is to continue around corners of the room, but is not to extend across doorways, archways, fireplaces, passageways, or other space unsuitable for having a flexible cord extended across it. The location of outlets for special appliances within 6 ft of the appliance [as per Section 210-50(c)] does not affect the spacing of general-use convenience outlets but merely adds a requirement for special-use outlets.

Figure 25 also shows two wall sections, 9 and 3 ft wide, extending from the same corner of the room. The receptacle located in the wider section of the wall will permit the plugging in of a lamp or appliance located within 6 ft of either side of the receptacle.

Receptacle outlets must be provided for all wall space within the room except individual isolated sections which are less than 2 ft in width. For example, a wall space 23 in wide and located between two doors would not need a receptacle outlet. Sliding panels in exterior walls are counted as regular wall space, and floor-type receptacles can be used to meet the required spacing. It is considered more desirable and safer to have outlets at these locations than to have occupants run extension cords which may, in many instances, cross a passageway used in connection with the sliding glass door or panel.

Fixed room dividers must be considered in spacing receptacles. This is illustrated by the sketch of Fig. 26. In effect, the two side faces of the room divider provide additional wall space, and a table lamp placed as shown would be more than 6 ft from both receptacles A and B. Also, even though no place on the wall is more than 6 ft from either A or B, a lamp or other appliance placed at a point such as C would be more than 6 ft from B and out of reach from A because of the divider.

FIG. 26 A fixed room divider must be counted as wall space requiring receptacles.

In kitchen and dining areas, a receptacle outlet must be installed at each counter space wider than 12 in (Fig. 27). Countertop spaces separated by range tops, sink, etc., must be considered as separate counter spaces. Receptacles rendered inaccessible by appliances fastened in place or occupying dedicated space (such as an outlet behind a refrigerator) are not considered as outlets relative to the required spacing.

At least one wall receptacle outlet must be installed in each bathroom adjacent to the basin location (Fig. 28).

At least one outdoor receptacle must be installed for each one-family dwelling and for each unit of a two-family dwelling (Fig. 29). No outdoor receptacles are required for apartment houses, motels, hotels, or other multioccupancy buildings.

At least one receptacle outlet must be installed in the basement of a "one-family dwelling." This requirement does not apply to two-family houses, apartment houses, hotels, motels, dor-

Receptacle required at each counter space wider than 12 in.

COUNTER SPACES in kitchen and dining rooms such as shown by arrows (above) must be supplied with receptacles if they are over 12 in. wide. Appliances are frequently used even on narrow counter widths; this requirement is designed to remove the dangerous practice of stretching cords across sinks, behind ranges, etc., to feed such appliances.

Inaccessible receptacles.

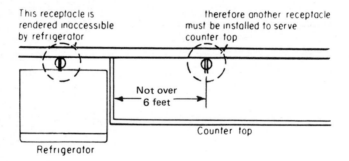

RECEPTACLE LOCATED behind an appliance, making the receptacle inaccessible, does not count as one of the required "counter-top" receptacles. (Neither does it count as one of the appliance-circuit receptacles required to be located every 12 ft.)

FIG. 27 Design requirements for kitchen receptacle layout.

mitories, and the like. A one-family dwelling is "a building consisting solely of one dwelling unit" (Fig. 30).

Any one-family dwelling having an attached garage also must have at least one receptacle outlet in the garage. There is no requirement for a receptacle outlet to be installed in a detached garage. But the rule of **NEC** Section 210-8 does require GFCI protection for a 125-V, single-phase, 15- or 20-A receptacle installed in either an attached or detached garage.

At least one 3-wire, 20-A, 240/120- or 208/120-V circuit shall be provided to serve only receptacles for the small-appliance load in the kitchen, pantry, dining room, and breakfast room of a dwelling unit. The 3-wire circuit must supply the receptacle outlets in these areas and must have no other outlets. Of course, two 2-wire, 20-A, 120-V circuits are equivalent to the 3-wire circuit and could be used.

FIG. 28 In a bathroom of a dwelling unit or hotel or motel guest room, a receptacle must be installed adjacent to the sink. The location of the receptacle will vary, depending on available wall space. The arrows show several possibilities. A receptacle in a medicine cabinet or in the bathroom lighting fixture does not satisfy this rule.

At least one receptacle must be installed out-doors for a one-family dwelling and for each dwelling unit of a two-family dwelling — with GFCI protection in or ahead of each receptacle.

FIG. 29 At least one outdoor receptacle outlet must be provided for *each* dwelling unit of these occupancies.

After many years of controversy and confusion about what a "family room" is and how the requirement for a 20-A small-appliance circuits relates to the receptacles, the phrase "family room" has been deleted from the rules of **NEC** Section 220-3(b). As a result, a room that is a so-called family room (whether it is called a den or a recreation room) may have its receptacle outlets supplied by 15- or 20-A general-purpose circuits that supply just the receptacles or the

At least one receptacle outdoors for one-family dwelling — with GFCI protection in receptacle or ahead of it

At least one receptacle in an *attached* garage — with GFCI protection

One-family dwelling unit

At least one receptacle in basement — for general use

A least one additional receptacle in basement—for a laundry area that might be located there. And a receptacle is required at the laundry no matter where it is located in any dwelling unit.

FIG. 30 Residential design *must* include all these receptacles.

receptacles and lighting outlets. But the rules still require the "two or more 20-A small appliance branch circuits" for all receptacles in the kitchen, pantry, breakfast room, and dining room of a dwelling unit.

As noted, the **NEC** requires that two or more 20-A branch circuits be provided to supply all the receptacle outlets in the kitchen, pantry, dining room, and breakfast room of any dwelling unit—one-family house, apartment unit, or motel or hotel suite with cooking facilities. Such a 20-A kitchen-appliance circuit must supply *only* receptacles and may not supply lighting outlets: but it may supply a clock-hanger outlet in the kitchen or an outdoor outlet, such as at a patio outside the kitchen or dining room. These circuits must not have outlets in any other rooms. The **NEC** states that at least two such circuits must supply countertop receptacle outlets in the kitchen itself [Section 220-3(b)]. See Fig. 31.

The two circuits feeding outlets in the kitchen may also feed outlets in the other designated areas (dining room, pantry, etc.). What the **NEC** prohibits is, say, one circuit feeding the kitchen outlets and the other circuit or circuits feeding the outlets in the other prescribed areas. At least two circuits must have outlets at the countertop in the kitchen, to supply the usual plug-connected appliances (toaster, coffee maker, etc.).

In a dwelling unit, at least one outlet must be installed for the laundry, and at least one 20-A branch circuit must be provided for the one or more laundry receptacles. Outlets in sections of the dwelling that take special appliances, such as laundry equipment, must be placed within 6 ft of the intended location of the appliance [Section 210-50(c)]. A laundry receptacle is not required in an apartment or other dwelling unit in a multifamily building where common laundry facilities are available to all occupants (Fig. 32).

All receptacles in these rooms must be spaced in accordance with Section 210-52(a)—i.e., 12 ft apart—but they DO NOT have to be fed by 20-amp small appliance circuits . . .

Both No.12 small-appliance circuits feed kitchen receptacles

In typical case, kitchen receptacles could be split-wired to 3-wire, 1Ø,240/120V circuit

Pantry   Breakfast room

Den

Kitchen   Family room

Dining room   Recreation room

. . . they may be fed from general-purpose 15- or 20-amp lighting and appliance circuits.

20-amp required kitchen appliance circuit

Convenience plug outlets in kitchen

Plug outlet behind gas wall oven for clock and controls

Range hood light and fan

VIOLATION—Small-appliance circuit must supply only receptacle outlets and not other loads

20-amp appliance circuits

Kitchen   Dining room

Laundry

VIOLATION—Circuit for kitchen, dining room, etc., may not serve laundry

Both circuits feed kitchen receptacles

No. 12 wire   Kitchen   Den

Dining room

VIOLATION—Small-appliance circuits may not have any outlets in other than designated rooms

FIG. 31   Kitchen and dining-room receptacles must be on at least two separate 20-A circuits.

NEC rules on the location and spacing of receptacle outlets in dwelling units also apply to dormitories and to guest rooms in hotels, motels, and similar occupancies that may not conform to the definition of "dwelling unit" but are covered by Section 210-60. An exception notes that receptacles in rooms of hotels and motels may be located so they are convenient for the permanent furniture layout rather than on 12-ft spacing.

In single-family dwellings, apartment units, and hotel and motel rooms, all receptacle outlets rated 20 A or less (except air-conditioning outlets and those outlets on the kitchen-appliance circuits) are considered as outlets for general illumination, and no additional branch-circuit capacity has to be provided for such outlets. They may be connected to separate receptacle circuits or to the circuits supplying lighting.

In a dwelling unit, any number of receptacle outlets may be connected on a circuit, whether or not that circuit also supplies lighting outlets. Recognizing that receptacle outlets have great diversity of usage, that loads are intermittent and/or light-duty cord-connected appliances, and

FIG. 32 A laundry must have at least one outlet on a 20-A circuit.

that the number of such receptacles simply provides convenience of connection, the **NEC** does not require each receptacle to be taken as a fixed 180-VA load—as is required in nonresidential occupancies.

In occupancies other than dwelling units, circuit capacity must be provided for all receptacle outlets and outlets serving loads separate from general illumination [Section 220-2(c)], as covered later under "Receptacle Circuits."

Figures 33 through 35 present recommended design characteristics for residential appliance circuits.

Residential lighting outlets

Certain lighting outlets are mandatory for all dwelling-type occupancies—one-family houses, apartment houses, hotels, motels, dormitories, etc. Section 210-70 requires that at least one wall-switch–controlled lighting outlet be installed in every habitable room, in hallways, stairways,

	Typical Load in Watts	Volts	Wires	Circuit Breaker or Fuse	Number of Outlets	Notes
RANGE	12000	120/240	3- #6	50-60A.	1	Use of more than one outlet is permitted, but not recommended.
OVEN (Built in)	4500	120/240	3- #10	30A.	1	Appliance may be direct connected.
RANGE TOP	6000	120/240	3- #10	30A.	1	Appliance may be direct connected.
RANGE TOP	3300	120/240	3- #12	20A.	1 or more	
DISHWASHER	1200	120	2 #12	20A.	1	These appliances may be direct connected on a single circuit. Grounded receptacles required, otherwise.
WASTE DISPOSER	300	120	2 #12	20A.	1	
BROILER	1500	120	2 #12			Heavy duty appliances regularly used at one location should have a separate circuit. Only one such unit should be attached to a single circuit at a time.
FRYER	1300	120	At Least Two Kitchen Appliance Circuits	20A.	2 or more	
COFFEEMAKER	1000	120				
REFRIGERATOR	300	120	2 #12	20A.	2	Separate circuit serving only refrigerator and freezer is recommended.
FREEZER	350	120	2 #12	20A.	2	

FIG. 33 Circuits for kitchen appliances.

and attached garages, and at outdoor entrances. The rule does not stipulate that these must be ceiling lighting outlets; they may be wall mounted.

In addition, at least one lighting outlet must be installed in every attic, underfloor space, utility room, and basement if it is used for storage or if it contains equipment requiring service. In these latter cases, the lighting outlet does not have to be controlled by a wall switch; it may be a lamp socket controlled by a pull chain, or a canopy switch could be used (Fig. 36).

Two exceptions are given to the basic requirements. Exception No. 1 notes that in habitable rooms other than kitchens or bathrooms, a wall-switch–controlled receptacle outlet may be used instead of a wall-switch–controlled lighting outlet. The receptacle outlet can serve to supply a portable lamp, which would give the necessary lighting for the room. Exception No. 2 states that in hallways, stairways, and at outdoor personnel entrances, remote, central, or automatic control of lighting shall be permitted. This recognition appears to accept remote, central, or automatic control as an alternative to the wall-switch control mentioned in the basic rules.

However, every kitchen and every bathroom must have at least one wall-switch–controlled lighting outlet. And Section 210-70(b) says that at least one wall-switch–controlled lighting outlet or wall-switch–controlled receptacle shall be installed in each guest room of a hotel, motel, or similar occupancy.

Load Devices	Typical Load in Watts	Volts	Wires	Circuit Breaker or Fuse	Number of Outlets	Notes
			Circuits for Laundry Areas			
IRONER	1650	120	2 #12	20A.	1	Grounding type receptacle required.
WASHING MACHINE	1200	120	2 #12	20A.	1	Grounding type receptacle required.
DRYER	5000	120/240	3 #10	30A.	1	Appliance may be direct connected — must be grounded.
			Circuits for Other Loads			
HAND IRON	1000	120	2 #12	20A.	2 or more	
WATER HEATER	3000					Consult Utility Co. for load requirements
WORKSHOP	1500	120	2 #12	20A.	2 or more	Separate circuit recommended.
PORTABLE HEATER	1300	120	2 #12	20A.	1	Should not be connected to circuit serving other heavy duty loads.
TELEVISION	300	120	2 #12	20A.	2 or more	Should not be connected to circuit serving appliances.

FIG. 34 Appliance circuits.

Size of Air Conditioner	Average Wattage	Circuits Required	Size of Each Circ.	Number of Outlets	Remarks
Air Conditioner ¾ hp	1200	Separate Circuit	2 #12 120-v	1	Use of 3-wire, 120/240-volt circuits to unit conditioners offers circuit flexibility for 120 or 240 volts
Air Conditioner 1½ hp	2400	Separate Circuit	3 #12 120/240-v	1	

FIG. 35 Circuits for unit air conditioners.

Outlet Loads Other than General Lighting

In all types of occupancies, branch-circuit capacity must be provided for specific outlets. For lighting other than general lighting, and for appliances other than motor-operated types, circuit capacity for each outlet must be provided as follows.

1. For each outlet supplying a specific appliance or load device, circuit capacity equal to the current rating of the appliance or device must be provided.

A wall-switched lighting outlet is required *in* garage and *at* doors intended only for *personnel entry* to garage. BUT . . .

Wall-switched lighting outlets (or switched receptacle in any room other than kitchen or bathrooms)

Attached garage

Attic

Room   Room   Hall   Room

Room   Hall   Room

Basement

Lighting outlet may be pull-chain socket or may have canopy switch— only in basement, attic, utility room or crawl space

Outdoor personnel entrance— front, side, rear—wall-switched lighting outlet must be used

. . . a lighting outlet is *not* required at garage "vehicle" doors.

EXCEPTION:

Wall switch   Receptacle   Portable lamp

Wall-switched receptacle(s) may be used instead of a lighting outlet in habitable rooms other than kitchens and bathrooms.

FIG. 36 Circuit design for residences *must* include required lighting outlets.

2. In computing the minimum required branch-circuit capacity for outlets not used for general illumination, the actual voltampere rating of a recessed lighting fixture must be taken as the amount of load that must be included in the branch-circuit capacity. Local and/or decorative recessed lighting fixtures must be taken at their actual load values rather than as a load allowance of "180 VA per outlet"—even if each fixture is lamped at, say, 25 W. (In the case where a recessed fixture contained a 300-W lamp, an allowance of only 180 VA would be inadequate.) The inclusion of excessive or inadequate load in determining branch-circuit and feeder capacity must be carefully avoided.

3. For each outlet supplying a heavy-duty lampholder, 600 VA of circuit capacity must be allowed.

4. For each other outlet, such as each general-purpose convenience receptacle outlet in other than one-family and multifamily dwellings and guest rooms of hotels and motels, at least 180 VA of circuit capacity must be provided. This is fully discussed later.

5. When a circuit supplies only motor-operated devices, the provisions of **NEC** Article 430 or 440 must be taken into consideration.

These minimum load allowances for receptacle outlets and outlets for local lighting and appliances are modified as follows:

NEC Table 220-19 establishes the basis for computing the required branch-circuit capacity for electric cooking appliances. For example, the branch circuit for a 16.6-kW household electric range that operates at 230/115 V would be sized as follows (refer to **NEC** Table 220-19):

1. Column A applies to ranges rated not over 12 kW, but the range in this example is rated 16.6 kW.

2. Note 1 below the table explains how to use the table for ranges over 12 kW and through 27 kW. For such ranges, the maximum demand in column A must be increased by 5 percent for each additional kilowatt of rating (or major fraction thereof) above 12 kW.

3. The 16.6-kW range exceeds 12 kW by 4.6 kW (16.6 − 12).

4. In this example there is only one range. The maximum demand in column A for one range is 8000 W. Five percent of this value is 400 W (8000 × 0.05).

5. The maximum demand for this 16.6-kW range must be increased above 8000 W by 2000 W (400 W × 5). (The 5 is derived as 4 kW above 12 kW plus 1 for the remaining 0.6 kW.)

6. The required branch circuit must be sized, therefore, for a total demand load of 10,000 W (8000 W + 2000 W).

7. The required size of the branch circuit is, then,

$$\frac{10,000 \text{ W}}{230 \text{ V}} = 43 \text{ A}$$

The minimum branch-circuit size is 45 A.

For a circuit to the three cooking appliances shown in Fig. 37, the sizing of branch-circuit wires must be based on the following load calculation:

1. Note 4 of Table 220-19 says that the branch-circuit load for a counter-mounted cooking unit and not more than two wall-mounted ovens, all supplied from a single branch circuit and located in the same room, shall be computed by adding the nameplate rating of the individual appliances and treating this total as being equivalent to one range.

2. Therefore, for purposes of computing the branch-circuit load, the three appliances shown may be considered to be a single range with an 18-kW rating (6 kW + 6 kW + 6 kW).

FIG. 37 For branch-circuit sizing, these three cooking units are considered to be an 18-kW range.

3. From Note 1 of Table 220-19, since an 18-kW range exceeds 12 kW by 6 kW (18 − 12), the 8-kW demand of column A for a single range must be increased by 400 W (5 percent of 8000 W) for each of the six additional kilowatts above 12 kW.

4. Thus, the branch-circuit load is

$$8000 \text{ W} + (6 \times 400 \text{ W}) = 10,400 \text{ W}$$

For show-window lighting, instead of calculating circuit capacity at the specified load per outlet, 200 W of load may be allowed for each linear foot of show window, measured along its base (Fig. 38). And it may be safely assumed that the total load determined from such a calculation is based on the reality that such lighting is a continuous load, so that the 125 percent multiplier is not needed.

Required branch-
circuit capacity } = 200 watts X 20 linear ft
for show window
= 4000 watts

FIG. 38 Design capacity for branch-circuit supply to show-window lighting is usually added without knowledge of the actual loads.

A capacity of 1.5 A (or 180 VA) must be allowed for each 5 ft or fraction thereof of each separate and continuous length of fixed multioutlet assembly (surface raceway with spaced receptacles in the cover). Each such 5-ft length is considered as one outlet of 1.5-A capacity. In those cases where it is likely that a number of appliances will be used simultaneously, each 1 ft or fraction thereof is considered as an outlet and requires a load allowance of 1.5 A (Fig. 39).

Load allowed for this bench = 18 × 1½ = **27 A**
At 120 volts, this represents a load of **3240 VA**

THEREFORE, 3240 VOLTAMPERES OF CIRCUIT
CAPACITY MUST BE PROVIDED IN THIS EXAMPLE

FIG. 39 Adequate capacity must be allowed for heavy use of cord-and-plug–connected loads.

No branch-circuit capacity has to be included for multioutlet assemblies in dwellings or guest rooms in motels and hotels. In such occupancies, the multioutlet assembly is considered part of the general lighting load—just as standard receptacle outlets are—and the **NEC** imposes no restriction on the number of receptacles that may be connected to any single branch circuit.

Branch-Circuit Conductors

Although the **NEC** establishes No. 14 wire as the minimum size to be used for branch-circuit wiring, good design practice dictates the use of No. 12 as the minimum size, protected at either 15 or 20 A. For electric ranges of 8¾ kW or more, No. 8 copper or No. 6 aluminum is the **NEC** minimum size, but No. 6 copper or No. 4 aluminum is the recommended size. In a 3-wire, single-phase branch circuit to a household electric range, the neutral may be sized on the basis of 70 percent of the current-carrying capacity of one of the ungrounded conductors, although it may not be smaller than No. 10 wire.

The load current to be used in selecting the correct sizes of conductors for branch circuits can be calculated readily with the simple formulas given in Fig. 40. The selection of insulated conductors for branch-circuit wiring should be based on all conditions involved in any given application. The size should be sufficient to minimize voltage drop, to compensate for reasonable energy losses, and to contain short-circuit heating. The size also is influenced by metallic composition (copper, aluminum, or copper-clad aluminum), construction (stranded or solid), and the duty cycles of loads (continuous or intermittent operation).

2-WIRE CIRCUITS:

$$\text{Line current} \quad = \quad \frac{\text{connected load va (or watts at unity pf)}}{\text{line voltage}}$$

SINGLE-PHASE, 3-WIRE CIRCUITS

Apply above formula, considering each line to neutral separately and using line-to-neutral voltage. Result gives current in line conductors.

THREE-PHASE, 3-WIRE CIRCUITS:

$$\text{Line current} \quad = \quad \frac{\text{balanced 3-phase load va}}{\text{line voltage} \quad \times \quad 1.732}$$

FIG. 40 Lighting and appliance branch-circuit calculations at any power factor.

- The type of insulation on conductors must be suited to operation at the ambient temperature of the area in which they are used.
- Where the ambient temperature exceeds 86°F (or 30°C, the room-temperature maximum for which insulated conductors are rated in **NEC** Tables 310-16 through 310-19), the ampacities of conductors should be derated according to correction factors given at the bottom of each of the ampacity tables (Tables 310-16 through 310-19).
- Where more than three conductors are used in a raceway or cable, the conductor ampacities must be derated from the ampacity values given in Table 210-16 or 310-18 to compensate for the increased heating effect due to reduced ventilation of an enclosed group of closely spaced conductors. (See Note 8 of Tables 310-16 through 310-19.)
- The type of insulation must be suited to the moisture content of its surroundings in the given application. Advantages and limitations related to the dryness or wetness of the area in which

conductors are to be used must be considered. **NEC** regulations regarding the moisture content of the place of installation conform to the following definitions:

Damp location: A location subject to a moderate degree of moisture, such as some basements, some barns, some cold-storage warehouses, under canopies, marquees, roofed open porches, and the like.

Dry location: A location not normally subject to dampness or wetness. A location classified as dry may be temporarily subject to dampness or wetness, as in the case of a building under construction.

Wet location: A location subject to saturation with water or other liquids, such as locations exposed to weather, washrooms in garages, and like locations. Installations underground or in concrete slabs or masonry in direct contact with the earth shall be considered as wet locations.

- **NEC** Section 310-8 says that insulated conductors used underground, in concrete slabs or other masonry in direct contact with earth, in wet locations, or where condensation or accumulation of moisture within the raceway is likely to occur shall be (1) lead-covered; (2) of Type RHW, RUW, TW, THW, THWN, or XHHW; or (3) of a type listed by a testing laboratory as suitable for use in such locations. Cables of one or more conductors used in wet locations must be listed for use in wet locations. Such conductors are not suitable for direct burial in the earth unless they are of a type specifically listed for direct-earth burial. MI cable [Section 330-3, item (9)] and ALS cable (Sections 334-3 and 334-4) are recognized for direct-earth burial by the **NEC**. Type USE, UF, or any other type of cable listed and approved for such use may also be directly buried in the ground. Section 300-5 covers the installation of underground circuits.

Conductor insulations are designed to operate reliably at the anticipated operating temperature (ambient plus temperature rise due to load-induced heat); to resist environmental conditions (moisture, fumes, chemicals, oil and grease, acids, alkalies); to promote installation ease: to keep the costs of installation, maintenance, and replacement within reasonable limits; and to ensure efficient, reliable service by providing proper physical and electrical properties.

Most general-purpose 600-V building wire contains materials that are members of the rubber and thermoplastic families. Rubber, which also includes various butyl and mineral bases, is available in many grades and thicknesses to comply with cost, application, and overloading conditions. It also is applicable, when provided with auxiliary jacketing, in locations subject to oil or heat.

Thermoplastic insulations can resist moisture, oils, and chemicals without additional protection. This makes them smaller in diameter and lighter in weight than comparable rubber-based insulations. They are durable, tough, and lubricated with silicone, and therefore easy to handle and install. Temperature ratings range from 60 to 90°C.

For higher-temperature applications, silicone asbestos insulation can be selected. Types are available for continuous operation at 125°C and for intermittent duty at 200°C.

Type FEP is a small-diameter wire insulated with fluorinated ethylene propylene. Type FEPB has a thinner layer of insulation but has an outer glass braid in sizes 14 to 8, and an asbestos braid in sizes 6 to 2. These types have a 90°C rating when used in dry locations for general wiring, and a 200°C maximum rating for use in special applications.

Wiring in air-handling ceilings

A common application which raises questions about the effects of temperature and moisture on wire insulation is that of wiring above suspended ceilings when such spaces are used for distributing air. **NEC** Section 300-22 says:

(c)   Other Space Used for Environmental Air. Only mineral-insulated, metal-sheathed cable, Type MC cable without an overall nonmetallic covering, and Type AC cable and other factory-assembled multiconductor control or power cable which is specifically listed for the use shall be used for wiring in systems installed in other space used for environmental air. Other type cables and conductors shall be installed in electrical metallic tubing, flexible metallic tubing, intermediate metal conduit, rigid metal conduit, metal surface raceway or wireway with metal covers where accessible, or flexible metal conduit. Electric equipment that is permitted within a building concealed space shall be permitted to be installed in other space used for environmental air if the associated wiring material and fixtures are suitable for the ambient temperature.

The above applies to other spaces such as spaces over hung ceilings which are used for environmental air-handling purposes.

Exception No. 1:   Liquidtight flexible metal conduit in single lengths not exceeding 6 feet (1.83 m).

Exception No. 2:   Integral fan systems specifically identified for such use.

Exception No. 3:   This section does not include habitable rooms or areas of buildings, the prime purpose of which is not air handling.

Exception No. 4:   Listed prefabricated cable assemblies of metallic manufactured wiring systems without nonmetallic sheath shall be permitted where listed for this use.

Exception No. 5:   This section does not include the joist or stud spaces in dwelling units when wiring or equipment passes through such spaces perpendicular to the long dimension of such spaces.

(d)   Data Processing Systems. Electric wiring in air-handling areas beneath raised floors for data processing systems shall comply with Article 645.

Section 300-22(b) covers wiring methods and equipment within "ducts or plenums," which are channels or chambers intended and used only for the supply or return of conditioned air. Such ducts or plenums are sheet metal or other types of enclosures which are provided expressly for air handling; they must be distinguished from "other space used for environmental air," such as the space between a suspended ceiling and the floor slab above it or the space between a raised floor (as used for data-processing wiring) and the slab below the raised floor, which may or may not be intended and used for air handling. Space of that type is covered by Section 300-22(c), although a raised floor used for data-processing circuits must also comply with Section 645-2(c). NFPA Standard 90A defines a duct system as "a continuous passageway for the transmission of air which, in addition to ducts, may include duct fittings, dampers, plenums, fans, and accessory air handling equipment." The word "duct" is not defined, but a plenum is defined as "an air compartment or chamber to which one or more ducts are connected and which forms part of an air distribution system."

NEC Sections 300-22(b) and (c) clearly limit acceptable wiring methods to the ones described (Fig. 41). The listing of acceptable raceway methods includes flexible metallic tubing, which is a raceway recognized by the **NEC** and covered in Article 349. But flexible metallic tubing may be used in air-handling ceilings only in single lengths not over 6 ft long. The same length limitation applies to liquid-tight flexible metal conduit used in air-handling ceilings.

The effect of the above rules is to exclude from use in all air-handling spaces any wiring that is not metal-jacketed or metal-enclosed, to minimize the creation of toxic fumes by burning plastic in case of fire. Section 800-3(d) basically requires telephone, intercom, and other communications circuits to be wired with one of the metal-type wiring methods covered in Section 300-22 when such circuits are located in ducts, plenums, or air-handling ceilings. However, the exception to that rule does recognize nonmetallic communication cables (without raceway) for use in such air-handling spaces provided that such conductors are "classified" (by UL or some other testing lab) as having "adequate fire-resistant and low-smoke producing characteristics." Only that type of cable is permitted for use without metal raceway or jacketing.

NEC Section 725-2(b) says essentially the same thing: Signal and/or control wiring located in an air-handling ceiling must also comply with the rules of Section 300-22(c). Signal and communication circuits must be run in metal raceway or metal cable assembly when used in an

Any length of MI cable, MC cable (which now includes ALS and CS cable), and/or Type AC cable (which is BX cable) may be used.

Any length of rigid metal conduit, IMC, EMT, flexible metal conduit

Warm-air return duct

Warm-air return through fixtures

Cool air supply

Metal surface raceway with metal covers may be used where accessible

But liquidtight flex and the new flexible metallic tubing may be used only in single lengths not over 6 feet.

IN CEILING SPACE: COMPLETE SYSTEM OF PREWIRED CABLE LENGTHS WITH SNAP-IN CONNECTORS FOR FIXTURES AND SWITCHES

Ordered-to-length branch cable set

Power-in, power-out connector assembly

Homerun conduit back to lighting panelboard

Fluorescent luminaire

System includes switch legs and receptacle feeds

Exposed T-bar grid

FIG. 41 In air-handling ceilings, branch-circuit design may utilize *only* these wiring methods.

air-handling ceiling, as required by Sec. 300-22. The exception to section 725-2(b) permits Class 2 and/or Class 3 circuits (as defined in Section 725-3) to be used without metal raceway or metal cable cover in ducts, plenums, or ceiling spaces used for environmental air *provided that the conductors are "listed" as "fire-resistant and low-smoke producing."* Because of the definition of the word "listed" (See Article 100, "Definitions"), this rule would require that any such nonmetallic assembly of conductors used for those circuits be specifically listed in the UL's *Elec-*

trical Construction Materials Directory (or have similar third-party certification) as having the specified characteristics for use without metal raceway or covering in ducts, plenums, and air-handling ceilings.

NOTE: For any such application, it is important to check that the circuit conductors are definitely listed by UL or another testing lab.

The entire subject of the use of wiring methods and equipment above suspended ceilings is broad and detailed. Although the rules of Section 300-22(c) apply only to air-handling spaces above suspended ceilings, such spaces are also subject to the general rules that apply to non-air-handling spaces.

It is generally understood that the foregoing establishes specific limitations on the types of wiring methods which can be used in the space above suspended ceilings when that space is used for air handling. And in various inspection jurisdictions, Section 300-22 has been taken to clearly require one of the designated wiring methods in air-handling ceiling spaces. However, some inspection authorities do not enforce the above interpretation and do permit other wiring methods in air-handling plenum chambers. But that must be carefully checked out with local inspection authorities.

Overcurrent Protection

According to the basic **NEC** rule, the rating or setting of an overcurrent device in any branch circuit must not exceed the current-carrying capacity of the circuit conductor. Figure 42 shows the basic rules that apply to overcurrent protection for branch circuits.

When a branch circuit supplies a load for which loss of power would create a hazard, Exception No. 8 of Section 240-3 states it is not necessary to provide "overload protection" for its conductors, *but* "short-circuit protection" must be provided. By "overload protection," this exception means protection at the conductors' ampacity—that is, protection that would prevent overload by opening the circuit at any value of load current *over* the ampacity of the wires.

Several points should be noted about this exception, which is shown at the bottom of Fig. 42.

1. Use of this exception is reserved only to applications where circuit opening on "overload" would be more objectionable than the overload itself, "such as in a material handling magnet circuit." In that example, mentioned in the exception, the loss of power to such a magnet while it is lifting a heavy load of steel would cause the steel to fall and would certainly be a serious hazard to personnel working below or near the lifting magnet. To minimize the hazard created by such power loss, the circuit to the magnet need not be protected at the conductor ampacity. A higher value of protection may be used, to let the circuit sustain an overload rather than open on overload and drop the steel. Because such lifting operations are usually short-duration, intermittent tasks, an occasional overload is far less of a safety concern than the dropping of the magnet's load.

2. The permission to eliminate "overload protection" is not limited to a lifting-magnet circuit, which is mentioned simply as an example. Other electrical applications that present similar "hazards" would be equally open to the use of this exception.

3. Although the exception permits the elimination of overload protection and requires short-circuit protection, it gives no guidance on the selection of the actual rating of the protection that must be used. For such circuits, fuses or a CB rated, say, at 200 to 400 percent of the full-load operating current would give freedom from overload opening. Of course, the protective device ought to be selected with as low a rating as would be compatible with the operating characteristics of the electrical load. And it must have sufficient interrupting capacity for the circuit's available short-circuit current.

Circuit conductors protected by fuse or CB poles in panelboard—with amp rating of fuse or CB not in excess of conductor ampacity, per Sec. 210-19. Conductor ampacity from Table 310-16.

Panelboard

NOTE THIS

Each fuse or CB in a panel must not carry a load current greater than 80% of fuse or CB amp rating—

BUT, THIS IS AN IMPORTANT EXCEPTION! If "overload protection" creates a hazard, it may be eliminated.

Loss of power to magnet would present hazard of falling weight to personnel

Protective device **may** have rating higher than ampacity of circuit conductors

Load

If load on circuit operates for any period of 3 or more hours, load current must not exceed 80% of fuse or CB rating.

Branch circuit

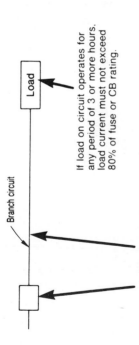

Fuse or CB with amp rating not greater than ampacity or load-current rating of circuit conductor. See **NEC** Sec. 210-19 and Tables 310-16 through -19 with notes.

NOTE: Although the rule of NEC Sec. 210-19(a) requires branch-circuit conductors to have an ampacity of not less than the rating of the branch circuit (which is the rating of the branch-circuit overcurrent device) for a multi-outlet branch circuit that supplies and readily accessible receptacles, conductors of a branch circuit supplying lighting outlets only is required to have ampacity adequate only for the load-current being supplied. The latter circuit may have an overcurrent device that is the next standard rating of device above the conductor ampacity where there is not a standard rating of a device that is exactly equal to conductor ampacity — such as where conductors have their ampacity derated because more than three conductors are installed in a conduit or EMT run.

Power circuit to lifting magnet of electromagnetic material-handling rig

FIG. 42 Basic design rules for branch circuits to other than motor loads.

4. Finally, this exception is not a mandatory rule but a permissible application. It says "overload protection shall not be required"; it does *not* say that overload protection shall not be used. Overload protection may be used, or it may be eliminated. Obviously, careful study should always precede the use of this exception.

A number of general and specific **NEC** rules apply to overcurrent protection for branch circuits (and the same rules apply to feeder circuits). These rules apply to both fuses and circuit breakers, as follows.

1. As shown in Fig. 43, an overcurrent device (fuse or CB trip unit) must be placed at the supply end of each ungrounded conductor of the circuit to be protected (except where the conductors are permitted to be used as tap conductors without overcurrent protection at the supply end—such as the 10-, 25-, and 100-ft tap conductors recognized by **NEC** Section 240-21).

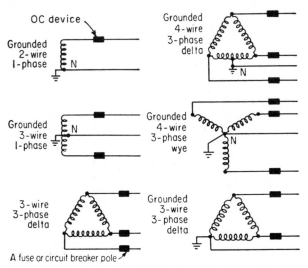

FIG. 43   An overcurrent device must be placed in each ungrounded conductor.

2. Where the device protecting a conductor is of a rating or setting that also provides protection for smaller conductors tapped from the larger conductor, there is no need to provide protection at the point where the smaller conductors are tapped from the larger conductor. Figure 44 shows this exception in relation to the basic rule above.

3. An overcurrent device must not be placed in any permanently grounded conductor, except (*a*) where the device simultaneously opens all conductors of the circuit, as a multipole CB could, or (*b*) where the device is used for motor running overload protection and is required by Sections 430-36 and 430-37 for the grounded conductor of a 3-phase, 3-wire circuit from a delta supply with a corner (one of the phase legs) grounded (Fig. 45).

4. Branch-circuit taps (the smaller tap conductors used to supply individual lighting outlets or appliances, as covered in Sections 210-19 and 210-20) are considered protected by the branch-circuit overcurrent devices (Fig. 46). For recessed incandescent fixtures that are wired with No. 16 or No. 18 fixture wire in flex "whips" that are between 4 and 6 ft long, the branch-circuit protective device, rated at 15 or 20 A is adequate protection for the fixture wires.

BASIC RULE

EXCEPTION

PROTECTING TWO SIZES OF CONDUCTORS

FIG. 44 An exception permits one set of protective devices to protect different sizes of wire.

5. Overcurrent devices must be located so they are readily accessible, with the following exceptions: overcurrent devices for circuits tapped from busway, which must be in the device for tapping the busway, in the cord plug of a fixed or semifixed luminaire supplied from troley busway, or mounted on a luminaire plugged into a busway; overcurrent devices that are supplementary overcurrent protection for subdivided electric-resistance heating elements used in ductwork above suspended ceilings, as required by Sections 424-22(b) and (c); and overcurrent devices that are supplementary to the protection required by the **NEC**, as recognized by Section 240-10 (Fig. 47). "Readily accessible" means "capable of being reached quickly, for operation, renewal, or inspections, without requiring those to whom ready access is requisite to climb over or remove obstacles or to resort to portable ladders, chairs, etc."

6. Plug fuses must not be used in circuits of more than 125 V between conductors, but they may be used in grounded-neutral systems where the circuits have more than 125 V between ungrounded conductors but not more than 150 V between any ungrounded conductor and ground. That means plug fuses may be used on circuits rated 120 V, derived from 240/120-V or 208Y/120-V systems. But an exception in that **NEC** rule does not permit the use of plug fuses for each hot leg of a 240-V, 2-wire or 3-wire circuit derived from a 240/120-V system: for each hot leg of a 208-V, 2-wire or 3-wire circuit derived from a 208Y/120-V system; or for each hot wire of a 208/120-V, 3-wire or 4-wire circuit derived from a 208Y/120-V system (Fig. 48). The screw shell of a plug fuseholder must be connected to

FIG. 45 Two cases where an overcurrent device is permitted in a grounded conductor.

Multi-outlet circuits

15A — #14 wire — May be No. 16 or No. 18 fixture wire.

20A — #12 wire

30A — #10 wire

15-amp taps to lighting units or individual outlets, with taps not over 18-in. long.

Taps to fixed or stationary cooking appliances, with taps rated not less than load and not less than 20 amps.

40A — #8 wire

Taps to fixed cooking appliances, with taps rated not less than load and not less than 20 amps.

50A — #6 wire

20-amp taps to lighting units or other outlets, with taps not over 18-in. long.

EXAMPLE:

Wall oven   Cook top

No. 6
Main 50–amp
circuit conductors — TAP CONDUCTORS RATED NOT LESS THAN 20 AMPS

.... BUT NOT LONGER THAN NEEDED FOR SERVICING THE UNITS

50 A fuses or CB   All appliances rated 115/230 volts and used in kitchen of residence   JB

FIG. 46   Sizes of tap wires protected by branch-circuit devices.

the load side of the circuit, and the **NEC** does not require a disconnecting means on the supply side of a plug fuse.

7. Fuseholders for plug fuses must be of Type S to accommodate Type S plug fuses. Fuseholders must be designed or equipped with adapters to take either a 0- to 15-A Type S fuse, a 16- to 20-A Type S fuse, or a 21- to 30-A Type S fuse. A 0- to 15-A fuseholder or adapter must not be able to accept a 20- or 30-A fuse and a 20-A fuseholder or adapter must not be able to take a 30-A fuse. (Fig. 49). The purpose of this rule is to prevent overfusing of 15- and 20-A circuits. Edison-base (non-Type S) plug fuses are recognized only for replacement use.

   NEC Section 240-60 recognizes the use of "300-volt type" cartridge fuses and fuseholders in circuits rated not over 300 V between conductors and in circuits supplied by a system with a grounded neutral and rated not over 300 V to ground from any ungrounded conductor. Such cartridge fuses are, therefore, recognized for use on circuits derived from a 480Y/277-V system.

8. Cartridge fuses and fuseholders must be such that a fuse of any given class cannot be used in a fuseholder of a lower current or higher voltage rating.

9. Fuseholders for current-limiting fuses must not accept fuses that are not current-limiting. Low-voltage current-limiting cartridge fuses up to 600 A are available with special ferrules or knife blades to provide installation in conformity with the requirements of Section 240-

60(b) (Fig. 50). Such fuses can fit into current-limiting fuseholders and also fit standard **NEC** fuseholders. But standard **NEC** fuses (Class H) cannot be inserted into the current-limiting holders. This arrangement maintains safety in applications where, say, the bracing of busbars in a panel or switchboard is based only on the amounts of short-circuit let-through current that a current-limiting fuse will pass but where the busbars could not take the higher current that would flow at that point in the event non-current-limiting fuses were used.

Class R fuses are UL-listed and meet the current **NEC**. They fit Class R fuseholders that accept only current-limiting fuses and reject fuses of low interrupting rating, such as Class H fuses. Class R fuses have the same overall physical dimensions as standard Class H (formerly **NEC**) fuses and will fit existing fused equipment, eliminating the need for duplicate stocking.

10. Fuses rated for 600 V may be used at any lower voltage.

11. A disconnecting means must be provided on the supply side of all fuses in circuits of more than 150 V to ground and of cartridge fuses in circuits of any voltage where the fuses would be accessible to other than qualified personnel (Fig. 51).

12. Circuit breakers must open simultaneously all ungrounded conductors of the circuits they protect; i.e., they must be multipole CB units, except that individual single-pole CBs may be used for the protection of each ungrounded conductor of certain types of circuits, including ungrounded 2-wire circuits, 3-wire single-phase circuits, and lighting or appliance branch circuits connected to 4-wire, 3-phase systems provided such lighting or appliance circuits are supplied from a grounded-neutral system and the loads are connected line to neutral (Fig. 52).

FEEDER TAP FROM BUSWAY

Plug-in connection for tapping off feeder or subfeeder must contain overcurrent protection, which does not have to be within reach of person standing on floor

Busway feeder

Panelboard, switchboard, motor control center, or trough with two or more branch circuits tapped off

BRANCH-CIRCUIT TAP FROM BUSWAY

Plug-in connection for tapping-off branch circuit must contain overcurrent protection, which does not have to be within reach of person standing on floor

Busway feeder

Branch circuit to motor (or lighting, etc.)

Motor

Starter

LUMINAIRE FED BY CORD FROM BUSWAY

Trolley busway

Fuse in cord plug out of **reach** from floor

Luminaire supplied by cord connection to busway

Floor level

FIG. 47 Overcurrent devices may not be readily accessible under these conditions.

Those are the rules of **NEC** Section 240-20(b). They coordinate with Section 210-10, which permits the use of 2-wire branch circuits tapped from the outside conductors of systems where the neutral is grounded on 3-wire dc or single-phase, 4-wire, 3-phase and 5-wire, 2-phase systems.

All poles of the disconnecting means used for branch circuits supplying permanently connected appliances over 300 VA or 1/8 hp must be operated at the same time. This requirement applies where the circuit is supplied through either circuit breakers or switches.

In the case of fuses and switches, when a fuse blows in one pole, the other pole may not

FIG. 48 Circuit voltage determines the acceptability of the use of plug fuses.

FIG. 49 Type S fuses provide safety through noninterchangeability.

Class RK-1 or Class RK-5 fuses

FIG. 50 Class R fuses and Class R fuseholders provide for rejection of inadequately rated fuses.

Disconnect required if fuse <u>is</u> accessible to non-qualified personnel

Cartridge fuse

Circuit of any voltage

Circuit not over 150 volts to ground

No disconnect required ahead of a plug fuse

...BUT THIS IS PERMISSIBLE

Panel with only cartridge fuses and no switching means, used where only qualified personnel will get at the fuses — such as at top of pole or structure for outdoor lighting

Fuse-only panelboard

Flood-lights in cluster at top

Pole

Grade line

Feeder to panel

FIG. 51   Cartridge fuses must generally be used in switches.

necessarily open; and the requirement to manually switch together involves only the manual operation of the switch. Similarly, when two circuit breakers are connected with handle ties, an overload of one of the conductors with the return circuit through the neutral may open only one of the circuit breakers; but the manual operation of the pair when used as a disconnecting means will open both poles. The words "manually switch together" should be considered as meaning "operating at the same time," i.e., during the same operating interval; they apply to the equipment when used as a disconnecting means and not when used as an overcurrent protective device.

Circuit breakers with handle ties are, therefore, considered as providing the disconnection required by this section. The requirement to "manually switch together" can be achieved by a "master handle" or "handle tie," since the operation is intended to be effected manually. The intent is not to require a common trip for the switching device but to require that it have the ability to disconnect ungrounded conductors with a single movement of the hand.

The basic rule of Section 210-4 requires that multiwire branch circuits (such as 240/120-V, 3-wire, single-phase and 3-phase, 4-wire circuits at 208/120 V or 480/277 V) be used only with loads connected from a hot or phase leg to the neutral conductor. The intent of the **NEC** is that line-to-line connected loads may be used, other than in Exception No. 1, only where the poles of the circuit protective device operate together—simultaneously and automatically—in response to overcurrent. A multipole CB satisfies the rule, but a fused multipole switch does not (Fig. 53).

BASIC RULE
3-phase 3-wire circuit
3-pole CB to open all ungrounded conductors of circuit
Single-pole CBs may not be used for a 3-phase, 3-wire circuit

EXCEPTIONS

2-wire ungrounded circuit
Two single-pole CBs may be used } Load

1-phase 3-wire circuit
Two single-pole CBs may be used } Load

3-phase 4-wire circuit
Single-pole CB may be used for each ungrounded conductor
Lighting and appliance branch circuits rated 120/208-v or 480Y/277-v depending upon the load

EXAMPLES:

Two 1-pole CBs may be used for a 240-volt or 208-volt circuit to an electric heating unit or air-conditioning unit if the CBs are not intended to provide the disconnect means required by Secs. 424-19(b)(1), 440-63, 426-20, 422-21(b), 430-85, or 430-103.

Two 1-pole CBs for 240-volt circuit (Also see Sec. 210-10)

2-pole thermostat with "OFF" position is disconnect (Sec. 424-20)

Heating unit

Two 1-pole CBs for 208-volt circuit

Plug is the disconnect (Sec. 440-63)

Air conditioner

NOTE: If multipole breakers are used instead of the single-pole units shown, the breakers could serve as disconnect means because they would satisfy Section 210-10 requiring poles of a disconnect to switch together.

NOTE: A 3-pole CB must always be used for a 3-phase, 3-wire circuit supplying phase-to-phase loads fed from an ungrounded delta system, such as 480-volt outdoor lighting for a parking lot, etc., as permitted by Section 210-6(b).

FIG. 52 Selecting between a multipole CB and multiple single-pole CBs.

The use of multipole switching (that is, the disconnect and not the "overcurrent device" itself) on such circuits provides safety to persons working on or replacing lighting fixtures or receptacle outlets connected line to line in existing systems. When a receptacle is fed by more than one hot wire, someone attempting to work on the receptacle may shut off only one of the branch-circuit breakers or switching devices in the branch-circuit panelboard, and believe that the circuit has been killed. However, the other leg to the receptacle will still be hot and will pose a shock hazard

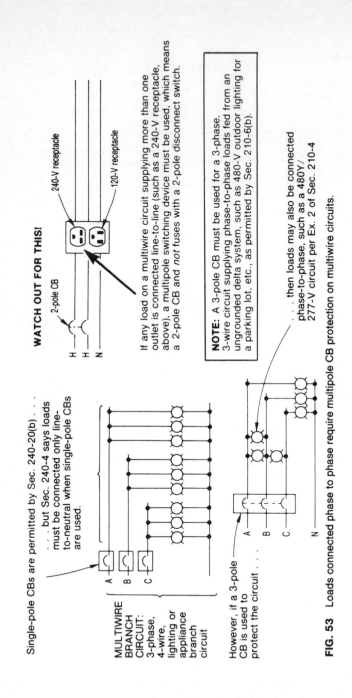

Single-pole CBs are permitted by Sec. 240-20(b) . . . . . . but Sec. 240-4 says loads must be connected only line-to-neutral when single-pole CBs are used.

MULTIWIRE BRANCH CIRCUIT: 3-phase, 4-wire, lighting or appliance branch circuit

However, if a 3-pole CB is used to protect the circuit . . . .

WATCH OUT FOR THIS!

2-pole CB

240-V receptacle

120-V receptacle

H
H
N

If any load on a multiwire circuit supplying more than one outlet is connected line-to-line (such as a 240-V receptacle, above), a multipole switching device must be used, which means a 2-pole CB and *not* fuses with a 2-pole disconnect switch.

NOTE: A 3-pole CB must be used for a 3-phase, 3-wire circuit supplying phase-to-phase loads fed from an ungrounded delta system, such as 480-V outdoor lighting for a parking lot, etc., as permitted by Sec. 210-6(b).

. . . then loads may also be connected phase-to-phase, such as a 480Y/277-V circuit per Ex. 2 of Sec. 210-4

A
B
C
N

FIG. 53 Loads connected phase to phase require multipole CB protection on multiwire circuits.

to that person. The same is true of lighting fixtures connected phase to phase at 240, 208, or 480 V.

The last sentence in Section 210-4 requires that a two-pole (simultaneously operating) circuit breaker or switch be used at the panelboard (supply) end of a 240/120-V, single-phase, 3-wire circuit that supplies one or more split-wired duplex receptacles or duplex switches (two switches on a single mounting yoke) in a dwelling unit (a one-family house, an apartment in an apartment house, a condominium unit, or any other occupancy that conforms to the definition of "dwelling unit" given in Article 100). Figure 54 shows that a two-pole CB, two single-pole CBs with a handle tie that enables them to be used as a two-pole disconnect, or a two-pole switch ahead of branch-circuit fuse protection will satisfy the requirement that both hot legs must be interrupted when the disconnect means is opened to deenergize a multiwire circuit to a split-wired receptacle. This provides the greater safety of disconnecting both hot conductors simultaneously to prevent shock hazard.

A shock hazard may exist when a piece of electrical equipment is replaced or maintained after only one of two hot supply conductors has been opened. If only one of the two single-pole CBs protecting a 240/120-V 3-wire circuit to a split-wired receptacle is shut off, the load device, say a lamp, connected to one of the receptacles will go out. Repair personnel may presume that they will not contact any hot terminals within the receptacle box, but they will receive an electric shock from the other hot leg.

ANY SPLIT-WIRED RECEPTACLE IN A
DWELLING UNIT MUST HAVE MULTIPOLE,
SIMULTANEOUS DISCONNECT MEANS AT
CIRCUIT ORIGIN . . .

. . . OR THIS MAY BE DONE . . .

A 2-pole switch ahead of plug fuses will satisfy as disconnect on circuit to one or more split-wired receptacles, provided there are no phase-to-phase-connected loads on the circuit.

FIG. 54 Two-pole disconnect must be used for switching residential split-wired receptacles.

1. "Single-pole CBs rated 120 volts ac are suitable for use in a single-phase multiwire circuit where the neutral is connected to the load."

2. "Single-pole circuit breakers rated 120/240 volts ac are suitable for use in a single-phase multiwire circuit *with or without* the neutral connected to the load."

FIG. 55 Single-pole CBs must be suited to their particular design function.

NOTE: Circuits supplying split-wired receptacles in commercial, industrial, and institutional occupancies (places that are not "dwelling units") may contain single-pole CBs, plug fuses without switches, or single-pole switches ahead of fuses.

It should also be noted that although a two-pole switch ahead of fuses may serve as the simultaneous disconnect required ahead of split-wired receptacles, such a switch does not satisfy the requirement of Exception No. 2 of section 210-4 for a simultaneous multipole "branch-circuit protective device" when a multiwire circuit supplies any loads connected phase to phase. In such a case, a two-pole CB must be used because fuses are single-pole devices and do not assure simultaneous opening of all hot legs on overcurrent or ground fault.

Figure 55 points out two specific UL limitations on the use of molded-case CBs: A CB marked "120 V ac" may be used only for a load that is connected phase to neutral. Two such CBs are not suitable for protecting, say, a 2-wire, 240 V circuit with the load connected phase to phase. For such a circuit, UL regulations require CBs marked "120/240 V ac," and these may be used for any phase-to-neutral load as well as for phase-to-phase loads.

Conductor Temperature Rating

An important step in the design of branch circuits is the selection of the type of conductor to be used: TW, THW, THWN, RHH, THHN, XHHW, etc. The various types of conductors are covered in **NEC** Article 310, and the ampacities of conductors with different insulations and temperature ratings are given in Tables 310-16 through 310-19 for varying conditions of use—in raceways, in open air, at normal or higher-than-normal ambient temperatures. Conductors must be used in accordance with all the data in those tables and in the detailed notes given with them.

The selection of the best wire or cable for a proposed application has become increasingly complex because of the hundreds of different types of wire and cable available. Modern building wire (rated 600-V) can have thermoplastic insulation rated for use at various ambient temperatures, in wet or dry locations, or with a somewhat smaller overall diameter (THHN, THWN) to permit greater conduit fill or easier pulling. Or, a wire with cross-linked polyethylene insulation (XHHW) can be selected for tough applications. Cable assemblies range from nonmetallic sheathed cable (Type NM) and service-entrance cable (Type SE), which have been in common use for many years, to the new flat-conductor cable assemblies (Type FCC) for surface wiring on floors under carpet squares.

In selecting wire or cable, it is important to check the UL *Electrical Construction Materials Directory (Green Book)*. **NEC** Section 110-2 says, "The conductors and equipment required or permitted by this Code shall be acceptable only when approved." This wording is being widely interpreted to require the use of only UL-listed products if UL listing is given. This is consistent with the OSHA law, which requires the use of products listed by a nationally recognized testing lab.

In selecting the type and temperature rating of wire for branch circuits, consideration must be given to a very important UL qualification regarding the temperature ratings of equipment terminations. Although application data on temperature ratings of conductors connected to equipment terminals are not given in the **NEC**, they nevertheless become part of the mandatory regulations of the **NEC** because of Section 110-3(b). This section incorporates the instructions in UL and other listings books as part of the **NEC** itself. It reads as follows: "Listed or labeled equipment shall be used or installed in accordance with any instructions included in the listing or labeling."

A basic rule in the UL *Green Book* states that, in general:

> The termination provisions on equipment are based on the use of 60°C conductors in circuits rated 100 A or less and the use of 75°C conductors in higher rated circuits.
>
> If the termination provisions on equipment are based on the use of other conductors, the equipment is either marked with both the size and temperature rating of the conductors to be used or with only the temperature rating of the conductors to be used. If the equipment is marked with only the temperature rating of the conductors to be used, that temperature rating is required for the ambient temperature in the equipment and the 60°C ampacity (100-A or less circuits) or 75°C ampacity (over 100-A circuits) should be used to determine the size of the conductors.
>
> Higher-temperature-rated conductors may be used, though not required, if the size of the conductors is determined on the basis of the 60°C ampacity (100-A or less circuits) or 75°C ampacity (over 100-A circuits).

This temperature limitation on terminals applies to the terminals on all equipment—circuit breakers, switches, motor starters, contactors—except where some other specific condition is recognized in the general information preceding the product category. Figure 56 illustrates this vitally important matter, which has been widely disregarded in general practice. When terminals are tested for suitability at 60°C or 75°C, the use of 90°C conductors operating at their higher current ratings poses a definite threat of heat damage to switches, breakers, etc. Many termination failures in equipment suggest overheating even where the load current did not exceed the current rating of the breaker, switch, or other equipment.

When a terminal that is rated at 60°C is fed by a conductor loaded to its 90°C ampacity and operating at 90°C, a substantial amount of heat will be conducted from the 90°C conductor metal to the 60°C terminal. Over a period of time, this can damage the termination even though

UNLESS A CIRCUIT BREAKER OR SWITCH IS MARKED OTHER-WISE, circuit conductors connected to the terminals must not operate at more than a 60 C ampacity for a breaker or switch rated 100 A or less and must not operate at more than a 75 C ampacity for a breaker or switch rated over 100 A (refer to **NEC** Tables 310-16 through -19).

THAT MEANS:

FOR CBs, SWITCHES, CONTACTORS, ETC. RATED 100 A OR LESS—Use TW wire (or use THW, THHN, RHH, XHHW or other higher-temperature wire at the ampacity of the corresponding size of TW wire).

FOR CBs, SWITCHES, CONTACTORS, ETC. RATED OVER 100A—Use TW, THW, THWN OR XHHW wire at their ampacities permitted up to 75 C (or use RHH, THHN or other higher-temperature wire at the ampacity of the corresponding size of 75 C wire).

SWITCH (fusible or nonfusible)

Load terminals

Circuit breaker

FIG. 56 UL specifies maximum temperature ratings for conductors connected to equipment terminals.

the load current does not exceed the equipment current rating and does not exceed the ampacity of the 90°C conductor. Whenever two metallic parts at different operating temperatures are tightly connected together, the higher-temperature part (say 75°C or 90°C wire) will transmit heat to the lower-temperature part (the 60°C terminal) and thereby raise its temperature.

The ampacity of any given size of conductor is established by the ability of the conductor insulation to withstand the I^2R heat produced by the current flowing through the conductor. But it must not be assumed that the equipment to which that conductor is connected also is capable of withstanding the heat that will be thermally conducted from the metal of the conductor to the metal of the terminal to which the conductor is tightly connected.

Although this limitation on the operating temperature of terminals in equipment might appear to eliminate any advantage that higher-temperature conductors have over lower-temperature conductors, there are still advantages to using the higher-temperature wires.

NEC Ampacities

The ampacity of a conductor is the amount of current, in amperes, that the conductor can carry continuously under specified conditions of use without developing a temperature in excess of the value that represents the maximum temperature that the conductor insulation can withstand. **NEC** Table 310-16, for instance, specifies ampacities for conductors where not more than three conductors are contained in a single raceway or cable or directly buried in the earth, provided that the ambient temperature is not in excess of 30°C (86°F). For higher ambient temperatures, the ampacity must be derated in accordance with the *correction factors* at the bottom of the table. For applications with more than three conductors in a conduit or cable, the ampacities of the conductors must be derated in accordance with Note 8 to Tables 310-16 to 310-19. And where there are both an elevated ambient temperature and more than three conductors in a cable or conduit, both deratings must be made; that is, the conductors must be derated twice.

NEC Table 310-16 clearly indicates that No. 14, 12, and 10 conductors rated at 90°C do, in fact, have higher ampacities than those of the corresponding sizes of 60°C and 75°C conductors. As shown in Fig. 57, No. 12 TW and No. 12 THW copper conductors are both assigned an ampacity of 25 A under the basic application conditions of the table. But, for a copper conductor, a No. 12 THHN, RHH, or XHHW (dry location) has an ampacity of 30 A. However, the

FIG. 57 Excerpt from **NEC** Table 310-16 shows higher ampacities for No. 14, No. 12, and No. 10 branch-circuit conductors.

> †The load current rating and the overcurrent protection for conductor types marked with an obelisk (†) shall not exceed 15 amperes for 14 AWG, 20 amperes for 12 AWG, and 30 amperes for 10 AWG copper; or 15 amperes for 12 AWG and 25 amperes for 10 AWG aluminum and copper-clad aluminum.

FIG. 58   This note below **NEC** Table 310-16 radically alters conductor applications for No. 14, No. 12, and No. 10 circuit wires.

footnote to Table 310-16 (shown in Fig. 58) requires that the "load-current rating" and "over-current protection" for No. 14, 12, and 10 copper conductors be taken as 15, 20, and 30 A, respectively, regardless of the type and temperature rating of the insulation on the conductors. When applied to the selection of branch-circuit wires in cases where conductor ampacity derating is required by Note 8 of Tables 310-16 through 310-19 for conduit fill (over three wires in a raceway), the footnote to Table 310-16 affords advantageous use of the 90°C wires for branch-circuit makeup. The reason is that, as stated in Note 8, the derating of ampacity is based on taking a percentage of the actual ampacity value shown in the table, and the ampacity values for 90°C conductors are higher than those for 60°C and 75°C conductors.

Where not more than three conductors are used in a conduit, copper and aluminum conductors may supply load currents up to the ampacities given in **NEC** Table 310-16 for the various sizes and insulations of conductors with temperature ratings up to 90°C, with the exceptions indicated in the footnote to Table 310-16. These ampacities apply when the ambient temperature does not exceed 30°C (86°F). If higher ambients exist, the ampacity must be derated in accordance with the table of correction factors given at the bottom of Table 310-16.

Note 8: ampacity derating

When there are more than three current-carrying conductors in a raceway or cable, their current-carrying capacities must be decreased to compensate for proximity heating effects and reduced heat dissipation due to reduced ventilation of the individual conductors, which are bunched or which form an enclosed group of closely placed conductors. In such cases, the ampacity of each conductor must be reduced as indicated in the table of Note 8, which appears after **NEC** Tables 310-16 to 310-19.

Note 8 to Table 310-16 says: "Where the number of conductors in a raceway or cable exceeds three, the ampacities given in Table 310-16 shall be reduced as shown in the following table." If, for instance, four No. 8 THHN conductors are used in a conduit, the ampacity of each No. 8 is reduced from the 50-A value shown in the table to 80 percent of that value. Each No. 8 then has a new (reduced) ampacity of 0.8 × 50 A or 40 A. Moreover, Section 240-3 of the **NEC** states, "Conductors shall be protected against overcurrent in accordance with their ampacities as specified in Tables 310-16 through 310-19 and all applicable notes to these tables." Thus, fuses or CB poles rated at 40 A are, as a general rule, required for overcurrent protection.

The application of these No. 8 conductors and their protection rating is based on the general concept behind the **NEC** tables of maximum allowable current-carrying capacities (now called *ampacities*). The **NEC** tables of ampacities of insulated conductors installed in raceways or cables have always set the maximum continuous current that a given size of conductor can carry continuously (for 3 hours or longer) without exceeding the temperature limitation of the insulation on the conductor—that is, the current above which the insulation would be damaged.

This concept has always been verified in Section 240-1, whose wording has been virtually unchanged for over 30 years, and which says:

Overcurrent protection for conductors and equipment is provided to open the circuit if the current reaches a value that will cause an excessive or dangerous temperature in conductors or conductor insulation.

NEC Table 310-16 gives ampacities under two conditions: that the raceway or cable containing the conductors is operating in an ambient temperature not over 30°C (86°F) and that there are not more than three current-carrying conductors in the raceway or cable. Under those conditions, the ampacities shown correspond to the thermal limit of each particular insulation. But in any case where either of the two conditions is exceeded, the ampacity of the conductors of a circuit must be reduced (and protection provided at the reduced ampacity) to ensure that the temperature limit of the insulation is not exceeded:

1. If the ambient temperature is above 30°C, the ampacity must be reduced in accordance with the correction factors given with Table 310-16.

2. If more than three current-carrying conductors are used in a single cable or raceway, the conductors tend to be bundled in such a way that their heat-dissipating capability is reduced, and excessive heating will occur at the ampacities shown in the table. As a result, Note 8 requires the reduction of ampacity, and conductors have to be protected at the reduced ampacity.

It should be clearly understood that any reduced ampacity, required because of a higher ambient and/or conductor bundling, has the same meaning as the value shown in the table: derated ampacity and normal (table) ampacity both represent a current value above which excessive heating would occur under the particular conditions. And if there are two conditions that lead to excessive heating, then more reduction of current is required than if only one such condition existed.

(In the 1975 and 1976 editions of the **NEC**, alterations were made in the text of Note 8 that seemed to depart from the long-standing concept. Without any explanation, the rewording no longer clearly referred to a "change" in ampacity when the percentage was applied for more than three conductors in a raceway. In the 1981 **NEC**, the radical change was completed—again without any explanation in the voluminous committee reports and technical documentation that preceded it. However, the **NEC** no longer contains the confusing and contradictory rules of the 1981 edition.)

The application of Note 8 depends upon how many current-carrying conductors are in a raceway. A true neutral conductor (a neutral carrying current only under conditions of unbalanced loading on the phase conductors) is not counted as a current-carrying conductor. If a 208Y/120-V circuit or a 480Y/277-V circuit is made up of three phase legs and a true neutral in a conduit, the circuit is counted as consisting of only three conductors in the conduit, and it is not necessary to derate for conduit fill, as described in Note 8 of Tables 310-16 to 310-19. But neutrals for circuits with these voltage ratings must be counted as current-carrying conductors if the major portion of the load consists of electric-discharge lighting [Note 10(c) of the tables]. Thus, if the circuit supplies fluorescent, mercury, or metal-halide lamps, the neutral is counted as the fourth current-carrying conductor because it carries third-harmonic current, which approximates the phase-leg current under balanced loading. Any such 4-wire circuit must have its ampacity derated to 80 percent of the ampacity given in Table 310-16, as required by Note 8 to the tables.

As shown in Fig. 59, the makeup of a branch circuit consists of first selecting the correct size of wire for the particular load current (based on the number of wires in the raceway, ambient temperature, and ampacity deratings) and then relating the rating of the overcurrent protective device to all the conditions.

In Table 310-16, which applies to conductors in raceways and in cables and covers the vast majority of conductors used in electrical systems for power and light, the ampacities for No. 14,

FIG. 59 Loading and protection of branch-circuit wires must account for derating of conductor "ampacity."

12, and 10 conductors are particularly significant because copper conductors of those sizes are involved in the vast majority of branch circuits in modern electrical systems. Number 14 wire has an ampacity of 20 A; No. 12 has an ampacity of 25 A: and No. 10 has an ampacity of 30 A. The typical impact on circuit makeup and loading is as follows.

1. Number 12 TW or THW copper is shown to have an ampacity of 25 A instead of 20 A; based on the general UL requirement that equipment terminals be limited to use with 60°C conductors up to a 100-A rating, No. 12 THHN or XHHW copper conductors must also be treated as having only a 25-A continuous rating and not a 30-A rating, as shown in Table 310-16. But the footnote to Table 310-16 limits all No. 12 copper wires to a maximum load of 20 A and requires that they be protected at not more than 20 A.

2. The ampacity of 25 A for No. 12 TW and THW copper wires interacts with Note 8 to Tables 310-16 to 310-19 where there are, say, six No. 12 TW current-carrying wires for the phase legs of two 3-phase, 4-wire branch circuits in one conduit supplying, say, receptacle loads. In such a case, the two neutrals of the branch circuits do not count in applying Note 8, and only the six phase legs are counted in determining the extent to which all circuit conductors must have their ampacities derated to the "percent of values in Tables 310-16 and 310-18," as given in the table in Note 8. In the case described here, that literally means that each No. 12 phase leg may be used at an ampacity of 0.8 $\times$ 25 A, or 20 A. And the footnote to Table 310-16 would require the use of a fuse or CB rated not over 20 A to protect each No. 12 phase leg. Each No. 12 would be protected at the current value that represents the maximum I^2R heat input that the conductor insulation can withstand. The only possible qualification is that Section 384-16(c) would require the load current on each of the phase legs to be further limited to no more than 80 percent of the 20-A rating of the overcurrent device—that is, to 16 A—if the load current is "continuous" (operates steadily for 3 hours or more), an unlikely condition for receptacle-fed loads. But the above-described application is different from what it would have been under previous **NEC**s.

3. If two 3-phase, 4-wire branch circuits of No. 12 TW or THW copper conductors were installed in a single conduit or EMT run to supply, say, fluorescent or other electric-dis-charge lighting, the two neutrals of the branch circuits would carry harmonic current even under balanced conditions and would have to be counted, along with the six phase legs, as current-carrying wires for ampacity derating in accordance with Note 8. In such a case, as the table of Note 8 shows for 7 to 24 conductors, their ampacities would have to be derated to 70 percent of the 25-A value shown in Table 310-16, which gives each conductor an ampacity of 17.5 A (0.7 $\times$ 25 = 17.5). If 20-A overcurrent protection were used for each No. 12 phase leg and the load were limited to no more than 17.5 A (or 16 A, which is 0.8

× 20 A, for a continuous load), the application would satisfy Note 8 in the **NEC**. And Section 210-19(a) permits the use of 20-A protection as the next higher standard rating of protective device above the conductor ampacity of 17.5 A, provided that the circuit supplies only lighting outlets—and not readily accessible receptacle outlets that might allow the conductors to be overloaded above 17.5 A.

Use of the 20-A protection on conductors rated at 17.5 A is recognized only for fixed circuit loading (like lighting-fixtures outlets) and not for the variable loading of receptacle outlets, because any increase in load current over 17.5 A would produce excessive heat input to the eight bundled No. 12 conductors in the conduit, which would damage and ultimately break down conductor insulation. Of course, the question arises: Over the operating life of the electrical system, how can excessive current be prevented? The practical, realistic answer is: It can't! It would be better to use 15-A protection on the No. 12 conductors or use conductors rated at 90°C.

4. Number 12 THHN or XHHW conductors, with their 90°C rating and consequently greater resistance to thermal damage, could be used for the two 3-phase, 4-wire, 20-A circuits to the electric-discharge lighting load; they would satisfy all **NEC** rules and would not be subject to insulation damage. With eight current-carrying wires in the conduit, the ampacity derating to 70 percent as required by Note 8 would be applied to the 30-A value shown in Table 310-16 as the ampacity of No. 12 THHN, RHH, or XHHW (dry locations). Then, because 0.7 × 30 A = 21 A, the maximum of 20-A protection required by the footnote to Table 310-16 would ensure that the conductors could never be subjected to excessive current and its damaging heat. And if the original loading on the conductors is set at 16 A [the 80 percent load limitation of Section 384-16(c)] for continuous operation of the lighting, any subsequent increase in load—even up to the full 20-A capacity—would not reach the 21-A maximum ampacity set by Note 8.

Figure 60 summarizes the application described in items 3 and 4 above.

Advantage of 90°C wires

If the four circuit wires in Fig. 59 are rated at 90°C, as are THHN, RHH, or XHHW conductors, then the loading and protection of the circuit must be related to the required ampacity derating as shown in Fig. 61. The application is based on these considerations:

1. As described in Figs. 57 and 58, each No. 12 THHN has an ampacity of 30 A from **NEC** Table 310-16, but the footnote to the table limits the load and overcurrent protection on any No. 12 THHN to not more than 20 A.

2. Because the neutral of the 3-phase, 4-wire circuit must be counted as a current-carrying wire, there are four conductors in the conduit; each conductor must therefore have its ampacity derated to 80 percent of its table ampacity value, as required by Note 8 of Tables 310-16 to 310-19. Each No. 12 then has a new (derated) ampacity of 0.8 × 30 A, or 24 A.

3. By using a 20-A, single-pole protective device (fuse, single-pole CB, or one pole of a three-pole CB), which is the maximum protection permitted by the footnote to Table 310-16, each No. 12 THHN is easily made to comply with the requirement of Section 210-19 that the branch-circuit wire have an ampacity "not less than" the maximum load current being supplied.

4. If the lighting load on the circuit is noncontinuous—that is, does not operate for any period of 3 hours or more—the circuit may be loaded up to its 20-A maximum rating.

5. If the load is continuous—full-load current flows for 3 hours or more—the load on the circuit must be limited to 80 percent of the rating of each 20-A fuse or CB pole, as required by

Eight wires for two 3-phase, 4-wire branch circuits to fluorescent lighting that will operate for periods of over 3 hrs

All 20-A, 1-pole CBs

CASE 1
With TW or THW No. 12 copper conductors:
No. 12 ampacity = 25 A, from Table 310-16
From Note 8, max load = 0.7 × 25 = 17.5 A
From Sec. 384-16 (c), max load = 0.8 × 20 = 16 A

CASE 2
With THHN, XHHW or RHH No. 12 copper conductors:
No. 12 ampacity = 30 A, from Table 310-16
From Note 8, max load = 0.7 × 30 = 21 A
From Sec. 384-16 (c), max load = 0.8 × 20 = 16 A

IN CASE 1, CONDUCTORS ARE NOT PROTECTED IN ACCORDANCE WITH THE 17.5-A MAXIMUM ALLOWABLE LOAD CURRENT. IN CASE 2, THEY *ARE* PROTECTED AGAINST EXCESSIVE LOAD CURRENT.

FIG. 60 Conductors with 90°C insulation eliminate the chance of conductor damage due to overload.

All 20-A, 1-pole CBs

Conduit

4 No. 12 THHN in conduit

Circuit to electric-discharge lighting

Each derated No. 12 has a load rating of 0.8 × 30 A, or **24 A**

NOTE: TW or THW wires would have to be derated from 25 A to 20 A.

FIG. 61 Derating of 90°C branch-circuit wires (No. 14, No. 12, and No. 10) is based on higher ampacities.

Section 384-16(c). Then 16 A is the maximum load. [Although **NEC** Section 210-22(c) at one time contained an Exception that dispensed with the 80 percent load limitation on conductors for continuous circuit operation, if conductors are derated in accordance with Note 8, a protective device in a panelboard must comply with Section 384-16(c), which does require the 80 percent load limitation in any case where the protective device is not recognized (by UL) for continuous operation at 100 percent load. UL rules state that "unless otherwise marked, circuit breakers should not be loaded to exceed 80 percent of their current rating, where in normal operation the load will continue for 3 hours or more."]

Figure 62 shows the use of two 3-phase, 4-wire circuits of THHN conductors in a single conduit. The 90°C wires offer distinct advantages (substantial economies) over the use of either 60°C (TW) or 75°C (THW) wires for the same application, as follows.

FIG. 62   90°C conductors can take derating without losing the full circuit load-current rating.

Resistive load: If the circuit shown feeds only incandescent lighting or other resistive loads (or electric-discharge lighting does not make up "a major portion of the load"), then Note 10(c) of Tables 310-16 to 310-19 does not require the neutral conductor to be counted as a current-carrying conductor. In such cases, circuit makeup and loading could include these considerations:

1. With the six phase legs as current-carrying wires in the conduit, Note 8 requires that the ampacity of each No. 12 be derated from its basic table value of 30 A to 80 percent of that value, or 24 A.
2. Then each No. 12 is properly protected by a 20-A CB or fuse, satisfying both Section 210-19 and the footnote to Table 310-16.
3. If the circuit load is not continuous, each phase leg may be loaded to 20 A.
4. If the load is continuous, a maximum of 16 A (80 percent) must be observed to satisfy Section 384-16(c).

Electric-discharge load: If the two circuits of Fig. 62 supply electric-discharge lighting (fluorescent, mercury vapor, metal halide, high-pressure sodium, or low-pressure sodium), the makeup and loading must be as shown in case 2 of Fig. 60. If that same makeup of circuits supplies noncontinuous loads, the circuit conductors may be loaded up to 20 A per pole.

In case 2 of Fig. 60, the only difference between such circuit makeups using XHHW or

RHH conductors and those using THHNs is the need for ¾-in conduit instead of ½-in conduit because of the larger cross-sectional area of RHH and XHHW (see Tables 3A and 3B in **NEC** Chapter 9).

Color Coding of Circuit Conductors

NEC rules on the color coding of conductors (Section 210-5) apply only to branch-circuit conductors and do not directly require the color coding of feeder conductors. However, Section 384-3(f) does require identification of the different phase legs of feeders to panelboards, switchboards, etc.—and that requires some technique for marking the phase legs. Many design engineers insist on color coding for feeder conductors, to afford effective balancing of loads on the different phase legs.

The color coding of branch-circuit conductors can be divided into three categories.

1. Grounded conductor: The grounded conductor of a branch circuit (the neutral of a wye system or a grounded phase of a delta) must be identified by a continuous white or natural-gray color for the entire length of conductors of size No. 6 or smaller. Where wires of different systems (such as 208/120 and 480/277 V) are installed in the same raceway, box or other enclosure, the neutral or grounded wire of one system must be white; the neutral of the other system must be gray or white with a color trace (stripe) along its insulation to distinguish it from the white neutral (Fig. 63). If there are three or more systems in the same raceway or enclosure, the additional neutrals must be white with color traces other than green. The point is that neutrals of different systems must be distinguished from each other when they are in the same enclosure.

FIG. 63 Neutrals of different systems must be distinguished where circuits are in the same raceway or enclosure.

Exceptions to Section 200-6 modify the basic rule that requires the use of continuous white or natural gray color along the entire length of any insulated grounded conductor (such as a grounded neutral) in sizes No. 6 or smaller. One exception permits the use of other colors (black, purple, yellow, etc.) for a grounded conductor in a multiconductor cable under certain conditions:

1. Such a conductor is used only where qualified persons supervise and perform service or maintenance on the cable—such as in industrial and mining applications.

2. Every grounded conductor of color other than white or gray is effectively and permanently identified at all terminations by distinctive white marking or other effective means applied at the time of installation.

This permission is intended to allow the practice in commercial and industrial facilities where multiconductor cables are commonly used.

2. Hot conductors: The **NEC** no longer requires that individual hot conductors of a multiwire circuit be identified. It is no longer necessary, for instance, to use black, red, and blue for the phase conductors of a circuit, with a white conductor as a neutral. The absence of a specific rule on required colors for the phase (or hot) legs of branch circuits permits all the phase conductors of a circuit to have the same color. That is, all circuit conductors may be black, or blue, or red, or any other color (Fig. 64).

NEC Section 215-8 does require positive identification of the "high" leg (the hot conductor with higher voltage to ground than the other hot legs) of a 240/120-V, 3-phase, 4-wire delta

Neutral must be white or
gray. . .

208Y/120V
or
480/277V
3-phase, 4-wire
circuit

BUT, ungrounded conductors may be all black, all red, or all of any color or combination of colors other than white or green.

NOTE: Although color coding of hot conductors is not *required*, such practice certainly is acceptable to the **NE Code** where designers and/or installers find additional safety and more effective installation provided by color identification of each different phase of a system.

IF THERE ARE TWO SYSTEM VOLTAGES:

208Y/120V
circuit

480Y/277V
circuit

White neutral

All of these hot legs may be black, or all of any color or combination of colors other than white, natural gray, or green.

Gray neutral
(or white with color stripe)

208Y/120V
circuit

Separate
raceways

480Y/277V
circuit

White neutral (or gray)

All of these hot legs may be black, or all of any color or combination of colors other than white, natural gray, or green.

White neutral (or gray)

FIG. 64 Color coding is not required for ungrounded (hot) conductors of multiwire branch circuits.

system, as shown in Fig. 65. The wording of this section recognizes orange as the preferred color for the high leg of a 4-wire delta supply without disturbing current practices in various local areas, where other colors (such as red, yellow, or blue) or other means of identification are required by electric-utility regulations or local code.

HIGH-LEG CONDUCTOR may be orange in color or may be some other color—such as red or yellow—as long as the color or tagging or other identification clearly distinguishes this as the one with higher voltage to ground at any connection point where the neutral is present.

FIG. 65   High-leg conductor (heavy arrow) is the only hot leg required by **NEC** to be identified.

Note that identification of the phase leg with 208 V to ground is required only at those points in the system where the neutral is present, such as in panelboards, motor-control centers, and other enclosures where circuits are connected. The purpose is to provide a warning that 208 V, and not 120 V, exist from the high leg to the neutral. Such a warning minimizes the chances that a 208-V circuit might be accidentally or unwittingly connected to a 120-V load, such as lamps or appliances or 120-V operating coils in motor starters. The connection would burn out the 120-V equipment and present a hazard to personnel.

3. Grounding conductor: An equipment grounding conductor of a branch circuit (if one is used) must be color-coded green or green with one or more yellow stripes; or, the conductor may be bare, as covered in Section 210-5(b). However, Exception No. 1 of Section 210-5(b) refers to Section 250-57(b), which now says that an equipment grounding conductor of size larger than No. 16 may be other than a green insulated conductor or green with yellow stripe(s). Section 250-57 permits an equipment grounding conductor with insulation that is black, blue, or any other color—provided that one of the three techniques specified in Section 250-57 is used to identify this conductor as an equipment grounding conductor.

The first technique consists in stripping the insulation from the insulated conductor for the entire length of the conductor appearing within a junction box, panel enclosure, switch enclosure, or any other enclosure. With the insulation stripped, the conductor appears as a bare conductor, which is recognized by the **NEC** for the purpose.

The second techinque that is acceptable is to paint the exposed insulation green for its entire length within the enclosure. If, say, a black insulated conductor is used in a conduit coming into a panelboard, the length of the black conductor within the panelboard can be painted green to identify it as an equipment grounding conductor.

Equipment grounding conductor up to No. 6 may be bare or covered to show a green color or green with one or more yellow stripes.

But, for a grounding conductor larger than No. 6, an insulated conductor of other than green color or green with yellow stripes may be used provided *one* **of the following steps is taken:**

1. Stripping the insulation from an insulated conductor of another color (say, black) for the entire length that is exposed in the box or other enclosure, so that the conductor appears as a bare conductor.

2. Painting the exposed insulation green for its entire length within the enclosure.

3. Wrapping the entire length of exposed insulation with green-colored tape or green-colored adhesive labels.

NOTE: See Section 250-57(b), Exception No. 3 and Section 310-10(b), Exception No. 2 for use of "multiconductor cables."

FIG. 66 Equipment grounding conductor for a branch circuit must be identified.

The third acceptable method is to mark the exposed insulation with green-colored tape or green-colored adhesive labels. Green-colored conductors must not be used for any purpose other than equipment grounding.

Figure 66 summarizes these rules regarding the identification of branch-circuit equipment grounding conductors.

Need for color coding

The color coding of circuit conductors (or the use of some other method of identifying them) deserves the close, careful, and complete attention of all those concerned with electric wiring. By allowing ready identification of the two or three phase legs and neutrals in a wiring system, color coding provides the easiest and surest way of balancing loads among the phase legs, thereby providing for full, safe, and effective use of circuit capacities. In circuits where color coding is missing or not effectively applied, loads or phases get unbalanced, many conductors are either badly underloaded or excessively loaded, and breakers or fuses often are increased in size to eliminate tripping due to overload on only one phase leg. For reasons of safety and energy conservation, as well as full, economical application of system equipment and materials, modern electric usage demands the many real benefits that color coding can provide.

For the greater part of its existence the **NEC** required the very clear, rigid color coding of branch circuits for good and obvious safety reasons. The color coding of hot legs to provide load balancing is a safety matter. Section 220-3(d) requires the balancing of loads from branch-circuit hot legs to neutral. Section 220-22 bases the sizing of feeder neutrals on clear knowledge of load balance to determine "maximum unbalance." And mandatory differentiation of voltage levels is in the safety interest of electricians and others maintaining or working on electric circuits, to warn of various levels of hazard.

In the interest of safety and the effective use of any electrical system, modern design demands the thorough, effective, and consistent color coding of circuit wires. For all branch circuits in a given system, all ungrounded wires of the same color must be connected to the same ungrounded feeder conductor. This means that all branch-circuit conductors connected to a given busbar in a panelboard must be of the same color. In a 3-phase panel, for instance, all red wires must connect to only one of the busbars; all black wires must connect to another busbar; etc. (Fig. 67). And if the red wires in one panel are connected to a busbar fed by, say, phase A of the feeder, all red wires in all other panels throughout the entire wiring system must also connect to busbars fed by a phase-A feeder conductor. It can readily be seen that such color coding of branch-circuit conductors would require that the various phases of feeder circuits be identified. Color coding is usually limited to conductors pulled in raceways. If cables had to meet color-coding rules, 2-wire cables would have to be made in three different color combinations—red and white, blue and white, and black and white—to provide for 2-wire circuits from the different busbars in panels and the 2-wire taps from multiwire circuits.

FIG. 67 Color coding of phases must be consistent throughout the entire system.

Because 99 percent of all electrical systems involve no more than two voltage configurations for circuits up to 600 V, and because there has been greater standardization in circuit voltage levels, there should be industry-wide standardization on circuit conductor identification. A clear, simple set of rules could cover the preponderant majority of installations, with exceptions made for the relatively small number of cases where unusual conditions exist and where the local inspector may authorize other techniques. Color coding should follow some basic pattern, such as the following.

120-V, 2-wire circuit: grounded neutral, white; ungrounded leg, black

240/120-V, 3-wire, single-phase circuit: grounded neutral, white; one hot leg, black; one hot leg, red

208Y/120-V, 3-phase, 4-wire circuit: grounded neutral, white; one hot leg, black; one hot leg, red; one hot leg, blue

240-V delta, 3-phase, 3-wire circuit: one hot leg, black; one hot leg, red; one hot leg, blue

240/120-V, 3-phase, 4-wire, high-leg delta circuit: grounded neutral, white; high leg (208 V to neutral), orange; one hot leg, black; one hot leg, red

480Y/277-V, 3-phase, 4-wire circuit: grounded neutral, gray; one hot leg, brown; one hot leg, orange; one hot leg, yellow

480-V delta, 3-phase, 3-wire circuit: one hot leg, brown; one hot leg, orange; one hot leg, yellow

The standardization effected by coding according to a set of simple, specific color designations will ensure all the safety and operating advantages of color coding to all electrical systems. Particularly today, with electrical systems being subjected to an unprecedented amount of alterations and additions because of the development and expansion of electric usage, conductor identification for safety is required over the entire life of each system.

A second step in the clarification and expansion of color coding would be required color coding for all circuits—multiwire branch circuits, branch circuits without a neutral or other grounded conductor, feeders, and even service conductors. As now indicated in **NEC** Section 384-3(f), the correct and effective loading of all circuits (to use the full capacity of all phases and to prevent unintentional imbalances, with the attendant chance of oversizing of protection and overloading of conductors) depends upon ready identification of all conductors at all points in a system.

Of course, there are alternatives to color identification throughout the lengths of conductors. Color differentiation is almost worthless for color-blind electricians. And it can be argued that color differentiation of conductors poses problems because electrical work is commonly performed in darkened areas where color perception is reduced, even for those with good eyesight. The **NEC** already recognizes the use of white tape or paint over conductor insulation ends at terminals, to identify neutrals where the conductor size is larger than No. 6 (Section 200-6). Number markings (1, 2, 3, etc.) spaced along the length of a conductor, on the insulation—say, white numerals on black insulation—might prove very effective for differentiating conductors. Or the letters *A*, *B*, and *C* could be used to designate specific phases. Or a combination of color and such markings could be used. But some kind of conductor identification is essential to the safe, effective hookup of the ever-expanding array of conductors used throughout buildings and systems today.

Receptacle Circuits

The design of branch circuits for plug receptacles requires careful determination of particular requirements. The type and size of occupancy and the nature of the work performed there will indicate the best manner of handling plug-connected loads. For known appliances, individual or multioutlet branch circuits should be used, depending on the size of the load.

Automatic appliances should always be provided with separate circuits to isolate them from the effects of faults or other disturbances in other load devices. Fans or heaters might be individually fed by branch circuits, grouped on their own branch circuits, or connected to receptacles on general lighting circuits.

Of course, the number of plug-in appliances in a particular area will greatly affect the circuiting. In general, plug-in devices should be supplied from receptacles on circuits other than general-lighting circuits. In this way, the loading of circuits can be kept under control, and spare-capacity provisions can be much more realistic.

The number of plug outlets connected on a single circuit should be related to the amount of load likely to be connected to any one receptacle in a particular occupancy. When plug-connected loads are known, determination of the proper number of plug outlets is relatively easy. And the number of circuits required in such cases follows directly from the loads and circuit capacities.

When plug-connected loads are not known, the type of occupancy will indicate the possible appliances to be provided for. If the possible appliances are relatively heavy-current devices, two or three outlets per circuit might be the maximum number to allow for efficient and convenient use of the circuits. If the possible appliances are low-current devices, up to 10 or even more plug outlets may be connected on the circuit without likelihood of overload.

NEC limitations regarding the number of plug outlets on any one branch circuit should be carefully observed in selecting the number of receptacle circuits, as follows.

Nonresidential occupancies

NEC Section 210-50 specifies where and when receptacle outlets are required on branch circuits. Note that there are no specific requirements for receptacle outlets in commercial, industrial, and institutional installations other than for store windows in Section 210-25(d). There is the general rule that receptacles do have to be installed where flexible cords are used. In nonresidential buildings, if flexible cords are not used, there is no requirement for receptacle outlets. They have to be installed only where they are needed, and the number and spacing of receptacles are completely up to the designer. The **NEC** takes the position that receptacles in nonresidential buildings have to be installed only where needed for connection of specific flexible cords and caps and demands that where such receptacles are installed, each must be taken as a load of 180 VA.

Section 220-2(c)(4) requires that every general-purpose single, duplex, or triplex convenience receptacle outlet be taken as a load of 180 VA, and that amount of circuit capacity must be provided for each such outlet, as shown in Fig. 68. If a single 15-A, 115-V circuit is used to supply only receptacle outlets, then the maximum number of general-purpose receptacle outlets that may be fed by that circuit is

$$15 \text{ A} \times 115 \text{ V} \div 180 \text{ VA or 9 outlets}$$

For a 20-A, 115-V circuit, the maximum number of general-purpose receptacle outlets is:

$$20 \text{ A} \times 115 \text{ V} \div 180 \text{ VA or 12 outlets}$$

NOTE: In these calculations, the actual results work out to be 9.58 receptacles on a 15-A circuit, and 12.77 on a 20-A circuit. Some inspectors round off these values to the nearest integer and permit 10 receptacle outlets on a 15-A circuit and 13 on a 20-A circuit.

The value of 115 V is used in these calculations, although, in a reference to "voltage" covering the sample calculations in its Chapter 9, the **NEC** does specify that for uniform application of the provisions of Articles 210, 215 and 220, a voltage of 120, 120/240, 240, or 208Y/120V be used in computing the ampere load on conductors.

Using 115 and 230 V instead of 120 and 240 V does produce higher current values when volts are divided into watts or voltamperes (for example, 6000 W ÷ 120 V = 50 A, and 6000 W ÷ 115 V = 52 A). Using higher current values for sizing conductors and other equipment gives greater assurance of adequate capacity and completely satisfies all **NEC** rules.

Based on experience and very clear evidence that electrical systems quickly tend to be overloaded, and because load growth is a certainty in this all-electric age, many designers resolve all questions of circuit loading and permitted capacity in favor of greater capacity. The use of 9 instead of 10 as the maximum number of receptacles on a circuit calls for more circuits and gives higher capacity to each receptacle outlet. "When in doubt, design up, not down."

For office areas, the **NEC** permits either of two methods for determining the required minimum branch-circuit capacity for receptacle outlets. The first is the method just described, in which a load of 180 VA is taken for each single or duplex receptacle outlet and sufficient branch-circuit capacity is provided for the total load (180 VA times the number of receptacle outlets).

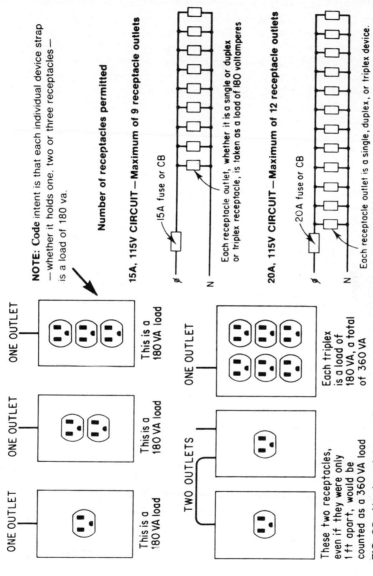

NOTE: Code intent is that each individual device strap —whether it holds one, two or three receptacles— is a load of 180 va.

Number of receptacles permitted

15A, 115V CIRCUIT—Maximum of 9 receptacle outlets

15A fuse or CB

Each receptacle outlet, whether it is a single or duplex or triplex receptacle, is taken as a load of 180 voltamperes

20A, 115V CIRCUIT—Maximum of 12 receptacle outlets

20A fuse or CB

Each receptacle outlet is a single, duplex, or triplex device.

ONE OUTLET

This is a 180 VA load

ONE OUTLET

This is a 180 VA load

ONE OUTLET

This is a 180 VA load

ONE OUTLET

Each triplex is a load of 180 VA, a total of 360 VA

TWO OUTLETS

These two receptacles, even if they were only 1 ft apart, would be counted as a 360 VA load

FIG. 68 Number of receptacles on a circuit is limited only for nonresidential circuits.

2.71

The second, or alternative, approach is described in the last footnote to **NEC** Table 220-2(b) (the double-asterisk note).

Based on extensive analysis of load densities for general lighting in office buildings, Table 220-2(b) now requires a minimum unit load of only 3.5 W/ft²—rather than the previous unit load of 5 W/ft²—for general illumination in "office buildings" and for "banks." And branch-circuit capacity for general lighting in such occupancies must not be less than 3.5 times the area (in square feet) of the space being supplied with the general lighting. But the double-asterisk note at the bottom of the table requires the addition of another 1 W/ft² to the 3.5-W/ft² value to cover the loading added by general-purpose receptacles in those cases where the actual number of receptacles is not known at the time feeder and branch-circuit capacities are being calculated. In such cases, a unit load of 4.5 W/ft² must be used; the calculation based on that figure will yield the minimum branch-circuit capacity for both general lighting and for all general-purpose receptacles that may later be installed. Whatever number of receptacles may later be installed, they may be divided among lighting circuits or separate receptacle circuits, with the loads "evenly proportioned" among the circuits. The allowance of 180 VA for each receptacle outlet may be disregarded.

But the rule does require that, where the actual number of general-purpose receptacles is known, the general lighting load be taken at 3.5 W/ft² for branch-circuit and feeder capacity, and each receptacle be taken as a load of 180 VA, in determining the total required branch-circuit capacity. The demand factors of Table 220-13 may be applied to get the minimum required feeder capacity for receptacle loads.

As noted previously, Exception No. 1 of Section 220-2(c) requires branch-circuit capacity to be calculated for multioutlet assemblies (prewired surface metal raceways with plug outlets spaced along their length). Exception No. 1 says that each 1-ft length of such strip must be taken as a load of 1.5 A when the strip is located where a number of appliances are likely to be used simultaneously. For instance, in industrial applications, on assembly lines involving frequent simultaneous use of plugged-in tools, a loading of 1.5 A/ft must be used. (A loading of 1.5 A for each 5-ft section may be used in commercial or institutional applications of multioutlet assemblies when the use of plug-in tools or appliances is not heavy.)

From the standpoint of design, the determination of the number of plug outlets required for various nonresidential occupancies is not a matter of easy or standard calculation. The following are suggestions for receptacle layouts in various areas:

1. In office buildings, the growing use of business machines dictates heavier load allowances for plug outlets. In separate offices less than 400 ft² in area, at least one plug outlet should be allowed for each 10 linear feet of wall space. In each office over 400 ft² in area, eight plug outlets should be allowed for the first 400 ft² of floor area and three plug outlets for each additional 400 ft² of floor area or fraction thereof. The number of outlets obtained in this way should be evenly distributed throughout the area.

2. In schoolrooms, common practice has been to provide at least one plug outlet on each wall of the room. Figure 69 shows a school panelboard with wide variation in the number of receptacle outlets connected to 120-V circuits. The layout and circuiting of receptacles were based on a study of likely loads, possible simultaneous use of appliances, and habits of use of electrical appliances. In a home-economics classroom, for example, heavy appliance loads and sewing-machine desks dictated the use of only two receptacles per No. 12 2-wire circuit.

3. In stores, at least one plug outlet should be provided for each 400 ft² of floor area or major fraction thereof.

4. In industrial areas, plug outlets should be provided on the basis of particular conditions. Figure 70 shows the circuit layout and supply to receptacles operating at 60 and 400 Hz in a modern airplane hangar.

FIG. 69 Variation in the number of receptacles per circuit is dictated by different applications.

Of course, the possible number of appliances to be used in any area will also affect the design of receptacle circuits. When special requirements for receptacle outlets arise after tenants move into an area, the spare circuits in the panelboard and extra capacity in the existing system of raceways or underfloor system will offer a solution to these requirements.

In addition to conventional receptacle circuits, modern buildings increasingly require new and often complex types of receptacle circuits. In particular, data-processing equipment and other special machines used in today's commercial and industrial buildings call for multiwire receptacle circuits of varying voltage, phase, and frequency characteristics.

Figure 71 shows a nine-pole receptacle used for dozens of outlets in a computer room. The receptacles were installed in a wireway under a raised floor and provided for flexibility in the movement and connection of various machines, each of which required single-phase, 120-V power and 3-phase, 4-wire, 120/208-V, 400-cycle power. Each machine is equipped with a nine-conductor cord and plug for ready connection to one of the receptacles. And an expanded color code was used for the circuit wiring.

Figure 72 shows the amperage loading schedule for the nine-pole receptacles—the single-phase, 60-Hz and the 3-phase, 400-Hz. There is a receptacle for each entry under the heading "Machine No." Note that the single-phase loads of the receptacles are balanced among the three phases of the 60-Hz, 3-phase, 4-wire branch circuits. And note the substantial spare capacity, enough for over 100 percent load growth in most circuits. The multioutlet branch circuits used here—50- and 70-A—are special and do not conform to **NEC** regulations regarding conventional receptacle circuits.

Residential occupancies

The **NEC** covers the maximum permitted number of receptacle outlets on a circuit in commercial, industrial, institutional, and other nonresidential installations, but there are no such limitations on the number of receptacle outlets on residential branch circuits. Instead, the **NEC** simply assumes that cord-connected appliances will always be used in all residential buildings, and it requires general-purpose receptacle outlets of the number and spacing indicated in Sections 210-52(a) and (c). These rules cover one-family houses, apartments in multifamily houses, guest rooms in hotels and motels, living quarters in dormitories, etc. But because so many receptacle

RECEPTACLE CIRCUITS in modern jet airplane hangar supply 60-cycle and 400-cycle energy for servicing of planes in the hangar bays. The circuits are protected at their origins and at the receptacles. Aluminum interlocked armor cable was selected for the 400-cycle circuits because of the non-magnetic character of the armor. This minimizes self-induction and consequent voltage drop. The armored cable was selected over busway with aluminum enclosure and conductors in aluminum conduit because of its unusual flexibility for this application.

FIG. 70 Large-capacity receptacle circuits are common in industrial locations.

FIG. 71 Branch-circuit wiring to special receptacles used for supplying computer equipment.

9-pole polarized receptacle in side of wireway

Tap conductors

Solderless connectors

Wire size varies – see panel schedules– Fig. 72

Wireway receptacles are connected to ϕA, ϕB, or ϕC of 60~ as per schedule–Fig. 72

A B C D E F G H J

No. 12 No. 8 No. 12 No. 12 No. 12 No. 8 No. 12 No. 8

Color code

Yellow	400~	ϕA
Purple	400~	ϕB
Brown	400~	ϕC
White w/marker	400~	neutral
Green	equip	ground
White	60~	neutral
Black	60~	ϕA
Red	60~	ϕB
Blue	60~	ϕC

PANEL SCHEDULES

	60 CYCLES—PANEL "B"								400 CYCLES—PANEL "A"								
Circuit No.	Circuit Breaker			Machine No.	60~ Kva	60~ Amps Per φ			400~ Kva	400~ Amps Per φ			Machine No.	Circuit Breaker			Circuit No.
	Frame	Poles	Trip			φA	φB	φC		φA	φB	φC		Frame	Poles	Trip	
1.	100A	3	50A	Console	0.6	5.0	....	....	0.5	1.4	1.4	1.4	Console	100A	3	50A	1.
				304	0.5	....	4.2	....	6.9	19.0	19.0	19.0	304				
4/8 RHW				Future	0.5	....	....	4.2	0.5	1.4	1.4	1.4	Future				4/8 RHW
				Total	1.6	5.0	4.2	4.2	7.9	21.8	21.8	21.8					
2.	100A	3	70A	332	1.3	11.0	....	....	0.3	0.9	0.9	0.9	332	100A	3	70A	2.
				332	1.3	....	11.0	....	0.3	0.9	0.9	0.9	332				
4/6 RHW				332	1.3	....	....	11.0	0.3	0.9	0.9	0.9	332				4/6 RHW
				332	1.3	11.0	....	....	0.3	0.9	0.9	0.9	332				
				332	1.3	....	11.0	....	0.3	0.9	0.9	0.9	332				
				332	1.3	....	....	11.0	0.3	0.9	0.9	0.9	332				
				332	1.3	11.0	....	....	0.3	0.9	0.9	0.9	332				
				332	1.3	....	11.0	....	0.3	0.9	0.9	0.9	332				
				332	1.3	....	....	11.0	0.3	0.9	0.9	0.9	332				
				330	0.3	2.8	....	....	3.9	10.8	10.8	10.8	330				
				Future	0.3	....	2.8	....	3.9	10.8	10.8	10.8	Future				
				"	1.3	11.0	....	....	0.3	0.9	0.9	0.9	"				
				"	1.3	....	11.0	....	0.3	0.9	0.9	0.9	"				
				"	1.3	....	....	11.0	0.3	0.9	0.9	0.9	"				
				Total	16.2	46.8	46.8	44.0	11.4	32.4	32.4	32.4					
3.	100A	3	70A	340B	0.2	1.7	....	....	3.9	10.8	10.8	10.8	340B	100A	3	70A	3.
				340T	5.8	....	48.0	....	0.4	1.1	1.1	1.1	340T				
4/6 RHW				320	0.3	....	....	2.5	4.4	12.2	12.2	12.2	320				4/6 RHW
				332	1.3	11.0	....	....	0.3	0.9	0.9	0.9	332				
				Maint. Rm.	2.0	16.7	....	....	1.0	2.8	2.8	2.8	Maint. Rm.				
				Future	0.2	1.7	....	....	3.9	10.8	10.8	10.8	Future				
				"	5.8	....	....	48.0	0.4	1.1	1.1	1.1	"				
				"	0.1	....	0.9	....	2.1	5.9	5.9	5.9	"				
				"	1.3	11.0	....	....	0.3	0.9	0.9	0.9	"				
				Total	17.0	42.1	48.9	50.5	16.7	46.5	46.5	46.5					
4.	100A	3	70A	360	2.5	21.0	....	....	1.0	2.8	2.8	2.8	360	100A	3	50A	4.
				370	1.3	....	11.0	....	0.7	1.9	1.9	1.9	370				
4/6 RHW				380	5.0	....	....	42.0	1.0	2.8	2.8	2.8	380				4/8 RHW
				Future	1.3	....	11.0	....	1.0	2.8	2.8	2.8	Future				
				Total	10.1	21.0	22.0	42.0	3.7	10.3	10.3	10.3					

FIG. 72 Load schedules for special receptacles.

outlets are required in such occupancies and because the use of plug-connected loads is intermittent and has great diversity of load values and operating cycles, the **NEC** notes at the bottom of Table 220-2(b) that the loads connected to such receptacles are adequately served by the branch-circuit capacity required by Section 220-3; no additional load calculations are required for such outlets.

In dwelling occupancies, it is first necessary to calculate the total "general lighting load" from Section 220-2(b) and Table 220-2(b) (at 3 W/ft^2 for dwellings or 2 W/ft^2 for hotels and motels, including apartment houses without provisions for cooking by tenants). Then the minimum required number and rating of 15 and/or 20-A general-purpose branch circuits must be provided to handle that load, as covered in Section 220-3(a). As long as the basic circuit capacity is provided, any number of lighting outlets may be connected to any general-purpose branch circuit, up to the rating of the branch circuit if loads are known. The lighting outlets should be evenly distributed among all the circuits. Although residential lamp wattages cannot be anticipated, the **NEC** method covers fairly heavy loading.

When the above rules on circuits and outlets for general lighting are satisfied, general-purpose convenience receptacle outlets may be connected on circuits supplying lighting outlets; or, receptacles only may be connected on one or more of the required branch circuits; or, additional circuits (over and above the required minimum) may be used to supply the receptacles. But no matter how general-purpose receptacle outlets are circuited, any number of general-purpose receptacle outlets may be connected on a residential branch circuit—with or without lighting outlets on the same circuit.

And, when small-appliance branch circuits (in kitchen, dining room, pantry, etc.) are provided in accordance with the requirements of Section 220-3(b), any number of small-appliance receptacle outlets may be connected on the 20-A small-appliance circuits. However, only receptacle outlets may be connected to these circuits, and only in the specified rooms.

Section 210-52(a) applies to the spacing of receptacles connected on the 20-A small-appliance circuits, as well as the spacing of general-purpose receptacle outlets. That section, therefore, establishes the minimum number of receptacles that must be installed. Of course, a greater number of receptacles may be installed for greater convenience of use.

Applying Plug Receptacles

The first rule on the use of receptacles is that any receptacle must have an ampere rating at least equal to the load it supplies. In addition, on circuits having two or more outlets, receptacles must be rated as follows:

On 15-A circuits: not over 15-A rating

On 20-A circuits: 15- or 20-A rating

On 30-A circuits: 30-A rating

On 40-A circuits: 40- or 50-A rating

On 50-A circuits: 50-A rating

The **NEC** requires that only grounding-type receptacles be installed on 15- and 20-A branch circuits. This means that all general-purpose, 120-V plug outlets must be of the three-pole type. The rule applies to all types of buildings and areas and to receptacles at any voltage level. The grounding terminals on such receptacle outlets must be connected to ground either by a specific grounding conductor run with the circuit conductors or by bonding (with a jumper or self-grounding screws) to the metal circuit outlet box which is, itself, grounded by the metallic jacketing on cable, by an equipment grounding conductor in nonmetallic sheathed cable, or by a metallic raceway system.

With a properly grounded metallic system (conduit or armored cable) run to a surface-mounted metal outlet box, receptacles are grounded through the mounting bracket that is metallically connected to the metal box—which, in turn, is grounded through the metal raceway or cable armor supplying it. At recessed boxes set into the wall, a separate jumper must be used from the green hex-head screw on the receptacle to a solid connection on the grounded metal box, unless the receptacle is a self-grounding type; that is, unless it has a yoke especially designed to provide a sure grounding path of low resistance through the screw to the threaded hole on the box (Fig. 73). A grounding wire run with the circuit conductors to provide for equipment grounding is mandatory in nonmetallic wiring systems.

NEC Section 410-56(g) requires that when aluminum conductors are connected directly to the terminals of any receptacle rated 20 A or less, the device must be marked "CO/ALR" (Fig. 74).

Section 210-21(b) contains two paragraphs of importance. The second paragraph reads: "A single receptacle installed on an individual branch circuit shall have an ampere rating of not less than that of the branch circuit." Since the branch-circuit overcurrent device determines the branch-circuit rating (or classification), a single receptacle (not a duplex receptacle) supplied by an individual branch circuit cannot have a rating below that of the branch-circuit overcurrent device (Figure 75).

Poke-through receptacles

Poke-through wiring—the technique in which floor outlets in commerical buildings are wired through holes in concrete-slab floors—is an acceptable wiring method if use is made of UL-

Special mounting screws

MOUNTING SCREW is equipped with a pressure clip to assure tight contact between the screw and the device yoke, eliminating the need for using a grounding jumper wire from the green, hex-head screw to the box.

Surface-mounted box grounded by conduit system

Bracket grounds receptacle to box

Grounding slots

Receptacle

A jumper wire to connect the grounding-screw terminal to the grounded box is not required with a surface-mounted box, but is required when receptacle is used in a recessed box (Section 250 – 74).

FIG. 73 Receptacle in a surface box does not require a bonding jumper, and self-grounding receptacles do not require such a jumper in a recessed *or* surface box.

15-amp or 20-amp
overcurrent device

Neutral

Raceway, cable armor or
separate grounding
conductor to ground the
grounding terminal and
mounting yoke

Ground terminal is
green hex head screw

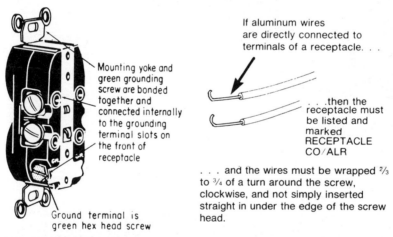

Mounting yoke and
green grounding
screw are bonded
together and
connected internally
to the grounding
terminal slots on
the front of
receptacle

Ground terminal is
green hex head screw

If aluminum wires
are directly connected to
terminals of a receptacle. . .

. . .then the
receptacle must
be listed and
marked
RECEPTACLE
CO/ALR

. . . and the wires must be wrapped ⅔
to ¾ of a turn around the screw,
clockwise, and not simply inserted
straight in under the edge of the screw
head.

FIG. 74 Only grounding receptacles may be used on all 15- and 20-A circuits and only CO/ALR receptacles with aluminum wire.

N

120 V

Individual
circuit

20 A

Single 20-amp
gr-type receptacle
(125 V)

Receptacle must be
rated at least the
same as the fuse
or CB

This is the only load
on the circuit

FIG. 75 A single receptacle on an individual circuit must
have an ampere rating equal to the circuit rating.

listed poke-through fittings that have been tested and found to preserve the fire rating of the concrete floor (Fig. 76). Poke-through wiring continues to be a popular and very effective method of wiring floor outlets in office areas and other commercial and industrial locations, as an alternative to underfloor raceway or wiring in cellular floor construction. Holes are cut or drilled in concrete floors at the desired locations of floor outlets, and floor-box assemblies are

THIS IS ACCEPTABLE—UL LISTED

One side of box is for power and light wiring

Other side is for telephone or signal wiring

Numbered components include (1) combination floor service box, (2) fire-rated center coupling, (3) concrete slab, (4) barriered extension, and (5) barriered junction box.

. . . OR THIS FITTING

Fire-rated concrete WALL OR FLOOR standard designs for 3 hour rating. Higher ratings available.

Socket head cap screws easy installation

Intumescent material

Swells during fire to seal void caused by melted cable insulation.

Insulation layer

On each side to protect against temperature build up.

Elastomeric seal Seals vapors, water and smoke. Designs available for sealing up to 100 psi. Also provides cable support.

Two steel pressure discs for uniform sealing. (Alternate materials and slotted designs available to prevent induction heating.)

Core drilled or cast

FIG. 76 Poke-through wiring of floor outlets must be carefully designed.

installed and wired from the ceiling space of the floor below. The method permits installation of each and every floor box at the precise location that best serves the layout of desks and other office equipment. But a serious question is posed about how the use of the poke-through wiring technique can be properly reconciled with **NEC** Section 300-21.

Each floor outlet at a poke-through location may be wired basically in either of two ways: by some job-fabricated assembly of pipe nipples and boxes or by means of a manufactured through-floor assembly made expressly for the purpose and tested and listed by a nationally recognized testing lab such as UL. Violation of the **NEC** has been indicated where floor receptacles are fed by using job-fabricated assemblies through holes in the floor, thereby destroying the fire rating of the floor. Figure 77 shows a typical objectionable method. Section 300-21 requires that electrical installations be made so as to substantially protect the integrity of rated fire walls and fire-resistant or fire-stopped walls, partitions, ceilings, and floors. There are UL-listed and UL-approved through-floor fittings that permit poke-through wiring that satisfies Section 300-21.

FIG. 77 Untested floor penetrations must not be used for poke-through wiring.

Designers must give careful consideration to the liabilities incurred through less-than-strict compliance with the letter and intent of **NEC** and OSHA regulations. Given the nature of today's electrical work—with much broader, more intensified enforcement of national codes and standards—the use of new types of products, such as fire-rated floor outlets, must be thoroughly evaluated. This also applies to cable and/or conduit penetrations of fire-rated walls, floors, and ceilings without altering the fire rating of the structural surface.

FCC wiring

Flat-conductor cable (FCC) is a surface-mounted cable for use on floors, under carpet squares not over 30 in square; it is the basis for a complete wiring system for supplying power, signal, and communications floor outlets. This alternative to underfloor raceway, cellular-floor wiring, and conventional pipe-and-wire and poke-through receptacles is a UL-listed, **NEC**-recognized (Article 328) branch-circuit wiring system that may be used to supply floor outlets in office areas and other commercial and institutional interiors. The system may be used for new buildings or for modernization or expansion in existing interiors. FCC wiring may be used on any hard, sound, smooth floor surface—concrete, wood, ceramic, etc. The great flexibility and ease of

FIG. 78 A typical flat-conductor-cable system includes the cable assembly and all accessory devices to make a complete layout.

installation of this surface-mounted wiring system are given added importance by the fact that the average floor power outlet in an office area is relocated every 2 years (Fig. 78).

Undercarpet wiring to floor outlets eliminates any need for core drilling of concrete floors, avoiding noise, dripping water, falling debris, and the disruption of normal activities in an office area. Alterations or additions to Type FCC circuit runs are neat, clean, and simple and may be made during office working hours; this eliminates the overtime labor rates incurred by floor drilling, which must be done at night or on weekends. The FCC method also eliminates the use of conduit or cable and the need to fish conductors.

Type FCC wiring offers versatile supply to floor outlets for power and communication at any location on the floor. The flat cable is inconspicuous under the carpet squares. And, the elimination of floor penetrations maintains the fire integrity of the floor, as required by Section 300-21.

Hospital-grade receptacles

The **NEC** does not contain a general requirement that "hospital-grade" receptacles (the UL-listed "green-dot" wiring devices) must be used in health-care facilities. However, it should be noted that Sections 517-101(b)(5) and 517-101(c)(2) do require the use of "receptacles and attachment plugs" that are "listed for hospital use" above "hazardous" anesthetizing locations and in "other-than-hazardous" anesthetizing locations. As described, those rules require that all two-pole, 3-wire grounding-type receptacles and plugs "for single-phase 125- or 250-V ac service" must be marked "Hospital Only" or "Hospital Grade," with a green dot on the face of each receptacle (Fig. 79). The relation between the phrase "listed for hospital use" and the phrase "hospital grade" is explained in the UL *Green Book,* under the heading "Attachment Plug Receptacles and Plugs." It says:

Receptacles listed for hospital use in other than hazardous locations in accordance with Article 517 of the **National Electrical Code** are identified (1) by the marking "Hospital Only" or (2) by the marking "Hospital Grade" and a green dot on the receptacle. The green dot is on the face of the receptacle where visible after installation.

But it should be noted that the requirement for hospital-grade wiring devices applies only with respect to anesthetizing locations. It does not apply to patient-care areas or other areas. And it does not apply to nursing homes, clinics, or medical offices.

Of course, in the defined hazardous areas where flammable anesthetics are used (Sections 517-100 and 517-101), receptacles must be of the explosion-proof type, listed for Class I, Division 1 areas.

UL has a special series of tests for devices

Nylon face resists impact cracking

Heavy-duty double-wipe blades

Special marking shows ground symbol

Separate grounding pigtail or cover shields over terminals can be special features

FIG. 79 Hospital-grade receptacles are *required* only in anesthetizing locations, but they offer long, reliable service in other areas.

intended for hospital use. These tests are substantially more abusive than those performed on general-purpose devices and are designed to ensure the reliability of the grounding connection in particular, when these devices are used in the hospital environment. Hospital-grade receptacles have stability and construction in excess of standard specifications and can stand up to abuse and hard usage. Devices which pass this test are listed as "Hospital Grade" and are identified with these words and a green dot, both of which are visible after installation. UL listings include 15- and 20-A, 125-V grounding, nonlocking-type plugs, receptacles, and connectors. This class of devices is acceptable for use in any nonhazardous anesthetizing location. Their use might not be acceptable above hazardous anesthetizing areas if they are not totally enclosed and permit the escape of sparks or hot metal particles which could fall into the hazardous area, as prohibited by **NEC** Section 517-101(b).

Residential GFCI

NEC Section 210-8 (a) is headed "Dwelling Units." The very clear and detailed definition of those words, as given in **NEC** Article 100, indicates that all the ground-fault circuit interruption rules apply to

- All one-family houses
- Each dwelling unit in a two-family house
- Each apartment in an apartment house
- Each dwelling unit in a condominium

GFCI protection is required for all 125-V, single-phase, 15- and 20-A receptacles installed in bathrooms of dwelling units and hotel or motel units and garages of dwelling units only (Fig. 80). The requirement for GFCI protection in "garages" is included because homeowners do use outdoor appliances (lawnmowers, hedge trimmers, etc.) plugged into garage receptacles. Such receptacles require GFCI protection for the same reason as "outdoor" receptacles. In both

GFCI is needed on—

Single-phase, 120V, 15A or 20A receptacles

Either GFCI CB in panel or GFCI receptacles at box locations

In bathrooms

In Garages

FIG. 80 GFCI protection is required for receptacles in bathrooms and garages of dwelling units.

places, GFCI protection may be provided by a GFCI circuit breaker that protects the whole circuit and any receptacles connected to it; or, the receptacle may be a GFCI type that incorporates the components that give it the necessary tripping capability on low-level ground faults.

A lot of the controversy that was generated by the question "What is a bathroom?" has been eliminated by the inclusion of the **NEC** of a definition of the word "bathroom." A "bathroom" is "an area" (which means it could be a room or a room plus another area) that contains, first, a "basin" (sometimes called a "sink") and, second, at least one more plumbing fixture—a toilet, a tub, and/or a shower. A small room with only a basin (a washroom) is not a "bathroom." Neither is a room that contains only a toilet and/or a tub or shower.

As noted above, GFCI protection is required by Section 210-8(a) in bathrooms of hotel or motel rooms but not in dormitory rooms or any other residential occupancy that does not conform to the definition of "dwelling unit" (Fig. 81).

In addition to bathroom receptacles, kitchen plug receptacles above countertops and within 6 ft of the kitchen sink must be provided with GFCI protection. Such receptacles, which are connected on the 20-A kitchen-appliance circuits, may be GFCI receptacles, or the entire circuit may be protected by a GFCI circuit breaker. And at least one GFCI-protected receptacle must be installed in the basement of a dwelling unit.

The rule requiring GFCI protection in garages applies to both attached garages and detached (or separate) garages associated with dwelling units—such as one-family houses or multifamily houses where each unit has its own garage. But Section 210-52(f) requires at least one receptacle only for "attached" garages. A receptacle is not required in a detached garage, but if one is used in such a garage, it must have GFCI protection.

The basic rule requires all 125-V, single-phase, 15- or 20-A receptacles installed in garages to have ground-fault circuit interrupter protection. However, Exception No. 1 excludes a ceiling-mounted receptacle that is used solely to supply a cord-connected garage-door operator. And, as worded, the exception excludes any receptacles that "are not readily accessible"—that is, any receptacle that requires the use of a portable ladder or a chair to get up to it.

Receptacles for "dedicated appliances" (those that are put in place and not normally moved owing to weight and size, such as refrigerators, freezers, and washing machines) are excluded from the need for GFCI because they will not normally be used to supply hand-held cord-connected appliances (lawnmowers, hedge trimmers, etc.) that are used outdoors (Fig. 82). The intent of this rule seems to be that any such receptacle for a "dedicated" appliance must be a *single* receptacle, and not a duplex receptacle with one of the two receptacles left unused.

NOTE: A garage receptacle that is excluded from the requirement for GFCI protection, as noted in the two exceptions of this section, may not be considered as the receptacle that is

TYPICAL BEDROOM SUITE IN ONE-FAMILY HOUSE OR APARTMENT UNIT

AND WATCH THIS ONE!

THIS IS A COMMON LAYOUT OF PLUMBING FIXTURES IN HOTEL AND MOTEL UNITS

Although this **area** with basin is **outside** room with tub and toilet, the intent of Section 210-52(c) requires a receptacle at basin; and Section 210-8(a) requires that it be GFCI-protected.

NOTE: It is important to understand that the Code meaning of "bathroom" refers to the total "area" made up of the basin in the alcove *plus* the "room" that contains the tub and toilet. Although a receptacle is *not* required in the "room" with the tub and toilet, if one is installed in that room, it must be GFCI-protected because such a receptacle is technically "in the bathroom," just as the one at the basin location is "in the bathroom."

Guest rooms in hotels and motels are required by Section 210-60 to have the same receptacle outlets required by Section 210-52 for "dwelling units." The requirement for a wall receptacle outlet at the "basin location" applies to bathrooms; and the anteroom area with only a basin is, by definition and Code intent, a "bathroom." A receptacle at the basin would, therefore, be required by Code. Section 210-8(a), which applies only to bathrooms and GFCI protection is required for this receptacle.

FIG. 81 Intent of the bathroom GFCI rule applies to an "area" (not simply a "room") in the "bathroom" of a "dwelling unit" or a hotel or motel unit.

If a receptacle is installed to supply a freezer in a garage, it does *not* have to be GFCI protected . . .

. . . **BUT AT LEAST ONE OTHER RECEPTACLE must be installed in the garage for using hand-held electric tools or appliances, and it MUST be GFCI-protected.**

FIG. 82 Garage receptacle does not require GFCI if it supplies a "dedicated" appliance (or if it is ceiling mounted or otherwise not "readily accessible").

required by Section 210-52(f) to be installed in an attached garage. Thus, if a non-GFCI-protected receptacle is installed in a garage at the ceiling for connection of a door operator, or if a non-GFCI-protected receptacle is installed in a garage for a freezer, at least one additional receptacle must be installed in the garage to satisfy Section 210-52(f) for general use of cord-connected appliances; such a receptacle must be installed not over 5½ ft above the floor, and it must have GFCI protection (either in the panel CB or in the receptacle itself).

Application of the two exceptions of Section 210-8(a) may prove troublesome from an inspection standpoint, because receptacle outlets are most commonly installed in a garage during the construction of a house and before it is known what appliances might be used. Under such conditions, GFCI receptacles would be required because "dedicated space" appliances are not in place. Then if, say, a freezer is later installed in the garage, the GFCI receptacle could be replaced with a non-GFCI device. But such replacement would not be acceptable for a receptacle that is the only one in the garage, because such a receptacle is required by Section 210-52(f), must be GFCI protected, and is not subject to Exception No. 2 of Section 210-8(a).

Section 210-8(a)(3), on outdoor receptacles, requires GFCI protection of "all 125-V, single-phase, 15- and 20-A receptacles installed outdoors" at dwelling units. The rule specifies that such protection of outdoor receptacles is required only "where there is direct grade-level access to the dwelling unit and to the receptacles." The phrase "direct grade-level access" is not defined and does lend itself to varied interpretations (Fig. 83). Because there is no "grade-level access" to apartment units constructed above ground level, there would be no need for GFCI protection of receptacles installed outdoors on balconies of such apartments or condominium units. Likewise, GFCI protection would not be required for any outdoor receptacle installed on a porch or other raised part of even a one-family house, provided that there was not grade-level access to the receptacle (Fig. 84).

Outdoor receptacles at a dwelling unit may be installed in the exterior surface of the building, in a post-light, or in a box fed out of the ground. Any box containing a receptacle fed out of the ground must be supported as shown in Fig. 85. The **NEC** considers a threaded box or fitting not over 100 in³ in volume to be adequately supported if two or more conduits are threaded into the

Receptacle on each outdoor apartment balcony does <u>not</u> have grade-level access

Duplex receptacle outdoors with weatherproof cover and grade-level access

Earth

Receptacles fed out of ground *must be* GFCI-protected

Apartment balcony receptacles do *not* require GFCI protection

NOTE: In any case of an outdoor receptacle at a dwelling unit, GFCI protection is not required if there is no access to the receptacle from the grade-level of ground around the building.

FIG. 83   Outdoor receptacles of "dwelling units" need GFCI protection only if they have "direct grade-level access."

box or fitting and the conduits are adequately supported within 18 in of the box or fitting. This makes it permissible to support elevated threaded junction boxes with conduits emerging up to 18 in from the earth, for example near swimming pools or other outdoor areas where small PAR (parabolic-reflector) lampholders or receptacles are attached to the box. Support by a single conduit is not recognized.

Branch-Circuit Controls

Branch-circuit design must include provision for control of the circuits. This may take the form of control over individual outlets or groups of outlets or may consist of control of the entire circuit.

Local control of one or more lighting outlets (less than the entire circuit) is generally effected with manual, direct-acting, wiring-device switches. Alternating-current snap switches feature a wide variety of modern designs. Such switches permit full loads for incandescent and fluorescent lighting and 80 percent loads for motors. The switches are available with exposed screw terminals or screwless pressure-locking terminals. Mercury switches have ac-dc snap-switch ratings (Fig. 86). Local control is not generally needed or required for energizing and deenergizing receptacle outlets, although such control is sometimes desirable and recommended. A commonly used technique for receptacle-outlet control is to split-wire a duplex receptacle (by breaking off the metal tab between the brass-colored screws on the ungrounded side of the device) so the top outlet is controlled by a wall switch and the bottom outlet is always hot (Fig. 87).

Although no **NEC** rules require switch control of general-purpose lighting or receptacle outlets in commercial, institutional, or industrial electrical systems, the **NEC** does make certain switched lighting outlets mandatory for all dwelling units (one-family houses, apartment units, etc.). Section 210-70 requires that at least one wall-switch–controlled lighting outlet be installed in every habitable room, in hallways, stairways, and attached garages, and at outdoor entrances. The rule does not stipulate that these must be ceiling lighting outlets; any or all of them may be wall-mounted lighting outlets (Fig. 88).

A vehicle door in an attached garage is not considered as an outdoor entrance. The **NEC** does not require a lighting outlet at any garage door that is provided as a vehicle entrance because the lights of the car provide adequate illumination when such a door is being used in darkness.

Grade-level patio or open porch with receptacle accessible from grade level

No grade-level access: no stairs to second-floor deck porch of one-family house —only access is door from house

Open-front porch of one-family house, with access by steps from grade level

Outdoor receptacle on second-floor deck porch would **not** require GFCI protection, but receptacle on porch below **would** need GFCI protection.

Outdoor receptacle on this porch might likely be used to connect appliance used on ground, and GFCI protection **would** seem to be required.

IMPORTANT: A hotel, motel or dormitory is *not* a "dwelling unit". . .

. . . and an outdoor receptacle does *not* require GFCI protection, even if it *does* have grade-level access.

FIG. 84 Required GFCI protection for outdoor receptacles applies only where there is direct grade-level access and only at "dwelling units."

PAR lamp and holder

2 conduits supported by earth

JB

Not over 18"

Outdoor receptacles

Ground

Threaded hubs

Only one supporting conduit

THIS IS OK   THIS IS OK   THIS IS A VIOLATION!

FIG. 85 Outdoor boxes fed from underground must be properly supported.

Toggle handle
15A or 20A
120V or
120/277V

Push handle
15A or 20A
120/277V

Lighted rocker
handle
15A, 120V

Rocker handle
15A or 20A
120/277V

Rotary handle
15A, 120/277V

Rocker handle
15A or 20A
w/special plate
120/277V

Rocker-type
operated by special
snap-on actuator plate
15A or 20A, 120/277V

FIG. 86 Wide variety of wiring-device switches offers great flexibility for design of branch-circuit control.

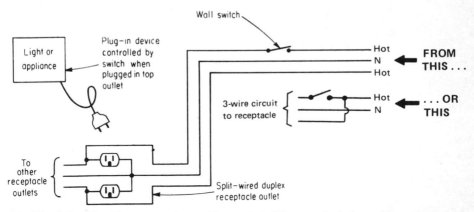

Light or
appliance

Plug-in device
controlled by
switch when
plugged in top
outlet

Wall switch

Hot
N
Hot

FROM THIS ...

3-wire circuit
to receptacle

Hot
N

... OR THIS

To
other
receptacle
outlets

Split-wired duplex
receptacle outlet

FIG. 87 Split wiring of a receptacle provides control of the receptacle from either a 2-wire or 3-wire circuit.

But the wording of the **NEC** does suggest that a rear or side door that is provided for personnel entry to an attached garage would be "considered as an outdoor entrance" because the note excludes only "vehicle" doors. Each such personnel entrance from outdoors to the garage would require a wall-switched lighting outlet.

In addition, at least one lighting outlet must be installed in every attic, underfloor space, utility room, and basement if it is used for storage or if it contains equipment requiring servicing. In these latter cases, the lighting outlet does not have to be controlled by a wall switch, although it may be. A lamp socket controlled by a pull chain or a canopy switch could be used.

Two exceptions are given to the basic requirements. Exception No. 1 notes that in habitable rooms other than kitchens or bathrooms, a wall-switch–controlled receptacle outlet may be used

A wall-switched lighting outlet is required *in* garage and *at* doors intended only for *personnel entry* to garage. BUT . . .

Wall-switched lighting outlets (or switched receptacle in any room other than kitchen or bathrooms)

Attached garage

Lighting outlet may be pull-chain socket or may have canopy switch— only in basement, attic, utility room or crawl space

Outdoor personnel entrance— front, side, rear – wall- switched lighting outlet must be used

. . . a lighting outlet is *not* required at garage "vehicle" doors.

EXCEPTION:

Wall switch   Receptacle   Portable lamp

Wall-switched receptacle(s) may be used instead of a lighting outlet in habitable rooms other than kitchens and bathrooms.

FIG. 88 In dwelling units only, specific switch-controlled lighting outlets are required.

instead of a wall-switch–controlled lighting outlet. The receptacle outlet can serve to supply a portable lamp, which would give the necessary lighting for the room. Exception No. 2 states that in hallways, stairways, and at outdoor entrances, remote, central, or automatic control of lighting shall be permitted. This latter recognition accepts remote, central, or automatic control as an alternative to the wall-switch control mentioned in the basic rules.

But note carefully that every kitchen and every bathroom must have at least one wall-switch–controlled lighting outlet.

Section 210-70(b) states that at least one wall-switch–controlled lighting outlet or wall-switch–controlled receptacle must be installed in guest rooms of hotels, motels, and similar occupancies.

An important rule on the use of wiring-device switches is contained in **NEC** Section 380-8(b), as follows:

Snap switches shall not be grouped or ganged in outlet boxes unless they can be so arranged that the voltage between adjacent switches does not exceed 300, or unless they are installed in boxes equipped with permanently installed barriers between adjacent swtiches.

This applies to switches with either screw-type connections or screwless connections (push-in type). It applies where 277-V switches, mounted in a common box (such as two- or three-gang switches), control 277-V loads, with the voltage between exposed line terminals of adjacent switches in the common box being 480 V. If the adjacent switches have exposed live terminals, anyone changing one of the switches without disconnecting the circuit at the panel could contact 480 V, as shown in Fig. 89. The rule requires permanent barriers between adjacent switches located in the same box when the voltage between such switches exceeds 300 V and terminals are exposed.

Screwless terminal switches (with no exposed live parts) would seem to satisfy the intent (and literal text) of Section 380-8 because the switches could be considered to be "arranged" to

Snap switches shall not be grouped or ganged in outlet boxes where voltage between adjacent switches exceeds 300 volts.

Installation shown would be acceptable if a separate single-gang box and plate are used for each switch; or a common wire from only one phase (A, B or C) supplies switches in 3-gang box or a box with permanent barriers between the switches would also comply with the Code.

FIG. 89 Snap switches on different phase legs of a circuit fed into a multigang box must not present a personnel hazard.

prevent exposure to 480 V. Of course, the hookup shown would be acceptable if a separate single-gang box and plate were used for each switch, or a common wire from only one phase (*A*, *B*, or *C*) supplied all three switches in the three-gang box. The inspector should be consulted concerning questionable cases.

Full-circuit control

When local switching is not necessary, the entire circuit can be controlled by the branch-circuit breaker or switch in the panelboard. Although panelboard control is a common and satisfactory practice in many occupancies—particularly industrial and large commercial areas—most areas should be provided with the convenience and economy of operation afforded by local switching. Switching at the panelboard requires a separate circuit for each group of outlets to be controlled together.

Such factors as distance from panelboard to lighting fixtures, partitioning of the area of lighting coverage fed from the panelboard, and need for three- or four-way switch control will determine the amount of local switching, the amount of panelboard switching, and the number of circuits needed.

In commercial and industrial electrical systems, the breakers in the lighting panel are commonly used for on-off control of lighting; this eliminates the need for local wiring-device switches. The UL states that "circuit breakers marked SWD are suitable for switching 120-V fluorescent lighting on a regular basis." Such listing indicates that any CB used for regular switching of 120-V fluorescents must be marked SWD (switching duty) and that breakers not so marked are not suitable for panel switching of lighting. **NEC** Section 240-83(d) requires SWD-marked circuit breakers for switching 120- and 277-V fluorescent lighting (Fig. 90). Both 120- and 277-V CBs are designed to be suitable for regular switching of lighting of any kind; these are marked SWD. Only those breakers bearing the designation SWD should be used as snap switches for lighting control. Type SWD breakers have been tested and found suitable for the greater frequency of on-off operations required for switching duty, as compared with strictly overcurrent protection, where the breaker is used only for infrequent disconnecting for circuit repair or maintenance.

FIG. 90 Circuit breakers used for on-off switching of full-circuit lighting load must be suitable.

Remote and/or automatic control

Automatic direct switching of lighting circuits is commonly accomplished by means of time switches, which may be used to control all or some of the loads fed by branch circuits. A time switch may provide direct control, with its contacts making and breaking the current to the load, as in the case of the double-pole switch at the left of Fig. 91; or, the time switch may be used to

Single-pole, double-throw, 35-amp time
switch with astronomic dial and 120-volt
motor-operating in coil circuit of
contactor to control lighting circuits in
top section of panel

Lighting panelboard

6-20A circuits-Sec. 2 →   4 No. 10

4-20A circuits-Sec.1 →   4 No. 10

1-30A 3P circuit

30A. 208v 3φ mechanically-
held contactor

To front parking
lights 2 cir.

120/208-volt feeder
from lighting switchboard

Double-pole, single-throw, 35-amp
time switch with astronomic dial
and 120-volt motor- directly
switching two lighting circuits for
parking area during darkness

FIG. 91 Time switches can provide automatic control of lighting
and other loads—either directly or through a contactor.

control the operating coil of a magnetic contactor, as shown at the right in that diagram. And
there is a growing trend toward the convenience and flexibility of low-voltage relay switching
and the use of various types of remote-control magnetic switching through relays and contactors.
Such applications are particularly effective for controlling individual lighting fixtures or groups
of fixtures when the ability to turn off lights not in use is desired for energy conservation and
reduced electric bills.

Low-voltage relay switching

Low-voltage relay switching is used where remote control or control from a number of spread-
out points is required for each of a number of small 120- or 277-V lighting or heating loads. In
this type of control, contacts operated by low-voltage relay coils are used to open and close the
hot conductor supplying the one or more luminaires or load devices controlled by the relay. The
relay is generally of the 3-wire, mechanically held, on-off type, energized from a step-down
control transformer.

In a typical installation, all the relays may be mounted in an enclosure near the panelboard
supplying the branch circuits which the relays switch, with a single transformer mounted there
to supply the low voltage (Fig. 92). Where a single panelboard serves a large number of lighting
branch circuits in a very large area, as in large office areas in commercial buildings, several
relays associated with each section of the overall area may be group mounted in an enclosure in
that section.

Figure 93 shows 24-V control of 277-V fixtures, with constantly illuminated switchplates
alongside doorways to define interior exit routes and practical hollow-partition clip-in switch
boxes in interior labs of a medical research center. Control relays are in compact boxes atop
luminaires. Relays are connected to switches and to 50-VA continuous-duty 120/24-V transformers by multiwire Class 2 remote-control circuits (**NEC** Article 725) routed through overhead
plenums and supported by insulator rings attached to fixture hangers by spring clips. No switching circuit controls more than four relays. Switching circuits and exit lights are served from an
emergency power source, with general-lighting fixtures served through normal 480/277-V feeders. Interior-area route-indicating lights and general-lighting switches are mounted together in
thin boxes set in partitions.

Figure 94 shows the schematic and physical hookup of the components of another low-voltage
relay control system.

Power supply is mounted on relay cabinet. (Only one power source is required for entire system.)

110/24 volt dc. momentary impulse voltage

Pilot light transformer

110-volt ac

Relays in cabinet

Load A   Load B

Relays

To master control panel, if used

To switches fed by other relays

No.18 low-voltage wiring

Switch-pilot light assemblies

Single-gang switches

FIG. 92 Typical low-voltage relay switching system provides great flexibility and convenience of control.

Tabs spot-welded to steel rim of box

Screw holes drilled and tapped to take switchplate screws

L.V. cable

Conduit

Exit route light

Standard shallow cutlet box

L.V. switch

MOUNTING DETAILS FOR HOLLOW PARTITIONS...

...AND ALSO FOR PLASTER PARTITIONS...

50-V.A. continuous-duty transformer located in ceiling

24-volt secondary

To 120-volt emergency power supply

Exit route light remains on continuously

Wiring grouped and installed on rings attached to building steel

Low-voltage control relays for lighting fixtures

24-volt momentary contact

3-wire on-off switch

WIRING DIAGRAM FOR EXIT ROUTE LIGHT AND L.V. SWITCHES

FIG. 93 Details of a low-voltage relay system.

EQUIPMENT LAYOUT

CIRCUIT HOOKUP

FIG. 94 Low-voltage relay may be located at a fixture or remote from a fixture.

Contactor control

One of the most popular types of modern branch-circuit control is contactor switching. Magnetic contactors—either electrically held or mechanically held, depending upon the stability of the control voltage and such factors as coil hum and coil power drain—may be used in individual enclosures for control of circuit or load devices. The basic difference between electrically (or magnetically) held contactors and mechanically held contactors is shown in Fig. 95.

Magnetic contactors: The basic magnetic contactor is magnetically operated by energizing its operating coil to close its contacts. It is magnetically held in position by maintaining current flow through the coil. If the voltage to the coil is removed, the contactor opens. The voltage supply to the coil is generally taken from the line side of the contactor itself. Magnetic contactors are made in a wide range of ratings up to about 1000 A, with one to four poles, and to 600 V.

① Maintained-contact
   switch or automatic
   pilot control device–or

② Momentary-contact
   pushbutton used with
   auxiliary holding contacts

Holding contact for use with
momentary pushbutton-connected
across "start" button

Single operating coil

MAGNETICALLY HELD

TWO-COIL
OR
LATCHING TYPE

Both the "close" and "open" switches
are momentary-contact type

Close

Open

Main closing and
latching solenoid

Unlatching solenoid
to open the contactor

Either
coil
energized
only during
operation

3-wire control circuit from
contactor to control switch

SINGLE-COIL TYPE
WITH
LOCKING LINKAGES

Momentary-contact or maintained-
contact switches (toggle or pushbutton,
or automatic device) may be used

Single coil operates main
contacts and auxiliary
contact to break control
circuit after main contacts
open or close. Coil is
energized only during
opening or closing operation

Open

Close

3-wire control circuit from
contactor to control switch

MECHANICALLY HELD

FIG. 95 Each type of magnetic contactor has its own operating features.

Magnetically held contactors provide effective remote switching control of power and lighting loads where stable circuit voltage will prevent objectionable low-voltage dropout and where the inherent ac hum of continually energized coils is not objectionable. They are particularly applicable to controlling motor loads where running-overload protection is not required or is provided elsewhere. In such use, the magnetically held contactor can be arranged for low-voltage protection or low-voltage release for the motor load.

Mechanically held contactors: A variation of the basic magnetic contactor is the mechanically held contactor, in which the operating coil is momentarily energized to close its contacts and momentarily energized to open its contacts.

Mechanically held contactors are best suited for use as remotely controlled safety switches. They are used for switching feeders, branch circuits, and individual loads—lighting, motor, or other power loads. They offer a particular advantage where a single contactor is to be controlled from a number of different remote locations. The mechanically held design is also highly desirable where continuity of circuit connection must be maintained during voltage dips and power outages and where severe voltage fluctuations would drop out a magnetically held contactor.

A second important advantage of the mechanically held contactor is the absence of ac hum. Because the coil is only momentarily energized for opening and for closing, the coil does not constantly carry current as long as the contacts are in the closed position. This suits the mechanically held contactor to application is such quiet areas as schools, hospitals, and libraries.

In keeping with present energy-conservation efforts in most facilities, contactors are being applied increasingly to control lighting panels, heating and ventilating equipment, and production lines furnished with continuously running electrical equipment. Mechanically held contactors are commonly chosen for this type of control.

The control voltage of the contactor operating coil must be carefully selected. Standard control voltages are 24, 120, 208, 240, 277, 480, and 550 V. Coils can be provided for any voltage, regardless of the voltage of the switched circuit. If the control-circuit voltage is the same as the phase-to-phase or phase-to-neutral voltage of the circuit being controlled by the contactor, the control circuit can be tapped from the line-side terminals of the contactor. Safety considerations frequently dictate the use of a lower control-circuit voltage when the voltage of the controlled circuit is 440 V or higher. In this event, a control transformer can be added to the contactor to obtain the desired voltage, or the control voltage may be derived from a separate ac or dc source and not from the controlled circuit.

Both the magnetic and the mechanically held contactor are operated via a control circuit, the operating action being initiated by a pilot device such as a pushbutton or time clock. The contactor is placed in the circuit to be switched. This may be in the middle of a split-bus panelboard, where the switch controls one section of bus (Fig. 96). It also may be adjacent to a panel, where the switch controls a branch circuit or feeder from the panel. Industrial applications of remote-control switching can follow a basic hookup, with the individual mechanically held contactors mounted in enclosures high up on columns in plant areas. Each contactor can be located at the approximate center of its lighting load, to keep circuit wiring as short as possible for minimum voltage drop, and can be tapped from the nearest point on the busway. Control circuits can then be carried to one or more control panels, as shown in Fig. 97. The operating coil of a contactor (or *remote-control switch,* as it also is known) may be switched either manually (using pushbuttons or momentary-contact switches) or automatically by a time switch or photoelectric device.

A typical contactor application for full-panel control might involve locating the contactors at widely spaced lighting panelboards supplying outdoor lighting, with all the control circuits brought to pilot switches at a common point of control. For the control of individual circuits supplying lighting loads, contactors may be located near the panelboard or near the loads, with the control circuits arranged—for maximum convenience of operation—at a central control point or at any number of points, depending upon the job requirements (Fig. 98). Where incandescent lighting is to be switched, the contactor must be tungsten-rated; where motors are involved, the contactor must be able to break the inductive load current.

FIG. 96 Contactors within a panelboard provide remote control of loads connected to the two sections of the split-bus panelboard.

FIG. 97 Remote-control switching for large-block control of lighting.

Figure 99 shows a magnetic contactor used for the control of heating circuits in a night-setback hookup. During the day, the electric heaters are controlled by room thermostats at their locations; the overcall key switch is in the closed position, thereby keeping the contactor coil energized and its contacts closed. At night, the overcall switch is opened and the magnetic contactor is controlled by the zone thermostat shown, which operates to maintain the temperature at the night setting of 50°F. The local thermostats are all closed at that temperature.

TYPES OF CONTACTORS USED IN THIS SYSTEM

Type I –
Electrically operated, mechanically held, 60-amp 250-volt ac contactor, with 1 NO contact and with 208-volt operating coil

Type II –
The same as Type I but with 2 NO contacts

Type III –
Electrically operated, electrically held, NEMA Size 4, 600-volt contactor, with 480/120-volt control transformer and 120-volt operating coil.

Astronomical timer in plant's protection office operates the control circuits of contactors at preset times of day. Can be over-ridden by master manual pushbutton control.

Control voltage 208 volts

Master manual control from push button station in plant's protection office

On
Off

2 No.8-$\frac{3}{4}$"C. to 35A
2 P. in L.P.

To L.P. 120 volts

To L.P. 1-25A. 1P. A.C.B. 2 No.10-$\frac{3}{4}$"C.

Type I contactor

To floodlights

120-v

3 No.10-$\frac{3}{4}$"C. to 2-25A 1P. A.C.B's

120-v

Type II contactor

To floodlights and obstruction light

480-v 3-φ

3 No. 2-1$\frac{1}{4}$"C to bus duct

Type III contactor

To parking lot floodlights

Type I contactor

3 No.8-$\frac{3}{4}$"C.

120-v

2 No.10-$\frac{3}{4}$"C. to L.P. 1-25A. 1P. A.C.B.

Type I contactor

To floodlights

120-v

3 No.10-$\frac{3}{4}$"C. to L.P. 2-25A.1P. A.C.B's

Type II contactor

To floodlights and obstruction light

Contactors spread throughout the plant are fed from different panelboards and supply widespread loads as shown.

FIG. 98 Central control of widespread lighting with contactors.

2.99

Typical coil-circuit control devices for contactors include maintained-contact and momentary-contact pushbuttons, selector switches, control- and master-type switches, pressure switches, float switches, limit switches, time switches, thermostats, plugging switches, and contacts of control relays. Which of these devices is used depends upon the nature and layout of the load; more than one type may be used. Contactors may be used for the control of branch circuits or feeders and for supplying lighting loads, heating loads, and/or motor or other power loads.

FIG. 99 Contactor control of electric-heating circuits.

The design and layout of magnetic-contactor control circuits and their components are regulated by **NEC** rules. Article 725 covers remote control. As shown in Fig. 100, protection of the coil-circuit conductors is set forth in Section 725-12.

Exception No. 3 to Section 725-12 covers protection of the remote-control circuit that energizes the operating coil of a magnetic contactor, as distinguished from a magnetic motor starter. Although it is true that a magnetic starter is a magnetic contactor with the addition of running-overload relays, Exception No. 3 covers only the coil circuit of a magnetic contactor. Exception No. 4 states that contactor remote-control conductors can be satisfactorily protected by overcurrent devices rated at not more than 300 percent of the carrying capacity of the control-circuit conductors. That applies to control wires for magnetic contactors used for the control of lighting and heating loads, but not motor loads. Section 430-72 covers requirements for control circuits of magnetic motor starters.

It should be noted that the overcurrent protection is required for the control conductors and not for the operating coil. Because of this, control conductors can be sized for use without separate overcurrent protection.

In Fig. 101, office lighting made up of continuous-row fluorescent luminaires is circuited from a remote-control contactor panelboard which is outside the office area, on the wall of the adjoining manufacturing area. The special panelboard is fed by conductors tapped from a busway plug switch, as indicated. The panel is equipped with a main contactor which switches the supply to the panel mains. Each branch circuit consists of a branch CB and a remote-control contactor. All control circuits are carried in conduits to a control panel on the wall at one end of the office area.

Branch-Circuit Panelboards

Lighting and appliance branch circuits originate in panelboards in which the overcurrent protective devices for the circuits are mounted. A panelboard usually contains a means for switching each circuit. Typical panelboard configurations are shown in Fig. 210. Some panelboards have a switch and a fuse for each circuit; other panelboards have circuit breakers that provide both protection and switching for the circuits. Overcurrent protection is an **NEC** requirement for branch circuits, but disconnect provisions are not required for branch circuits (although a means

Feeder or branch-circuit protective device rated not more than three times the current rating of the control conductors

Magnetic contactor

Line

To load
(e.g., fluorescent
lighting panel)

Neutral

Opening and
closing coil

Overcurrent protection
not required for
above condition of
circuit protection

Class 1 remote-
control circuit

Remote-control
station

EXAMPLE: 30-A fuses at "A" would be adequate protection if No. 14 wire, rated at 20 A, is used for the remote-control circuit, because 30 A is *less than* 3 × 2|0 A. If fuses at "A" were rated over 60 A, 15 -A protection would be required at "B" for No. 14 wire.

FIG. 100   Design of contactor control for branch circuits includes sizing of coil-circuit wires and provision of overcurrent protection for such wires.

is required for disconnecting cartridge fuses in circuits at any voltage if they are accessible to other than qualified persons). A panelboard, therefore, may contain only fuses. In the great majority of cases, however, the advantages of switch or CB control of each circuit at the panelboard dictate use of such control.

The term *loadcenter* is commonly used to describe the light-duty type of panelboard used for residential and light commercial and industrial systems. However, both the UL and the **NEC** refer only to "panelboards" and include both loadcenters and the heavier-duty panelboards within this single category.

The selection of a panelboard is based first on the number of circuits it must serve. Then the designer must ensure that the busbars in the panelboard for any application have sufficient current rating for the demand load of all of the branch circuits.

Standards of UL and NEMA base the ratings of main switches, overcurrent devices, and busbars feeding lighting and/or appliance branch devices within the panel on an assumed average load current of not more than 10 A per branch circuit. Where the continuous loading will be in excess of an average of 10 A per branch circuit, busbars, integral main switches, and overcurrent devices of greater capacity are required.

All panelboards—for lighting and power—are required by **NEC** Section 384-13 to have a rating (the current capacity of the busbars) not less than the **NEC** minimum feeder-conductor capacity for the entire load served by the panel. That is, the panel busbars must have a nameplate ampere rating at least equal to the required ampere capacity of the conductors feeding the panel (Fig. 103). A panel may have a busbar current rating greater than the current rating of its feeder but must never have a current rating lower than that required for its feeder. Although

1. Size of feeder from bus plug switch to main contactor is based on total volt-amperes of fluorescent–lighting load at 100% demand.

2. Additional feeder capacity is included for load growth.

3. Further upsizing of feeder conductors keeps voltage drop under 1% from service to branch circuit panelboard and reduces the heating effect of continuous load operation.

4. Main contactor is sized to match feeder capacity.

5. Lighting panelboard is rated for the total feeder capacity.

6. Branch circuit protection consists of 20-amp, 3-pole CB's for multi-wire branch circuits to 277-volt fluorescent ballast.

7. Individual branch circuit contactors are 3-pole, 30-amp remote-control switches connected in the lighting panelboard on the load side of the branch CB's.

8. Each branch circuit is loaded to 50% of its capacity (80% loading is the NEC maximum for continuously operating circuits).

FIG. 101 Major considerations in circuit design for contactor control.

Section 220-10(b) notes that a feeder for a continuous load must be rated at least 125 percent of the load current, there are no safety reasons for having the panel busbars rated at more than 100 percent of the load current.

In addition to the rule of Section 384-13, the rule of Section 220-3(d) applies to panelboard ampacity ratings; it requires any panelboard to have busbar ampacity at least equal to the min-

CIRCUIT-BREAKER PANEL

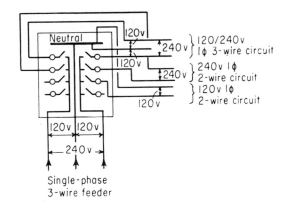

SWITCH-AND-FUSE PANEL

FIG. 102 Typical circuit makeups in panelboards.

imum calculated branch-circuit loads from Section 220-2. For a panelboard supplying general lighting, the rule of Section 220-3(d) requires busbars to have an ampere rating of "not less than the calculated load," as determined from the minimum watts-per-square-foot value for the branch-circuit load and the area served—even if the result is less than the actual connected branch-circuit loading.

A panelboard must have a minimum ampacity sufficient to serve the calculated total load of lighting, plus appliances, motors, and other loads supplied. Of course, if the connected lighting load is greater than that calculated on the watts-per-square-foot basis, the greater value of load must be used in determining the number of branch circuits, the panelboard capacity, and the feeder capacity.

Feeder conductors

Busbars

Neutral

Branch circuits

Minimum ampere rating of busbars in any panel

must not be less than

the required minimum ampere rating of the feeder conductors

as determined from

the sum of all branch-circuit loads calculated according to NEC Sec. 220-10

FIG. 103 Panelboard bus must have a rating equal to the calculated minimum feeder ampacity.

EXAMPLE

For a 200- × 200-ft office area, a 3-phase, 4-wire, 460/265-V feeder and branch-circuit panelboard must be selected to supply 265-V HID lighting that is a 92,000-W connected load and will operate continuously (3 hours or more). What are the minimum size of feeder conductors and minimum panelboard rating that must be used to satisfy Section 220-3?

SOLUTION

For the branch-circuit 92,000-W connected load, the panelboard would be sized for the total calculated load of 140,000 W as shown in Fig. 104.

The actual connected lighting load for the area, calculated from the lighting design, is 92,000 W ÷ (460 V × 1.732) = 116 A per phase. Sizing of the feeder and panelboard, however, must be based on 176 A, not 116 A, to satisfy Section 220-3(d). But then, other **NEC** rules require adjustment of that figure of 176 A to get the usable capacity for a 176-A continuous load.

The next step is to correlate the rules of Sections 220-3(a) and (d) with those of Section 220-10(a). The latter requires a feeder to be sized for the "computed load" as determined according to Part **A** [Section 220-2(b)]. The feeder to the continuous calculated load of 176 A must have an ampacity at least equal to 125 percent of the calculated load; and, the feeder protective device must be sized on the same basis, as required by Section 220-10(b):

$$176 \text{ A} \times 1.25 = 220 \text{ A} \qquad [\text{Section 220-10(b)}]$$

Thus the feeder conductors must have that much ampacity, and, say, No. 4/0 THW copper conductors might be used. And a 225-A protective device would be used at the supply end of the feeder.

Section 384-13 requires this panelboard to have a rating not lower than the minimum required ampacity of the feeder conductors—which, in this case, means the panel must have a busbar rating not lower than 220 A. A 225-A panelboard is therefore required, even though a 125-A panel would be adequate for the actual load current of 116 A.

As noted previously, Section 220-3(d) specifically eliminates the need to install enough panelboard protective devices and circuit wires to handle the calculated load of 140,000 W. The number of branch-circuit protective devices required in the panel (the number of branch

Fixture layout for 265-V lighting system is a 92,000-W actual load that draws 116 A per phase . . .

200 ft

200 ft

. . . but calculated load is 140,000 W (176 A per phase); and Secs. 220-2, 220-3(d) and 220-10(b) require a minimum of 225-A panel . . .

. . . and feeder conductors with at least 220-A capacity.

200 ft × 200 ft = 40,000 sq ft
40,000 sq ft @ minimum of 3.5 W/sq ft = 140,000 W

The minimum computed load for the feeder for the lighting is

140,000 W ÷ (460 × 1.732) = 176 A per phase

FIG. 104 Panelboard busbar capacity must be rated for the "calculated load."

circuits) is based on the size of the branch circuits used and their capacity as related to the connected load. If, say, all circuits are to be 20-A, 265-V, phase-to-neutral circuits, then each pole may be loaded at no more than 16 A, because Section 220-2(a) requires the continuous circuit load to be limited to 80 percent of the 20-A protection rating. With the 116 A of connected load per phase, a single-circuit load of 16 A calls for a minimum of 116 A ÷ 16 A or 8 poles per phase leg. Thus a 225-A panelboard with 24 breaker poles would satisfy the rule of Section 220-3(d). But, because the panelboard must have capacity for the calculated continuous load of 176 A per busbar, the required panel must have space for a total number of poles (spare pole spaces plus the eight active poles per phase) at least equal to 176 A ÷ 16 A (per 20-A pole) or 11 poles per phase leg. Thus the required panel would have to contain at least 33 pole-places (3 × 11), although only 8 pole-places would have to be used per phase leg. Section 220-3(a) also requires that the minimum number of branch-circuit poles be based on the "total computed load" and not on a lesser value of connected load.

Panel main protection

Rules in **NEC** Section 384-16 concern the selection of lighting and appliance branch-circuit panelboards. Any panelboard that meets the **NEC** definition (Section 384-14) of a lighting and appliance branch-circuit panel must be individually protected on its supply side by not more than two main circuit breakers or two sets of fuses having a combined rating not greater than that of the panelboard (Fig. 105). In the vast majority of cases, a single main protective device (CB or set of fuses), rated at not more than the panel's rated busbar ampacity, is used to provide

ONE main protective device, either in panel or just ahead of it, rated not greater than panel busbar rating.

TWO main protective devices, dividing supply into two circuits, have combined rating not greater than panel busbar rating.

FIG. 105 Lighting and appliance panelboard must have one or two main protective devices (that is, one or two CBs or sets of fuses).

panelboard main protection. Individual protection is not required when a lighting and appliance branch-circuit panelboard is connected to a feeder that has overcurrent protection not greater than that of the panelboard.

NEC Section 384-14 defines a lighting and appliance branch-circuit panelboard as "one having more than 10 percent of its overcurrent devices rated 30 amperes or less, for which neutral connections are provided."

Watch out for this definition! Many panels have makeups that appear to make them power panels or distribution panels, yet they are technically lighting and appliance panels in accordance with the above definition, and they must have protection for the busbars. Figure 106 shows an example of how it is determined whether a panelboard is or is not "a lighting and appliance branch-circuit panelboard." The determination is important because it indicates whether or not main protection is required for a particular panel. Figure 107 shows panels that do not need main protection because they are not lighting and appliance panels, which are the only panels required by Section 384-16 to have main protection.

Because of the wording of the definition in Section 384-14, it is important to evaluate a panel carefully to determine if protection is required. And just as it is strange, in Fig. 106, to identify a panel that supplies no lighting as a lighting panel, it is also strange that some panels that supply only lighting are technically not lighting panels. Because of the definition of Section 384-14, if the protective devices in a panel are all rated over 30 A, or if no neutral connections are provided in the panel, then the panel is not a lighting and appliance panel, and it does not require main overcurrent protection.

Figure 108 shows a panelboard which just meets **NEC** rules for classification as a lighting and appliance panel, as follows:

1. There is a total of 24 overcurrent devices in the panel.

2. Three of the overcurrent devices are rated 30 A or less, with neutral connections provided for them.

GIVEN: This 208Y/120-V, 3φ, 4-wire panel in an industrial plant.

Based on NEC Sec. 384-14:

1. 42 poles in panel × 10% = 4.2 poles.
2. This panel has 6 poles rated 30 A or less.
3. This means that *more than 10%* of the panel's devices are rated 30 A or less, and neutral connections are provided.

Therefore:

This *is* a "lighting and appliance branch-circuit panelboard," and main overcurrent protection must be provided for the panel busbars.

3φ motor

3φ motor

3φ feeder

3φ motor

3φ motor

Two 3φ, 4-W 20-A branch circuits (6 20-A poles) to 120-V receptacle outlets at work benches

Neutral

3φ motor

3φ feeder

3φ motor

3φ motor

3φ motor

3φ feeder

3φ welder

FIG. 106 Main protection is required for any panel that meets the definition of Section 384-14.

EXAMPLE: Overcurrent devices rated over 30 A

Main protection is not required.

This is not a lighting and appliance panelboard as defined in Sec. 384-14.

37.5 A per phase

Typical 9-light pole on baseball field with 3 1500-W floodlights per phase [Sec. 210-23(c)]

4 single-conductor No. 6 UF cables underground from panel to pole

Neutral

Panel with all 50-A, single-pole CBs

EXAMPLE: No neutral connections provided

Main protection is not required.

All circuits feed lighting: ungrounded 3-wire circuits to fluorescent or other electric-discharge sources with luminaires connected phase-to-phase (no neutral connections)

Supply from ungrounded. 480-V, 3ϕ, 3-wire transformer secondary (no neutral)

3-pole CB for each 3-wire circuit per NEC Sec. 240-20(b) rated not over 20 A for fluorescent fixtures, up to 50 A for other electric-discharge lighting [Secs. 210-23 and 210-21(a)]

NOTE: Fusible equipment is also permissible for such applications.

FIG. 107 Some panels that supply only lighting are not lighting and appliance panels and do not require main protection.

FIG. 108 Panel with more than 10 percent of its overcurrent devices rated 30 A or less, for which neutral connections are provided, is a lighting and appliance panelboard.

3. Ten percent of the number of overcurrent devices (24) is 2.4.

4. The panelboard therefore has more than 10 percent of its overcurrent devices rated 30 A or less, with neutral connections.

5. Such a panelboard is designated by the **NEC** as a "lighting and appliance" panelboard, and it must be protected in accordance with **NEC** Section 384-16.

The overcurrent devices referred to in the foregoing determination may be fuses or CB poles. A two-pole CB is considered to be two overcurrent devices; a three-pole CB is three overcurrent devices. And the **NEC** says that not more than 42 overcurrent devices, "other than those provided for in the mains," may be installed in any one cabinet or cutout box for a lighting and appliance panelboard. Thus, a panel may contain 42 overcurrent poles plus main protection—such as a three-pole main CB ahead of the 42 devices.

Figure 109 shows the three locations at which panelboard main protection may be placed. Where a number of panels are tapped from a single feeder protected at a current rating higher than that of the busbars in any of the panels, the main protection may be installed as a separate device just ahead of the panel or as a device within the panel feeding the busbars (cases 2 and 3 in Fig. 109). The main protection would normally be a circuit breaker or fused switch with the number of poles corresponding to the number of busbars in the panel.

Figure 110 shows two panels that require main protection because they are both fed by the same feeder, which is protected at a current rating greater than the busbar rating of the panels. Figure 111 shows how feeder protection can protect panelboards.

Other protection rules

Section 384-16(b) says, "Panelboards equipped with snap switches rated at 30 amperes or less shall have overcurrent protection not in excess of 200 amperes." Thus any panel (lighting or power) which contains snap switches rated 30 A or less must have overcurrent protection not in

CASE 1

Fused switch or CB at supply end of feeder to panel

Any distance

Feeder to panel

CASE 2

Fused switch or CB just ahead of panel

Lugs only

CASE 3

Fused switch or CB within panel enclosure

IN ALL CASES CB or fuses are rated *not over* panel busbar ampacity.

FIG. 109 Main panel protection may be placed at any one of these locations.

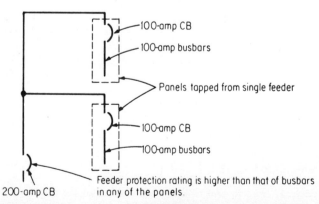

100-amp CB

100-amp busbars

Panels tapped from single feeder

100-amp CB

100-amp busbars

Feeder protection rating is higher than that of busbars in any of the panels.

200-amp CB

FIG. 110 Single main protects each panel in accordance with the basic rule of Section 384-16(a).

FIG. 111 Panels do not require main protection because feeder protection protects panels [Exception No. 1 of Section 384-16(a)].

excess of 200 A. Panels which are not lighting and appliance panels and do not contain snap switches rated 30 A or less do not have to be equipped with main protection and may be tapped from any size of feeder. Figure 112 shows two examples of these overcurrent protection requirements for panelboards.

Section 384-16(d) applies to a panelboard fed from a transformer. The rule requires that overcurrent protection for such a panel, as required in parts (a) and (b) of the same section, must be located on the secondary side of the transformer.

But the primary-circuit protection is not acceptable as suitable protection for a panelboard on the secondary of a transformer with a 3-wire or 4-wire secondary—even if the panel busbars have an ampacity equal to the ampacity of the primary protection times the primary-to-secondary voltage ratio.

Primary protection of a panel on the secondary of a single-phase, 2-wire to 2-wire transformer is permitted if the primary overcurrent protection does not exceed the value determined by multiplying the panel busbar ampacity by the secondary-to-primary voltage ratio (Fig. 113).

Residential panel mains

Main overcurrent protection is generally required for any lighting and appliance panelboard. However, Exception No. 2 of **NEC** Section 384-16(a) notes that a panelboard used as residential service equipment for an "existing" installation may have up to six main protective devices [as permitted by Exception No. 3 of Section 230-90(a)]. Such usage is limited to "individual residential occupancy," such as a private house or an apartment in a multifamily dwelling, where the panel is truly service equipment and not a subpanel fed from service equipment in the basement or from a meter bank on the load side of a building's service.

But it must be noted that the phrase "for existing installations" makes it a violation to use a service panelboard (loadcenter) with more than two main service disconnect-and-protection devices for a residential occupancy in a new building (Fig. 114). That applies to one-family houses and to "dwelling units" in apartment houses, condominiums, and the like. The use of, say, a split-bus panelboard with four or six main protective devices (such as two-pole CBs) in the busbar section that is fed by the service-entrance conductors is limited to a service panel in "an individual residential occupancy" in an *existing* building, for example, for service modernization (Fig. 115). For any new job, any panelboard—for service or otherwise, and in any type

FIG. 112 Power panels do not require main overcurrent protection except with snap switches rated 30 A or less.

of occupancy—may have no more than two main protective devices if the panel is a lighting and appliance panelboard.

A lighting and appliance panelboard containing six single-pole breakers or six two-pole breakers (or even six three-pole breakers or fuses) of any rating may not be used as new residential service equipment without a main protective device ahead. For an existing service, where it is permissible to use a split-bus panel with six main devices in the section fed by the service conductors, it is permissible to use 15- and/or 20-A protective devices in that main section. Of course, 15- and/or 20-A protective devices may be used in the bottom section of the panel, which is protected by one of the mains in the top part of the panel.

Other design concerns

NEC Section 384-16(c) states that "the total load on any overcurrent device located in any panelboard shall not exceed 80 percent of its rating where in normal operation that load will continue for three hours or more." An exception to this would be an assembly, together with the overcurrent device, which has been found suitable for continuous duty at 100 percent of its rating and is UL-listed (or otherwise listed) for such application. The rule of Section 384-16(c) parallels that of Section 210-23(b), which has the same effect of limiting load on the protective device.

Panelboards may be equipped with a main switch or circuit breaker feeding the entire bus or a section of bus in the board. In cases where a lighting panel is supplied by a tap from a feeder with overcurrent protection greater than the main busbar rating of the panel, a main CB

Primary circuit protection... ...does not qualify as suitable protection for secondary conductors

...or a lighting and appliance panelboard

Conductors also rated 125% TPFLC

Max rating or setting of CB or fuses is 125% TPFLC

Δ 480V   Y 208/120V

Panel requiring main protection

... But this is OK.

2-wire

Panel

240 V

120 V

50-A primary CB ...

... protects secondary panel with busbars rated at 100 A

FIG. 113 Main protection for a lighting and appliance panel is not generally permitted on the primary side of the supply transformer.

This application is prohibited for new installations, which may have no more than two main protective devices.

Not more than six fused switches or CBs

240/120-V service

No main protection ahead of six service OC protection devices

N

Panelboard used as service equipment for individual residential occupancy

Any or all branches rated 15 A or 20 A or higher

FIG. 114 For new installations, main overcurrent protection is required for a lighting and appliance panel used as service equipment for a residence.

FIG. 115   Split-bus loadcenter with more than two
main overcurrent devices is permitted for residential
service equipment only in "existing" installations.

or fused switch required for the panel provides simultaneous disconnect of circuits from the feeder. In many cases, a remote-control switch, operated by a manual pilot switch or by some automatic pilot device like a time clock, is used to control the circuits connected to the bus in the panel by switching the feed to the bus, as shown in Fig. 116.

To supplement the spare capacity allowed during branch-circuit design, spare circuit capacity should be allowed in panelboards. One spare circuit should be allowed in the panel for each five active circuits. Inasmuch as panelboards are made in multiples of four switch-and-fuse units or circuit breakers, the spare circuits can often be provided without using a panelboard larger than is required for the active circuits.

When spare circuits are provided in flush-type panelboards, the required conduit capacity to wire these circuits also should be provided, to avoid the need to tear out walls in the future. This extra conduit capacity may consist of empty conduit runs to the ceiling and floor, terminated in covered outlet boxes, spare capacity in other raceways for branch circuits from panelboards, etc.

In addition to standard panelboard hookups, there are cases which call for careful design of special panel arrangements. A typical special panel application is commonly used in gas stations. **NEC** Section 514-5 (in Article 514, "Gasoline Dispensing and Service Stations") contains a specific requirement regarding "circuit disconnects." It states:

FIG. 116 Panelboard control may include contactor switching of separate busbar sections.

Each circuit leading to or through a dispensing pump shall be provided with a switch or other acceptable means to disconnect simultaneously from the source of supply all conductors of the circuit, including the grounded neutral, if any.

Figure 117 shows how this can be accomplished using a gas-station-type panelboard, which has its bussing arranged to permit the hookup of standard solid-neutral circuits in addition to the required switched-neutral circuits. Another way of supplying such switched-neutral circuits is with circuit-breaker-type panelboards for which there are standard accessory breaker units,

FIG. 117 Branch circuits to gasoline dispensing pumps must have switched neutrals.

FIG. 118 Two-pole, switched-neutral CB has a pigtail for connecting one pole to the neutral block.

which have a trip element in the ungrounded conductor and only a switching mechanism in the other pole of the common-trip breaker, as shown in Fig. 118. Either two- or three-pole units may be used for 2-wire or 3-wire circuits, rated 15, 20, or 30 A. No electrical connection is made to panel busbar by the plug-in grip on the neutral breaker unit. A wire lead connects the line side of the neutral breaker to the neutral block in the panel, or two clamp terminals are used for the neutral.

Branch-Circuit Voltage Drop

In laying out circuits, the loading and lengths of homeruns and runs between outlets must be related to voltage drop and the need for spare capacity in the circuits for possible future increases in load. Each lamp, appliance, or other utilization device on the circuit is designed for best performance at a particular operating voltage. Although such devices will operate at voltages on either side of the design value, adverse effects will generally result from operation at voltages lower than the specified value.

A 1 percent drop in voltage to an incandescent lamp produces about a 3 percent decrease in light output; a 10 percent voltage drop will decrease the output about 30 percent. In heating devices of the resistance-element type, voltage drop has a similar effect on heat output. In motor-operated appliances, low voltage to the device will affect the starting and pullout torques, and the current drawn from the line will increase with drops in voltage. As a result, the heat rise in the motor windings will be above normal.

Voltage drop in the conductors is due to the resistance of the conductors plus, in ac circuits, reactance. And the heat developed by the dissipation of power (the I^2R losses) in the wiring, which itself costs money, deteriorates the conductor insulation. In good design, the voltage drop is held to no more than 1 percent.

Figure 119 shows the basic formulas used to calculate voltage drop, with examples of their application. Figure 120 shows how copper loss (the I^2R loss) is calculated for circuits with power factors of 100 percent and less than 100 percent.

In Section 210-19(a), the **NEC** comments (but does not make a mandatory rule) on branch-circuit voltage drop, as follows:

Conductors for branch circuits as defined in Article 100, sized to prevent a voltage drop exceeding 3 percent at the farthest outlet of power, heating, and lighting loads, or combinations of such loads and where the maximum total voltage drop on both feeders and branch circuits to the farthest outlet does not exceed 5 percent, will provide reasonable efficiency of operation. See Section 215-2 for voltage drop on feeder conductors.

1 **Two-wire, single-phase circuits (inductance negligible):**

$$V = \frac{2k \times L \times I}{d^2} = 2R \times L \times I$$

$$d^2 = \frac{2k \times I \times L}{V}$$

V = drop in circuit voltage (volts)
R = resistance per ft of conductor (ohms/ft)
I = current in conductor (amps)
L = one-way length of circuit (ft)
d^2 = cross section area of conductor (circular mils)
k = resistivity of conductor metal (cir mil-ohms/ft) (k = 12 for circuits loaded to more than 50% of allowable carrying capacity; k = 11 for circuits loaded less than 50%; k = 18 for aluminum conductors) at 30 C

EXAMPLE:
What is the voltage drop in this circuit?

$$V = \frac{2k \times L \times I}{d^2} = \frac{2 \times 12 \times 50 \times 60}{26,240 \ (\ \text{NE Code} \ \ \text{Table 8, Ch. 9})}$$

$$V = \frac{72,000}{26,240} = 2.7 \ volts$$

2 **Three-wire, single-phase circuits (inductance negligible):**

$$V = \frac{2k \times L \times I}{d^2}$$

V = drop between outside conductors (volts)
I = current in more heavily loaded outside conductor (amps)

EXAMPLE:

$$V = \frac{2 \times 12 \times 100 \times 24}{10,380} = \frac{57,600}{10,380}$$

$$V = 5.5 \ volts$$

3 **Three-wire, 3-phase circuits (inductance negligible):**

$$V = \frac{2k \times I \times L}{d^2} \times 0.866$$

V = voltage drop of 3-phase circuit

EXAMPLE:

$$V = \frac{2 \times 18 \times 10 \times 100}{6530} \times 0.866$$

$$V = \frac{36,000}{6530} \times 0.866 = 4.76 \ volts$$

4 **Four-wire, 3-phase balanced circuits (inductance negligible):**
For lighting loads: Voltage drop between one outside conductor and neutral equals one-half of drop calculated by formula for 2-wire circuits.
For motor loads: Voltage drop between any two outside conductors equals 0.866 times drop determined by formula for 2-wire circuits.

EXAMPLE:

What size of copper wire must be used here to keep voltage drop to 1% under balanced conditions (i.e., 1% of 277 or 2.8 volts)?

For a 2-wire circuit, $d^2 = \dfrac{2k \times I \times L}{V}$

For a 3-ϕ, 4-wire circuit, $d^2 = \dfrac{2k \times I \times L}{V} \times \dfrac{1}{2}$

$$d^2 = \frac{2 \times 12 \times 15 \times 200}{2.8} \times \frac{1}{2} = \frac{72,000}{5.6}$$

$$= 12,857 \ circular \ mils$$

From Table 8, Ch. 9, NE Code —
No. 10 = 10,380 CM   No. 8 = 16,510 CM
No. 8 must be used.

FIG. 119   Basic formulas for branch-circuit voltage drop.

That wording is only a recommendation, and the voltage drop should be held to the above-recommended 1 percent in lighting branch-circuit conductors to minimize energy losses in the conductors and ensure optimal operation and efficiency for load devices. Figure 121 is a tabulation of circuit loads and the maximum circuit lengths from the panelboard to the first outlet (on a balanced 3-wire, single-phase circuit or a 4-wire, 3-phase circuit) that will hold the voltage

where E = circuit voltage (volts)
       R_1 = total resistance of circuit conductors (ohms)
       R_2 = resistance of noninductive load (ohms)
       I = current flowing in circuit (amps)

With 100% load power factor and negligible conductor reactance

When load power factor is 100% and conductor reactance due to self-induction is negligible, calculation of voltage drop in the circuit and calculation of copper loss due to heating effect of current in the circuit conductors follows the standard relations between current and voltage. The supply voltage is equal to the arithmetic sum of the voltage drop in the conductors and the voltage across the load. Copper loss—energy wasted as heat produced in the circuit conductors—is a simple "I^2R" loss.

$$I = \frac{E}{R_1 + R_2} = \frac{480}{1 + 15} = 30 \text{ amps}$$

Voltage drop in conductors $= I \times R_1 = 30 \times 1$
$= 30 \text{ volts}$

Copper loss in conductors $= I^2 \times R_1 = (30)^2 \times 1$
$= 900 \text{ watts}$

Percent voltage drop $= \dfrac{30}{480} \times 100 = 6\%$

With less than 100% load power factor and negligible conductor reactance

When the power factor of the load is less than 100% and the conductor reactance is negligible, the sum of the voltage drops across the load and the resistance of the circuit conductors is no longer equal to the circuit supply voltage. In the following example, note that although the voltage drop in the circuit conductors is 30.4 volts and might appear to leave only 449.6 volts for the load (480 − 30.4), there is actually 456 volts across the load (30.4 amps x 15 ohms of load impedance). The real loss in voltage to the load is, therefore, only 24 volts (480 −456). The 30.4-volt drop in the conductors is only the voltage across the resistive conductor load, which differs in phase from the voltage across the reactive load device fed by the circuit. Although the arithmetic sum of the two voltages (across the conductor resistance and across the load) is equal to 486.4 (30.4 + 456), correct vector addition of the two voltages gives the 480-volt value of the supply. And note that the significant percent voltage drop is different from the apparent value. Copper loss, however, remains a simple "I^2R" value.

$$Z_1 = \sqrt{(R_2)^2 + (X_L)^2} \quad (Z_1)^2 = (R_2)^2 + (X_L)^2$$
then, $(15)^2 = (R_2)^2 + (X_L)^2$

but, POWER FACTOR $= 0.8 = \dfrac{R_2}{Z_1} = \dfrac{R_2}{15}$

thus, $R_2 = 0.8 \times 15 = 12 \text{ ohms}$
then from $(X_L)^2 = (Z_1)^2 - (R_2)^2$
$= (15)^2 - (12)^2$
$= 225 - 144 = 81$
$X_L = \sqrt{81} = 9 \text{ ohms}$

The total impedance of the circuit, Z_2, includes conductor resistance and must be calculated in order to find circuit current.

$$Z_2 = \sqrt{(R_1 + R_2)^2 + (X_L)^2} = \sqrt{(1 + 12)^2 + (9)^2}$$
$= \sqrt{169 + 81} = \sqrt{250} = 15.8 \text{ ohms}$

Then $I = \dfrac{480}{15.8} = 30.4 \text{ amps}$

Voltage drop in conductors $= I \times R_1 = 30.4 \times 1$
$= 30.4 \text{ volts}$

Copper loss in conductors $= I^2 \times R_1 = (30.4)^2 \times 1$
$= 924 \text{ watts}$

Apparent % voltage drop $= \dfrac{30.4}{480} \times 100 = 6.3\%$

Voltage delivered to load $= I \times Z_1 = 30.4 \times 15$
$= 456 \text{ volts}$

Significant % voltage drop

$$= \frac{480 - 456}{480} \times 100 = \frac{24}{480} \times 100 = 5\%$$

where X_L = inductive reactance component of load (ohms)
Z_1 = impedance of load (ohms)
Z_2 = total impedance of circuit (ohms)

FIG. 120 Calculating voltage drop and copper loss.

drop to a 1 percent maximum. Figure 122 contains a graph that may be used to directly select the wire size required to hold the voltage drop to 1 percent.

Although 50 percent loading of circuits is recommended for substantial protection against excessive voltage drop, the sizes of conductors for long runs should always be carefully selected to ensure that if spare capacity is put to use in the future, it will not increase the voltage drop beyond the limit. For this reason, when the design intent is to use 50 percent loading to provide

Loads and lengths in feet for 1% drop on 3- and 4-wire, 115-volt circuits

Amp load	#10 wire	#12 wire	#14 wire
1	946	596	374
2	474	298	188
3	316	198	124
4	236	148	94
5	190	120	76
6	158	100	62
7	136	86	54
8	118	74	46
9	106	66	42
10	94	60	38
11	86	54	34
12	78	50	32
13	72	46	28
14	68	42	26
15	64	40	24
16	60	38	
17	56	36	
18	52	34	
19	50	32	
20	48	30	
21	46		
22	44		
23	42		
24	40		
25	38		
26	36		
27	36		
28	34		
29	32		
30	32		

Calculations based on copper resistance of 13 ohms per CM-ft at 60C (140F).

For a circuit made up of two phase wires and the neutral of a 3-phase, 4-wire wye system, multiply given lengths by 0.67.

Balanced 3-wire single-phase circuit

Balanced 4-wire 3-phase circuit

NOTE:
This table is based on a 1% voltage drop in each 115-volt circuit which is part of a 3-wire, single-phase circuit or part of a 4-wire, 3-phase circuit—provided the circuit is operating under balanced conditions. This is important because under balanced conditions, the neutral associated with each 115-volt circuit does not carry current. As a result, each 115-volt circuit is, in effect, fed by only one conductor. This produces half the voltage drop that would occur under unbalanced conditions, with both the phase leg and neutral carrying current. The distances given are, therefore, twice what they would be for the same voltage drop under unbalanced conditions.

EXAMPLE:

Q. In making up a 115/230-volt, 3-wire circuit to carry 15 amps on each hot leg for a total circuit length of 57 ft, what size wire must be used to keep the voltage drop to 1% with a balanced load?

A. Come down the left hand column, "Amp load," to the circuit load of 15 amps. Move across and note maximum distances for each wire size.

It can be seen that No. 12 wire reaches a drop of 1% (1.15 volts) at a length of 40 ft. No. 10 wire can carry a 15-amp circuit for up to 64 ft before reaching the voltage drop level.

Because the required circuit length of 57 ft is greater than 40, No. 12 cannot be used. But because it is less than 64 ft, *No. 10* wire will keep voltage drop within limits.

FIG. 121 Sizing branch-circuit wires for 1 percent voltage drop in the homerun (the wires between the panelboard and the first outlet).

spare capacity in the circuit, the conductors used in long runs should be sized for voltage drop on the basis of the maximum possible loading.

For lighting and appliance branch circuits, calculations of voltage drop should be made on a single-phase, 2-wire basis. That is, the voltage drop in the wires should be limited to some percentage of the voltage from the hot leg to the neutral—whether the circuit is 2-wire or 3-wire, single-phase or 4-wire, 3-phase. Since most branch-circuit loads are connected hot leg to neutral, the major concern should be with the voltage delivered to the loads, rather than with the phase-to-phase voltage when that voltage is not supplied to any load device. Of course, 2-wire circuits made up of ungrounded legs of single-phase or 3-phase systems should be designed

VOLTAGE DROP AND CAPACITIES FOR COPPER CONDUCTORS

GENERAL NOTES

60 cycle—single-phase, 2 or 3 wires in steel conduit (49C temp.).

Curves are for a 1-volt drop and 90% PF.

For any other voltage drop, ratio ordinates.

TYPICAL EXAMPLE

Required: to carry 10 amps 100 ft with a maximum of 2-volt drop.

What is the wire size?

a. 100 ft for 2-volt drop is the same as 50 ft for 1-volt drop.

(Convert to distance per volt) Note arrows on graph.

b. Intersection of 10 amps and 50 ft is just about a No. 10 wire.

Balanced 3-wire single-phase circuit

Basic 2-wire single-phase circuit

NOTE:

In both of the types of circuits shown, there are two conductors carrying current—one out to the load and one back from the load. For the single-phase circuit derived from the 3-phase, 4-wire wye, it is best to select conductors for a 3-phase, 4-wire circuit on the basis of the voltage drop which would exist if only one phase was operating. This gives the worst condition for voltage drop. When such a circuit is truly balanced, there is no current in the neutral, and the phase-to-neutral connected load is being served by one conductor—which produces only half the voltage drop that will be present under conditions of maximum unbalance.

EXAMPLES:

1. **Q.** What wire size should be used to make up a 2-wire, 480-volt circuit to carry 15 amps a distance of 100 ft with a drop of 5 volts?

A. A drop of 5 volts over a 100-ft circuit is equal to a 1-volt drop for each 20 ft (100 ÷ 5). By converting to this value of distance per volt, we can use the table directly. On the horizontal scale, locate the load of *15 amps*. On the vertical scale, locate the distance of *20 ft*. The intersection of co-ordinates shows that *No. 12* conductors would be O.K.

2. **Q.** What wire size should be used for a 3-phase, 4-wire, 480/277-volt circuit that supplies a balanced lighting load of 16 amps per phase—to keep voltage drop at not more than 3 volts over the circuit length of 120 ft, even under conditions of maximum unbalance?

A. A drop of 3 volts over 120 ft is the same as 1 volt for 40 ft. Using the coordinates of 16 amps and 40 ft indicates that No. 10 conductors would do the job.

FIG. 122 Graph shows branch-circuit voltage drop and minimum wire size for a given load current.

for a voltage drop that is some percentage of the voltage between them when they supply loads at other than phase-to-neutral connection.

The voltage drop of any 2-wire circuit (either phase to neutral or phase to phase) can be taken simply as the *IR* drop of the conductors, using the total dc resistance of the two circuit wires. For conductors up to No. 3 AWG—which are used in vast majority of branch circuits—the ac resistance of the wire is equal to the dc resistance; the latter is readily obtained from the right-hand columns of Table 8 in **NEC** Chapter 9. In such wire sizes, the inductive reactance of the circuit is negligible. Figure 123 shows an example of this approach.

The maximum tolerable voltage drop in a motor circuit usually depends upon the speed and torque requirements of the motor and how they would be affected by low voltage. In Fig. 124, for example, the characteristics of the dc motor and its load-torque requirements indicate that the motor needs at least 223 V for proper performance. For operation at normal ambient temperature, the branch-circuit conductors would be sized as follows to ensure at least the minimum allowable voltage at the motor.

The branch-circuit conductors must provide the voltage drop along with the required carrying capacity. **NEC** Section 430-22 requires branch-circuit conductors supplying a single motor to have an ampacity not less than 125 percent of the motor full-load current rating:

$$1.25 \times 20 \text{ A} = 25 \text{ A}$$

From **NEC** Table 310-16, a No. 12 TW copper conductor has an ampacity of 25 A; this conductor would therefore satisfy Section 430-22 as regards the minimum ampacity for the motor circuit. And because the actual full-load current of the motor is 20 A, there is no conflict with

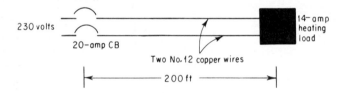

Total length of No. 12 conductor:
$$200 \times 2 = 400 \text{ ft}$$
From Table 8, Chapter 9, NE Code, No. 12 solid copper conductor has a dc resistance of 1.62 ohms per 1000 ft.

Then the 400-ft circuit length has a total resistance of

$$1.62 \times \frac{400}{1000} = 0.648 \text{ ohm}$$

Voltage drop = I × R = 14 amps × 0.648 ohm
= *9 volts.*

In the preceding problem, the conductors were assumed to be exposed to an ambient temperature of 25C (77F). The resistance of the conductors, and thus the voltage drop, would increase under conditions of higher ambient temperature.

FIG. 123 Calculating voltage drop from **NEC** table of wire resistances.

FIG. 124 Voltage drop for dc motor branch circuit involves only resistance and current.

the footnote to Table 310-16 requiring that the "load current rating" of No. 12 copper must not exceed 20 A.

Although the same footnote also sets 20 A as the maximum rating of overcurrent protection for No. 12 copper conductors, Exception No. 3 of Section 240-3 exempts "motor-circuit conductors" from the basic requirement of Section 240-3 that "conductors . . . shall be protected against overcurrent in accordance with their ampacities as specified in Tables 310-16 through 310-19 and all applicable notes to these tables."

To ensure a voltage of at least 223 V at the motor, the maximum permitted voltage drop in the 250-ft circuit run from the panelboard to the motor must be

$$228 \text{ V} - 223 \text{ V} = 5 \text{ V}$$

Table 8 in **NEC** Chapter 9 shows that a No. 12 copper conductor has a dc resistance of 1.62 Ω per 1000 ft at a temperature of 77°F. Because resistance is the only cause of voltage drop in a dc circuit (there is no inductive reactance as in ac circuits), the voltage drop of this circuit, using No. 12 copper wire, can be readily calculated.

With two conductors for the 250-ft-long circuit, the total length of No. 12 wire is 2 × 250 ft or 500 ft. At 1.62 Ω per 1000 ft, the total resistance of the circuit would be half of 1.62 Ω, or 0.81 Ω.

At a full-load current of 20 A, the drop in voltage from the panel to the motor would be

$$V_d = I \times R = 20 \text{ A} \times 0.18 \text{ Ω} = 16.2 \text{ V}$$

That would produce a motor voltage of

$$228 \text{ V} - 16.2 \text{ V} = 211.8 \text{ V}$$

Obviously, this would not satisfy the requirement of 223 V at the motor. A larger conductor, with less resistance, must be used.

Because the drop of 16.2 V with No. 12 wire is over three times the maximum permitted drop of 5 V, the required conductor must have a resistance less than ⅓ that of No. 12 wire.

Reference to the table of dc resistances shows that No. 8 wire has a resistance of 0.6404 Ω per 1000 ft, which is more than 1.62 Ω/3. But No. 6 wire, with a resistance of 0.410 Ω per 1000 ft, looks like it might do.

The calculation is repeated for the No. 6 wire: At a resistance of 0.410 Ω per 1000 ft, the required 500 ft of No. 6 wire for the 250-ft-long 2-wire circuit would have a total resistance equal to half of 0.410 Ω, or 0.205 Ω.

Then, at the 20-A load current, the voltage drop would be

$$V_d = I \times R = 20 \text{ A} \times 0.205 \text{ }\Omega = 4.1 \text{ V}$$

The motor supply voltage then would be

$$228 \text{ V} - 4.1 \text{ V} = 223.9 \text{ V}$$

This satisfies the need for at least 223 V at the motor, and No. 6 copper conductor (TW, THW, RHH, THHN, or XHHW) must be used.

As shown in Table 9 in **NEC** Chapter 9, conductors smaller than No. 2 AWG have negligible self-inductance, and therefore there is no difference between the resistance of such conductors to direct current and their resistance to alternating current. In larger conductors, the "skin effect" (self-inductance forcing current to flow in the outer part of the conductor cross-sectional area) produces an ac resistance that is higher than the dc resistance. For the vast majority of branch-circuit designs, the dc resistance of wire is entirely satisfactory for use in voltage-drop calculations.

In the smaller sizes of conductors, the inductive reactance produced under ac conditions is also generally negligible. However, it is important to distinguish between the increase in ac resistance due to self-inductance and that due to inductive reactance, which is also produced by the self-inductance of the conductor.

As noted above, ac flow in conductors is subject to the skin effect, which produces an apparent increase in resistance over the resistance value which would obtain for dc flow. This is due to a reduction in the effective conductor cross section resulting from the tendency of ac to flow close to the surface (or "skin") of the conductor. Generally, this increase in resistance to ac is of little consequence in conductors smaller than No. 4/0 AWG.

Reactance in conductors carrying ac power depends upon the size of the conductor, the spacing between it and other conductors carrying current, the position of the conductor with respect to conductors close to it, the frequency of the alternating current, and the presence of magnetic materials close to the conductor. In an ac circuit, the reactance of the conductors may be reduced by placing the conductors close together and/or by placing them in nonmagnetic raceway instead of steel conduit or raceway. In many large-size or long ac circuits, the voltage drop due to impedance is often far greater than the drop due simply to the resistance of the conductors.

When current flows in a conductor in which the reactance due to self-induction is negligible, the voltage drop is equal to the product of the current in amperes and the total resistance of the conductor in ohms. But when the reactance of the conductor is not negligible, the voltage drop is equal to the product of the current in amperes and the total impedance of the conductor, which is determined from the formula

$$Z = \sqrt{R^2 + X^2}$$

where Z = total impedance, Ω
R = total ac resistance of conductor, Ω
X = reactance of conductor, Ω

The voltage drop in such a conductor is

where V = voltage drop, V
$\quad I$ = current flowing in conductor, A
$\quad Z$ = total impedance of conductor, Ω

Resistance and reactance data on wires and cables are given in literature made available by the manufacturers. Tables and graphs are also available for quickly and easily computing the voltage drops in large, heavily loaded feeders operating at less than unity power factor and with considerable conductor reactance. But, for branch-circuit calculations of voltage drop, inductive reactance is not nearly as significant as it is for the larger conductors used for feeders.

SUMMARY ON CIRCUIT DESIGN

The requirements and recommendations of the foregoing discussion lead to a set of basic standards for modern branch-circuit design. Of course, in many cases, practical considerations such as size of area or type of load devices will require deviation from the letter of these standards. However, general standards for the majority of applications may be listed as follows:

- Separate branch circuits should be provided for general lighting, for automatic appliances, for fixed appliances, and for plug receptacles. Generally, each automatic or fixed appliance should be served by an individual circuit.

- Branch circuits with more than one outlet should not be loaded in excess of 50 percent of their carrying capacity.

- An individual branch circuit should have spare capacity to permit a 20 percent increase in load before reaching the level of maximum continuous load current permitted by the **NEC** for that circuit.

- At least one spare circuit should be allowed for each five circuits in use.

- The smallest wire size used in branch circuiting should be No. 12.

- The wire used in a branch-circuit homerun should be at least one size larger than that computed from the loading when the distance from the overcurrent protective device to the first outlet is over 50 ft.

- When the distance from the overcurrent protective device to the first plug outlet on a receptacle circuit is over 100 ft, the size of the circuit homerun should not be less than No. 10 for a 20-A branch circuit; it may be larger, depending upon the rating of the particular circuit, the actual distance, the voltage drop, and the load conditions.

- Homeruns on lighting circuits should be limited to a maximum of 100 ft, unless the load on the circuit is so small that the voltage drop between the overcurrent protective device and any outlet is under 1 percent. Careful layout of panelboard locations and use of sufficient number of panelboards will avoid this problem of long homeruns.

General design standards for the selection and layout of panelboards are as follows:

- No more than 42 branch-circuit phase conductors (for example, 42 single-pole protective devices) may originate from a single lighting and appliance panelboard. And, again, in determining the number of poles in a panel, a two-pole CB is considered to be two single poles, and a three-pole CB is considered to be three single poles.

- No branch circuit in a panelboard should run more than 100 ft to the first outlet of the circuit.

- All panelboards should be readily accessible.

- Each panelboard should be placed as near as possible to the center of the layout of load outlets that it handles.

- If circuit switching from the panelboard is desired, a switch-and-fuse or CB panelboard must be used, and such a panelboard should be used generally.
- Panelboard locations should be selected to conform as much as possible to the routing of feeders, ensuring the shortest possible feeder runs and a minimum number of bends.
- Every panelboard must have a rating that is not lower than the minimum feeder capacity required to serve the load, as determined from **NEC** Article 220.
- At least one lighting and appliance branch-circuit panelboard should be provided on each active floor of a building.

LAYOUT ON PLANS

The actual design procedure for lighting and appliance branch circuits involves working on a set of plans for the building. Such plans are usually available or can be made. The design work consists in using the foregoing information and whatever other data are known to lay out the outlets, circuit runs, and control legs. Of course, this step depends upon a decision having been made as to the general type of distribution and the characteristics and voltage of the feeders from which the branch circuits will be supplied. Figure 125 shows the kind of one-line diagram that can be included on electrical plans to show the overall relationship of branch circuits to the feeders supplying them.

The first step in circuit design is to indicate the locations of general-lighting outlets on floor plans. This layout of general-lighting outlets is part of overall lighting design. As a follow-up to this first part of the work, the wattage of each outlet is determined, the total load is determined, and the number of circuits is selected.

Design standards for lighting and appliance circuits are aimed at providing convenience, flexibility, operating efficiency, and reliability in using the available energy. To ensure all the advantages of sufficient circuit capacity plus spare capacity for load growth, modern design practice dictates the separation of loads into known, approximated, and unknown loads.

General illumination is a known load—whether derived from a detailed lighting layout or developed from a watts-per-square-foot calculation. The number, rating, and layout of outlets for general illumination can easily and accurately be apportioned among a number of branch circuits. Such circuits can be carefully loaded with due regard to voltage drop, operating cycle, possible future increases in lighting level, and required control.

Figure 126 shows branch-circuit loading schedules for two different installations. In Fig. 126a, the number of outlets and total load for each branch circuit in each panelboard are indicated. The relative balancing of loads on the two busbars is also shown. The top part of the panel in Fig. 126b shows the loading of 277-V fluorescent lighting on 20-A circuits, each of which has a load capacity of 5500 W. The schedule in Fig. 126b is actually for a loadcenter in a 480/277-V distribution system serving a modern school. Classroom and corridor lighting is supplied by 277-V fluorescent-lighting circuits from the 480/277-V panel (circuits 1 to 8 in the schedule). The 30-kVA transformer fed by circuit 20 in this panel supplies 120/208-V power to a general-purpose lighting and appliance panelboard. Note the spare circuits.

In the design of many commercial and industrial buildings, specific lighting requirements may not be known until after it is known what type of work will be performed and tenants decide upon their working layout. In such cases, the provision of sufficient capacity and circuits in the panelboard is the initial limit of circuit design work, pending determination of the loads.

After lighting outlets are laid out, the next usual step is to indicate convenience and special receptacle outlets on the plans. Depending upon the occupancy, receptacle circuits may be run in conduit to wall-mounted or column-mounted outlet boxes or fed up from the floor slab or

FIG. 125 Light and power branch-circuit layout served from a 480Y/277-V system.

Feeder

Distribution panel air circuit breaker type 480Y/277v 3-phase, 4-wire

480v 4-wire subfeeder

Sub-distribution panel with current-limiting fuses 277v,3-phase, 4-wire with solid neutral

277v lighting branch circuit

Control relay or contactor using 120v.

Fixtures

Fluorescent lighting using 277v ballasts

277-24v control transformer for relay switching of 277v circuits

Motor subfeeder

Single-phase dry-type distribution transformer 480-240/120v, 10 or 15kva

Comb. fused starter

Motor branch circuit

Vent fan motor

M

Sub-distribution panel for plug outlets and incandescent lighting 240/120v 3-wire with solid neutral

G Lighting Panelboard				K Lighting Panelboard			
Circ. No.	No. of Outlets	Wattage Bus A	Bus B	Circ. No.	No. of Outlets	Wattage Bus A	Bus B
1	4	1000		1	4	480	
2	4		1000	2	3		1500
3	4	1000		3	3	1500	
4	4		1000	4	5		750
5	4	1000		5	5	750	
6	4		1000	6	2		800

(a)

480Y/277 volt mains

800 W 1
2700 W 3
3000 W 5
3600 W 7
9
11
13
Immersion heaters 15
2-5 kw, 2 No.10, ¾"C {

5hp blower
Heating unit No.1 {

2 2100 W
4 3600W   277-volt
6 3600W   lighting
8 3300W

10
12
14
16 Immersion heaters, 2-5kw, 2 No.10, ¾"C
20 Transformer 30 kva 3 No.6, 1¼"C

Service: 4 No. 2/0, 2½"C

(b)

FIG. 126 Examples of circuit loadings in lighting panelboards.

down from the ceiling; the circuits may be run in underfloor raceway or a cellular metal floor system; the receptacles may be fed by a flat-conductor cable system under carpet squares, or they may be fed by poke-through floor fittings.

Figure 127 shows the standard symbols used to designate the elements of an electrical-system layout on a set of plans. Because electrical plans, along with a set of specifications, constitute the instructions to the electricians installing the system (that is, the persons exccuting the design concepts set forth in the plans), as much detail as possible should be included in the job drawings.

Graphical interpretation of electrical design involves indication of the types of electrical equipment on scaled representations of the building areas served by the equipment. Such indication includes the wire and raceway interconnections among the various elements and must convey all the design concepts to the mind of the installer who will work from the plans. Figure 128 shows the branch circuiting for a small school building with an outside eating area. A legend clarifies the symbols, and a schedule for the panelboard serving the circuits shows the various loads. For the lighting circuits, these loads are known from the lighting design which established the number and ratings of the various lighting units. For the ½-hp motor on circuits 8 and 10, the load is known. For receptacle circuits, loads are determined either by allowing 1.5 A for

FIG. 127 Electrical symbols for architectural plans.

each outlet or by using load figures for known or estimated ratings of appliances. Ratings of overcurrent protection devices for the branch circuits are given under "Remarks" in the schedule.

When lighting and appliance branch circuits, with their controls and panelboards, have been laid out, a panelboard schedule should be made up and included in the plans and specifications.

SCHEDULE – LIGHTING PANEL "A"				
Ckt No.	No outlets	Watts bus 1	Watts bus 2	Remarks
1	6	900		15 A Ltg. Reception Rm.
2	6	700		Closets and O.S.
3	4		800	Work Rm
4	4		800	Principal's and Nurse
5	8	1050		Heads. Htr. Rm and Fans
6	3	1000		20A Rec. Nurse's Room
7	7		1050	Reception Room
8	½ Hp		400	15A 15A-2P
9	6	900		20A Rec Work Rm ½ Hp
10	½ Hp	400		15A Motor
11	1		500	P. A. System
12	1		500	20A. Intercom. System
13	1	500		Clock, Bell & F.A.System
14	2	900		20A Recep's.
15	5		500	15A Snack Bar & Outside
16	1		500	20A Recep. Snack Bar
17	2	500		20A Receps.Work Rm.
18	–	1000		15A Spare
19	–		1000	
20	–		1000	
Totals		7850	7050	
Main lugs only				

Plan – Covered Eating Area

LEGEND

Fluorescent fixtures – capital letter and number indicate panel and circuit number per schedule. Lower case letter corresponds to letter on switch controlling unit

Duplex convenience receptacle – letter and number show panel and circuit

Incandescent lighting fixture – letter in circle gives fixture type from fixture schedule, "A2" gives panel and circuit, "e" indicates switch controlling unit and "100" is unit wattage

Floor Plan – Light and Power

Se   Switch – lower case letter indicates lighting units controlled by switch

Fan motor

Concealed branch circuit home run

Concealed branch circuit conduit

Underfloor branch circuit home run

Underfloor branch circuit conduit

CIRCUITING DESIGNATIONS

One circuit, two wires, ½" conduit

Two circuits, three wires, ½" conduit

Two circuits, four wires, ¾" conduit

Two wires   ½" conduit

Three wires

Four wires   ¾" conduit

FIG. 128   Details on electrical plans must convey clear, easily understood, and complete details of the branch-circuit design concepts.

This schedule should assign some code-letter designation to each panelboard (such as L1a for lighting panel *a* on the first floor, L2b for lighting panel *b* on the second floor, etc.) and should include a tabulation of panel locations, number and size of circuits in each panel, types and ratings of circuit protective devices, capacities of mains, size and type of main protection and disconnect, and any pertinent remarks about each panelboard which might clarify the design intent for the installer (Fig. 129).

Depending upon the type of occupancy and how the interior is to be laid out from a work standpoint, outlets for local and/or special lighting units represent approximated loads. Such outlets are included in the overall lighting design, either as fixed unit loads or estimated plug-in loads. Outlets—either lampholders or plug receptacles—for such lighting units may be con-

LIGHTING PANELBOARD SCHEDULE

Panel Designation	Location	Equip.	No. Active	No. Spares	Total No.	Poles	Trip Calibr.	Type	Cap	Phase	Feed	Voltg.	Remarks
			Branch Circuits					Mains			Service		
LP1A	1st Floor Stair No. 5	Cct Bkr	25	7	32	SP	20A	Cct Bkr.	225/125	3	4-300 MCM	120/208	
LPIC	1st Floor Main Stair		27	7	34	SP	20A	"	225/150	3	4-4/0	"	
LP2A	2nd Floor Stair No. 6		29	7	36	SP	20A	"	225/150	3	4-300 MCM	"	
LP2B	2nd Floor Stair No.1		1 / 22	1 / 8	2 / 30	2P / SP	20A / 20A	"	225/125	3	4-4/0	"	
LP2C	2nd Floor Main Stair		14	6	20	SP	20A	"	100/70	3	4-4/0	"	
LP3A	3rd Floor Stair No. 6		39	3	42	SP	20A	"	225/175	3	4-350 MCM	"	
LP3B	3rd Floor Stair No.1		1 / 20	1 / 6	2 / 26	2P / SP	20A / 20A	"	100/100	3	4-4/0	"	
LP3C	3rd Floor Main Stair		12	4	16	SP	20A	"	100/100	1	—	"	Connect to Existing 3 Wire Feed
LP4A	4th Floor Stair No. 6		18	4	22	SP	20A	"	100/100	3	4-350 MCM	"	
LP4B	4th Floor Stair No.1		1 / 13	0 / 5	1 / 18	2P / SP	20A / 20A	"	100/70	3	4-300 MCM	"	
LP5A	5th Floor Stair No. 6		11	9	20	SP	20A	"	100/100	3	4-400 MCM	"	
LP5B	5th Floor Stair No.1		1 / 11	1 / 5	2 / 16	2P / SP	20A / 20A	"	100/70	3	4-300 MCM	"	
LP6A	6th Floor Stair No. 6		14	8	22	SP	20A	"	100/100	3	4-400 MCM	"	
LP6B	6th Floor Stair No.1		1 / 15	1 / 5	2 / 20	2P / SP	20A / 20A	"	100/70	3	4-300 MCM	"	
LPPH-1A	Penthouse		3	3	6	SP	20A	Lugs Only	NEMA STD	1	3 No.12	"	
LPBA	Basement Near Stair No.1		9	3	12	SP	20A	Lugs Only	NEMA STD	3	4 No 2	"	
LPBB	Basement Near Stair No.1		11	3	14	SP	20A	Lugs Only	NEMA STD	3	4 No 2	"	
LPBC	Basement Near Col. 32	▶	5 / 0	5 / 1	10 / 1	SP / 2P	20A / 20A	Lugs Only	NEMA STD	3	4 No 6	"	
LPBD	Basement Near Col. 8		6 / 0	4 / -	10 / 1	SP / 2P	20A / 20A	Lugs Only	NEMA STD	3	4 No 8	"	

FIG. 129 Typical detailed schedule of lighting panelboards gives selection and installation data for all panels.

2.131

FIG. 130 Diagrams on plans should show adequate details of special hookups, such as emergency lighting and night lighting.

nected to general-lighting circuits, provided with separate circuiting, or included in circuits allowed for plug receptacles.

Local or special lighting outlets include corridor lights, exit lights, entrance lights, washroom lights, emergency lights, and other known lighting-outlet requirements. Figure 130 shows a typical layout of special lighting outlets for the continuous operation of emergency lights which also serve as night lights in a hospital corridor. The circuiting arrangement for the two-lamp corridor lighting fixtures provides separate conduit and conductors for normal-circuit lighting with a 150-W lamp in each fixture and for emergency-circuit lighting with a 25-W night lamp in each fixture. In this diagram, *D* circuits 12 and 14 supply the 150-W lamps and may be switched off at night by means of the wall switch at the upper left. Panel D is fed from the normal (or nonessential) section of bus in the main switchboard. The 25-W night lamps are fed from panel 1 EM (circuits 3, 5, and 7 shown here) and operate day and night, with no local switching available. Panel 1 EM is fed from the section of main switchboard bus which is fed from the normal supply, and from the emergency generator when the normal supply fails.

In addition to one-line diagrams showing the basic power distribution and full wiring schematics, to detail the actual hookups and circuit wiring layouts, electrical plans should include isometric (three-dimensional) drawings to show the physical relationships between electrical equipment and building interior. Such drawings indicate the installed positions of the electrical equipment in relation to architectural and/or structural members of the building. They are particularly useful for the layout of wiring methods—which extend over very large areas of a building's floor or ceiling space. An example of this type of diagrammatic elaboration is presented in Fig. 131, which shows details from the plans of an industrial building in which fluorescent lighting is fed by cord-and-plug connection to trolley busway. As shown, the trolley busway serves as a lighting branch circuit and is fed from four 50-A CBs in a lighting panel. The panel is supplied from a transformer secondary, powered from a tap to a busway feeder or subfeeder. The lighting fixtures are suspended from the trolley busway, from beams, or from the ceiling. The trolley busway is supported by the ceiling, beams, messenger cable, or braces supported by the beams.

Figures 132 and 133 illustrate the advantage of isometric drawings in clarifying design concepts of electrical wiring methods with respect to floor construction.

In the design of many systems, the general construction of the building, including architectural features and the type of floor-slab construction, will affect the way the circuits are run and the locations of outlets. Detailed, effective drawings can greatly facilitate the task of making

Messenger cable support

Hanger

Trolley busway

Chain

Lighting fixture

Clamp

Beam

Trolley busway

Chain

$\frac{3}{8}$" rod

Hanger

Plug and cord

Lighting fixture

Trolley perpendicular to beams

Beams

"C" clamp

Brace

Trolleys parallel to beams

Trolley busway

SUSPENSION DETAILS FOR TROLLEY BUSWAY

Trolley busway

$\frac{1}{2}$" steel rod

10'

4'±

8'

50A 250 vac

Connecting strap

Feed-in sectionalizer

8'-4 tube slimline fixtures

No. 10 AWG flexible rubber cord

℄

Steel column

25 kva 1φ transformer

100 A

240 v 1φ primary

3φ 3P 600A feeder busway

From main switchboard

Distribution panel 4-50A circuit breakers Single pole

10'

3'-6"

Floor line

ONE TYPE OF TROLLEY BUSWAY DESIGN

FIG. 131 Isometric drawings on electrical plans can greatly clarify the installed relationship of electrical components.

Underfloor Raceway for Use in Conventional Concrete Slab Floors

FIG. 132 High-quality, detailed drawings on electrical plans assure fast and reliable communication of the designer's concepts to the installing electrician.

known to the installer exactly what is to be done. An example is the drawing in Fig. 134, for a data-processing center. The locations of the various types of receptacle outlets are laid out on the floor plan, and a legend is part of the plan. The numbers in the rectangles identify various types of IBM machines. Neoprene-jacketed cables with separable cord connectors run under the raised floor from the panels to the machines.

Special attention should be given to receptacle circuits serving special loads. Figure 135 shows the layout (on the electrical plans) for special receptacle outlets mounted in wireways to supply 9-wire circuits to computing machines in a data-processing center. The floor plan of the area shows the layout of the wireways and the locations of the receptacle outlets. The phase

FIG. 133 Layout of electrical components embedded in concrete must be fully detailed.

designation alongside each receptacle symbol (ϕA, ϕB, or ϕC) indicates which phase of the 60-Hz circuit is tapped to that particular receptacle. The arrow from each receptacle symbol points to the machine supplied by that receptacle. This grid layout of wireways provides for the installation of receptacles at any location, as necessary to serve machines anywhere in the area. The nine conductors for each circuit include three phases and a neutral for 60-Hz supply, three phases and a neutral for 400-Hz supply, and a separate ground wire (either No. 8 or No. 6).

FIG. 134 Floor-plan layout specifies installed location of a wide range of different receptacles for computers.

FIG. 135 This kind of layout ensures that the complex 60- and 400-Hz system will be installed correctly.

2.137

section

Motor Branch Circuits and Control Circuits

Motor loads vary widely in size and electrical characteristics, but all motor circuits require careful wiring and the protection of conductors and equipment to ensure safe and reliable operation. In any plant or building, of course, the problem of providing maximum safety and reliability must be solved along with other problems—minimizing voltage drop, avoiding excessive copper loss, providing sufficient flexibility for changes in the locations of equipment, designing for ease and economy of maintenance of the motors and equipment, and providing spare capacity for future load increases.

Although the effective application of motor controllers is based primarily on thorough engineering analysis, careful consideration should also be given to the **National Electrical Code (NEC)**, which sets forth minimum safety provisions for the control of motors.

In no way is the **NEC** a substitute for the intelligent design of motor control circuits suited to the particular characteristics of each individual application. However, because it does represent the accumulation of years of experience with motor circuits, it presents an excellent general outline of motor-circuit design. Within this basic framework, designers can add specific equipment features and circuit techniques to meet their own needs.

The minimum requirements that must be observed in designing branch circuits for motor loads are set out in two **NEC** articles which are directed specifically to motor applications:

1. Article 430 of the **NEC** covers the application and installation of motor circuits and motor control hookups, including conductors, short-circuit and ground-fault protection, starters, disconnects, and running overload protection.
2. Article 440, "Air-Conditioning and Refrigerating Equipment," contains provisions for such motor-driven equipment and for branch circuits and controllers for this equipment, tak-

ing into account the special considerations involved with the use of sealed (hermetic-type) motor-compressors, in which the motor operates under the cooling effect of the refrigeration.

It must be noted that the rules of Article 440 apply in addition to or are amendments to the rules given in Article 430 for motors in general. The basic rules of Article 430 apply to air-conditioning and refrigerating equipment unless exceptions are indicated in Article 440. Article 440 further clarifies the application of **NEC** rules to air-conditioning and refrigeration equipment as follows:

1. Air-conditioning and refrigerating equipment which does not incorporate a sealed (hermetic-type) motor-compressor must satisfy the rules of Article 422 (appliances), Article 424 (space-heating equipment), or Article 430 (conventional motors), as they apply. For instance, where refrigeration compressors are driven by conventional motors, the motors and controls are subject to Article 430, not Article 440. Furnaces with air-conditioning evaporator coils installed must satisfy Article 424. Other equipment in which the motor is not a sealed compressor and which must be covered by Article 422, 424, or 430 includes fan-coil units, remote forced-air-cooled condensers, remote commercial refrigerators, and similar equipment.

2. Room air conditioners are covered in Part **G** of Article 440 (Sections 440-60 through 440-64 but must also comply with the rules of Article 422).

3. Household refrigerators and freezers, drinking-water coolers, and beverage dispensers are considered by the **NEC** to be appliances; their application must comply with Article 422 and must satisfy the rules of Article 440, because such devices contain sealed motor-compressors.

Figure 136 shows the six basic elements which the **NEC** requires the designer to account for in any motor circuit. Although these elements are shown separately here, in certain cases the **NEC** permits a single device to serve more than one function. For instance, in some cases, one switch can serve as both disconnecting means and controller. In other cases, short-circuit protection and overload protection can be combined in a single circuit breaker (CB) or set of fuses.

Throughout this discussion of motor and control circuits, it should be clearly understood that the **NEC** rules provide only the foundation for modern design. Many equipment applications, circuit layouts, and features must be incorporated in the design, over and above **NEC** rules, to provide convenience, flexibility, and fulfillment of the objectives of a specific installation.

Motor-Circuit Conductors

The basic **NEC** rule says that the conductors carrying load current to a single-speed motor used for continuous duty must have a current-carrying capacity of not less than 125 percent of the motor full-load current rating. In the case of a multispeed motor, the selection of branch-circuit conductors on the line side of the controller must be based on the highest of the full-load current ratings shown on the motor nameplate. The selection of branch-circuit conductors between the controller and the motor, which are energized for a particular speed, must be based on the current rating for that speed.

Figure 137 shows the sizing of branch-circuit conductors to four different motors fed from a panel. The full-load current for each motor is taken from **NEC** Table 430-150. Running overload protection is sized on the basis that nameplate values of motor full-load currents are the same as the values in Table 430-150. If the nameplate and table values are not the same, overload protection is sized according to the nameplate value. The conductor sizes shown are for copper. The given current values and **NEC** Table 310-16 may be used to select the correct sizes of aluminum conductors. (Figure 137 also shows the sizing of branch-circuit protection and running overload protection, as discussed below. Refer to **NEC** Table 430-150 for motor full-load currents, and Table 430-152 for maximum ratings of fuses.)

Used with
wound-rotor motors

Secondary
resistors   Secondary
controller

Two elements commonly
in single enclosure

Remote control circuit
of contactor

Motor

PROVIDE

For a motor of given
horsepower, voltage,
torque, temperature
rise—driving a load
of known torque
requirements

Running overload
protection to open
the circuit on
overloads up to
and including
stalled-rotor
current

Controller
suited to the
load and to
the supply
system

Remote control
circuit for manual
or automatic
operation of
contactor coil,
if magnetic
controller
is used

Disconnecting
means to
disconnect both
motor and
controller from
all ungrounded
supply conductors

Branch circuit
conductors
rated at least
125% of motor
full-load
current

Branch circuit
protective device
to protect against
short circuits
and grounds in
motor circuit

FIG. 136 Design of any motor branch circuit must account for each of these elements.

3.3

Branch-circuit conductors
125% x 40 = 50A
No 6 TW, RHW, THW
RHH, THHN

460V, 3-phase
motor feeder

Branch-circuit
fuses 300% x 65
= 195A (200A fuses)

Squirrel-cage
65A full load
full-voltage
starting

Motor panel

Motor OC
protection
125% x 40 = 50A

Maximum rating
of branch-circuit
fuse 150% x 40
= 60A (60A fuses)

50hp

30hp

Branch-circuit conductors
65 x 125% = 82A

No 2 TW
THW, THHN, RHH or XHHW

Wound-rotor
40 A full load

Motor OC protection
125% x 65 = 82A
(90A rating)

Branch-circuit fuses
300% x 14 = 42A (45A fuses)

Branch-circuit conductors
125% x 14 = 17.5 A (No. 12)

Motor OC protection
125% x 14 = 17.5 to 20A rating

Squirrel-cage 14A full-load
full-voltage starting

10hp

10hp

FIG. 137 Basic design steps in sizing motor branch-circuit conductors and overcurrent devices.

Conductors supplying two or more motors must have a current rating of not less than 125 percent of the full-load current rating of the largest motor supplied plus the sum of the full-load current ratings of all other motors supplied.

It is important to note that these are minimum conductor ratings based on temperature rise only; they do not take into account voltage drop or power loss in the conductors. Such considerations frequently require an increase in the size of the branch-circuit conductors.

Section 430-22(a) includes requirements for sizing individual branch-circuit wires serving motors used for short-time, intermittent, periodic, or other varying duty. For such types of duty, the frequency of starting and duration of operating cycles impose varying heat loads on conductors. Conductor sizing, therefore, varies with the application. But it should be noted that any motor is considered to be wired for continuous duty unless the nature of the apparatus that it drives is such that the motor cannot operate continuously with load under any condition of use.

Conductors connecting the secondary of a wound-rotor induction motor to the controller must have a carrying capacity at least equal to 125 percent of the motor's full-load secondary current if the motor is used for continuous duty. If the motor is used for less than continuous duty, the conductors must have a capacity not lower than the percentage of full-load secondary nameplate current given in **NEC** Table 430-22(a). Conductors from the controller of a wound-rotor induction motor to its starting resistors must have an ampacity in accordance with Table 430-23(c). (See Fig. 138.)

What current to use?

The **NEC** contains definite provisions for determining current-carrying capacities:

Section 430-6: For general motor applications (excluding applications of torque motors and sealed hermetic-type refrigeration motor-compressors), whenever the current rating of a motor

FIG. 138 Circuit design for wound-rotor motors must include the sizing of both primary and secondary conductors.

is used to determine the current-carrying capacities of conductors, switches, fuses, or circuit breakers, the values given in Tables 430-147 through 430-150 must be used instead of the current rating marked on the motor nameplate. However, the selection of separate motor running overcurrent protection must be based on the actual motor nameplate current rating.

For torque motors, shaded-pole motors, permanent split-capacitor motors, and ac adjustable-voltage motors, refer to Section 430-6.

Section 440-5: For sealed (hermetic-type) refrigeration motor-compressors, the actual nameplate full-load running current of the motor must be used in determining the current rating of the disconnecting means, controller, branch-circuit conductors, short-circuit and ground-fault protective devices, and motor running overload protection. When such equipment is marked with a branch-circuit selection current, it must be used instead of the rated-load current to determine the rating or ampacity of the disconnecting means, branch-circuit conductors, controller, and branch-circuit, short-circuit, and ground-fault protection. But the nameplate rated load current must still be used for sizing separate running overload protection.

EXAMPLE

The circuit components for a sealed hermetic motor-compressor with a nameplate current rating of 10 A are sized as follows:

1. The full-load current used for all calculations for a sealed (hermetic) motor-compressor is the current marked on the unit's nameplate [**NEC** Section 440-5(a)].

2. For continuous operation, the branch-circuit conductors must have a rating at least equal to 125 percent of 10 A or 13 A (**NEC** Section 440-32).

3. Conductors for the branch circuit must, therefore, be not smaller than No. 14 copper or No. 12 aluminum (Type TW, THW, RHW, RHH, THHN or XHHW).

4. Running overload protection from overload relays in the starter must have a rating or setting not over 140 percent of the full-load (nameplate) current rating. This calls for an overload relay set to trip at 1.4 × 10 A, or 14 A [**NEC** Section 440-52(a)].

5. The maximum fuse rating for short-circuit protection for a hermetic motor-compressor is 175 percent of the motor nameplate full-load current rating, which is 1.75 × 10 A or 17.5 A. Section 440-22 does not indicate that the next higher standard rating of fuse may be used, but it does permit up to 225 percent of the motor current where needed for starting of the motor. That would indicate the use of 20-A fuses.

Short-Circuit Protection

The **NEC** requires that the branch-circuit protection for motor circuits must protect the circuit conductors, the control apparatus, and the motor itself against overcurrent due to short circuits or grounds (Sections 430-51 through 430-58).

The first, and obviously necessary, rule is that the branch-circuit protective device for an individual branch circuit to a motor must be capable of carrying the motor starting current without opening the circuit. Given this condition, the **NEC** places maximum values on the ratings or settings of such overcurrent devices. It says that such devices must not be rated in excess of values determined from Table 430-152. If such values do not correspond to standard sizes or ratings of fuses, nonadjustable CBs, or thermal devices or to possible settings of adjustable CBs adequate to carry the load, the next higher size, rating, or setting may be used. Where absolutely necessary for motor starting, the device may be rated as follows:

1. The rating of a non-time-delay fuse not exceeding 600 A may be increased but may in no case exceed 400 percent of the full-load current.

2. The rating of a time-delay (dual-element) fuse may be increased but must never exceed 225 percent of the full-load current.

3. The setting of an instantaneous-trip circuit breaker (without time delay) may be increased, but never over 1300 percent of the motor full-load current.

4. The rating of an inverse-time (time-delay) circuit breaker may be increased but must not exceed 400 percent of a full-load current of 100 A or less and must not exceed 300 percent of a full-load current over 100 A.

5. The rating of a fuse rated 601 to 6000 A may be increased but must not exceed 300 percent of the full-load current.

6. Torque motors must be protected at the motor nameplate current rating; if a standard overcurrent device is not made in that rating, the next higher standard rating of protective device may be used.

For a multispeed motor, a single short-circuit and ground-fault protective device may be used for one or more windings of the motor, provided the rating of the protective device does not exceed the above applicable percentage of the nameplate rating of the smallest winding protected.

The **NEC** establishes maximum values for branch-circuit protection, setting the limit of safe application. However, the use of branch-circuit protective devices of smaller sizes is obviously permitted, and this does offer the opportunity for substantial economy in the selection of circuit breakers, fuses and the switches used with them, panelboards, etc. However, a branch-circuit device with a lower rating than the maximum permitted rating must have sufficient time delay in operation to permit the motor starting current to flow without opening the circuit. But a circuit breaker for branch-circuit protection must have a continuous current rating of not less than 115 percent of the motor full-load current.

Where maximum protective-device ratings are shown in the manufacturer's heater table for use with a marked controller or are otherwise marked with the equipment, they must not be exceeded, even if higher values are allowed by the rules given above. (See Fig. 137 for the sizing of overcurrent protection in accordance with Table 430-152.)

Instantaneous-only circuit breakers

The **NEC** recognizes the use of instantaneous-trip circuit breakers (without time delay) for short-circuit protection of motor circuits. Such breakers (also called *magnetic-only* breakers) may be used only if they are adjustable and only when combined with motor starters in combination assemblies.

Molded-case circuit breakers containing only magnetic instantaneous-trip elements are available in almost all sizes. The use of such a device requires careful accounting for the absence of overload protection in the CB, up to the short-circuit trip setting.

This type of CB is designed for use as shown in Fig. 139. The circuit conductors are sized for at least 125 percent of the motor current. The thermal overload relays in the starter protect the entire circuit and all equipment against operating overloads up to and including stalled-rotor current. They are commonly set at 125 percent of the motor current. In such a circuit, a CB with an adjustable magnetic trip element can be set to take over the interrupting task at currents above stalled-rotor current and up to the short-circuit duty of the supply system at that point of installation. The magnetic trip in a typical unit might be adjustable to operate at from 3 to 13 times the breaker current rating; that is, a 100-A CB can be adjusted to trip anywhere between 300 and 1300 A. Thus, the CB serves as motor-circuit disconnect and short-circuit protection.

The selection of an instantaneous-only CB requires the choice of a nominal size to provide a current rating at least equal to 115 percent of the motor full-load current, to carry the motor current and to qualify as a disconnect means according to the **NEC**. Then the adjustable magnetic trip is set to provide the short-circuit protection (the value of current at which instantaneous circuit opening takes place, which should be just above the starting current of the motor involved), using a value of about 1.5 times the locked-rotor current to account for asymmetry in the starting current. Manufacturers of equipment of this type provide very simple instructions that make proper selection of combination starters a quick, easy matter.

FIG. 139 Instantaneous-only CBs are permitted only in combination starters.

The combination motor starter with instantaneous-trip breaker offers the use of a CB that is smaller than the standard thermal-magnetic CB. And the smaller CB offers faster operation for greater protection against grounds and short circuits, in addition to greater economy. The following example shows the basic idea.

EXAMPLE

A 30-hp, 230-V, 3-phase squirrel-cage motor is marked with the code letter M, indicating that the motor has a locked-rotor current of 10 to 11.19 kVA/hp. A full-voltage controller is also provided, with running overload protection in the controller to protect the motor within its heating-damage curve on overload. Select a circuit breaker which will provide short-circuit protection and will qualify as the motor-circuit disconnect means.

SOLUTION

The motor has a full-load current of 80 A (Table 430-150). A CB suitable for use as the disconnect must have a current rating of at least 115 percent of 80 A. Table 430-152 permits the use of an inverse-time (the usual thermal-magnetic) CB rated not more than 250 percent of the motor full-load current (although a CB could be rated as high as 400 percent of full-load current if such size were necessary to pass motor starting current without opening). Based on 2.5 × 80 A or 200 A, a 225-A frame size with 200-A trip setting could be selected. This CB is large enough to generally take the starting current of the motor without tripping either the thermal element or the magnetic element in the CB. The starting current of the motor will initially be about 882 A [30 hp × 11.19 kVA/hp ÷ (220 V × 1.73)]. The instantaneous trip setting of the 200-A CB will be about 200 A × 10 or 2000 A. Such a CB will provide protection for grounds and shorts without interfering with the motor running overload protection.

But consider the use of a 100-A CB with thermal and adjustable magnetic trips. The instantaneous trip setting at 10 times the current rating would be 1000 A, which is above the 882-A locked-rotor current. But starting current would probably trip the thermal element and open the CB. This problem can be solved by removing the CB thermal element and leaving only the magnetic element in the CB. Then conditions of operating overload can be cleared by the running overload devices in the motor starter, right up to stalled-rotor current, with the magnetic trip adjusted to open the circuit instantaneously on currents above, say, 1300 A (882 A × 1.5). But because the value of 1300 A is greater than 1300 percent of the motor full-load current (80 A × 13 = 1040 A), the unnumbered exception in Section 430-52 would prohibit setting the CB at 1300 A. The maximum setting would be 1000 A.

Because a magnetic-only CB does not protect against low-level grounds and shorts in the circuit conductors on the line side of the running overload relays, this type of CB may be used only where the CB and starter are installed as a combination starter in a single enclosure.

Motor short-circuit protectors

A motor short-circuit protector (MSCP) is a fuse-like device designed for use only in its own type of fusible-switch combination motor starter. The combination offers short-circuit protection, running overload protection, disconnect means, and motor control—all with assured coordination between the short-circuit interrupter (the motor short-circuit protector) and the running overload devices. It involves the simplest method of selecting the correct short-circuit protector for a given motor circuit. This packaged assembly is a third type of available combination motor starter, along with the conventional fusible-switch and circuit-breaker types.

The **NEC** recognizes the use of motor short-circuit protectors in Sections 430-40 and 430-52, provided the combination is especially approved for the purpose. Practically speaking, this means a combination starter equipped with motor short-circuit protectors should be listed by UL as a package called an MSCP starter.

One circuit for two or more motors

A single branch circuit may be used to supply two or more motors as follows.

Small motors: Two or more motors, each rated not more than 1 hp and each drawing not over 6 A full-load current, may be used on a branch circuit protected at not more than 20 A at 125 V or less, or 15 A at 600 V or less. The rating of the branch-circuit protective device marked on any of the controllers must not be exceeded. [See Section 430-53(a).] Individual running overload protection is necessary in such circuits, unless the motor (1) is not permanently installed, or (2) is manually started and is within sight from the controller location, or (3) has sufficient winding impedance to prevent overheating due to stalled-rotor current, or (4) is part of an approved assembly which does not subject the motor to overloads and which incorporates protection for the motor against stalled rotor, or (5) the motor cannot operate continuously under load.

Motors of any rating: Two or more motors of any rating, each having individual running overload protection, may be connected to a branch circuit which is protected by a short-circuit protective device selected in accordance with the maximum rating or setting of a device which could protect an individual circuit to the motor of the smallest rating. This may be done only where it can be determined that the branch-circuit device so selected will not open under the most severe normal conditions of service which might be encountered [Section 430-53(b)].

This permission of Section 430-53(b) offers wide application of more than one motor on a single circuit, particularly small, integral-horsepower motors installed on 440-V, 3-phase systems. The application primarily concerns the use of small, integral-horsepower 3-phase motors in 208-V, 220-V, and 440-V industrial and commercial systems. Only such 3-phase motors have full-load operating currents low enough to permit more than one motor on circuits fed from 15-A protective devices.

Figure 140 covers the use of more than one motor on a branch circuit, where small integral-horsepower and/or fractional-horsepower motors are used in accordance with Section 430-53(b).

In case I, with a three-pole CB used as the branch-circuit protective device, application is made in accordance with Section 430-53(b) as follows:

1. The full-load current for each motor is taken from **NEC** Table 430-150 [as required by Section 430-6(a)].

2. A circuit breaker is used instead of fuses for branch-circuit protection. The rating of the branch-circuit protective device, 15 A, does not exceed the maximum short-circuit protection required by Section 430-52 and Table 430-152 for the smallest motor of the group, which is the 1½-hp motor. Although 15 A is greater than the maximum value of 250 percent of the motor full-load current (2.5×2.6 A = 6.5 A) set by Table 430-152, the 15-A breaker is the "next higher size, rating or setting" for a standard circuit breaker as permitted by Section 430-52. A 15-A CB is the smallest standard rating recognized by the **NEC** in Section 240-6.

3. The total load of motor currents is

$$4.8 \text{ A} + 3.4 \text{ A} + 2.6 \text{ A} = 10.8 \text{ A}$$

CASE I—USING A CIRCUIT BREAKER FOR PROTECTION

One 15-amp, 3-pole CB

> **HERE IS THE KEY:** A 15-amp, 3-pole CB is used, based on Section 430-52 and Table 430-152. This is the "next higher size" of standard protective device above 250% × 2.6 amps (the required rating for the smallest motor of the group). The 15-amp CB makes this application possible, because the 15-amp CB is the smallest standard rating of CB and is suitable as the branch-circuit protective device for the 1½-hp motor.

CASE II—USING A CIRCUIT BREAKER FOR PROTECTION

FIG. 140 Branch circuit supplying two or more motors must be carefully designed.

This is well within the rating of the 15-A CB, which has sufficient time delay in its operation to permit starting of any one of these motors with the other two already operating. The torque characteristics of the loads on starting are not high, and it was determined that the CB will not open under the most severe normal service.

4. Each motor is provided with individual running overload protection in its starter.

5. The branch-circuit conductors are sized in accordance with Section 430-24:

$$4.8 \text{ A} + 3.4 \text{ A} + 2.6 \text{ A} + 0.25 \times 4.8 \text{ A} = 12 \text{ A}$$

The conductors must have an ampacity at least equal to 12 A, and No. 14 THW, TW, RHW, RHH, THHN, or XHHW conductors will fully satisfy this requirement.

In case II, a similar hookup is used to supply three motors—also with a CB for branch-circuit protection:

1. Section 430-53(b) requires the branch-circuit protection rating to be not higher than the maximum current set by Section 430-52 for the lowest-rated motor of the group.
2. From Section 430-52 and Table 430-152, that maximum protection rating for a circuit breaker is 250 percent of 1 A (the rating of the lowest-rated motor) or 2.5 A. But, 2.5 A is not a "standard rating" of circuit breaker from Section 240-6; the third paragraph of Section 430-52 permits use of the "next higher size, rating or setting" of standard protective device.
3. Because 15 A is the lowest standard rating of CB, it is the "next higher" device rating above 2.5 A and satisfies **NEC** rules on the rating of the branch-circuit protection.

The applications shown in cases I and II permit the use of several motors up to circuit capacity, based on Sections 430-24 and 430-53(b) and on the starting-torque characteristics, the operating duty cycles of the motors and their loads, and the time-delay of the CB. Such applications greatly reduce the number of CB poles, number of panels, and amount of wire used in the system. However, one limitation is placed on this practice in the next to the last sentence of Section 430-52, as follows:

> Where maximum branch-circuit short-circuit and ground-fault protective device ratings are shown in the manufacturer's overload relay table for use with a motor controller or are otherwise marked on the equipment, they shall not be exceeded even if higher values are allowed as shown above.

In case III, shown in Fig. 141, the three motors of case II would be hooked up differently to comply with the rules of Section 430-53(b) if fuses, instead of a circuit breaker, were used for branch-circuit protection:

BUT, WATCH OUT!!!

CASE III—USING FUSES FOR CIRCUIT PROTECTION

> Interpretation of **NE Code** rules of Section 430-53(b) in conjunction with the "standard" ratings of fuses in Section 240-6 may require different circuit makeup when fuses are used to protect the branch circuit to several motors.

FIG. 141   Fuses, instead of CBs, alter the design characteristics of a multimotor branch circuit.

1. To comply with Section 430-53(b), fuses used as branch-circuit protection must have a rating not in excess of the value permitted by Section 430-52 and Table 430-152 for the smallest motor of the group—here, one of the ½-hp motors.

2. Table 430-152 shows that the maximum permitted rating for non-time-delay fuses is 300 percent of the full-load current for 3-phase squirrel-cage motors. Applying that rule to one of the ½-hp motors gives a maximum fuse rating of 3.0 × 1 A = 3 A.

Fuses without individual disconnects might be acceptable under Section 240-40;or a single disconnect switch, fused at 3 amps, could be installed at point "A," eliminating the need for fuses in the two starters for the ½-hp motors.

. . . **AND A HOOKUP LIKE THIS MIGHT BE REQUIRED TO MAKE CASE 1 OF FIG. 140 COMPLY WITH** NE CODE **RULES**

FIG. 142   Other designs for multimotor circuits.

3. But there is no permission for the fuses to be rated higher than 3 A, because 3 A is a "standard" rating of fuse (but not a standard rating of circuit breaker). Section 240-6 considers fuses rated at 1, 3, 6, and 10 A to be "standard."

4. The maximum branch-circuit protection permitted by Section 430-53(b) for a ½-hp motor is therefore 3 A.

5. The two ½-hp motors may be fed from a single branch circuit with three 3-A fuses in a 3-pole switch.

6. Following the same **NEC** rules, the 2-hp motor would require fuse protection rated not over 10 A (3.0 × 3.4 A = 10.2 A).

   NOTE: Because there are standard fuse ratings below 15 A, fuses have a different relationship to the applicable **NEC** rules than CBs; interpretation of the rules will be required to resolve the question of acceptable application in case II versus case III. Such interpretation will be necessary to determine if circuit breakers are excluded as circuit protection in those cases where the use of fuses, in accordance with the precise wording of the **NEC**, provides lower rated protection than circuit breakers when the rule of the third paragraph of Section 430-52 is applied. And if the motors of case I are fed from a circuit protected by fuses, the literal effect of the rules is to require different circuiting for those motors.

   Figure 142 shows one way of combining cases II and III to satisfy Sections 430-53(b), 430-52, and 240-6; however, the 15-A CB would then be feeder protection, because the fuses would be serving as the "branch-circuit protective devices" as required by Section 430-53(b). Those fuses might be acceptable in each starter, without a disconnect switch, in accordance with Section

240-40, which allows the use of cartridge fuses at any voltage without an individual disconnect for each set of fuses, provided only qualified persons have access to the fuses. But Section 430-112 would have to be satisfied if the single CB were to be used as a disconnect for the group of motors. And part *b* of the exception to that section recognizes one common disconnect in accordance with Section 430-53(a) but not with Section 430-53(b). Certainly, the use of a fusible-switch-type combination starter for each motor would fully satisfy all rules.

FIG. 143 Economy often dictates the use of multiple fractional-horsepower or small integral-horsepower motors on a single circuit.

   Figure 143 shows another hookup of several motors on one branch circuit—an actual job installation which was based on application of Section 430-53(b). The installation was studied as follows.

EXAMPLE

A factory has one hundred 1½-hp, 3-phase motors with individual motor starters incorporating overcurrent protection, rated for 460 V. Provide circuits.

SOLUTION

Prior to 1965, the **NEC** would not permit several motors on one branch circuit fed from a three-pole CB. Each of the 100 motors would have had to have its own individual 3-phase circuit fed from a 15-amp, three-pole CB in a panel. As a result, a total of 300 CB poles would have been required, calling for seven panels of 42 circuits each plus a smaller panel (or special panels of more than 42 poles per panel).

   Under present rules, depending upon the starting-torque characteristics and operating duty of the motors and their loads, with each motor rated for 2.6 A, three or four motors could be connected on each 3-phase, 15-A circuit; this greatly reduces the number of panelboards and overcurrent devices and the amount of wire involved in the system. The time delay of the CB influences the number of motors on each circuit. However, an extremely important requirement is given in the next to last sentence of Section 430-52, which says:

Where maximum branch-circuit short-circuit and ground-fault protective device ratings are shown in the manufacturer's overload relay table for use with a motor controller or are otherwise marked on the equipment, they shall not be exceeded even if higher values are allowed as shown above.

Because Section 430-53(b) includes the condition that "the branch-circuit short-circuit and ground-fault protective device is selected not to exceed that allowed by Section 430-52," it is clear that all of Section 430-52 (including the sentence quoted above) must be fully satisfied.

Motors for "group installation": Two or more motors of any rating may be connected to one branch circuit if each motor has running overload protection, if the overload devices and controllers are approved for group installation, and if the branch-circuit fuse or time-delay CB rating is in accordance with Section 430-52 for the largest motor plus the sum of the full-load current ratings of the other motors. The branch-circuit fuses or circuit breaker must not be larger than the rating or setting of short-circuit protection permitted by Section 430-52 for the smallest motor of the group, unless the thermal device is approved for group installation with a given maximum size of fuse or time-delay CB for short-circuit protection. [See Sections 430-53(c) and 430-40.]

Air Conditioning and Refrigeration

NEC Section 440-22(a) covers the rating or setting of the branch-circuit short-circuit and ground-fault protective device for a circuit to an individual sealed hermetic motor-compressor. The rule says that the device "shall be capable of carrying the starting current of the motor." The required protection is considered to be obtained when the device has a rating or setting not exceeding 175 percent of the compressor rated load current or branch-circuit selection current, whichever is greater (with a 15-A size minimum); however, where the specified protection is not sufficient for the starting current of the motor, it may be increased but may not exceed 225 percent of the motor rated load current or branch-circuit selection current, whichever is greater.

Section 440-22(b) covers sizing of the short-circuit and ground-fault protective device for a branch circuit to equipment which incorporates more than one sealed hermetic motor-compressor or one sealed motor-compressor and other motors or other loads. This extensive coverage of branch-circuit protection for such complicated motor loads should be carefully studied to ensure effective compliance with the **NEC** on such work. In all such cases, where more than one motor is supplied by a single branch circuit, the rules of Section 430-53 ("Several Motors or Loads on One Branch Circuit") must be applied.

Section 440-62(a) points out that a room air conditioner must be treated as a single motor unit in determining its branch-circuit requirements when all the following conditions are met:

1. The unit is cord-and-plug connected.
2. Its total rating is not more than 40 A and 250 V, single phase.
3. The total rated load current is shown on the unit nameplate, rather than individual motor currents.
4. The rating of the branch-circuit, short-circuit, and ground-fault protective device does not exceed the ampacity of the branch-circuit conductors or the rating of the receptacle, whichever is less.

Section 440-60 describes a room air conditioner as an ac appliance of the air-cooled window, console, or in-wall type, with or without provisions for heating, installed in the conditioned room, and incorporating one or more hermetic refrigerant motor-compressors.

Figure 144 calls attention to the fact that branch-circuit protection must always be capable of interrupting the amount of short-circuit current which might flow through it. In the diagram, a short-circuit fault at C will draw current until the circuit is opened by the protection at B. The value of the short-circuit current available at C depends upon the apparent-power (kVA) rating of the supply transformer A, percentage reactance of the transformer, secondary voltage, and effective impedance of the current path from the transformer to the point of fault. In application, therefore, motor controllers must be carefully coordinated with the characteristics of the branch-circuit protective device, which must be able to safely interrupt the short-circuit current. And not only must the device be rated to interrupt the fault current, but it must also act quickly enough to open the circuit before let-through current can damage the controller. The speed in clearing the circuit must be compared to the abilities of the various circuit elements to withstand the damaging effects of short-circuit current flow dur-

FIG. 144 Branch-circuit protection must be properly rated for the maximum current drawn by a fault anywhere on its load side.

ing the time it takes the protective device to operate. Figure 145 shows a typical example of the problem that can arise when the operating speed of the branch-circuit overcurrent device does not protect other circuit elements from the effects of let-through current. The damage to the motor controller could have been avoided by use of a faster-opening CB or fast-acting fuses which would limit the fault current to a value that the controller could withstand.

Motor Controllers

As used in the **NEC**, the term "controller" includes any switch or device normally used to start and stop a motor, in addition to motor controllers as such. The basic requirements of Sections 430-81 and 430-83 regarding sizes and types of motor controllers are as follows (refer to Fig. 146, where controller types are numbered to correspond to the following list):

① Short circuit develops in motor winding.

② Circuit breaker clears fault in several cycles of the fault current. Breaker operates well within its rating and is not damaged by the opening of the fault current.

③ But the high level of fault current that flowed for the several cycles the CB took to open welded the starter contacts closed and damaged the overload relays.

FIG. 145 Effective design can prevent this kind of common fault.

THIS IS THE BASIC RULE

THESE ARE EXCEPTIONS TO THE BASIC RULE

FIG. 146 Some type of motor controller must be provided.

1. A *controller* must be capable of starting and stopping the motor which it controls, must be able to interrupt the stalled-rotor current of the motor, and must have a horsepower rating not lower than the rating of the motor, except that

2. The *branch-circuit protective device* may serve as the controller for a stationary motor rated ⅛ hp or less which is normally left running and is not subject to damage from overload or failure to start. (Clock motors are typical of this application.)

3. A *plug-and-receptacle connection* may serve as the controller for portable motors up to ⅛ hp.

4. A *general-use switch* rated at not less than twice the full-load motor current may be used as the controller for stationary motors up to 2 hp, rated 300 V or less. On ac circuits, a general-use snap switch suitable only for use on ac may be used to control a motor having a full-load current rating not over 80 percent of the ampere rating of the switch.

5. A *branch-circuit circuit breaker,* rated in amperes only, may be used as a controller. If the same circuit breaker is used both as a controller and to provide overload protection for the motor circuit, it must be rated accordingly.

6. Exception No. 3 of **NEC** Section 430-83 covers controllers for torque motors.

Although Section 430-83 permits the use of horsepower-rated switches as controllers and the UL lists horsepower-rated switches up to 500 hp, the UL states in its *Green Book* that "enclosed switches rated higher than 100 hp are restricted to use as motor disconnect means and are not for use as motor controllers." But a horsepower-rated switch up to 100 hp may be used as both a controller and a disconnect if it breaks all ungrounded legs to the motor.

For sealed (hermetic-type) refrigeration motor-compressors, selection of the size of controller is slightly more involved than for standard applications. Because of their low-temperature operating conditions, hermetic motors can handle heavier loads than general-purpose motors of equivalent size and rotor-stator construction. And because the capabilities of such motors cannot be accurately defined in terms of horsepower, they are rated in terms of full-load current and locked-rotor current for polyphase motors and larger single-phase motors. Accordingly, the selection of controller size for a hermetic motor is different from that for a general-purpose motor, for which horsepower ratings must be matched.

NEC rules on controllers for motor-compressors are covered in Section 440-41. The controller must have a continuous-duty full-load current rating and a locked-rotor current rating that are not less than the full-load and locked-rotor currents of the motor.

For controllers rated in horsepower, the size required for a particular hermetic motor can be selected after the nameplate rated load current or branch-circuit selection current (whichever is greater) and locked-rotor current of the motor have been converted to an equivalent horsepower rating. To get this equivalent horsepower rating, which is the required size of controller, the tables in Article 430 must be used. First, the nameplate full-load current of the motor is located in one of Tables 430-148 to 430-150, and the horsepower rating which corresponds to it is noted. Then the nameplate locked-rotor current of the motor is found in Table 430-151, and again the corresponding horsepower is noted. In all tables, if the exact value of current is not listed, the next higher value should be used. If the two horsepower ratings obtained in this way are not the same, the larger value is taken as the required size of controller.

EXAMPLE

A 230-V, 3-phase, squirrel-cage induction motor in a compressor has a nameplate full-load current of 25.8 A and a nameplate locked-rotor current of 90 A. What size of horsepower-rated controller should be used?

SOLUTION

From Table 430-150, 28 A is the next higher current than the nameplate current of 25.8 A, and the corresponding horsepower rating for a 230-V, 3-phase motor is 10 hp.

From Table 430-151, a locked-rotor current rating of 90 A for a 230-V, 3-phase motor requires a controller rated at 5 hp. The two values of horsepower obtained are not the same, so the higher rating is selected as acceptable for the given conditions. A 10-hp motor controller must be used.

Some controllers may be rated in full-load current and locked-rotor current rather than in horsepower. For use with a hermetic motor, such a controller must have current ratings equal to or greater than the nameplate full-load current and locked-rotor current of the motor.

Starter poles

It is interesting to note that the **NEC** says that a controller need not open all conductors to a motor, except when the controller serves also as the required disconnecting means. For instance, a two-pole starter could be used for a 3-phase motor if running overload protection is provided in all three circuit legs by devices separate from the starter. The controller must interrupt only enough conductors to start and stop the motor.

However, when the controller is a manual (nonmagnetic) starter or is a manually operated switch or CB (as permitted by the **NEC**), the controller itself may also serve as the disconnect means if it opens all ungrounded conductors to the motor. This eliminates the need for another switch or CB to serve as the disconnecting means. But it should be noted that only a manually operated switch or circuit breaker may serve such a dual function. A magnetic starter cannot also serve as the disconnecting means, even if it does open all ungrounded conductors to the motor. These conditions are shown in Figs. 147 and 148.

The word "ungrounded" above refers to the condition that none of the circuit conductors is grounded. They may be the ungrounded conductors of grounded systems.

Installed location

Basically, the **NEC** requires that the motor and its driven machinery be within sight from the controller for the motor. When the controller is out of sight (and the **NEC** considers a distance of 50 ft or more to be equivalent to "out of sight," even though the motor and its load might

FIG. 147   Manual starter that opens all ungrounded conductors to a motor may serve as *both* controller and disconnect.

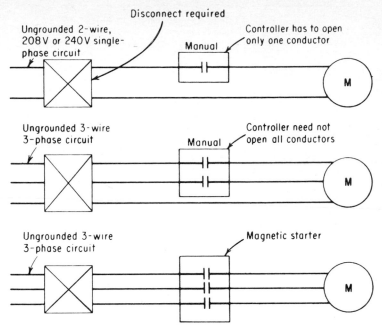

FIG. 148 If a manual starter or switch does not open all ungrounded conductors, a separate disconnect is required.

actually be visible from the controller location), the controller must comply with one of the following conditions:

1. The controller disconnecting means must be capable of being locked in the open position, or

2. A manually operable switch, which will provide disconnection of the motor from its power-supply conductors, must be placed within sight from the motor location. And this switch may not be a switch in the control circuit of a magnetic starter.

These requirements are shown in Fig. 149. Specific layouts of the two conditions are shown in Fig. 150.

NOTE: The **NEC** provisions shown in these sketches are minimum safety requirements. The use of additional disconnects, with or without lock-open means, may be made necessary or desirable by job conditions.

The **NEC** at one time permitted a controller to be installed out of sight of the motor and its load if a switch in the coil circuit of the starter was installed within sight of the motor. However, this is no longer acceptable; one of the two conditions shown in Fig. 150 must be met.

Generally, an individual motor controller is required for each motor. However, for motors rated not over 600 V, a single controller rated at not less than the sum of the horsepower ratings of all the motors in the group may be used with a group of motors in any one of the following cases:

1. Where a number of motors drive several parts of a single machine or piece of apparatus: metal- and wood-working machines, cranes, etc.

2. Where two or more motors are under the protection of one overcurrent device, as in the case of small motors supplied from a single branch circuit. This use of a single controller applies only to cases involving motors of 1 hp or less, as provided in Section 430-53(a).

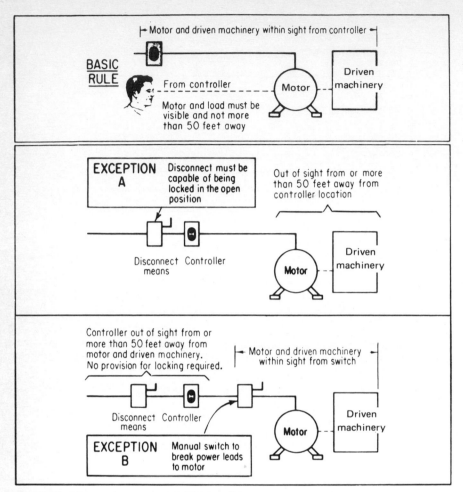

FIG. 149   These rules cover controller location.

3. Where a group of motors is located in one room and all are within sight from the controller location.

On the subject of motor controllers, the **NEC** further requires that speed-limiting devices be used with separately excited dc motors, with series motors, and with motor-generators and converters which can be driven at excessive speed from the direct-current end, as by a reversal of current or decrease in load. Certain exceptions are made in Section 430-89.

Adjustable-speed motors that are controlled by means of field regulation must be designed or equipped and connected so that they cannot be started under a condition of weakened field.

Types of controllers

Controllers for ac motors can readily be divided into two types according to the condition of starting voltage: full-voltage controllers and reduced-voltage controllers. A *full-voltage* or *across-the-line* controller is one which connects its controlled motor directly to the full value of the

EXCEPTION A

Combination fused-switch magnetic starter unit in control center has **lock-open provision** on switch.

Motor and driven load are out of sight from controller location

Motor control center

Note: Pushbutton station is not the controller. The magnetic starter in the control center is the controller.

EXCEPTION B

Circuit breaker in panel provides branch-circuit protection and is nonlock-open disconnect in sight from controller

Pushbutton station

Motor and driven load are out of sight from controller location

Magnetic starter controlled by pushbuttons at motor location

A manual switch in sight from motor in circuit to motor windings eliminates need for lock-open disconnect. But this is an additional switch and is not needed if disconnect ahead of starter is lock-open type.

FIG. 150   One of these design measures will permit a controller to be out of sight from its motor load.

motor-circuit voltage. A *reduced-voltage* controller, as the name implies, initially connects the motor to a value of voltage less than that of the supply circuit and then increases the voltage gradually until full circuit voltage is impressed across the motor terminals.

Any polyphase induction motor can be started safely using a full-voltage controller, without doing damage to the motor. Under proper conditions, any size of motor on any voltage can be started at full voltage. However, when the full voltage is impressed, the initial or starting-current surge drawn from the line might be as much as eight times the normal running current of the motor. The motor itself can handle the current and will start developing rotation; but the driven load may be damaged by the shock of the starting torque of the motor, and severe voltage disturbances may be set up in the distribution system supplying the motor. In such cases, it might be better (or necessary) to start the motor on reduced voltage.

In small and moderate-sized motor applications the torque, speed, and power requirements of the driven load generally permit full-voltage starting without objectionable results.

Across-the-line or full-voltage motor controllers can be divided into two categories according to the manner in which the contacts in the starter are closed or opened. In the manual starter, the contacts are operated by a mechanical linkage from the toggle handle or pushbuttons provided with the unit. In the magnetic starter, the contacts are operated by an electromagnetic coil which is controlled by switching energy to the coil.

The simplest type of manual starting switch is the one- or two-pole fractional-horsepower toggle switch used for infrequent starting and stopping of single-phase motors, up to a maximum of 1 hp at 120 or 240 V. This switch consists of a basic snap-action mechanism which connects the motor to the line in the ON position and disconnects it in the OFF position. To provide running overload protection, the small assembly contains a thermal device to open the circuit on overload, as shown in Fig. 151.

Manual starting switches for use with single-phase and polyphase integral-horsepower motors are across-the-line starters containing electrical contact assemblies which are opened and closed by mechanical action. Manual operation may be effected either by pushbuttons or by a toggle handle, mechanically coupled to the contact assembly and protruding through the starter enclosure. Starters of this type may be used on motor circuits rated up to 600 V ac, for single-phase motors rated up to 5 hp, and for polyphase motors rated up to 7½ hp. Running overload

FIG. 151 Manual starters are mechanical switches that include running overload trip devices.

protection is provided by thermal-relay assemblies similar to those used in fractional-horsepower starters.

Although manual starters are equipped with running overload protection, they do not have the form of protection known as *low-voltage* or *undervoltage* protection. As a result, the motor is not protected against the overheating of low-voltage operation, and a power failure or other loss of voltage to the motor circuit will stop the controlled motor but will not cause it to be disconnected from the supply circuit because the controller is mechanically fixed in the closed position. The starter contacts will remain closed, and the motor will restart immediately upon restoration of the circuit voltage. This can be hazardous when machine operators or maintenance personnel who are working on the motor are surprised by the return of power. To avoid such a hazard, manual starters must always be opened on power failure. As will be seen later, magnetic starters can provide "low-voltage protection" and "low-voltage release," which often dictate their use in cases where manual starters could otherwise be used.

The typical magnetic across-the-line starter is generally similar to the manual starter in the construction of the contact assembly; but, instead of requiring mechanical, hand-applied force to open and close the contacts, the magnetic starter uses electromagnetic energy to actuate the contacts. This starter connects the motor to its power-supply conductors at full line voltage. And the unit is equipped with running overload protection and can provide protection against undervoltage. Magnetic across-the-line starters are made for single-phase and polyphase motors.

The most common type of control switch used for the coil in a magnetic starter is a pushbutton station—a set of START and STOP pushbuttons. The pushbutton station is either in the front of the starter enclosure or installed separately from the starter at some remote location and wired to the starter, thereby providing remote control of the starter. Such remote control is one of the major advantages of magnetic starters.

When a magnetic starter is controlled by a remote pushbutton station, the coil circuit is called a *3-wire control circuit.* In such a hookup, three wires run from the starter assembly to the remote pushbutton station.

Other switching devices which may be used to provide remote and automatic control of the starter include float switches, pressure switches, limit switches, thermostats, and control relays. With any of these devices, when the switching contacts close, the coil in the starter is energized. Opening the pilot contacts breaks the circuit to the coil, and the starter disconnects the motor from its supply. Connection of these devices to the starter is made with two wires. Coil circuits which are controlled by on-off automatic pilot devices (float switches, limit switches, etc.) are called *2-wire control circuits.*

Motor running overload protection is a standard provision in magnetic across-the-line starters. Several different types of overload relays are used on the different types and sizes of controllers. In construction and operation, these devices are generally similar to the overload relays in manual starters.

Overload Protection

The **NEC** contains specific requirements regarding motor running overcurrent (overload protection); these are intended to protect the elements of the branch circuit—the motor itself, the motor control apparatus, and the branch-circuit conductors—against excessive heating due to motor overloads. Such an overload is considered to be an operating overload up to and including stalled-rotor current. When the overload persists for a sufficient length of time, it will cause damage or dangerous overheating of the apparatus. This overload type does not include fault current due to shorts or grounds.

Typical **NEC** requirements for running overcurrent protection, shown in Fig. 152, are as follows:

1. MOTORS RATED MORE THAN 1 HP

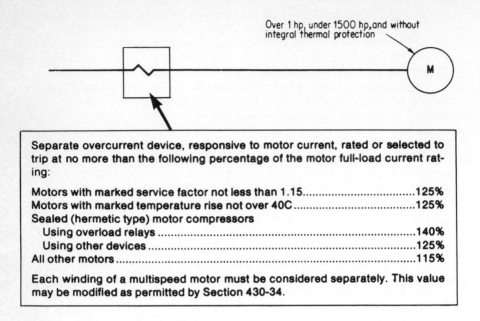

Over 1 hp, under 1500 hp, and without integral thermal protection

M

Separate overcurrent device, responsive to motor current, rated or selected to trip at no more than the following percentage of the motor full-load current rating:

Motors with marked service factor not less than 1.15	125%
Motors with marked temperature rise not over 40C	125%

Sealed (hermetic type) motor compressors

Using overload relays	140%
Using other devices	125%
All other motors	115%

Each winding of a multispeed motor must be considered separately. This value may be modified as permitted by Section 430-34.

2. MOTORS RATED 1 HP OR LESS

Portable motor 1 hp or less, manually started, and within sight from starter

M

Branch circuit protection (fuse or CB) is considered to be sufficient running OC protection

Plug connection may be the starter

or

Manual starter may be on the machine

Motor 1 hp or less, automatically started or out of sight from starter or not permanently installed

M

For motors without integral thermal protectors, this OC device must be the same as in 1 (above) for motors over 1 hp

FIG. 152 Circuit design must include some form of protection against overloads, up to locked-rotor current.

3.24

1. Running overload protection must be provided for motors of more than 1 hp, if used for continuous duty. This protection may be an external overcurrent device actuated by the motor running current and set to open at not more than 125 percent of the motor full-load current for motors marked with a service factor of not less than 1.15 and for motors with a temperature rise not over 40°C. Sealed (hermetic-type) refrigeration motor-compressors must be protected against overload and failures to start, as specified in Section 440-52, by one of the following:

   a. An overload relay set at up to 140 percent of the motor full-load current

   b. An approved, integral thermal protector

   c. A branch-circuit fuse or CB rated at not over 125 percent of the rated-load current

   d. A special protective system

   The overload device must be rated or set to trip at not more than 115 percent of the motor full-load current for all other motors, such as motors with a 1.0 service factor or a 55°C rise.

   Section 430-32 presents a thorough guide to the application of motor and branch-circuit overcurrent (overload) protection for conventional motors of the nonhermetic type. Section 440-52 gives detailed regulations on running overload protection for motor-compressors.

2. Motors of 1 hp or less which are not permanently installed and are manually started are considered protected against overload by the branch-circuit protection if the motors are within sight from the starter. Running overload devices are not required in such cases. A distance of over 50 ft is considered out of sight.

   Any motor of 1 hp or less which is not portable, is not manually started, and/or is not within sight from its starter location must have specific running overload protection. Automatically started motors of 1 hp or less must be protected against running overload in the same way as motors rated over 1 hp. That is, a separate or integral overload device must be used.

Basic **NEC** requirements are concerned with the rating or setting of overcurrent devices separate from motors. However, the rules permit the use of thermal protectors integral with motors, provided that such devices are approved for their particular applications and that they prevent dangerous overheating of the motors. Exceptions to the basic rules on providing specific running overload protective devices are as follows:

1. Where the values specified for motor running overload protection do not permit the motor to start or to carry the load, the next higher size of overload relay may be used, but not higher than the following percentages of motor full-load current rating:

   - Motors with a marked service factor not less than 1.5: 140 percent
   - Motors with a marked temperature rise not over 40°C: 140 percent
   - Sealed (hermetic-type) motor-compressors: 140 percent
   - All other motors: 130 percent

   Fuses or circuit breakers may be used for running overload protection but may not be rated or set up to those values. Fuses and CBs must have a maximum rating as listed above for motors of more than 1 hp. If the value determined as indicated there does not correspond to a standard rating of fuse or CB, the next smaller size must be used. A rating of 125 percent of the full-load current is the absolute maximum for fuses and circuit breakers.

2. Under certain conditions, no specific running overload protection need be used: The motor is considered to be properly protected if it is part of an approved assembly which does not normally subject the motor to overloads and which has controls to protect against stalled

rotor. Or, if the impedance of the motor windings is sufficient to prevent overheating due to failure to start, the branch-circuit protection is considered adequate.

3. A motor used for service which is inherently short-time, intermittent, periodic, or varying duty (see **NEC** Table 430-22) is considered as protected against overcurrent by the branch-circuit overcurrent device. A motor is considered to be wired for continuous duty unless the motor cannot operate continuously with load under any condition of use.

Complete data on the required number and location of overcurrent devices are given in Table 430-37. Table 430-37 requires three running overload devices (trip coils, relays, thermal cut-outs, etc.) for all 3-phase motors unless protected by other approved means, such as specially designed embedded detectors with or without supplementary external protective devices (Fig. 153).

OL relay required in each leg

3-phase motor

Stop   Start

FIG. 153 Three-phase starter must have three overload devices.

The usual acceptable rating of the overload protection in a motor starter is 125 percent of the motor full-load current (or as specifically required in **NEC** Sections 430-32 and 430-34). But when a power-factor capacitor is installed on the load side of the motor starter or at the motor itself, a correction must be made in the rating or setting of the overload device. The device must be rated to take into account the fact that the magnetizing current for the motor is being supplied by the capacitor, and the total current flowing from the supply circuit through the starter is lower than it would be without the power-factor capacitor.

Figure 154 explains the need for correcting the size of the running overload protection in motor controllers when power-factor capacitors are used on the load side of the controller, as required by Section 460-9 on the use of capacitors to correct power factor. And, as required by the **NEC**, the rating of the capacitors should not exceed the value required to raise the no-load power factor of the motor to unity. Capacitors of these maximum ratings usually result in a full-load power factor of 95 to 98 percent.

Although the **NEC** contains all those requirements on the running overload protection of motors, Section 430-44 does recognize that there are cases when automatic opening of a motor circuit due to overload may be objectionable from a safety standpoint. In recognition of the circumstances of many industrial applications, Section 430-44 permits alternatives to automatic opening of a circuit in the event of overload. This permission to eliminate overload protection is similar to the permission given in Section 240-12 to eliminate overload protection when automatic opening of the circuit on an overload would constitute a more serious hazard than the overload itself. As the rule notes, "if immediate automatic shutdown of a motor by a motor overload protective device(s) would introduce additional or increased hazard(s) to a person(s) and continued motor operation is necessary for safe shutdown of equipment or process" then automatic overload opening is not required. However, as shown in Fig. 155, the circuit must be provided with a motor overload sensing device conforming to the **NEC** requirement on overload protection, to indicate the presence of the overload by means of a supervised alarm. Such overload indication (instead of automatic opening) will alert personnel to the objectionable condition and will permit corrective action for an orderly shutdown, either immediately or at some more convenient time, to resolve the difficulty. But, as is required in Section 240-12, short-circuit

A (total motor current) = vector sum of B and C
where B = in-phase, working current
C = reactive, magnetizing current

EXAMPLE:

A motor with 70% power factor has a full-load current rating of 143 amps. Normally, the OL relay would be set for, say, 125% of 143 or 179 amps. *BUT*, because a PF capacitor is installed at the motor, the OL relay no longer will have 143 amps flowing through it at full load. If the capacitor corrects to 100% PF, the effect will be as follows:

I_{x_c} cancels I_{x_L} at 100% PF, leaving only the working or in-phase current tô be supplied ˜rom the circuit.

WORKING CURRENT =

TOTAL MOTOR CURRENT × POWER FACTOR

= 143 × 0.70 = 100 AMPS

THAT IS THE CURRENT THAT WILL BE FLOWING THROUGH THE OL RELAY AT FULL LOAD.

THE OL RELAY, THEREFORE, MUST BE SET AT 125% × 100 AMPS = 125 AMPS

FIG. 154 Setting of overload protection must be corrected for reduced line current due to capacitors.

FIG. 155   Overload protection may be eliminated under these conditions.

protection on the motor branch circuit must be provided to protect against those high-level ground faults and short circuits that would be more hazardous than a simple overload.

Motor Disconnect

The **NEC** specifically requires that a disconnecting means—basically, a motor-circuit switch rated in horsepower, or a CB—be provided in each motor circuit to disconnect the motor and its controller from all ungrounded supply conductors. In a motor branch circuit, every switch in the circuit in sight from the controller must satisfy the requirements regarding the type and rating of the disconnect means. And the disconnect switch or CB must be rated to carry at least 115 percent of the nameplate current rating of the motor for circuits up to 600 V. Figure 156 sets forth the basic requirements on types of disconnects.

The **NEC** includes a basic requirement that the disconnecting means for a motor and its controller be a motor-circuit switch rated in horsepower. For motors rated up to 500 hp, this rule is readily complied with, inasmuch as the UL lists motor-circuit switches up to 500 hp and manufacturers mark switches to conform. But for motors rated over 100 hp, the **NEC** does not require that the disconnect have a horsepower rating. An exception to the basic rule permits the use of ampere-rated switches or isolation switches, provided the switches have a carrying capacity of at least 115 percent of the nameplate current rating of the motor. The UL notes that a horsepower-rated switch that is rated over 100 hp must not be used as both a disconnect *and* a controller. But switches rated up to 100 hp may serve the dual function of disconnect and controller.

As noted above, although Exception No. 4 to Section 430-109 sets the maximum horsepower rating required for motor-circuit switches at 100 hp, higher-rated switches are now available, will provide additional safety, and should be used to assure adequate interrupting ability.

Isolation switches for motors over 100 hp, if they are not capable of interrupting stalled-rotor currents, must be plainly marked "Do not open under load" (Fig. 157).

EXAMPLE

Provide a disconnect for a 125-hp, 3-phase, 460-V motor. Use a nonfusible switch, inasmuch as short-circuit protection is provided at the supply end of the branch circuit.

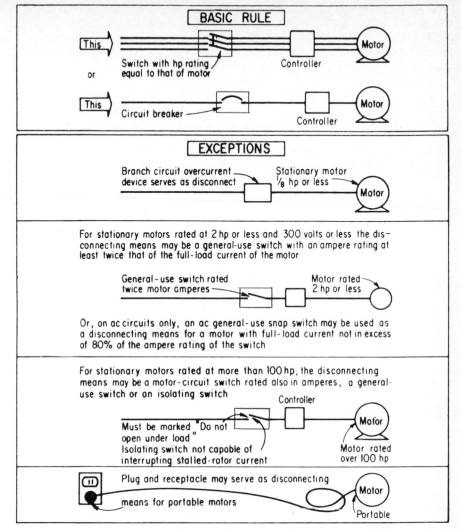

FIG. 156 Design of motor branch circuit must include a suitable disconnecting means.

SOLUTION

The full-load running current of the motor is 156 A. A suitable disconnect must have a continuous carrying capacity of 156 A × 1.15 or 179 A. This calls for a 200-A, three-pole switch rated for 480 V. The switch may be a general-use switch, a current-and-horsepower marked motor-circuit switch, or an isolation switch. A motor-circuit switch with the required current and voltage rating for this case would be marked for 50 hp, but the horsepower rating is of no concern because the switch does not have to be horsepower-rated for motors larger than 100 hp.

If the 50-hp switch were of the heavy-duty type, it would have an interrupting rating of 10 × 65 A (the full-load current of a 440-V, 50-hp motor), or 650 A. But the locked-rotor current of the 125-hp motor might run as high as 900 A. In such a case, the switch should be marked "Do not open under load."

If a fusible switch had to be provided for this motor for disconnect and short-circuit protection, the size of the switch would be determined by the size and type of fuses used. For a fuse rating of 250 percent of the motor current (which does not exceed the 300 percent maximum in Table 430-152) for standard fuses, the application would call for 400-A fuses in a 400-A switch. This switch would certainly qualify as the motor disconnect. However, if time-delay fuses were used, a 200-A switch would be large enough to take the time-delay fuses and could be used as the disconnect (because it is rated at 115 percent of motor current).

In the foregoing, the 400-A switch might have an interrupting rating high enough to handle the locked-rotor current of the motor. Or the 200-A switch might be of the type that has an

FIG. 157 Motor disconnect selected on the basis of ampere rating must be carefully applied.

interrupting rating up to 12 times the rated load current of the switch itself. In either of these cases, there would probably be no need for marking the switch "Do not open under load."

Up to 100 hp, a switch which satisfies the **NEC** rule on rating for use as a motor controller may also provide the required disconnect means—the two functions being performed by the one switch—provided it opens all ungrounded conductors to the motor, is protected by an overcurrent device (which may be the branch-circuit protection or may be fuses in the switch itself), and is a manually operated air-break switch or an oil switch not rated over 600 V or 100 A.

As described under "Controllers," a manual starting switch capable of starting and stopping a given motor, capable of interrupting the stalled-rotor current of the motor, and having the same horsepower rating as the motor may serve the functions of controller and disconnecting means in many motor circuits, if the switch opens all ungrounded conductors to the motor. A single circuit breaker may also serve as controller and disconnect; this is permitted by Section 420-111. However, an autotransformer-type controller, even if manual, may not also serve as the disconnecting means. Such controllers must be provided with a separate means for disconnecting the controller and motor.

The acceptability of a single switch used as both the controller and disconnecting means is based on the single switch satisfying **NEC** requirements for a controller and for a disconnect. This "combination" finds application where general-use switches or horsepower-rated switches are used, as permitted by the **NEC**, in conjunction with time-delay fuses which are rated low enough to provide both running overload protection and branch-circuit (short-circuit) protection. In such cases, a single fused switch may serve a total of four functions: (1) controller, (2) disconnect, (3) branch-circuit protection, and (4) running overload protection. And it is also possible for a single circuit breaker to serve these four functions.

For sealed refrigeration compressors, Section 440-12 gives the procedure for determining the disconnect rating, based on the nameplate rated load current or branch-circuit selection current (whichever is greater) and the locked-rotor current of the motor-compressor. As an example, suppose a 3-phase, 460-V hermetic motor rated at 11 A branch-circuit selection current and 60 A locked-rotor current is to be supplied with a disconnect switch rated in horsepower. The first step is to determine the equivalent horsepower rating of the motor by referring to **NEC** Table 430-150. This table lists 7½ hp as the horsepower size of a 460-V, 11-A motor. To ensure adequate interrupting capacity, Table 430-151 is now used. This table shows 7½ hp as the equivalent horsepower rating for any locked-rotor current over 45 A and up to 66 A for a 460-V motor. Both tables thus establish a 7½-hp disconnect as adequate for the given motor. If the ratings obtained from the two tables had been different, the higher rating would have been chosen.

In general, each motor is provided with a separate disconnecting means. However, a single disconnect sometimes may serve a group of motors. Such a disconnect must have a rating sufficient to handle a single load equal to the sum of the horsepower ratings or current ratings of all the motors it serves. The single disconnect may be used for a group of motors driving different parts of a single piece of apparatus, for several motors on one branch circuit, or for a group of motors in a single room within sight from the disconnect location (Section 430-112).

An important rule in **NEC** Section 430-102 states simply and clearly that "a disconnecting means shall be located in sight from the controller location." This applies always, for all motor loads rated up to 600 V—even if an "out-of-sight" disconnect can be locked in the open position (Fig. 158). There are, however, two exceptions to this rule:

Exception No. 1 permits the disconnect for a high-voltage (over 600 V) motor to be out of sight from the controller location if the controller is marked with a warning label giving the location and identification of the disconnecting means to be locked in the open position.

Exception No. 2 is aimed at permitting practical, realistic disconnect means for large and complex industrial machinery utilizing a number of motors to power various interrelated

THIS IS THE BASIC RULE

The disconnect means, whether it is a switch or CB, individually mounted or installed in a panel or switchboard, whether lock-open type or nonlock-open type...

... must be within sight and not more than 50 ft away from the motor controller.

Lock-open or nonlock-open type fused switch for motor branch-circuit protection out of sight from starter

Magnetic starter

Motor

EXAMPLE

A disconnect means must be placed somewhere in this part of circuit so it is in sight from the starter. Or, a nonfusible combination starter could be used in place of the starter shown.

FIG. 158 Basic motor-circuit design includes an "in-sight" disconnect.

machine parts. This exception recognizes that a single common disconnect for a number of controllers, as permitted by part *a* of the exception to Section 430-112, often cannot be installed "within sight" of all the controllers, even though the controllers are "adjacent one to the other." On much industrial process equipment, the components of the overall structure obstruct the view of many controllers. Exception No. 2 permits the single disconnect to be technically out of sight from some or even all of the controllers if the disconnect is simply "adjacent" to them—that is, nearby on the equipment structure.

Those two exceptions are shown in Fig. 159.

CONTROL CIRCUITS

A control circuit, as discussed here, is any circuit which has as its load device the operating coil of a magnetic motor starter, a magnetic contactor, or a relay. Strictly speaking, it is a circuit that exercises control over one or more other circuits. And these other circuits may themselves be *control* circuits, or they may be *load* circuits carrying utilization current to a lighting, heating, power, or signal device. Figure 160 clarifies the distinction between control circuits and load circuits.

A **lock-open type** switch or CB may be an out-of-sight disconnect for a . . .

. . . controller for a high-voltage motor (over 600 volts) if the controller is marked to warn personnel about the disconnect.

Warning sign or label tells where the lock-open disconnect is and how it may be identified.

For a "multi-motor continuous process machine" . . .

. . . a single disconnect that is located "adjacent to a group of coordinated controllers" . . .

. . . does not have to be within sight of each and every individual controller.

FIG. 159 These "out-of-sight" disconnects may be used.

Load circuit

Operating coil

Control circuit

Control switch

FIG. 160 Control circuit provides on-off operation of load-circuit current.

The elements of a control circuit include all the equipment and devices concerned with the function of the circuit: conductors, raceway, contactor operating coil, source of energy supply to the circuit, overcurrent protective devices, and all switching devices that govern the energizing of the operating coil.

NEC rules on motor-starter control circuits must be evaluated against a number of very important background facts:

1. The **NEC** has no specific rule that covers the ampacity of control-circuit wires with respect to the amount of current that a starter operating coil draws. It simply presumes that the wires of a control circuit will have ampacity sufficient for the coil current.

2. **NEC** rules require the protection of control-circuit wires against current in excess of their ampacity or spell out very specific conditions under which overcurrent devices rated higher than wire ampacity may be used or under which the protection of control wires may be eliminated or must be eliminated.

3. The **NEC** rules on control-circuit overcurrent protection are not concerned with protecting the operating coil itself against overcurrent. Because of that, it is permissible to select the size of control-circuit wires solely in relation to the rating of branch-circuit protection when the use of larger control wires would eliminate the need for a control-circuit fuse block in the starter [i.e., control-wire size selected so that the branch-circuit fuses or CB are not more than 300 percent of the wire ampacity, as in Exception No. 2 to Section 430-72(b)].

The **NEC** covers the application of control circuits in Article 725 and in Sections 240-3 and 430-71 through 430-74.

Section 430-72 covers motor control circuits that are derived within a motor starter from the branch-circuit wires that connect to the line terminals of the starter. When a control circuit derives its current supply from the same branch circuit that supplies power to the motor (either directly, by being tapped from the line terminals within the starter, or indirectly by being fed from the secondary of a control transformer with its primary tapped from the starter line terminals and located within the starter enclosure), all the rules of Section 430-72 must be observed. The rules refer to such a control circuit as one "tapped from the load side" of the fuses or circuit breaker that provide branch-circuit protection for the conductors that supply the starter (Fig. 161). Motor control circuits that are *not* tapped from the line terminals within a starter must be protected against overcurrent in accordance with Section 725-12 or 725-35. Such control circuits would be those that are derived from a panelboard or a control transformer—as

FIG. 161 Control circuit derived within a motor controller must satisfy **NEC** Section 430-72 on overcurrent protection.

where, say, 120-V circuits are derived from a source external to the starters and are typically run to provide lower-voltage control for 230-, 460-, or 575-V motors (Fig. 162).

Section 725-2(e) states that the remote-control circuit of a motor starter must comply with the rules of Part **F** of Article 430 (Sections 430-71 through 430-74), as well as with any applicable rules of Article 725, in all cases where the control circuit is derived from the line terminals of a starter. When a control circuit for the operating coil of one or more magnetic motor starters is derived from a separate control transformer (one that is not within the starter housing and is not fed from a motor branch cirucit), the control circuit or circuits and all components are covered by the rules of Article 725. In the same way, a control circuit that is taken from a panelboard for power supply to the operating coil of one or more motor starters is also covered by Article 725, and not by the rules of Section 430-72.

The design requirements for the coil circuit of a magnetic contactor—as distinguished from a motor starter—are based solely on the rules of Article 725. For any magnetic contactor used for remote control of power and/or lighting loads, Exception No. 3 to Section 725-12(b) governs the overcurrent protection of the control-circuit wires. The rule is basically the same as that of Exception No. 2 to Section 430-72(b). Or, if a magnetic contactor has its coil circuit fed from the secondary of a control transformer, Exception No. 2 of Section 725-12(b) applies.

FIG. 162 Coil circuit with external voltage supply must satisfy Section 725-12 on overcurrent protection.

The design and installation of control circuits are basically divided into three classes (in Article 725) according to the energy available in the circuit. Class 2 and Class 3 control circuits have low energy-handling capabilities; to qualify as a Class 2 or 3 control circuit, a circuit must have open-circuit voltage and overcurrent protection that is limited to the conditions given in Section 725-31.

The vast majority of control circuits for magnetic starters and contactors could not qualify as Class 2 or 3 circuits because of the relatively high energy required for operating coils. And any control circuit rated over 150 V (such as 220- or 440-V coil circuits) can never qualify, regardless of energy.

Class 1 control circuits include all operating-coil circuits for magnetic starters that do not meet the requirements for Class 2 or 3 circuits. Class 1 circuits must be wired in accordance with Sections 725-11 to 725-20. The rules may be summarized as follows:

1. In general, the wiring of Class 1 control circuits must be the same as general-purpose power and lighting wiring.

2. Conductor sizes generally are limited to a minimum of No. 14, but No. 16 or No. 18 fixture wires may be used if installed in raceway or approved cable or flexible cord, loaded up to the ampacities shown in Section 402-5. Conductors of sizes No. 18 and No. 16 are considered as protected by overcurrent devices rated not over 7 and 10 A, respectively.

3. Wires larger than No. 16 must be of Type TW, THW, THHN, or some other approved type. Conductors of size No. 18 or No. 16 in raceway or cable must be of Types RFH-2,

FFH-2, etc. [See Section 725-16(b).] Other conductors may be used with specific approval for the particular purpose.

4. As specified in **NEC** Section 725-17, the number of Class 1 control-circuit conductors in a conduit must be determined from Tables 1 through 5 in **NEC** Chapter 9. There is a definite limit on the number of wires permitted in a raceway; if these wires carry continuous loads, the derating factors of Note 8 to Tables 310-16 through 310-19 must be applied to their current ratings. But Section 430-72(b)(1) exempts motor-starter coil-circuit wires from such derating.

5. When power-supply conductors and Class 1 remote-control conductors are used in a single conduit or EMT run (as permitted by Section 725-15), the derating factors of Note 8 must be applied as follows:

a. Note 8 must be applied to all conductors in the conduit when the remote-control conductors carry continuous loads and the total number of conductors (remote-control and power wires) is more than three. In Fig. 163*a,* the conduit size must be selected according to the number and sizes of the wires. Because two of the control wires to the pushbutton and the power wires to the motor will carry a continuous load, a derating factor of 80 percent (from Note 8) must be applied to all the wires in the conduit.

b. Note 8 must be applied only to the power wires when the remote-control wires do not carry a continuous load and when the number of power wires is more than three. In Fig. 163*b,* no derating at all is applied because the control wires do not carry continuous current (only for the instant of switching operation), and there are only three power wires.

c. Conductors for two or more Class 1 control circuits (ac or dc or both) may be run in the same raceway if all conductors are insulated for the maximum voltage of any conductor in the raceway. Another permitted installation is shown in Fig. 163*c.*

Control Wires in Raceways

Class 1 control wires (starter-coil circuit wires) may be run in raceways by themselves in accordance with the first sentences of Section 725-15. A given conduit, for instance, may carry one or several sets of Class 1 control-circuit wires. And Section 725-14 says

FIG. 163   Design of every contactor or motor starter-coil circuit must satisfy certain basic rules.

Class 1 circuit wires must conform to the same basic rules from **NEC** Chapter 3 that apply to standard power- and lighting-circuit wiring. Conduit fill must be made in accordance with Section 725-17, which cites the usual rules on number of wires in rigid metal conduit, in IMC, in EMT, etc. When more than three control wires are installed in a raceway, the ampacity derating factors of Note 8 to Tables 310-16 to 310-19 must be applied only if the conductors carry continuous loads (that is, carry their load current continuously for 3 hours or more), as required by Section 725-17. But that applies only to wires of control circuits that are derived external to the starter, as in those cases where the coil circuit is energized from a separate (external) control transformer or from a panelboard. Section 430-72(b), however, applies to coil circuits for starters in which control power is tapped from the starter line terminals; it requires basically that No. 14 wires and larger be protected at their ampacities as given in Tables 310-16 through 310-19, without derating factors.

Elimination of the requirement to derate conductor ampacities (or limit the load) according to Note 8 of Tables 310-16 to 310-19 and Section 725-17(a) has been a controversial point in the past. Although the derating required by Section 725-17 is still part of the 1981 **NEC**, that rule literally applies only to control circuits that are derived from an external source—such as a transformer or panelboard. The elimination of derating factors in Section 430-72(b) applies only to control circuits tapped within starters.

Two sections of the **NEC** cover the placement of Class 1 control-circuit conductors in the same raceway, cable, or enclosure that contains the circuit wires carrying power to the motor windings. Section 300-3 covers the general use of "conductors of different systems" in raceways as well as in cable assemblies and in equipment wiring enclosures, (that is, cabinets, housings, starter enclosures, junction boxes, etc.). But Class 1 control wires are also regulated by the rules of Article 725, covering "remote-control" wiring.

CASE I—SECTION 725-15 (LAST SENTENCE) PERMITS THIS

FIG. 164   Control and power wires may always be in the same raceway for a single motor.

Figures 164 to 166 show three cases where Class 1 control wires are run in raceway with the motor power conductors. Although the sketches show specific hookups of equipment, the rules described apply to any hookup or equipment layout. The cases shown are intended only to focus on the placement of control wires in raceway with power wires—no matter how the hookup is made.

CASE II—SECTION 300-3(a) PERMITS COMMON RACEWAY TO CONTAIN POWER AND CONTROL-CIRCUIT WIRES FOR TWO OR MORE MOTORS

This common raceway (conduit, EMT, wireway, or etc.) may contain **all** power conductors and **all** control conductors for two or more motors . . .

Motors operate together and are functionally associated as integral parts of a machine or process

Individual conduit runs for power and control wires to each motor   Pushbuttons

MOTOR CONTROL CENTER

. . . **but**, the intent of this **Code** rule, along with that of Section 725-15, permits a common raceway **only** where the power and control conductors are for a number of motors that operate integrally—such as a number of motors powering different stages or sections of a multi-motor process or production machine. Such usage complies with Section 725-15 (last sentence), which permits power and control wires in the same raceway, cable, or other enclosure when the equipment powered is "functionally associated"—that is, the motors have to run together to perform their task.

FIG. 165 Specific conditions permit all power and control wires for two or more motors to be installed in a single raceway.

CASE III—BUT THE LETTER AND INTENT OF SECTION 725-15 PROHIBITS INTERMIXING OF POWER AND CLASS 1 (STARTER COIL-CIRCUIT) WIRES WHEN MOTORS ARE NOT "FUNCTIONALLY ASSOCIATED"

Separate conduit or raceway is required for power and control wires to each motor . . .

Pushbutton stations

3 power wires plus 3 control wires in each conduit

MOTOR CONTROL CENTER

. . . When these are individual motors that do not operate together as parts of a machine or process—that is, each motor has a separate, independent load that may operate by itself.

FIG. 166 Power and control wires must be segregated for motors that are not "associated."

Case I: Locating the power and control wires for an individual motor in the same raceway, as here, is acceptable under Section 725-15.

Case II: The wording of Section 300-3(a) recognizes the placement of power and control wires in a single raceway to supply more than one motor, but such usage must conform with the last sentence of Section 725-15. As explained in the accompanying sketches, the two **NEC** rules must be combined carefully.

A common raceway, as shown, may be used only where the two or more motors are required to be operated together to serve their load function. Many industrial and commercial installations have machines, manufacturing operations, or processes in which a number of motors operate in unison to perform the various parts of a task. In such cases, either all the motors operate or none do. The placement of all the control and power wires for such a group of motors in the same raceway would not produce a situation where a fault in one motor circuit could disable another circuit to a motor that might otherwise continue to operate.

But when a common raceway is used for power and control wires to separate, independent motors, a fault in one circuit could knock out all the others—and motors would be shut down unnecessarily. Because motor circuits are so closely associated with vital equipment such as elevators, fans, and pumps in modern buildings, it is a matter of safety to separate such circuits and minimize the chance of a large-scale outage due to a fault in a single circuit. For safety's sake, the **NEC** says, in effect, "Do not put all your eggs in one basket." But, again, the loss of more than one motor because of a single fault is not objectionable where all motors must be shut down when any one is stopped—as in multimotor machines and processes.

Case III: In cases where each of several motors serves a separate, independent load (with no interconnection of their control circuits and no mechanical interlocking of their driven loads), the use of a separate raceway for each motor is required by the last sentence of Section 725-15—but only when control wires are carried in the raceways. It would be acceptable to run the power conductors for all three motors in a single raceway, and all the control-circuit wires in another raceway. Such a hookup would not violate Section 725-15. However, the conductors would have to be derated, and there is a definite chance of losing more than one motor on a fault in only one of the circuits in either the power raceway or the control raceway.

Figure 167 shows the details of Section 300-3(b) as they apply to high-voltage motor starters and their control circuits.

Overcurrent Protection

In general, remote-control conductors must be protected against overcurrent. Section 430-72 covers overcurrent protection for remote-control conductors that are tapped from the line terminals within a motor starter. Section 725-12 covers overcurrent protection for remote-control conductors of motor starter-coil circuits that are derived from separate control transformers or from panelboards, and overcurrent protection for the coil circuits of magnetic contactors (as used for the control of lighting and/or power). Exception No. 3 of Section 725-12 states that remote-control conductors other than those for motor control circuits can be considered to be satisfactorily protected by the branch-circuit overcurrent device if it is rated at not more than 300 percent of the carrying capacity of the control-circuit conductors. That applies to control wires for magnetic contactors used for the control of lighting or heating loads, but not for motor loads. Section 430-72 includes a similar requirement for motor control circuits.

For motor starter-coil circuits tapped from line terminals within the starter enclosure, a number of points must be observed.

SECTION 300-3(b) PROHIBITS THIS—

Same conduit carries 5-kv power conductors and control wires under 600 volts

High-voltage motor (e.g. 4160V)

BUT, EXCEPTION NO. 3 PERMITS THIS—

Power circuit in raceway to motor

Enclosure of motor controller

High-voltage motor (e.g. 4160 V)

2. Any **raceway** containing conductors operating at over 600 volts **may not** also contain **any** conductors operating at 600 volts or less.

For any **individual** motor or starter—excitation, control, relay and/or ammeter conductors operating at 600 volts or less **may** occupy the same **starter or motor enclosure** as the conductors operating at over 600 volts.

FIG. 167   Control wires for high-voltage starters may be used in the starter enclosure, but not in raceway with power conductors.

1. Depending upon conditions set in Section 430-72, the conductors of the control circuit may be considered as protected either by the branch-circuit protective device ahead of the starter or by supplementary protection (usually fuses) installed in the starter enclosure.
2. The control circuit that is tapped from the line terminals within a starter is not a branch circuit itself.
3. Overcurrent protection is required for the control conductors and not for the operating coil. Because of this, the size of the control conductors can be selected to allow application without separate overcurrent protection.

Exception No. 1 to Section 430-72(b) specifies that the branch-circuit protective device may serve as sufficient protection to control-circuit conductors where the control conductors do not actually leave the enclosure of a magnetic motor starter, such as when the START-STOP push-button station is in the cover of the starter housing. But the rule adds that this is permissible only where the branch-circuit protective device has a rating not over 400 percent of the ampacity of the control-circuit wires. Obviously, this requires that the size and current rating of the control-circuit wires be known and evaluated against the rating of the branch-circuit protective device before it can be assumed that protection for the control circuit is not required within the starter. And, if such protection for the internal control circuit is required, a suitable fuse must be inserted in the controller housing (Fig. 168).

In using Exception No. 1, the branch-circuit protection must be compared to, and must be not over 400 percent of, the ampacity of the control-circuit wires that are factory installed in the starter and connected to the START and STOP buttons in the cover. The 400 percent rating may be taken on the free-air ampacities of Table 310-17 for 60°C wire.

Because the vast majority of starters are the smaller ones using No. 18 and No. 16 wires for their coil circuits, special reference is made to those wire sizes. For No. 16 control wires, branch-

**WHEN CONTROL WIRES
DO NOT LEAVE STARTER:**

Controller

Motor

. . . if branch-circuit protection
is rated:

Fuse protection of the coil
circuit wires is not needed
in the starter . . .

1. Not over 4 times the
   free-air ampacity of No. 14
   or larger control wires, or
2. Not over 40 A for No. 16
   control wires, or
3. Not over 25 A for No. 18
   control wires

FIG. 168 Overcurrent protection is usually not required for coil-circuit wires of
a starter with pushbuttons in the cover.

circuit protection rated up to 40 A would eliminate any need for a separate control-circuit fuse
in the starter. And, for No. 18 control wires, separate coil-circuit protection is not needed for a
starter with branch-circuit protection rated not over 25 A. Those values—40 A and 25 A—
represent 400 percent of the ampacities of No. 16 and No. 18 wires—10 and 7 A, respectively.

Overcurrent protection for remote-control conductors that run from a motor starter to external remote-control devices (pushbutton stations, float switches, limit switches, thermostats, etc.)
is specified by the **NEC** in several ways, as follows.

Figure 169 shows a basic approach to motor circuits with either 15- or 20-A fuses or CBs
for branch-circuit protection, to satisfy Exception No. 2 of Section 430-72(b). Note that separate

Branch-circuit protection
up to 20A . . .

No other protection here

Stop

Start

C

No. 14 control wires to
remote-control device.

Starter
enclosure

Motor

FIG. 169 This is a basic rule on the protection of remote-control wires.

WHEN REMOTE CONTROL IS USED:

If branch-circuit protection "A" is rated not over 3 times the ampacity of No. 14 or larger coil-circuit wires that run to a remote-control device . . . .

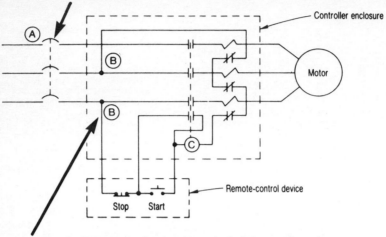

. . . . then separate protection is *not* required at the supply ends "B" of the control wires.

If protection at "A" exceeds that value, protection at "B" must protect control wire at its ampacity from Table 310-16 for 60 C wire.

FIG. 170 Separate control-wire protection may not be required for a circuit to remote pilot devices.

fuses do not have to be provided in the starter for protection of the remote-control conductors. As noted in **NEC** Section 430-72(b), the rating of overcurrent protection must not exceed 7 A for No. 18 control wire or 10 A for No. 16 control wire. That permission is also in Section 725-12(b).

Exception No. 2 to Section 430-72(b) covers the protection of No. 14 and larger control wires that run from a starter to a remote-control device (pushbuttons, limit switch, etc.).

Exception No. 2 applies only to No. 14 and larger wires that "extend beyond the motor control equipment"—that is, wires that run to a remote-control device. This rule is shown in Fig. 170. The control wires must be taken as having the following ampacities: No. 14, 15 A; No. 12, 20 A; and No. 10, 30 A. If No. 16 or No. 18 wires are used, they must be protected by 10- or 7-A fuses, respectively, in the starter.

When control power for motor starter-coil circuits is derived from a control transformer within the starter enclosure, protection must be provided; Exception No. 3 of Section 430-72(b) permits protection on the primary side of the control transformer to protect both the transformer and the secondary conductors in accordance with their ampacity, but this use is limited to transformers with 2-wire secondaries (Fig. 171). Because Section 430-72(a) states that the rules of Section 420-72 apply to control circuits tapped from the motor branch circuit, the rule of Exception No. 3 must be taken as applying to a control transformer installed within the starter enclosure—although the general application may be used for any transformer because it conforms both to Exception No. 5 of 240-3 and to Section 450-3.

The rating of the overcurrent protection must not be greater than the secondary conductor ampacity times the secondary-to-primary voltage ratio, and not in excess of 125, 167, or 300 percent of the transformer rated primary current, as specified in Section 450-3(b)(1).

WHEN CONTROL POWER IS DERIVED FROM A CONTROL TRANSFORMER:

Control wires fed by 2-wire transformer secondary may be protected
by primary protective devices sized, at transformer turns ratio, to
ampacity of secondary wires—and primary protection must be sized
not over 125%, 167% or 300% of transformer primary current.

FIG. 171 Coil-circuit wires from a control transformer may use primary
protection.

Section 430-72(c) covers the protection of control transformers—and, again, it must be taken
to apply specifically to such transformers that are connected to the motor branch circuit.
Although the basic rule calls for each control transformer to be protected in accordance with
Section 450-3 (usually by a primary-side protective device rated not over 125, 167, or 300 percent of the primary current, depending upon the transformer primary current rating), exceptions
are given. To eliminate fuse blowing on inrush current, Exception No. 1 permits primary transformer protection rated up to 500 percent of the rated primary current where the transformer
rated primary current is below 2 A (Fig. 172). That exception applies to a control transformer
that is an integral part of the motor controller. It should be noted that Exception No. 2 to Section

WHEN CONTROL TRANSFORMER IS WITHIN THE STARTER ENCLOSURE:

For transformer with rated primary current less than 2 amps, transformer protection may be provided by an overcurrent device in each ungrounded leg,
rated or set at not more than 5 times (500%) rated transformer primary current.

FIG. 172 Control transformer may have 500 percent primary protection.

450-1 exempts dry-type transformers that are component parts of other apparatus from the rules of Section 450-3 requiring overcurrent protection for the transformer. This has been taken to exempt control transformers within starters from the need for primary-side overcurrent protection.

The use of secondary protection for control wires is suited to applications in which the transformer is within the starter enclosure. Primary protection, without secondary protection, is suited to use with either internal or external control transformers with 2-wire primaries and 2-wire secondaries and constitutes more effective protection because *both* the transformer and the primary and secondary control wires are protected.

Exception No. 2 of Section 725-12 applies to the protection of remote-control conductors fed by the 2-wire secondary of a separate control transformer supplying the coils of one or more motor starters or magnetic contactors. The rule is the same as that of Exception No. 3 of Section 430-72(b), as described above. Figure 173 shows an example of the application of that rule to a 3000-VA control transformer supplying starter-coil circuits. A properly sized circuit breaker or fuses may be used at the supply end of the circuit that feeds the transformer primary and may provide overcurrent protection for the primary conductors, for the transformer itself, and for conductors of the control circuit that is run from the transformer secondary to supply power to motor starters or other control equipment, as follows:

1. The primary-side protection must not be rated greater than the protection required by Section 450-3(b)(1) for transformers rated up to 600 V. For a transformer rated 9 A or more, the rating of the primary CB or fuses must not be greater than 125 percent of the rated transformer primary current. However, if 125 percent of the rated primary current does not yield a standard rating of fuse or CB, the next higher rated standard protective device may be used. Where the transformer rated primary current is less than 9 A (as it would be for all the usual control transformers rated up to 5000 VA and stepping 480 V down to 120 V), the maximum permitted rating of the primary protection is 167 percent of the rated primary current. For a transformer with a primary rated less than 2 A, the primary protection must never exceed 300 percent of the rated primary current, except as noted above for control-circuit transformers within motor starters. For the vast majority of control transformers—

No. 14 copper wires are adequate for primary circuit and are protected.

Secondary control circuit wires must have ampacity of at least (480 ÷ 120) × 10 A, or 40 A. No. 8 copper wires are adequate.

3000-VA transformer

10A

480 V   120 V

10A

Power to starters, contactors, and/or relays: Class 1 circuit

Rated primary current is 6.25 A

For transformer protection, fuses must be rated not over 1.67 × 6.25, or 10 A [Sec. 450-3(b) (1), Ex. No. 1].

FIG. 173 Protection must be provided for a separate control transformer *and* the circuit wires.

with primary ratings well below 10 A—fuse protection will be required on the primary because the smallest standard CB rating is 15 A, and that will generally exceed the maximum values of primary protection permitted by Section 450-3(b)(1).

2. The primary protection must not exceed the ampere rating of the primary-circuit conductors. When protection is sized for the transformer as described above, No. 14 copper primary conductors will be protected well within their 15-A rating.

3. Secondary conductors for the control circuit can then be selected to have an ampacity at least equal to the rating of the primary protection times the primary-to-secondary transformer voltage ratio. Of course, larger conductors may be used if needed to keep the voltage drop within limits.

Exception No. 4 of Section 430-72(b) says that the overcurrent protection for a control circuit (other than branch-circuit protection) must be eliminated where opening of the control circuit would create a hazard. An example would be the control circuits for fire-pump motors.

Control Transformers

In the majority of magnetic motor controllers and contactors, the voltage of the operating coil is the voltage provided between two of the conductors supplying the load, or between one conductor and the neutral. Conventional starters are factory wired with coils of the same voltage rating as the phase voltage to the motor. However, in many cases it is desirable or necessary to use control circuits and devices of lower voltage rating than the motor. Such would be the case with high-voltage (over 600 V) controllers, for instance, when it is necessary to provide a source of low voltage for practical operation of magnetic coils. And even with motor controllers and contactors using less than 600 V, safety requirements dictate the use of control circuits of lower voltage than the load circuit.

Although contactor coils and pilot devices are available and are effectively used for motor controllers with up to 550-V control circuits, this practice has been prohibited where atmospheric and other working conditions make such high control-circuit voltages dangerous for operating personnel. And certain OSHA regulations require 120- or 240-V coil circuits for 460-V motors. In such cases, control transformers are used to step the voltage down to permit the use of lower-voltage coil circuits.

Control transformers are relatively small, compact, dry-type potential transformers. They are available in many ratings to meet any common motor control-circuit application. For a very wide range of motor controllers, control transformers are available as accessary equipment to the basic starter types. They can be supplied by manufacturers as separate units with provisions for mounting external to the controller or can be incorporated in the controller enclosure by the manufacturer, wired in with an operating coil of the proper voltage rating. Such transformers can be obtained with fused or otherwise protected secondaries to meet **NEC** requirements on control-circuit overcurrent protection. And extra transformer capacity can be included to permit the operation of a local lighting unit.

For low-voltage motor controllers, typical control transformers have single or double primary and secondary windings to give either a basic transformation (from, say, 480 to 120 V) or a selection of transformations (as 480/240-V primary to 240/120-V secondary). These units range in capacity from 25 VA to as high as 8000 VA. Control transformers for high-voltage controllers for 2300- and 4160-V motors are generally built into controller enclosures.

Selection of the proper control transformer for a controller is a simple matter of matching the characteristics of the control circuit to the specs of the transformer. The line voltage of the supply to the motor determines the required primary rating of the transformer. The transformer secondary must be rated to provide the desired control-circuit voltage to match the voltage of the controller operating coil. The continuous secondary-current rating of the transformer must be sufficient for the magnetizing current of the operating coil and must also be able to handle the inrush current. Of course, if other devices are to be energized from the transformer, additional capacity must be provided. Complete coordination between coil circuit and transformer is provided in factory-assembled controllers.

Grounding of control-transformer circuit

When a control transformer is used to derive control power, the transformer secondary may have to be operated with one conductor grounded. This may be either one of the conductors where the transformer has a 2-wire secondary, or the neutral conductor of a 3-wire, single-phase secondary.

NEC Section 250-5(b) applies to the secondaries of control transformers. According to the rules of this section:

1. Any 120-V, 2-wire circuit must normally have one of its conductors grounded.
2. The neutral conductor of any 240/120-V, 3-wire, single-phase circuit must be grounded.
3. The neutral of a 208/120-V, 3-phase, 4-wire circuit must be grounded.

Those requirements have often caused difficulty when applied to control circuits derived from a control transformer. For instance, there are cases where a ground fault on the hot leg of a grounded control circuit can cause a hazard to personnel by actuating the control-circuit fuse or CB and shutting down an industrial process in a sudden, unexpected, nonorderly way. A metal-casting facility is an example of an installation where sudden shutdown due to a ground fault in the hot leg of a grounded control circuit could be objectionable. Use of a 120-V control circuit without an intentional grounding connection for one of the two circuit conductors eliminates any chance of a control-circuit protective device being activated when a single accidental (fault) ground occurs on the circuit. Because designers often wish to operate such 120-V control circuits ungrounded, Exception No. 3 of Section 250-5(b) permits ungrounded control circuits under certain specified conditions. A 120-V control circuit may be operated ungrounded when all the following exist:

1. The circuit is derived from a transformer that has a primary rating below 1000 V.
2. In a commercial, institutional, or industrial facility, supervision will ensure that only persons qualified in electrical work will maintain and service the control circuits.
3. There is a need for preventing circuit opening on a ground fault—that is, continuity of power is required for safety or for operating reliability.
4. Some type of ground detector is used on the ungrounded system to alert personnel to the presence of a ground fault, enabling them to clear the ground fault in the normal downtime of the system (Fig. 174).

Although no mention is made of secondary voltage in this **NEC** rule, the exception permitting ungrounded control circuits is primarily significant for 120-V control circuits. The **NEC** has long permitted 240- and 480-V control circuits to be operated ungrounded. This exception can be applied to a 120-V control circuit derived from a control transformer in an individual motor starter or to a separate control transformer that supplies control power for a number of motor starters or magnetic contactors.

The previously described rules also apply to the hookup shown in Fig. 174. Overcurrent protection in each ungrounded leg of the circuit supplying a separate control transformer must be used on the primary side of the transformer (at *A*) and must be sized at not over 125 percent of the rated primary current [or as permitted by Section 450-3(b)]. The primary conductors

FOR SEPARATE OR BUILT-IN TRANSFORMER—

1. If transformer primary is rated less than 1000 volts . . .

2. . . . and the secondary supplies only control circuits where sudden shutdown would add or increase hazards . . .

3. . . . it is **not** necessary to ground one leg of the 120-volt secondary . . .

4. . . . **but**, some type of ground detector must be used to indicate when a ground fault occurs on the control system.

To other control circuits if separate transformer is used

FIG. 174   Under these conditions a 120-V control circuit may be used ungrounded.

must be protected at their ampacity by the primary protection. All conductors on the secondary side (that is, all control-circuit conductors) may be protected by the primary protection if the control conductors are sized so that the rating of the primary protection does not exceed the ampacity of the control conductors as reflected from the secondary to the primary. The primary-protection rating must not exceed the value obtained by multiplying the control-conductor ampacity by the secondary-to-primary voltage ratio. [See Exception No. 1 to Section 430-72(b).]

If the control transformer is supplied as part of an individual starter, within the starter enclosure, that transformer is a component part of the starter and is excluded by Exception No. 2 of Section 450-1 from the requirement for primary protection in Section 450-3(b). However, protection is required on the secondary side of the transformer by Section 430-72(b), to protect the control conductors against overload in excess of their ampacity. Of course, protection could be used on the primary instead of the secondary.

Grounding and bonding

According to the basic rule of Section 250-26(a), when one of the secondary conductors of a control transformer is grounded, it is necessary to (1) bond the grounded conductor to the metal case of the transformer and (2) run a grounding-electrode conductor from the grounded transformer secondary terminal to nearby grounded building steel or a grounded metal water pipe. But an exception exempts small control transformers from the second of these basic requirements. A Class 1 remote-control transformer that is rated not over 1000 VA simply has to have a grounded secondary conductor bonded to the metal case of the transformer; no grounding-electrode conductor is needed, provided that the metal transformer case itself is properly grounded by grounded metal raceway that supplies its primary or by means of a suitable (Section 250-57) equipment-grounding conductor that ties the case back to the grounding electrode for the primary system.

Control or signaling transformer not over 1000 VA

Frame

Fuse will operate Ground fault

Conduit

Grounded at source

Ground-fault current

Grounding electrode conductor not required

No. 14 Cu bonding jumper used to bond one leg of 120-V secondary to frame, which is grounded through primary conduit

FIG. 175 Small control transformers do not require a grounding-electrode conductor.

As shown in Fig. 175, the hookup of a control transformer under the two new exceptions consists in simply using a No. 14 copper conductor to bond the grounded leg of the transformer secondary to the transformer frame, leaving the supply conduit to the transformer to provide the path back to the main service ground, but with the connection between neutral and frame providing effective return for clearing faults, as shown.

Transformer housings must be grounded by connection to grounded cable armor or metal raceway or by use of a grounded conductor run with the circuit conductors (either a bare conductor or a conductor with green-colored covering).

Control Disconnects

NEC Section 430-74(a), on the disconnection of coil control circuits of magnetic motor starters, says:

> Motor control circuits shall be so arranged that they will be disconnected from all sources of supply when the disconnecting means is in the open position. The disconnecting means shall be permitted to consist of two or more separate devices, one of which disconnects the motor and the controller from the source(s) of power supply for the motor, and the other(s), the control circuit(s) from its power supply. Where separate devices are used, they shall be located immediately adjacent one to each other.

Remote disconnects

In recognition of the unusual and complex control conditions that exist in many industrial applications—particularly process industries and manufacturing facilities—Exception No. 1 to Section 430-74(a) alters the basic rule that disconnecting means for control circuits must "be located immediately adjacent one to each other." When a piece of motor control equipment has more than 12 motor control conductors associated with it, remote locating of the disconnect means is permitted under Exception No. 1. As shown in Fig. 176, this permission is applicable only where (1) only qualified persons have access to the live parts and (2) sufficient warning signs are posted on the equipment to locate and identify the various disconnects associated with the control-circuit conductors.

FOR CONTROL CONDUCTORS

Where an assembly of motor control equipment or a machine or process layout has **more than 12** control conductors coming into it and requiring disconnect means . . .

A warning sign must indicate location and identification of remote control disconnects

Control center or machine with motor power-circuit disconnects but not control disconnects

To remote disconnects for control circuits

. . . the disconnect devices required by Section 430-74(a) for the control conductors may be remote from, instead of adjacent to, the disconnects for the power circuits to the motor controllers.

FIG. 176   Remote control-circuit disconnects may be used under guarded conditions.

Exception No. 2 presents another instance in which control-circuit disconnects may be mounted other than immediately adjacent to each other. It notes that where the opening of one or more motor control-circuit disconnects might result in a hazard to personnel or property, remote mounting may be used under the conditions specified in Exception No. 1—that is, that access is limited to qualified persons and that warning signs are located on the outside of the equipment to indicate the location and the identification of each remote control-circuit disconnect.

Similar recognition of the need for remote disconnects in complex industrial layouts is indicated in Section 430-113. The basic rule of this section requires that a disconnecting means be provided for each source of electric energy input to equipment with more than one circuit supplying power to it. And each source is permitted to have a separate disconnecting means. But an exception to this rule states that where a motor receives electric energy from more than one source (such as a synchronous motor receiving both ac and dc energy input), the disconnecting means for the main power supply to the motor is not required to be immediately adjacent to the motor—provided that the controller disconnecting means, which is the disconnect ahead of the motor starter in the main power circuit, is capable of being locked in the open position. If, for instance, the motor control disconnect can be locked in the open position, it may be remote; but the disconnect for the other energy input circuit would have to be adjacent to the machine itself, as indicated in Fig. 177.

Protection against Accidental Starting

Combinations of ground faults often occur in motor control circuits, shorting the pilot starting device (pushbutton, limit switch, pressure switch, etc.) and accidentally starting the motor even though the pilot device is in the OFF position. And because many remote-control circuits are long, possible faults may occur at many points in those circuits. Insulation breakdowns, contact

FOR MULTIPLE POWER SOURCES

If the disconnect for the main power supply to a motor can be locked open, it may be installed remote from the motor . . .

. . . **but**, disconnects for any other power inputs to the motor must be "immediately adjacent" to the motor.

FIG. 177   One power-source disconnect may be installed remote from a motor.

shorts due to the accumulation of foreign matter or moisture, and grounds to conduit are common fault conditions responsible for the accidental starting of motor controllers.

As shown in Fig. 178, a magnetic motor controller used on a 3-phase, 3-wire ungrounded system always presents the possibility of accidental starting of the motor. If, for instance, an undetected ground fault exists on one phase of the 3-phase system—even if this ground fault is a long distance from the controller—a second ground fault in the remote-control circuit for the

FIG. 178   Ungrounded electrical systems present the hazard of unexpected motor starting due to faults.

FIG. 179 Control transformer can isolate coil-circuit wires from accidental starting connection.

operating coil of the starter can start the motor. Effective circuit design must always be utilized to prevent such accidental occurrences.

Figure 179 shows the use of a control transformer to isolate the control circuit and keep it from responding to the combination of ground faults shown in Fig. 178. This transformer may be a one-to-one isolating transformer, with the same primary and secondary voltages, or the transformer can step the motor-circuit voltage down to a lower level for the control circuit.

In the hookup shown in Fig. 180, a two-pole START button is used in conjunction with two

FIG. 180 Two-pole START button protects against accidental starting due to ground faults.

FIG. 181 Use of a two-pole START button and a two-pole STOP button provides the greatest safety.

sets of holding contacts in the motor starter. This hookup protects against accidental starting of the motor under the fault conditions shown in Fig. 178. The hookup also protects against accidental starting due to two ground faults in the control circuit simply shorting out the START button and energizing the operating coil. This could happen in the circuit of Fig. 178 or that of Fig. 179.

Another type of motor control-circuit fault can produce a current path through the coil of a closed contactor to hold it closed even if the pilot device operates to open the coil circuit. Again, this can result when a combination of ground faults shorts the STOP device. Failure to open can do serious damage to motors in some applications and can be a hazard to personnel. The operating characteristics of contactor coils contribute to the possible failure of a controller to respond to the opening of the STOP contacts. It takes about 85 percent of rated coil voltage to operate the armature associated with the coil; but it takes only about 50 percent of the rated voltage to enable the coil to hold the contactor closed once it is closed. Thus, even partial grounds and shorts on control contact assemblies can produce paths for sufficient current flow to cause shorting of the STOP position of pilot devices. And faults can short out running overload relays, negating the overcurrent protection of the motor, its associated control equipment, and conductors.

Figure 181 is a modification of the circuit of Fig. 180, using a two-pole START button and a two-pole STOP button; it protects against both accidental starting and accidental failure to stop when the STOP button is pressed. Both effects of ground faults are eliminated.

Figure 182 shows another example of a control-circuit installation that should be carefully designed—one that is required by the second sentence of Section 430-73 for any control circuit which has one leg grounded. Whenever the coil is fed from a circuit made up of a hot conductor and a grounded conductor (as when the coil is fed from a panelboard or separate control transformer, instead of from the supply conductors to the motor), care must be taken to place the pushbutton station or other switching control device in the hot leg to the coil, and not in the grounded leg to the coil. By switching in the hot leg, starting of the motor through an accidental ground fault can be effectively eliminated.

Section 430-74(a), on disconnection of coil control circuits of magnetic motor starters, was quoted under "Control Disconnects" above. Figures 183 and 184 show two methods of providing for the disconnection of control power. Note that, in Fig. 183, the single disconnect switch kills power to the motor and to the entire control circuit—because the step-down control transformer is installed on the load side of the disconnect. In Fig. 184, however, it is necessary to provide

RIGHT

Grounded neutral

Hot leg

Stop Start

Switching device is in hot leg of control circuit

HOOKUP SATISFIES NE CODE SECTION 430-73. GROUND FAULTS CANNOT START MOTOR

Any one of these ground faults will start motor

WRONG

Control circuit from panelboard, separate control transformer, or other separate energy source

Pushbutton station or other control switching device connected in the grounded conductor to operating coil of starter

Note : OL relays are not shown in diagrams.

FIG. 182 Switching must be done in the hot leg of a grounded control circuit.

WHERE TRANSFORMER IS IN STARTER

Control transformer in starter does not require overcurrent protection

Overcurrent protection for control circuit not needed if control circuit does not leave the controller

Disconnect switch or circuit breaker kills power circuit and control circuit as required by Section 430-74

3-phase 440V

Transformer secondary grounded as required by Section 250-5

Stop

Start

Operating coil

Ground fault could short out OL relays without stopping motor

Motor

FIG. 183 One disconnect kills all power to the starter.

WHERE TRANSFORMER IS OUTSIDE STARTER

FIG. 184 External control power must have a disconnect.

some means for ensuring that the control circuit is deenergized when the power circuit to the motor is open. Control circuits operating contactor coils and other devices within controllers present a shock hazard if they are allowed to remain energized when the disconnect is in the OFF position. Therefore, the control circuit either must be designed in such a way that it is disconnected from the source of supply by the controller disconnecting means or must be equipped with a separate disconnect immediately adjacent to the controller disconnect (Section 430-74) for opening both circuits. The four-pole disconnect shown in the sketch may be a four-pole switch or a three-pole switch with an auxiliary set of contacts to open the hot leg of the control circuit.

All the foregoing is based on rules given in **NEC** Article 430. For the most part, these rules are straightforward and clear. In any case where a question arises about the interpretation of **NEC** rules, the authority responsible for enforcement in the particular area should be consulted. Where interpretations are given in this presentation, they are offered to provide a concrete concept which can be checked out with the local inspector. Such consultation provides an easy way to find out exactly how an inspector will rule on specific fine points.

Voltage Regulation

To ensure the proper and efficient operation of motors, the matter of voltage regulation must be considered carefully. Many factors (e.g., number of motors, sizes and types of motors, duty cycles, load densities, type of distribution system, loading of various feeders, and power factor) are related to the design problem of assuring the necessary level and stability of motor voltages.

Voltage regulation follows through every step in design and must be accounted for in sizing conductors.

The voltage drop from the sources of voltage to any motor in the system must not exceed 5 percent. Normally, the proper power feeder design will limit the voltage drop to 3 percent, leaving a maximum permissible voltage drop of 2 percent in any motor branch circuit under full-load conditions. However, a 1 percent maximum circuit drop is recommended.

The power factor should be taken into consideration in all calculations; the known power-factor value or an 80 percent assumed value should be used. By raising the power factor, for the same actual power delivered, the current is decreased in the generator, transformers, and lines, up to the point where the capacitor is connected. Power-factor capacitors can be connected into motor circuits to neutralize the effect of lagging-power-factor loads, thereby reducing the current drawn for a given kilowatt load. In a distribution system, small capacitor units may be connected at the individual loads, or the total required capacitance (in kilovars) may be grouped at one point and connected to the main. Although the kilovar total is the same in both cases, the use of small capacitors at the individual loads reduces the current all the way from the loads back to the source; this method therefore has a greater power-factor corrective effect than the use of one big unit on the main, which reduces the current only from the point of installation back to the source.

Figure 185 shows the basic relationships involved in power-factor correction. A typical calculation would be made as follows.

EXAMPLE

A 3-phase, 460-V, 50-hp motor has a power factor of 70 percent. What rating of capacitor is needed to improve the power factor to 90 percent?

SOLUTION

From **NEC** Table 430-150, the full-load current of the motor is 65 A. That gives a kilowatt load of

$$L = \frac{460 \times 65 \times 1.73}{1000} PF_1$$

$$= \frac{51{,}740}{1000} \times 0.7 = 36.2 \text{ kW}$$

If $\cos \theta_1 = 0.70$, then $\theta_1 = 45.58°$ and $\tan \theta_1 = 1.0203$; and if $\cos \theta_2 = 0.90$, then $\theta_2 = 25.84°$ and $\tan \theta_2 = 0.4844$. Then

$$KVAR_R = L(\tan \theta_1 - \tan \theta_2)$$

$$= 36.2(1.0203 - 0.4844)$$

$$= 36.2 \times 0.5359 = 19.4 \text{ kvar}$$

Depending upon the relationship between the voltage drop and the power factor, measures for improving the power factor may be designed into motor branch circuits. At individual motor locations, power-factor-correcting capacitors offer improved voltage regulation. As shown in Fig. 186, power-factor capacitors installed at terminals of motors provide maximum relief from reactive currents, reducing the required current-carrying capacities of conductors from their point of application all the way back to the supply system. Such application also eliminates extra switching devices, since each capacitor can be switched with the motor it serves. Figure 187 shows details included on a typical set of electrical plans to convey the designer's concept for

TO IMPROVE THE POWER FACTOR OF A CIRCUIT FROM PF_1 TO PF_2 THE REQUIRED RATING OF A CAPACITOR MUST BE

$$KVAR_R = KW(\tan \theta_1 - \tan \theta_2)$$

where, in the sketch

$KVAR_R$ = required kilovars rating of capacitor to change from PF_1 to PF_2

KW = the kilowatt value of the circuit load

θ_1 = original phase angle

θ_2 = improved phase angle

PF_1 = power factor before correction

PF_2 = power factor after correction

$KVAR_1$ = reactive kilovolt-amperes at PF_1

$KVAR_2$ = reactive kilovolt-amperes at PF_2

NOTE: The phase angles θ_1 and θ_2 can be determined from a table of trigonometric functions using the following relationships:

θ_1 = The angle which has its cosine equal to the decimal value of the original power factor (e.g., 0.70 for 70% PF; 0.65 for 65%; etc.)

θ_2 = The angle which has its cosine equal to the decimal value of the improved power factor

FIG. 185 Motor-circuit power factor is raised with this procedure.

FIG. 186 Installation of the power-factor capacitors at the individual motors offers maximum benefit from the correction.

FIG. 187   Design must provide a place for connecting a capacitor unit between motor and starter.

power-factor correction of a large motor. For small, numerous motors that are operated inter-mittently, however, it is often economically more desirable to install the required capacitor kil-ovars at the motor load center, as at the right in Fig. 186.

In most applications, power-factor capacitors are installed to raise the system power factor for increased circuit or system current-carrying capacity, reduced power losses, and lower reac-tive-power charges (most utility companies include a power-factor penalty clause in their indus-trial billing). Additional benefits derived from power-factor-capacitor installation are reduced voltage drop and increased voltage stability. Capacitor manufacturers provide various tables and graphs to facilitate the selection of the proper capacitor for a given motor load.

The **NEC** limits power-factor correction to unity (100 percent or 1.0) when there is no load on the motor. This results in a power factor of 95 percent or better when the motor is fully loaded. Section 460-7 recognizes the use of capacitors either sized at the value that will produce 100 percent power factor when the motor is running at no load or sized at a value equal to 50 percent of the kilovoltampere rating of the motor input (as noted in the Exception), for motors up to 50 hp and 600 V, as shown in Fig. 188. Most motor-associated capacitors are used with

FIG. 188   The NEC sets a maximum value on capacitor rating for safety purposes.

Desired power factor in percentage

Original power factor in percentage	80%	81	82	83	84	85	86	87	88	89	90	91	92	93	94	95	96	97	98	99	100
50%	.982	1.008	1.034	1.060	1.086	1.112	1.139	1.165	1.192	1.220	1.248	1.276	1.303	1.337	1.369	1.402	1.441	1.481	1.529	1.590	1.732
51	.936	.962	.988	1.014	1.040	1.066	1.093	1.119	1.146	1.174	1.202	1.230	1.257	1.291	1.320	1.357	1.395	1.435	1.483	1.544	1.686
52	.894	.920	.946	.972	.998	1.024	1.051	1.077	1.104	1.132	1.160	1.188	1.215	1.249	1.281	1.315	1.353	1.393	1.441	1.502	1.644
53	.850	.876	.902	.928	.954	.980	1.007	1.033	1.060	1.088	1.116	1.144	1.171	1.205	1.237	1.271	1.309	1.349	1.397	1.458	1.600
54	.809	.835	.861	.887	.913	.939	.966	.992	1.019	1.047	1.075	1.103	1.130	1.164	1.196	1.230	1.268	1.308	1.356	1.417	1.559
55	.769	.795	.821	.847	.873	.899	.926	.952	.979	1.007	1.035	1.063	1.090	1.124	1.156	1.190	1.228	1.268	1.316	1.377	1.519
56	.730	.756	.782	.808	.834	.860	.887	.913	.940	.968	.996	1.024	1.051	1.085	1.117	1.151	1.189	1.229	1.277	1.338	1.480
57	.692	.718	.744	.770	.796	.822	.849	.875	.902	.930	.958	.986	1.013	1.047	1.079	1.113	1.151	1.191	1.239	1.300	1.442
58	.655	.681	.707	.733	.759	.785	.812	.838	.865	.893	.921	.949	.976	1.010	1.042	1.076	1.114	1.154	1.202	1.263	1.405
59	.618	.644	.670	.696	.722	.748	.775	.801	.828	.856	.884	.912	.939	.973	1.005	1.039	1.077	1.117	1.165	1.226	1.368
60	.584	.610	.636	.662	.688	.714	.741	.767	.794	.822	.849	.878	.905	.939	.971	1.005	1.043	1.083	1.131	1.192	1.334
61	.549	.575	.601	.627	.653	.679	.706	.732	.759	.787	.815	.843	.870	.904	.936	.970	1.008	1.048	1.096	1.157	1.299
62	.515	.541	.567	.593	.619	.645	.672	.698	.725	.753	.781	.809	.836	.870	.902	.936	.974	1.014	1.062	1.123	1.265
63	.483	.509	.535	.561	.587	.613	.640	.666	.693	.721	.749	.777	.804	.838	.870	.904	.942	.982	1.030	1.091	1.233
64	.450	.476	.502	.528	.554	.580	.607	.633	.660	.688	.716	.744	.771	.805	.837	.871	.909	.949	.997	1.058	1.200
65	.419	.445	.471	.497	.523	.549	.576	.602	.629	.657	.685	.713	.740	.774	.806	.840	.878	.918	.966	1.027	1.169
66	.388	.414	.440	.466	.492	.518	.545	.571	.598	.626	.654	.682	.709	.743	.775	.809	.847	.887	.935	.996	1.138
67	.358	.384	.410	.436	.462	.488	.515	.541	.568	.596	.624	.652	.679	.713	.745	.779	.817	.857	.905	.966	1.108
68	.329	.355	.381	.407	.433	.459	.486	.512	.539	.567	.595	.623	.650	.684	.716	.750	.788	.828	.876	.937	1.079
69	.299	.325	.351	.377	.403	.429	.456	.482	.509	.537	.565	.593	.620	.654	.686	.720	.758	.798	.840	.907	1.049
70	.270	.296	.322	.348	.374	.400	.427	.453	.480	.508	.536	.564	.591	.625	.657	.691	.729	.769	.811	.878	1.020
71	.242	.268	.294	.320	.346	.372	.399	.425	.452	.480	.508	.536	.563	.597	.629	.663	.701	.741	.783	.850	.992
72	.213	.239	.265	.291	.317	.343	.370	.396	.423	.451	.479	.507	.534	.568	.600	.634	.672	.712	.754	.821	.963
73	.186	.212	.238	.264	.290	.316	.343	.369	.396	.424	.452	.480	.507	.541	.573	.607	.645	.685	.727	.794	.936
74	.159	.185	.211	.237	.263	.289	.316	.342	.369	.397	.425	.453	.480	.514	.546	.580	.618	.658	.700	.767	.909
75	.132	.158	.184	.210	.236	.262	.289	.315	.342	.370	.398	.426	.453	.487	.519	.553	.591	.631	.673	.740	.882
76	.105	.131	.157	.183	.209	.235	.262	.288	.315	.343	.371	.399	.426	.460	.492	.526	.564	.604	.652	.713	.855
77	.079	.105	.131	.157	.183	.209	.236	.262	.289	.317	.345	.373	.400	.434	.466	.500	.538	.578	.620	.687	.829
78	.053	.079	.105	.131	.157	.183	.210	.236	.263	.291	.319	.347	.374	.408	.440	.474	.512	.552	.594	.661	.803
79	.026	.052	.078	.104	.130	.156	.183	.209	.236	.264	.292	.320	.347	.381	.413	.447	.485	.525	.567	.634	.776
80	.000	.026	.052	.078	.104	.130	.157	.183	.210	.238	.266	.294	.321	.355	.387	.421	.459	.499	.541	.608	.750
81	—.	.000	.026	.052	.078	.104	.131	.157	.184	.212	.240	.268	.295	.329	.361	.395	.433	.473	.515	.582	.724
82	—	—	.000	.026	.052	.078	.105	.131	.158	.186	.214	.242	.269	.303	.335	.369	.407	.447	.489	.556	.698
83	—	—	—	.000	.026	.052	.079	.105	.132	.160	.188	.216	.243	.277	.309	.343	.381	.421	.463	.530	.672
84	—	—	—	—	.000	.026	.053	.079	.106	.134	.162	.190	.217	.251	.283	.317	.355	.395	.437	.504	.645
85	—	—	—	—	—	.000	.027	.053	.080	.108	.136	.164	.191	.225	.257	.291	.329	.369	.417	.478	.620

FIG. 189   Multipliers for calculating capacitor kilovars required to raise the power factors of motor circuits.

low-voltage, 5- to 50-hp, 1800- and 1200-rpm, across-the-line-start motors. In this range, the no-load rule for determining the maximum capacitor kilovars restricts capacitors to the equivalent of less than 50 percent of the horsepower. Noticeable economy can be effected by applying larger capacitors, up to 50 percent of the motor-input power, to such motors. This has been done for years, with excellent results and no field trouble.

The no-load power factor of a motor is a design constant for the motor; it may be obtained from the manufacturer of the motor, or it may be measured or calculated. With reference to Fig. 185, the known no-load power factor (PF) of the motor may be used to calculate the required power in kilowatts as $kW = PF \times kVA_1$, where kVA_1 is calculated from the circuit voltage and current, as measured with a clamp-on ammeter. Then the kilovar rating of the capacitor required to raise the no-load power factor to 100 percent equals the square root of $(kVA_1)^2 - (kW)^2$.

Capacitors installed as a group or bank at some central point, such as a switchboard, load-center, busway, or outdoor substation, usually serve only to reduce the utility-company penalty charges. However, in many instances, installation costs also will be lowered.

Capacitor installations may consist of an individual unit connected as close as possible to the inductive load (e.g., at the terminals of a motor) or of a bank of many units connected in multiple

I. THE CONTROL CENTER LAYOUT

Contains branch circuit overcurrent protection and disconnect means

(Motor control centers may contain branch circuit overcurrent protection, disconnects and controllers)

Conductors with carrying capacity of at least 125% of motor full-load running current

Individual branch circuits to motors

Power panel

Feeder

Motors

Disconnects here if power panel is out of sight or over 50 ft away from controller

Motor controllers (disconnect switch may be incorporated in the starter enclosure if required as above)

2. THE CABLE TAP-BOX METHOD

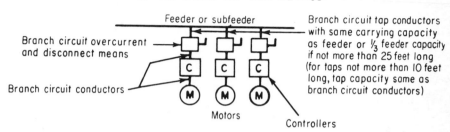

Branch circuit overcurrent and disconnect means

Branch circuit conductors

Feeder or subfeeder

Motors

Controllers

Branch circuit tap conductors with same carrying capacity as feeder or ⅓ feeder capacity if not more than 25 feet long (for taps not more than 10 feet long, tap capacity same as branch circuit conductors)

3. THE BUSWAY TAP-SWITCH METHOD

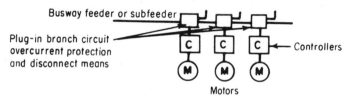

Busway feeder or subfeeder

Plug-in branch circuit overcurrent protection and disconnect means

Controllers

Motors

FIG. 190 Motor-circuit layout must suit the particular needs of a given installation.

across a main feeder. Units are available in specific kilovar and voltage ratings. Standard low-voltage capacitor units are rated from about 0.5 to 25 kvar at voltages from 216 to 600 V. For high-voltage applications, standard ratings are 15, 25, 50, and 100 kvar. Available in single-, 2-, or 3-phase configurations, power capacitors may be supplied either unfused or equipped with current-limiting or high-capacity fuses (single-phase units are furnished with one fuse; 3-phase capacitors usually have two fuses). On low-voltage units, fuses may be mounted on the capacitor bushings inside the terminal compartment.

Figure 189 is a table for calculating the capacitor kilovar rating required to raise the power factor from an original value to a desired higher value. Its use is illustrated in the following example.

EXAMPLE

A circuit supplies a load that has a 100-kW value (obtained either by calculation or from a wattmeter) and operates at a power factor of 68 percent. What capacitor kilovar rating is needed to raise the power factor to 96 percent?

SOLUTION

Find the required multiplier in the table of Fig. 189 by locating 68 percent in the vertical column at the left and then moving across to the column of multipiiers headed 96 percent. The multiplier is 0.788. Then the required rating is

$$100 \text{ kW} \times 0.788 = 78.8 \text{ kvar}$$

Circuit Layout

Based on the foregoing rules and considerations, motor branch circuits can be designed for any type of occupancy. Generally, motor-circuit requirements for commercial and institutional buildings are concerned with known loads which are fairly permanent in size and location. These loads include refrigeration and air-conditioning compressors, pumps, elevators, escalators, blowers, and fans. Such loads can be served by many types of wiring systems without special provisions for flexibility of power circuiting. Industrial plants, on the other hand, require considerable power-circuiting flexibility to accommodate the moving of motors, shifting of production lines, and expansion of motor loads. To provide this flexibility, motor circuits may be fed from motor control centers, from plug-in busway, from auxiliary gutters, from junction boxes

ROOF PLAN

FIG. 191   This kind of layout detail fully conveys to the installer the designer's intent regarding the circuiting of motors.

FIG. 192 One-line diagrams on electrical plans should fully detail the power-supply sequence for motor circuits at all voltages.

POWER PANEL SCHEDULE

Panel Designation	Location	Branch Cct. Equip.	Mains Type	Mains Cap	No.	Poles	Frame	Trip Calibr.	Phase	Voltg.	Cond. No.	Cond. Size	Cond. Type	Equipment	H P
PHP	Pent House	Cct. Bkr.	Lugs Only	NEMA Std.	—	—	—	—	3	120/208	3		R		
					1	3	225	125	3	"	3	2	R	Inter. Zone Fan	25
					2	3	100	50	3	"	3	8	R	Exter. Zone Fan	10
					3	3	50	15	3	"	3	12	R	Booster Pump	2
					4	3	50	15	3	"	3	12	R	Toilet Exh. Fan	3/4
					5	3	50	15	3	"	3	12	R	Dehumid. Spray Pump	1/2
					6	3	50	15	3	"	3	12	R	Power Roof Ventilator	1/2
					7	3	50	15	1	"	3	12	R	" "	1
					8	2	50	20	1	"	3		R	Ltg. Panel LPPH1A	1
					9	3	50	15	3	120/208	3	12	R	Roof Vent	1 3/8
					10,11,12	3	50	Spares			1				
PP3	3rd Fl. Mech. Rm.	Cct. Bkr.	Lugs Only	NEMA Std.	—	—	—	—	3	120/208			R		
					1	3	50	15	3	"	3	12	R	Dehum. Spray Pump	1/2
					2	3	225	100	3	"	3	4	R	A C Unit "C"	20
					3	1	50	15	1	"	2	12	R	Filter Motor	1/6
					4	1	50	15	1	"	2	12	R	Bd. Rm. Exh. Fan	1/4
					5	3	50	15	3	120/208	3	12	R	A C Unit "J"	1/2
					6	3	50	Spare							
					7	3	50	Spare							
					8	3	50	Spare							
BPP	Bsm't Power	Cct. Bkr.	Lugs Only	NEMA Std.	—	—	—	—	3	120/208			R		
					1	3	50	15	3	"	3	12	R	Evap. Cond. Fan	1
					2	3	50	15	3	"	3	12	R	Ch. D.W. Pump	1 1/2
					3	3	50	15	3	"	3	12	R	Ch. D.W. Pump	2
					4	3	50	30	3	"	3	12	R	D.W. Compr.	3
					5	3	50	30	3	"	3	12	R	D.W. Compr.	3
					6	1	50	20	1	"	2	12	R	Evap. Cond. Spray Pump	1/3
					7	1	50	20	1	"	2	12	R	D.W. Pump	1/4
					8	1	50	20	1	"	2	12	R	" "	1/4
					9	1	50	20	1	120/208	2	12	R	Boiler Cont.	1/4
					10,11,12	3	50	Spares							

FIG. 193 Power-panel schedule should summarize the ratings and layout of all motor-circuit equipment.

connected in conduit and raceways which allow changes and expansion in the wiring, or, in the case where many small motors are to be supplied, from an underfloor raceway system. Spare capacity for increased motor loads is a design consideration for motor feeders and subfeeders. Spare capacity should be included in the motor-circuit panelboard where required. By using sectional-type panelboards with interchangeable units and oversized raceway for branch-circuit wiring, future increases in the size of the motors on the circuit and in associated branch-circuit equipment are greatly facilitated.

As shown in Fig. 190, wiring to serve two or more motors may be laid out according to several plans:

1. Each motor may be served by a separate branch circuit from a panel or distribution center. Branch-circuit protective devices for motor circuits are commonly grouped with required controller and disconnect equipment. The grouping may be contained in a single enclosure, such as a motor control center (a so-called "MCC"), or it may consist of a number of individual enclosures at one location fed from a trough (an auxiliary gutter).

2. Individual circuits to motors may be tapped from a feeder without individual overcurrent protection for the taps, provided they are carried directly to the disconnecting means or combination starter for each motor. Such taps must have the same carrying capacity as the feeder if they are over 25 ft long. If they are under 25 ft long, they must have a carrying capacity at least equal to one-third the carrying capacity of the feeder. In such cases, the motor branch-circuit protection is included with the disconnect means (switch and fuses, or CB) or in a combination starter, and the branch circuit originates at that point. Taps not over 10 ft long need have only the same capacity as the branch-circuit wires. Taps not over 10 ft long are applied where two or more combination motor starters are nippled out of a gutter where the motor-circuit wires are tapped to the feeder conductors supplying the gutter.

3. Individual branch circuits to motors may originate at disconnects fed directly by a feeder or subfeeder (e.g., tap units on a busway), with disconnect and branch-circuit protection provided by the tap switch (with fuses) or CB.

The diagrams of motor branch circuits in Figs. 191 and 192 cover the most common layouts of motor circuits, supplied either from panelboards or switchboards or from taps to a feeder or subfeeder as shown. Figure 191, a typical layout of motor branch circuits in a fan room on top of a hospital, shows the circuit symbols and design information which must guide the installer in carrying out the plan. Other data on the motors and controls are incorporated in the job specifications. In Fig. 192, the distribution diagram for the boiler plant of a large medical research building, the motor branch circuits are derived from a 3-phase, 4-wire, 120/208-V motor control center for fan motors; from a 120/208-V switchboard for general-purpose motor applications; and from a 2400-V, 3-phase circuit-breaker switchboard supplying 2400-V circuits to motors for pumps, compressors, air-conditioning systems, etc. The successful design of motor branch circuits depends upon a thorough understanding of the ways in which the required types of protection may be combined.

Where a particular installation involves a large number of motors, the electrical plans should include one or more clear, detailed motor-circuit schedules. As shown in Fig. 193, power panels or motor control centers supplying motor branch circuits should be described as to both their feeds and the circuits they serve. Data on branch-circuit protection include frame size and trip calibration for CBs.

Feeders for Lighting and Power

In any electrical system, the distribution system is comprised of the equipment and methods used to carry power from the service equipment to the overcurrent devices protecting the branch circuits. This system is a layout of equipment designed to provide the right amount of current at the right voltage to each utilization outlet. The distribution system carries power to lighting panelboards, power panelboards, motor control centers, and the branch-circuit protective devices for individual motor or power loads.

Depending upon the type of building, the size and nature of the total load, various economic factors, and local conditions, a distribution system may operate at a single voltage level or may involve one or more transformations of voltage. A distribution system may also incorporate changes in the frequency of ac power or rectification from ac to dc power.

Figure 194 shows the basic layout of a distribution system and the terms used to refer to various parts of the system. Typical distribution systems can vary widely in size and complexity and will commonly involve the range of system components shown in Fig. 195. Because the terminology of distribution systems is frequently misused or misunderstood, it is best to begin this section with several definitions:

- *Mains* are the conductors extending from the utility service terminals at the building wall (or generator or converter bus) to the service switch or to the main distribution center.

- A *feeder* is a set of conductors originating at a main distribution center and feeding one or more subdistribution centers, one or more branch-circuit distribution centers, one or more branch circuits (as in the case of plug-in busway or motor-circuit taps to a feeder) or a combination of these. It may be a primary- or secondary-voltage circuit, but its function is always to deliver a

FIG. 194   Basic elements of electrical distribution at utilization voltage.

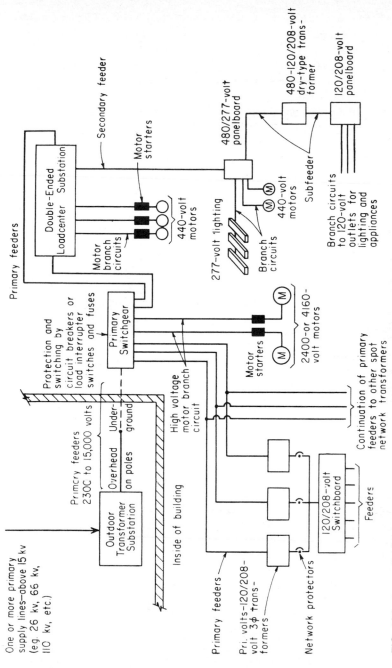

One or more primary supply lines—above 15 kv (eg. 26 kv, 66 kv, 110 kv, etc.)

Outdoor Transformer Substation

Primcry feeders 230C to 15,000 volts

Overhead on poles

Underground

Inside of building

Protection and switching by circuit breakers or load interrupter switches and fuses

Primary feeders

Double-Ended Loadcenter Substation

Secondary feeder

Motor starters

440-volt motors

Motor branch circuits

Primary Switchgear

High voltage motor branch circuit

Motor starters

2400- or 4160-volt motors

480/277-volt panelboard

277-volt lighting

Branch circuits

440-volt motors

480-120/208-volt dry-type transformer

120/208-volt panelboard

Subfeeder

Branch circuits to 120-volt outlets for lighting and appliances

Continuation of primary feeders to other spot network transformers

Primary feeders

Pri. volts—120/208-volt 3φ transformers

Network protectors

120/208-volt Switchboard

Feeders

FIG. 195 Large distribution systems may include a wide range of circuit configurations and voltages.

block of power from one point to another point at which the power capacity is apportioned among a number of other circuits.

- A *lighting feeder* is a feeder to a load which is made up primarily of lighting circuits.
- A *power feeder* is a feeder to a load of branch circuits for motors, heating, or other power loads.
- A *subfeeder* is a set of conductors originating at a distribution center other than the main distribution center and supplying one or more other distribution panelboards, branch-circuit panelboards, or branch circuits.
- A *switchboard* is a large single panel, frame, or assembly of panels with switches, overcurrent and other protective devices, and usually instruments mounted on the front and/or the back. Switchboards are generally accessible from the back as well as from the front and are not intended to be installed in cabinets.
- A *panelboard* is a single panel or a group of panel units assembled in the form of a single panel. It contains busses tapped by fuse holders, with or without switches, or by circuit breakers, providing protection for—and, if switches or circuit breakers are used, also providing control of—circuits for light, heat, or power. These circuits may be branch circuits or subfeeders. A panelboard is designed to be placed in a cabinet or cutout box placed in or against a wall or partition and accessible only from the front.

It should be noted that manufacturers commonly refer to light-duty panelboards as *loadcenters*. But **NEC** and UL regulations do not distinguish between light-duty and heavy-duty panelboards. All those regulations apply identically to panelboards and so-called loadcenters.

Distribution Voltage Levels

Distribution systems are basically classified according to the voltage levels used to carry the power either directly to the branch circuits or to loadcenter transformers or substations at which feeders to branch circuits originate. Figure 196 shows three basic layouts utilizing the loadcenter concept—carrying a feeder to the center of a layout of utilization loads and then subdividing the feeder capacity among branch circuits for motor and/or lighting loads.

Common distribution systems operate at the following voltage levels:

1. 120/240-V, 3-wire, single-phase combination light and power distribution (also called the 115/230-V system). This system is restricted to applications where the total load is small and is primarily lighting and receptacle circuits. It operates with the neutral wire grounded.

   The most common application is in residential occupancies—individual homes or multifamily dwellings. This system is also used in stores, small schools, and other small commercial occupancies. In most small commercial buildings, however, the use of 120/208-V, 3-phase distribution offers greater economy owing to the higher operating efficiency of 3-phase circuits.

   In those cases where 120/240-V feeders are used as the basic distribution method, the service to the premises is made at that voltage. Of course, 120/240-V distribution is frequently an effective and economical system for lighting subfeeder distribution in electrical systems which use a higher-voltage basic distribution system with loadcenter step-down to utilization voltages for local and incidental lighting and receptacle circuits.

2. 120/208-V wye, 3-phase, 4-wire light and power distribution. This is the most common type of system for commercial buildings, institutional occupancies, and industrial shops with limited electrical loads. Such a system operates with the neutral conductor grounded; it offers substantial economy over the 120/240-V system with regard to the amount of con-

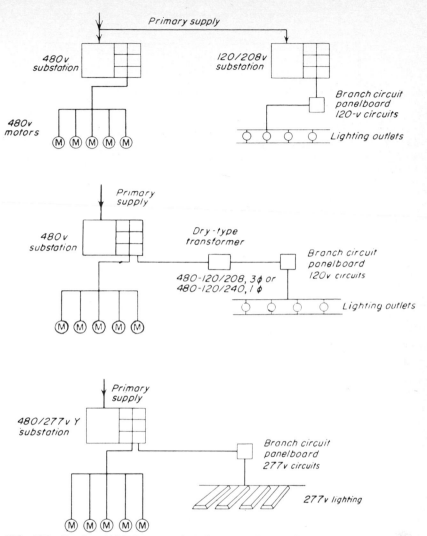

FIG. 196 Feeders to loadcenter substations, transformer distribution centers, and/or panels provide an effective, economical design technique.

ductor material required to carry a given amount of power to a load. This combination system provides 120 V phase to neutral for lighting and single-phase loads, and 208 V phase to phase for single- or 3-phase motor or other power loads.

The system is used as the basic distribution system in occupancies in which the service to the building is of this voltage. It is also the most common subdistribution system for lighting and receptacle circuits in occupancies using higher-voltage distribution to loadcenters. Figure 197 shows two examples of 120/208-V systems.

3. 240-V, 3-phase, 3-wire distribution. This is a common system for power loads in commercial and industrial buildings. It may operate ungrounded or with one of the phase legs grounded—in which case it is called a *corner-grounded delta system*. In such cases, service

Power and light feeders in 60,000-sq-ft factory

Transformation to utilization voltage is made in the outdoor transformer yard. Here a Y-Y connection of three 167-kva, oil-filled distribution transformers steps the incoming supply to a 120/208-volt utilization level. The 120/208-volt level was dictated by the ratings of existing motor loads on all of the production machinery which had been moved from the old building. As a result of this, there was no opportunity to take advantage of the distribution economies of 480/277-volt circuits for lighting and motor loads.

Subdistribution for vital hospital circuits

Plan shows three alternative circuits for supplying essential circuits to "iron lungs" in a research hospital. The 120/208-volt panelboard may be fed from: a step-down transformer fed by the 480-volt distribution system, supplied from a dual-primary network hookup; from Eng. Gen. 1; or from Eng. Gen. 2. Although not shown here, the panelboard for emergency lighting and the panelboard for isolated, ungrounded circuits in surgical areas are supplied from the same normal-and-emergency hookup, with the ungrounded circuits transformer-isolated.

FIG. 197 Utilization-equipment voltage requirements dictate extensive use of 120/208-V distribution.

to the premises is 240-V, 3-phase power. Feeders carry the power to panelboards or gutters supplying branch circuits for motor loads. Lighting loads are then handled by a separate single-phase service to the building or through a step-down transformer.

This system offers economical application where the power load is large compared to the lighting load. In some 240-V, 3-phase systems, a tap on one of the phases is brought out as a neutral to provide 120 V for lighting and receptacle circuits, forming a 4-wire system.

4. 480-V, 3-phase, 3-wire distribution. This system is commonly used in commercial and industrial buildings with substantial motor loads. It is generally operated ungrounded. Such a system provides a high level of power continuity because a ground fault on one phase will not cause automatic operation of a protective device to open the circuit, as in grounded systems. But it can be operated with one phase leg grounded—as a corner-grounded delta system. Service to the building may be made at this voltage, with the 480-V feeders carried to motor loads and to step-down transformers for lighting and receptacle circuits.

In many cases, 480-V feeders will be derived from loadcenter substations within the building and carried to motor loads or power panels (Fig. 198).

5. 480Y/277-V (460Y/265-V), 3-phase, 4-wire combination power and light distribution.

This is a very popular basic system for use in commercial and industrial buildings. In office and other commercial buildings, the 480-V, 3-phase, 4-wire feeders are carried to each floor, where 480-V, 3-phase, 3-wire power is tapped to a power panel or to motors; general area fluorescent lighting using 277-V ballasts is connected between each phase leg and the neutral. Required 120/208-V, 3-phase, 4-wire circuits are derived from step-down transformers for local lighting and appliance and receptacle circuits.

This system offers the advantage of economy over a 120/208-V system when fewer than about half the load devices require 120- or 208-V power. Where the 480Y system can be used, it costs less than the 120/208-V system, owing to savings on the smaller sizes of conductors, reduced raceway needs, reduced installation labor costs, and the lower cost of system elements, which have lower current capacities.

If the required amount of 120- or 208-V power is over half the total load in a building, the cost of the step-down transformers required to supply these circuits will generally offset the savings in the 480-V circuiting. The costs of switchgear and panelboards also affect the cost comparison between 120/208- and 480/277-V systems.

The 480Y system is generally more advantageous in multifloor buildings than in buildings of only a few floors, and more advantageous in large or long single-level areas than in small areas. The savings are directly related to reductions in the sizes of long feeders, allowing more kVA to be delivered per pound of installed system. (See Figs. 199 and 200.)

6. 2400-V, 3-phase delta distribution. This is an industrial-type system used to feed heavy motor loads directly, and motor and lighting loads through loadcenter substations and lighting transformers.

7. 4160Y/2400-V, 3-phase, 4-wire distribution with a grounded neutral. This system is more common than the 2400-V delta-connected system. Originally developed primarily for industrial applications, this system of high-voltage distribution has gained wide use for large-area, spread-out commercial and institutional buildings, such as shopping centers, schools, and motels.

The 4160Y/2400-V system also appears to have become the most popular high-voltage system for inside distribution. Economy in the use of cables and switchgear seems to be greatest for this particular primary system. In many installations, it offers the advantages of loadcenter primary distribution for power and lighting loads with direct connection of heavy motor loads operating at 4160 V.

This system is widely used to supply loadcenter substations in which the voltage is stepped to 480 V to feed motors and lighting transformers for 120/240- and/or 120/208-V circuits. It may also be used for distribution to substations stepping the voltage directly to 120/208 V. (See Figs. 201 and 202.)

8. 4800-V distribution. This is a delta-connected industrial system for layouts of substations supplying motors and lighting transformers.

9. 7200-V, 3-phase distribution. This is another industrial system used with substations for stepping voltage to lower levels for power and lighting.

10. 13.2Y/7.2-kV (or 13.8-kV), 3-phase, 4-wire distribution. This is a modern distribution system, widely used for large industrial plants. Power at this voltage is delivered to substations which step it to 480 V for motor loads and which supply 480-to-120/240-V or 480-to-120/208-V transformers for lighting. Or, 480Y/277-V substations may be used to supply motor loads and 277-V incandescent, fluorescent, or mercury-vapor lighting for office and industrial areas. Lighting transformers are then used to supply 120 V for lighting and receptacles. Figure 203 shows how this voltage level is used for distribution in a high-rise apartment building.

Power feeders for modernized distribution in existing plant

The outdoor enclosure houses a 4000-amp, load-break, 480-volt, 3-pole disconnect switch.

A 4000-amp bus-stub on the line side of the switch provides connection to the utility owned substation. The 4000-amp busway on the load side feeds into the building at the same elevation as the system inside.

Adequate fault and overcurrent protection is offered by the use of current-limiting fuses in the main and branch feeder circuit bus-plug switches. Proper selection and location in the system of these fuses provides a coordinated protective arrangement that will isolate faulted circuits without affecting the operation of equipment on other circuits in the system.

Plug-in busway distribution centers have not been used exclusively throughout this plant. In areas of fixed load, such as air conditioning machinery, motor-generator sets and plant lighting transformers, fusible-switch distribution panelboards have been used. These are wall-mounted types having capacity for future load growth.

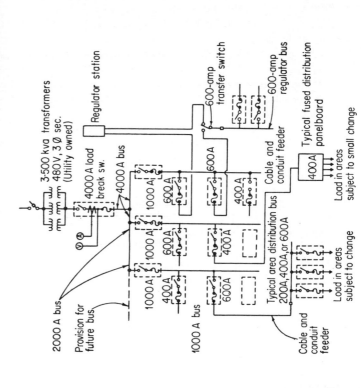

3-500 kva transformers
480V, 3 Ø sec.
(Utility owned)

Regulator station

4000 A load break sw.

4000 A bus

2000 A bus

Provision for future bus.

1000 A bus

1000 A

400 A

600 A

Cable and conduit feeder

1000 A

1000 A

600 A

400 A

600 A

400 A

600 A

600 A

400 A

600-amp transfer switch

600-amp regulator bus

Cable and conduit feeder

Typical fused distribution panelboard

400 A

Typical area distribution bus 200A, 400A, or 600A

Load in areas subject to change

Load in areas subject to change

Load in areas subject to small change

Underground feeders for 480-volt mercury-vapor ballasts

At the main distribution center, combination fused-switch motor starters are used to control and protect the underground feeder runs to the branch-circuit panels. These combination starters are used without overload relay elements and each functions simply as a fused switch and magnetic contactor. Each of two of the distribution feeders supplies three single-phase, 2-wire, 480-volt branch-circuit loadcenters. Each of the other two 480-volt distribution feeders supplies one 3-phase CB distribution center.

At each branch-circuit distribution center, a two-pole circuit breaker is used to control and protect Type UF branch-circuit conductors run underground to the lighting poles.

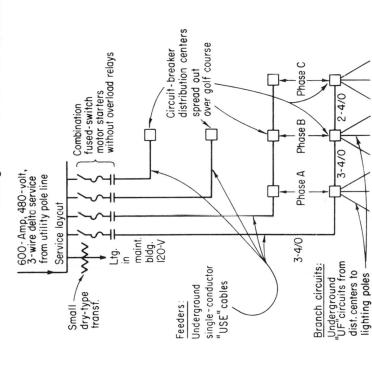

FIG. 198 480-V, 3-phase, 3-wire ungrounded distribution offers a high degree of power continuity.

POWER AND LIGHT FEEDERS
IN NEW INDUSTRIAL PLANT

Two of the transformers in the outdoor substation step 13.2 kv down to 480/277.
The third outdoor transformer steps 13.2 down to 2400 volts, 3-phase, 3-wire delta for circuits to three large machine motors.

FIG. 199 480/277-V distribution plus 2400-V delta serves lighting and heavy power loads.

Distribution layout for multibuilding computer center

One of the 13.8-kv feeders is routed to the 120/208-volt unit substation in the basement of the cafeteria building, and the second feeder is routed to the outdoor transformer mat for the 1000-kva and 1500-kva transformers rated for 480/277 volts on their secondary. Although the major lighting and power load in the cafeteria building was designed for 480/277 volts, 3-phase, 4-wire, the very substantial load of electric kitchen cooking equipment was rated at 120/208 volts, which would require large, high-current circuits with large conductor sizes to minimize voltage drop in the very long runs from the outdoor transformer mat to the cafeteria building. To eliminate the costly runs of secondary feeders for the cafeteria cooking equipment, the plan called for installing a 13.8-kv-to-120/208-volt transformer substation in the basement of the cafeteria building to supply the kitchen loads.

FIG. 200   480/277-V plus 13.8-kV distribution provide effective design for long feeders.

The voltage values given for these distribution systems are subject, of course, to the usual variation or spreads due to distance of transmission and distribution, local conditions of utility supply, and settings of transformer taps. Figure 204 shows how a high-voltage distribution system can be derived from a 120/208-V service. This is the service arrangement for a multibuilding suburban hospital facility. The utility company would supply power only at 120/208 V for this modernization project, but the engineer provided for immediate step-up of the 120/208 V to 4160Y/2400 V to supply an underground distribution system. The cost of the step-up transformers was more than offset by the savings due to high-voltage distribution.

There are times when the selection of distribution secondary voltages other than those described offers a better application, or is required. (See Figs. 205 and 206.) Other high-voltage systems may operate at 6.6, 8.3, 11, and 12 kV.

In all systems, operating at various voltage levels, there may be a need for equipment to modify the electrical characteristics of circuits. The characteristics of the distribution voltage often have to be modified or refined for such purposes as those indicated in Fig. 207. In such cases, the designer must analyze the load served from the various types of equipment and determine the magnitude and characteristics of each load as it is reflected back into the distribution system. Manufacturers' instructions should be carefully followed, and the engineering assistance of the manufacturer should be sought in extensive or complex installations.

Of the high-voltage (over 600 V) distribution systems, 4160- and 13,200-V systems are the most common; they represent good design selection and economy of application for most installations. However, selection of the primary voltage depends upon the size and layout of the load and the supply voltages which the utility makes available. Consultation with the utility must precede any design for high-voltage distribution.

Feeders to basement and roof vaults in high-rise apartment building

The basement vault supplies basement loads—oil burners, air-conditioning equipment, a 75-hp fire pump, garage loads, pool loads, laundry equipment, etc.—and risers to apartment panels, up to and including the first 21 floors of the building. The roof vault supplies typical roof loads, elevator motors, exhaust fans, cooling tower fans and mechanical auxiliary motors—and feeders running down to apartment panels from the 33rd floor down to, and including, the 22nd floor. Thus, the entire building load is clearly divided between the two vaults. The only secondary-voltage connection between the basement service and the roof service consists of two small conduit circuits—one to connect the secondary-voltage electric meter on the roof to a totalizing meter at the basement service, and one control conduit between air-conditioning equipment in the basement and the cooling tower on the roof.

FIG. 201 4160-V distribution minimizes voltage drop and affords great equipment economy.

FIG. 202   4160-V risers to loadcenter transformers provide low losses and low voltage drop.

Modern design practice has favored the use of 13-kV systems over other high-voltage distribution systems for large industrial plants, and the use of 4160/2400-V systems for spread-out commercial and institutional buildings. The total system load, layout, and load density will dictate the best high voltage for any specific system.

For any given system voltage configuration, the industry has more or less standardized on three different voltage values which may be used to describe the system: (1) the transformer voltage, (2) the nominal system voltage, and (3) the utilization voltage. For a given system these three levels account for the voltage drop from the source to the load. As shown in Fig. 208, for instance, the 480/277-V system has 480 V at the source or transformer; 460 V is the nominal system voltage; and 440 V is delivered to the 440-V-rated motor loads.

Distribution Design

The design of a distribution system is a matter of selecting circuit layouts and equipment to accomplish electrical actions and operations necessary for the conditions of voltage, current, and frequency required by the loads. This means that such factors as service voltage, distribution voltage or voltages, conductors, transformers, converters, switches, protective devices, regulators, and power-factor corrective means must be related to economy, load conditions, continuity of

Top and bottom feed for 776-unit apartment building

To provide high reliability of service, a dual 13.2-kv supply is provided. If service is interrupted on either utility line, a tie circuit breaker in incoming switchgear allows the alternate feeder to power entire building. Six high-voltage feeders carry 13.2-kv to unit substations, which are located in the basement and on the roof of each building wing. Total substation transformer capacity is 4250 kva. Basement subs supply bottom ten floors of building. Roof subs supply top ten floors.

FIG. 203   13.2-kV distribution supplies upper and lower halves of high-rise building.

service, operating efficiency, and future power requirements. The factors of capacity, accessibility, flexibility, and safety, which were discussed previously, must be carefully considered in designing for distribution.

Basic rules for distribution design are as follows:

1. Determine the magnitude and characteristics of all individual loads and load groupings.
2. Locate the one or more electric supply points as near as possible to building loadcenters.
3. Select and arrange feeders and other distribution equipment to provide the power continuity required by the commercial or industrial functions of the building.
4. Constantly relate the building's electrical requirements to the system characteristics of flexibility, accessibility, and regulation.
5. Provide a calculated amount of spare capacity in all system components from the supply to the load devices, carefully correlating feeder and subfeeder spare capacities to realistic demand expectations.
6. Use modern loadcenter layouts where possible.
7. In design calculations, observe minimum and maximum **NEC** values for conductors, conduit, protective devices, switches, and control equipment. Note that **NEC** standards must frequently be surpassed to obtain required convenience, flexibility, and effectiveness, and greater safety.

In an electrical system, the *feeders* are those conductors (usually insulated conductors in conduit, EMT, or busway) that carry electric power from the service equipment (or generator switchboard, where power is generated on the premises) to the overcurrent protective devices for

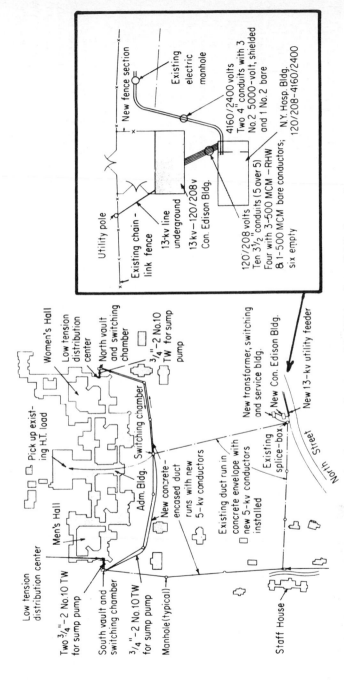

New fence section

Existing electric manhole

4160/2400 volts
Two 4" conduits with 3
No.2 5000-volt, shielded
and 1 No.2 bare

Utility pole

Existing chain-
link fence

13-kv line
underground

13kv—120/208v
Con. Edison Bldg.

120/208 volts
Ten 3½" conduits (5 over 5)
Four with 3–500 MCM – RHW
& 1–500 MCM bare conductors;
six empty

N.Y. Hosp. Bldg.
120/208–4160/2400

Women's Hall

Low tension
distribution center

North vault
and switching
chamber

³/₄"–2 No.10
TW for sump
pump

Pick up exist-
ing H.T. load

Switching chamber

Adm. Bldg.

New concrete-
encased duct
runs with new
5–kv conductors

New transformer, switching
and service bldg.

New Con. Edison Bldg.

New 13–kv utility feeder

North Street

Existing
splice-box

Existing duct run in
concrete envelope with
new 5–kv conductors
installed

Men's Hall

Low tension
distribution center

Two ³/₄"–2 No.10 TW
for sump pump

South vault and
switching chamber

³/₄"–2 No.10 TW
for sump pump

Manhole (typical)

Staff House

FIG. 204 Low-voltage service can often be stepped up to provide a high-voltage distribution system.

Two-stage step-down transformer hookup serves the ungrounded circuits required in a hospital operating room. The NE CODE generally requires that circuits in areas used for anesthetizing—operating rooms, delivery rooms, anesthesia rooms and corridors and utility rooms used for administering flammable anesthetics—must be supplied from an ungrounded distribution system which is isolated from any distribution system supplying areas other than anesthetizing locations [NE CODE Section 517-104.]. Such isolation is generally provided by means of two-winding transformers with no electrical connection between primary and secondary (that is, primary to secondary energy transfer must be made only by magnetic induction). It would seem, just from this, that the circuits could be provided by simply installing single-phase transformers for 480 to 120/240 volts or 3-phase transformers for 480-volt delta to 120/208-volt wye. But there is a further complication.

Section 517-104 of the NE CODE states that: "Circuits supplying primaries of isolating transformers [as required by Section 517-104.] shall operate at not more than 300 volts between conductors. . . ." This rules out the use of any transformer with a 480-volt primary for stepping down to the ungrounded distribution system for anesthetizing areas. As a result, a two-stage transformation is necessary to obtain the ungrounded system from the 480-volt delta system. This transformation is made in one of the roof penthouses where a 480-volt 3-wire feeder from the main switchboard comes into a hookup of three 5-kva, single-phase transformers connected for 480-volt delta down to 240-volt delta. The secondary conductors are brought into a trough from which single-phase, 2-wire subfeeds are

tapped to single-phase transformers (2 and 3 kva sizes) connected for 240 volts down to 120 volts. These small transformers provide required isolation of their secondary, 2-wire ungrounded subfeeds to panels for the ungrounded branch circuits. The small single-phase transformers are divided among the three phases of the 240-volt delta secondary of the 3-phase transformer bank.

FIG. 205 Special distribution voltage levels are sometimes dictated by job conditions, such as a hospital's need for 240-V, 3-phase ungrounded feeds to isolation transformers.

Output of emergency generator was selected at 600 volts to obtain the advantages of 600-volt distribution for the long feeder runs to widespread distribution centers. Use of 600-volt, 3-phase, 3-wire makeup permits use of smaller wires (due to reduced current) and smaller conduits than would be needed to deliver the same power at 208 volts or even 480 volts.

FIG. 206 600-V distribution offers maximum economy and minimum current by utilizing conductors and system equipment at their 600-V operating limit.

FIG. 207 Characteristics of the building distribution system must often be altered to supply special load equipment.

FIG. 208 Nominal voltage levels vary throughout a distribution system.

branch circuits supplying the various loads. *Main feeders* originate at the service equipment and may supply switchboards or panelboards where either subfeeders or branch circuits originate. *Subfeeders* originate at a switchboard, panelboard, or distribution center other than the service equipment or generator switchboard and supply one or more other distribution panelboards, branch-circuit panelboards, or branch circuits.

Feeders and subfeeders are sized to provide sufficient power to the circuits they supply. For the given circuit voltage, they must be capable of carrying the amount of current required by the load, plus any current which may be required in the future. The selection of the size of a feeder depends upon the size and nature of the known load, computed from branch-circuit data, anticipated future load requirements, and voltage drop. This represents one of the most important engineering tasks in electrical design work. The economy and efficiency of system operation and maintenance depend heavily on the selection of the proper size of feeders.

Experience with today's electrical modernization work has revealed feeder capacity as the big bottleneck in rewiring old buildings—and even some that are not so old. In those buildings, the cost of raising the capacity of the electrical system to meet modern load requirements would be considerably less if the original design of the feeders had been based on sound study of known and anticipated future loads. All this experience confirms the importance of careful sizing of the feeders, through calculations related to the particular conditions of the job and not just by a mechanical procedure of adding up load watts and dividing by volts to get the required current-carrying capacity.

MINIMUM FEEDER CAPACITY

According to the **NEC**, each feeder must be sized to carry a computed load current that is not less than the sum of the branch-circuit load currents it supplies, with certain qualifications:

1. For general illumination, a feeder must have sufficient capacity to carry the total load of the lighting branch circuits, determined as part of the lighting design; however, regardless of the actual connected branch-circuit loads, the feeder conductors must have an ampacity that is not less than the minimum branch-circuit load determined on a watts-per-square-foot basis from the table given in **NEC** Section 220-2(b), as shown in Fig. 209. (See Section 10 of this handbook for tables of unit loads in watts per square foot for hundreds of general lighting applications in commercial and industrial buildings.) Of course, if the actual connected load supplied by a feeder is greater than the total minimum required branch-circuit load computed on a watts-per-square-foot basis, then the feeder conductors must have ampacity at least equal to the actual connected load.

    When a feeder supplies a continuous load or a combination of continuous and noncontinuous loads, the rating of the feeder overcurrent device must not be less than the noncontinuous load plus 125 percent of the continuous load. However, if the overcurrent protection for the feeder is an assembly approved for operation at 100 percent of its rating, the ampacity of the feeder must simply be at least equal to the sum of the continuous load and the noncontinuous load.

    NEC Section 220-3 makes clear that a feeder supplying a branch-circuit panelboard must have a minimum ampacity sufficient to serve the calculated total load of lighting, appliances, motors, and other loads supplied. And the amount of feeder and panel ampacity required for the general lighting load must not be less than the current value determined from the circuit voltage and the total power (watts) resulting from multiplying the minimum unit load from Table 220-2(b) (in watts per square foot) by the area of the occupancy supplied by the feeder—even if the actual connected load is less than the calculated load determined on a watts-per-square-foot basis. But, as noted above, if the connected load is greater than that calculated on a watts-per-square-foot basis, the greater value of load must be used in determining the number of branch circuits, the panelboard capacity, and the feeder capacity. (Refer back to the discussion, under "Branch-circuit panelboards," covering the minimum sizing of feeder conductors where the total "computed" load is greater than the actual connected load.)

    It should be carefully noted that the first sentence of Section 220-3(d) states, "Where the load is computed on a watts-per-square-foot (0.093 sq m) basis, the wiring system up to and including the branch-circuit panelboard(s) shall be provided to serve not less than

Table 220-2(b).   **General Lighting Loads by Occupancies**

Type of Occupancy	Unit Load per Sq. Ft. (Watts)
Armories and Auditoriums	1
Banks	3½**
Barber Shops and Beauty Parlors	3
Churches	1
Clubs	2
Court Rooms	2
*Dwelling Units	3
Garages — Commercial (storage)	½
Hospitals	2
*Hotels and Motels, including apartment houses without provisions for cooking by tenants	2
Industrial Commercial (Loft) Buildings	2
Lodge Rooms	1½
Office Buildings	3½**
Restaurants	2
Schools	3
Stores	3
Warehouses (storage)	¼
In any of the above occupancies except one-family dwellings and individual dwelling units of multifamily dwellings:	
Assembly Halls and Auditoriums	1
Halls, Corridors, Closets	½
Storage Spaces	¼

For SI units: one square foot = 0.093 square meter.

* All receptacle outlets of 20-ampere or less rating in one-family and multifamily dwellings and in guest rooms of hotels and motels [except those connected to the receptacle circuits specified in Section 220-3(b)] shall be considered as outlets for general illumination, and no additional load calculations shall be required for such outlets.

** In addition a unit load of 1 watt per square foot shall be included for general purpose receptacle outlets when the actual number of general purpose receptacle outlets is unknown.

FIG. 209 Feeder conductors must always have capacity for the NEC minimum branch-circuit load.

the calculated load." Use of the phrase "wiring system up to and including" requires that a feeder must have capacity for the total minimum branch-circuit load, determined from the area in square feet times the minimum unit load [watts per square foot from Table 220-2(b)]. And that phrase clearly requires that amount of capacity to be allowed in every part of the distribution system supplying the load. The required capacity would, for instance, be required in a subfeeder to the panel, in the main feeder from which the subfeeder is tapped, and in the service conductors supplying the whole system.

Reference to the "wiring system" in Section 220-3(d) presents a requirement that goes beyond the heading "Branch Circuits Required" of Section 220-3 and, in fact, constitutes a requirement on feeder capacity that supplements the rule of the second sentence of Section 220-10(a). This requires a feeder to be sized to have enough capacity for "the computed load" as determined by Part **A** of this article (which means as computed in accordance with Section 220-2).

2. In sizing feeder conductors on the basis of a watts-per-square-foot calculation of the minimum required ampacity of a feeder, the demand factors given in **NEC** Table 220-11 may be applied to the total calculated branch-circuit load to get a reduced value of minimum required feeder capacity for lighting (but those demand factors must not be used in calculating branch-circuit capacity; Fig. 210).

Table 220-11.   Lighting Load Feeder Demand Factors

Type of Occupancy	Portion of Lighting Load to Which Demand Factor Applies (wattage)	Demand Factor Percent
Dwelling Units	First 3000 or less at....................... Next 3001 to 120,000 at .... Remainder over 120,000 at	100 35 25
*Hospitals	First 50,000 or less at .... Remainder over 50,000 at..............	40 20
*Hotels and Motels — Including Apartment Houses without Provision for Cooking by Tenants	First 20,000 or less at .... Next 20,001 to 100,000 at .... Remainder over 100,000 at	50 40 30
Warehouses (Storage)	First 12,500 or less at .... Remainder over 12,500 at..............	100 50
All Others	Total Wattage.............................	100

* The demand factors of this table shall not apply to the computed load of feeders to areas in hospitals, hotels, and motels where the entire lighting is likely to be used at one time, as in operating rooms, ballrooms, or dining rooms.

FIG. 210 Demand factors may be applied to the total branch-circuit load in sizing feeders.

3. If show-window lighting is supplied by the feeder, capacity must be included in the feeder to handle 200 W per linear foot of show-window length.

4. In single-family dwellings, in individual apartments of multifamily dwellings with provisions for cooking by tenants, or in a hotel or motel suite with cooking facilities or a serving pantry, at least 1500 W for each 2-wire, 20-A small-appliance circuit (to handle the small-appliance load in kitchen, pantry, and dining areas) must be added to the general lighting load on each feeder. The total small-appliance load may be added to the general lighting load and may be reduced by the demand factors given in **NEC** Table 220-11.

5. A feeder load of at least 1500 W must be added for each 2-wire, 20-A laundry circuit installed as required by Section 220-3(c). And that load may also be added to the general lighting load and reduced according to the demand factors of Table 220-11.

6. In dwelling units, receptacle outlets of 20 A or less, other than receptacles on the two or more required 20-A small-appliance circuits in kitchens and dining rooms, are considered to be outlets for general lighting. In other than dwelling occupancies, the branch-circuit load

for receptacle outlets that are taken as a load of 180 VA per outlet may be added to the general lighting load and may also be reduced by the demand factors in Table 220-11.

Prior to the 1978 **NEC**, Section 220-13 required a feeder to have ampacity for a load equal to 180 VA times the number of general-purpose receptacle outlets that the feeder supplies, and a feeder supplying such a load had to have a capacity for 100 percent of that total load in occupancies that come under the heading "All Others" at the bottom of Table 220-11 (such as office buildings, stores, and schools). But the **NEC** no longer requires that approach. Recognizing that there is great diversity in the use of receptacles in office buildings, stores, schools, and the occupancies that come under "All Others" in Table 220-11, Section 220-13 contains a table to permit reduction of the feeder capacity for receptacle loads on feeders in such occupancies (Fig. 211). The minimum required feeder capacity for a typical case where, say, an office-building feeder supplies two or more panelboards feeding a total of 500 receptacles is shown in Fig. 212.

Table 220-13
Demand Factors for Nondwelling Receptacle Loads

Portion of Receptacle Load to which demand factor applies (wattage)	Demand Factor Percent
First 10 kW or less	100
Remainder over 10 kW at	50

FIG. 211 Calculated receptacle load may be reduced by demand factors.

Although the calculations of Fig. 212 cannot always be taken as realistically related to usage of receptacles, it is relief from the 100 percent demand factor, which presumed that all receptacles were supplying 180-VA loads simultaneously in those occupancies that came under "All Others."

If the 500 receptacles in Fig. 212 were in a hospital instead of an office building, the capacity of a feeder supplying that total receptacle load could be determined with the demand factors of Table 220-11. As shown, the feeder capacity for that load in a hospital would have to be only 28,000 W, instead of 50,000 W.

7. Feeder capacity must be allowed for "household" electric cooking appliances rated over 1¾ kW, in accordance with **NEC** Table 220-19. Feeder demand loads for a number of cooking appliances on a feeder may be obtained from Table 220-19.

8. "Commercial" electric cooking loads must comply with Section 220-20 and its table of demand factors.

9. For fixed appliances (fastened in place) other than ranges, clothes dryers, air-conditioning equipment, and space-heating equipment, feeder capacity in dwelling occupancies must be provided for the sum of their loads, and the total load of four or more such appliances may be reduced with a demand factor of 75 percent (**NEC** Section 220-17).

10. **NEC** Article 430 (Sections 430-24 to 430-26) gives the required capacity of a feeder that supplies motor loads as well as appliance and lighting loads as described above. The feeder capacity for continuous-duty motor loads is taken at 125 percent of the full-load current rating of the largest motor supplied plus the sum of the full-load currents of the other motors supplied. For a motor load in which some or all of the motors do not operate continuously (such as motors on short-time, intermittent, periodic, or varying-duty cycles), the feeder capacity must be sized according to the procedure set forth in Section 430-24 or 430-25.

Take the total number of general-purpose receptacle outlets fed by a given feeder. . .

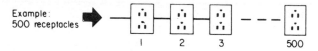

Example:
500 receptacles

. . . multiply the total by 180 voltamperes [required load of Section 220-2(c) (4) for each receptacle]. . .

500 × 180 VA = 90,000 VA

. . .and take the result as a load of unity power factor. (This is not mentioned in the Code rule.)

90,000 VA = 90,000 watts

Then apply the demand factors from Table 220-13:

First 10 kw or less @ 100% demand   **= 10,000 watts**
Remainder over 10 kw @ 50% demand
  = (90,000 – 10,000) × 50%
  = 80,000 × 0.5   **= 40,000 watts**
                Minimum demand-load total **= 50,000 watts**

Therefore, any feeder (or service-entrance conductors) that supplies those receptacles as part of its total load must include a current-carrying capacity of at least 50 kW for the total receptacle load. Required minimum ampacity for that load is then determined from the voltage and phase makeup (single or 3 phase) of the feeder.

But if the 500 receptacles are in a hospital:

500 receptacles × 180 VA = 90,000 VA = 90,000 watts

Then apply the demand factors from Table 220-11:

First 50 kW or less @ 40% demand   **= 20,000 watts**
Remainder over 50 kW @ 20% demand
  = (90,000 – 50,000) × 20% = 40,000 × 20%   **= 8,000 watts**
                Minimum demand-load total = **28,000 watts**

FIG. 212 How demand factors are applied to office or other nondwelling areas.

11. The capacity of a feeder supplying fixed electric space-heating equipment is determined on the basis of a load equal to the total connected load on all branch circuits served from the feeder. Under conditions of intermittent operation or where all units cannot operate at the same time, permission may be granted for use of less than a 100 percent demand factor in sizing the feeder. Sections 220-30 to 220-32 permit alternative calculations of electric heating loads in dwelling units.

12. Feeder capacity must be included for air-condiitoning equipment loads. Figure 213 shows unit loads that can be used in determining the total load, and that is usually taken as a continuous load at 100 percent demand factor. When dissimilar loads (such as space heating and air cooling) are supplied by the same feeder, the smaller of the two loads may be omitted

Type of Building	Watts per Square Foot of Floor Area of Conditioned Space		
	Low	Average	High
Banks	4	5	7
Department Stores			
Basement	4	5	10
Main Floor	5	7	12
Upper Floors	3	4	6
Hotel Guest Rooms			
(per room)	1	1.5	2
Office Buildings	3	4	6
Office Suites	3	3.5	5
Stores	2 to 6	4 to 7	5 to 10
Restaurants	10	15	20
Theatres (per seat)	65	75	80

FIG. 213 Air-conditioning unit loads can facilitate load calculations and feeder sizing.

from the total capacity required for the feeder if it is "unlikely" that the two loads will operate at the same time. Note that the rule of **NEC** Section 220-21, which permits elimination of the smaller of two loads in calculating minimum required feeder capacity, does not require that a double-throw switch or similar control arrangement be provided to prevent simultaneous connection of, say, a heating load and an air-conditioning load to the feeder where the feeder is sized for only the larger of the two loads. The rule simply refers to "dissimilar loads" and requires only that it be "unlikely" that both loads are connected at the same time. Of course an "either-or" control arrangement can prevent overload and undesired opening of the feeder protective device, as shown in Fig. 214.

13. The feeder load for each household electric clothes dryer in a dwelling unit must be 5000 W or the nameplate rating, whichever is larger. The demand factors in Table 220-18 may be used where a feeder supplies a number of such dryers.

Panelboard supplying electric heating loads

122-A load   83-A load

Manual or automatic transfer switch . . .

. . . but this could be simply a junction box to satisfy NEC rule in Sec. 220-21.

Feeder control and protection

Single feeder sized for larger (122 A) of two dissimilar loads

FIG. 214 Transfer switching hookup (such as "heating/air conditioning") ensures proper circuit operation where a single feeder supplies dissimilar loads that should not be energized simultaneously.

It should be noted that all the foregoing are general **NEC** requirements that cover minimum load conditions. This is particularly true of the watts-per-square-foot unit loads given in the table, which are based on 100 percent power factor. Feeder sizes determined solely on the basis of **NEC** requirements include no provision for the particular operating conditions of an occupancy, such as voltage stability and power losses, and no allowance for future growth in the load served by the feeder.

Although certain feeder applications can be satisfactorily sized with the **NEC** method, modern feeder design must carefully incorporate such factors as voltage drop, power factor, detailed analysis of watts-per-square-foot loads, realistic analysis and application of demand factors, and spare-capacity requirements.

Adding Branch-Circuit Loads

Based on the foregoing **NEC** rules, the required capacity of any feeder can be determined by adding up the branch-circuit loads supplied by the feeder. In earlier discussions of branch circuits, procedures were given for determining branch-circuit loads—for general lighting, fixed appliances, special receptacles, and general-use convenience receptacles—and for determining the number and characteristics of circuits required to handle the loads.

Whether general lighting loads are determined on a watts-per-square-foot basis or from a lighting design layout, the load for general illumination is established, and capacity for this load must be designed into the feeder, modified by the applicable demand factor from Table 220-11. Other circuit loads (such as for local and/or special lighting, motors, electric cooking appliances, and fixed appliances) are also generally known and can be accounted for in a feeder. Loads for convenience receptacle circuits and spare circuits in the panelboard, however, are not known and must be estimated in sizing a feeder.

From the known and estimated branch-circuit loads, the minimum capacity for a feeder to a number of branch circuits can be determined as follows:

1. For each multioutlet branch circuit supplying general lighting, allow voltamperes equal to the load on the circuit and not less than the minimum branch-circuit load set by the minimum watts per square foot of branch-circuit capacity—as described previously for **NEC** Sections 220-2 and 220-3(d).

2. For each multioutlet plug-receptacle branch circuit, allow voltamperes equal to the load on the circuit as used in branch-circuit design. The minimum allowance for each 120-V receptacle circuit must equal the number of receptacles (count a duplex receptacle as one) times 1.5 A times 120 V, unless a demand factor is considered acceptable for the expected use of the receptacles and is applied in accordance with Section 220-13 as described above.

    Where multioutlet assemblies (plug-in strip or raceway) are used, each 5 ft or fraction thereof of each separate and continuous length is taken as a load equal to 1.5 A. And when such assemblies are located where a number of appliances are likely to be used simultaneously, each 1 ft or fraction thereof is taken as a load of 1.5 A.

3. For each multioutlet branch circuit supplying heavy-duty lampholders, allow voltamperes equal to the load on the circuit as used in branch-circuit design. The minimum allowance for each heavy-duty lampholder circuit must equal the number of outlets times 600 W (which is equal to 600 VA at unity power factor).

4. For each multioutlet branch circuit supplying local or special lighting, allow voltamperes equal to the load on the circuit. **NEC** Section 220-2(c) establishes the minimum loads that must be allowed in computing the minimum required branch-circuit capacity for general-use receptacles and "outlets not used for general illumination." Item (3) of that section requires that the actual voltampere rating of a recessed lighting fixture be taken as the amount of load

that must be included in branch-circuit capacity. This permits local and/or decorative recessed lighting fixtures to be included at their actual load values rather than as "other outlets," which would require a load allowance of "180 VA per outlet"—even if such a fixture were lamped at, say, 25 W. (On the other hand, where a recessed fixture contained a 300-W lamp, an allowance of only 180 VA would be inadequate.)

5. For each individual branch circuit supplying a fixed appliance, allow voltamperes equal to the rating of the appliance.

6. For the sum of the loads of individual branch circuits supplying electric cooking appliances, allow kilowatts equal to the values given in Table 220-19 for the particular number and sizes of cooking appliances.

7. For each individual branch circuit supplying a motor, allow a voltampere load equal to 1.25 times the motor full-load current (from Tables 430-147 to 430-150 according to the horsepower of the motor) times the circuit voltage. For 3-phase motors, this product must be multiplied by 1.732 to obtain the correct load. In sizing conductors for required load currents, the voltampere load should always be used for devices operating at less than unity power factor; this will ensure the required current capacity, which is greater than would be required to supply only the wattage (effective power) represented by such loads.

8. For each branch circuit supplying more than one motor, allow voltamperes by proceeding as follows: Take 1.25 times the full-load current rating of the highest-rated motor; add to this the full-load current ratings of all the other motors; then multiply the sum by the circuit voltage. Again, this product must be multiplied by 1.732 to obtain the correct voltampere product for 3-phase motors.

9. For each branch circuit supplying show-window lighting, allow 200 W for each linear foot (measured horizontally) of show window served by the circuit.

By adding up these computed watts and voltamperes, a total number of volt-amperes is obtained which is the initial minimum feeder load. Of course, the majority of feeders used in commercial, institutional, and industrial buildings will serve only a few of the above-described loads. And feeders in different buildings will handle different combinations of loads.

For each spare branch circuit supplied by a feeder, voltamperes equal to the values given above for multioutlet general lighting branch circuits should be added to the feeder load.

The table in **NEC** Section 220-11 allows limited use of demand factors for general lighting circuits and multioutlet plug-receptacle circuits to reduce the feeder design load; however, most occupancies require a demand factor of 100 percent of the total load as obtained by adding all the circuit loads.

Using Demand Factors

Two terms constantly used in the literature of modern electrical design are *demand factor* and *diversity factor*. Because there is a very fine difference between the meanings for the words, the terms are often confused.

- The *demand factor* is the ratio of the maximum demand of a system (or part of a system) to the total connected load on the system (or part of the system). This factor is always less than unity.

- The *diversity factor* is the ratio of the sum of the individual maximum demands of the various subdivisions of a system (or part of a system) to the maximum demand of the whole system (or the part of the system under consideration). This factor generally varies between 1.00 and 2.00.

Demand factors and diversity factors are used in system design. For instance, the sum of the connected loads supplied by a feeder is multiplied by the demand factor to determine the load which the feeder must be sized to serve. This load is termed the *maximum demand* of the feeder. The sum of the maximum demand loads for a number of subfeeders divided by the diversity factor for the subfeeders gives the maximum demand load to be supplied by the feeder from which the subfeeders are derived.

Tables of demand and diversity factors have been developed from experience with various types of load concentrations and various layouts of feeders and subfeeders supplying such loads. **NEC** Table 220-11 presents common demand factors for feeders to general lighting loads in various types of buildings, as shown in Fig. 210.

It is common and preferred practice in modern electrical design to take unity as the diversity factor in main feeders to loadcenter substations, to provide a measure of spare capacity. Main secondary feeders are also commonly sized on the full value of the sum of the demand loads of the subfeeders supplied.

From power-distribution practice, however, basic diversity factors have been developed. These provide a general indication of the way in which main feeders can be reduced in capacity below the sum of the demands of the subfeeders they supply. On a radial feeder system, the diversity of the demands of a number of transformers reduces the maximum load which the feeder must supply to some value less than the sum of the transformer loads. Typical diversity factors for main feeders are as follows:

- Lighting feeders: 1.10 to 1.50
- Power and light feeders: 1.50 to 2.00 (or higher)

The selection of a diversity factor in any particular case must be based on a study of load characteristics and operating cycles. Very accurate data on demand diversity are recorded by operating personnel in many types of commercial and industrial buildings.

Another factor used to analyze systems is the *load factor,* which is the ratio of the average load over a particular period of time to the peak load occurring in that period of time.

Figure 215 illustrates the use of demand and diversity factors in sizing system components. It should be noted that the kilovoltampere rating of each of the subfeeders must have the minimum value as described in this section. And any feeder or subfeeder (even the main feeder in the diagram) can be sized directly by totaling all connected loads fed by that particular circuit and applying demand factors where conditions make it necessary or desirable.

Feeder Spare Capacity for Growth

The total feeder load obtained by summing all the branch-circuit loads fed by the feeder represents only the initial provision for branch-circuit loads. Future utilization of the spare capacity designed into the branch circuits (they are loaded only to 50 percent) is not accounted for in that total. And there is no provision for the future possibility of exceeding the maximum capacity of the panelboard itself, so that additional panelboard capacity—either a second panel or a new larger panel—is required.

Present experience with electrical modernization indicates that a little forethought and study will often reveal the advisability of providing spare feeder capacity—perhaps not in all or even most cases, but certainly in many cases. And only careful consideration of the nature of electrical utilization in a particular occupancy will accurately indicate possible future load requirements. But in all electrical designs, at least some spare feeder capacity must be provided.

Allowance for load growth in feeders should begin with the spare capacity designed into the branch circuits. In each circuit loaded to 50 percent of capacity, it can be assumed that there is

1. Sum of individual demands = 240 + 100 + 350 = 690 kva.
2. Sizing the substation at unity diversity, the required

$$kva = \frac{690}{1.00} = 690 \ kva.$$

3. To meet this load, use a 750-kva substation.
4. If analysis dictates the use of a diversity factor of 1.4 , the

$$required \ kva = \frac{690}{1.40} = 492 \ kva.$$

5. To meet this load, use a 500-kva substation.
6. Primary feeder to unit substation must have capacity to match the substation load.

FIG. 215 Application of demand factors and diversity factors must not conflict with provision of spare capacity.

an allowance for load growth equal to 30 percent of the circuit capacity. This allowance is based on the **NEC** limitation of 80 percent load on circuits that are in operation for long periods of time (that is, in continuous loading, for 3 hours or more). On circuits that supply motor-operated appliances in addition to other appliances and/or lighting, if an initial noncontinuous load is limited to no more than 50 percent of the circuit capacity (its ampere rating), there is at least spare capacity equal to 30 percent. The 30 percent growth allowance in all branch circuits should be converted to voltamperes and added to the load on the feeder as calculated above.

This grand total—the sum of the actual or calculated branch-circuit loads plus the allowable load growth—then represents the required feeder capacity to handle the full-circuit loads on the panelboard. The next step is to provide capacity in the feeder for anticipated load growth beyond the capacity of the circuits already provided (plus some amount for possible unforeseen future requirements). Based on experience, modern design practice dictates sizing to allow an increase of 25 to 50 percent in the load on a feeder, to provide for the addition of branch circuits where analysis reveals any load-growth possibilities.

In general, spare-capacity allowances for feeders depend upon the type of building, the work performed, the plans or expectations of management with respect to expansion of facilities or growth of business, the type of distribution system used, the locations of centers of loads, the permanence of various load conditions, and economic conditions. Thorough study of all pertinent factors should indicate the advisability of providing spare capacity for each feeder in a distribution system. Where such study indicates extra feeder capacity, a substantial growth allowance—50 percent—is generally essential to provide economy in future electrical expansion. Skimpy upsizing of feeder conductors or raceways has proved to be a major shortcoming of past electrical design work. This is particularly true in tall office buildings, apartment houses, and other commercial buildings in which the elimination of riser bottlenecks represents a large part of the total modernization cost.

Spare feeder capacity may be provided in one or more of several ways. If it is anticipated that the feeder load will be increased in the near future, the extra capacity should be included in the conductor sizing and installed as part of the initial electrical system. In many small office or commercial buildings and apartment houses, extra capacity for load growth should automatically be included in the feeder conductors when they are initially installed. Here are other ways of providing for growth in feeder loads:

1. Selecting raceways larger than required by the initial size and number of feeder conductors. This allows the conductors to be replaced with larger sizes if required at a later date. This method should be carefully considered for risers or underground or concealed feeder raceway runs and similar permanent installations, where it would be very costly to add conduit runs at a later date. In some cases, a compromise can be made between providing spare capacity in the feeder conductor size and increasing the size of the feeder raceway.

2. Including spare raceways in which conductors may be installed at a later date. Arrangements of multiple raceways must be carefully laid out and related to the existing overall system. The types of feeder distribution centers or main switchboards and the layout of local branch-circuit panelboards must be able to accommodate future expansion of distribution capacity based on the use of spare raceways, with modification and regrouping of feeder loads.

Figure 216 shows the several elements involved in sizing feeder conductors for the initial load plus two types of spare capacity. The feeder, consisting of three phase legs and the neutral, is run in conduit, so the neutral must be counted as a current-carrying conductor because of the third-harmonic current present in a circuit supplying fluorescent lighting [Note 10(c) to **NEC** Table 310-16]. Then Note 8 to the same table requires wire ampacity derating because there are four current-carrying conductors in the conduit. Although No. 4/0 THW copper wire has an allowable ampacity of 230 A (Table 310-16) and might appear satisfactory for the ultimate feeder load of 224 A, Note 8 derates ampacity of No. 4/0 to only 0.8 230 A or 184 A—which is inadequate. The minimum size of conductor that will have a derated ampacity at least equal to the required value of 224 A is 300-MCM THW copper, with an ampacity of 228 A (0.8 × 285 A) under the given conditions. And note how the continuous nature (3 hours or more of operation) of the feeder load involves Sections 220-10(b) and 384-16(a) on panelboard main protection.

Feeder Sizing versus Panelboard Rating

The sizing of feeder conductors for continuous loading (such as lighting in schools, office buildings, stores, and other occupancies where lighting is on all day long) raises problems in panelboard sizing. As discussed under the heading "Branch-circuit panelboards" in Section 2 of this handbook, **NEC** Section 384-13 requires that any panelboard have busbar ampacity at least equal to the **NEC** minimum feeder conductor capacity, required by Article 220. In effect, the clear, straightforward rule of Section 384-13 calls for a panelboard to have a busbar ampacity that is not less than the minimum ampacity required for the feeder conductors that supply the sum of

All 20-A
1-pole CBs in
42-circuit
225-A panel

Total branch-circuit load to
continuous fluorescent lighting
is supplied by 36 CB poles,
each loaded to 10 A.

1. Initial feeder load is 10 A × 12
poles per phase, or
120 A per feeder phase leg.

2. Spare feeder capacity
to handle a maximum allowable
continuous load of 16 A per circuit
pole is 6 A × 12 poles per phase
or **72 A** per phase

Six spare
pole spaces

3. Spare feeder capacity to handle
later addition of 2 poles per phase
loaded to 16 A is 16 A × 2 poles
per phase, or **32 A** per phase

N

Required feeder total capacity:
120 A + 72 A + 32 A = **224 A** per phase.

Use 300MCM THW copper conductors
rated at 285 A but limited to loading
of 80% × 285, or 228 A.

For ultimate loading:
To protect 228-A feeder conductors and to satisfy as main protection for
225-A panel, feeder protection must be rated at 225 A and UL-listed for
continuous load of 225 A (100% rated) [**NEC** Sec. 220-10(b)]

FIG. 216 For real spare capacity, the feeder and all circuit components must be sized to carry the ultimate load.

the branch-circuit loads fed from the panel. When the minimum required feeder ampacity is
calculated for the total of branch-circuit loads in accordance with Section 220-10, the panelboard
supplied by that feeder must have busbars with an ampacity rating at least equal to the minimum feeder ampacity. But here is where a controversy starts for feeders with continuous (3
hours or more) loading:

1. Section 220-10(b) sets two different possible minimum values for feeder ampacity, depending
   upon the type of overcurrent device protecting the feeder.

2. If the circuit breaker or fusible switch protecting the feeder conductors is not "listed" (such
   as by UL) for continuous operation at 100 percent of its current rating, the CB or fuses in
   the switch must be rated at least at 125 percent of the continuous load current; the feeder
   conductors must also have ampacity not less than 125 percent of the continuous load current.
   If the feeder protective device is listed for 100 percent continuous loading, then the minimum
   required feeder conductor ampacity is simply the feeder load current.

3. As a result of the interaction of Sections 384-13 and 220-10(b), the minimum required panel
   busbar rating depends on the type of protective device used in the feeder that supplies the
   panel.

4. As shown previously in Fig. 104, a continuous minimum feeder load current of 176 A must have feeder conductors (and therefore a panelboard busbar rating) of not less than 125 percent of 176 A, or 220 A, if the CB or fusible switch to the panel is not listed for 100 percent continuous loading. Only a very small percentage of feeder protective devices are so listed, and then only in CB and switch ratings of 600 A and above.

5. As shown in Fig. 217, the impact of these rules on panelboard busbar rating is to produce illogical conditions, because panelboard sizing is not related to the reason behind the 125 percent multiplier in Section 220-10(b). A circuit breaker or fusible-switch assembly that is not listed for 100 percent continuous loading would generate excessive heat within its own parts if it were operated at its full current rating continuously, and such heat would damage conductor insulation. For that reason, the 125 percent multiplier of Section 220-10(b) establishes the same limitation as the general UL rule that (unless marked otherwise) a circuit breaker must not be loaded continuously to more than 80 percent of its rating (which is the reciprocal of having a rating of at least 125 percent of the continuous load current). Circuit breakers and fusible switches that are listed for 100 percent continuous loading are constructed and/or ventilated to prevent the generation of excessive heat.

Branch-circuit loading of 224 A per phase busbar
in panel for continuous lighting on 480/277-V system

Minimum panel rating
of 225 A (Sec. 384-13)

Minimum panel rating
of 400 A (Sec. 384-13)

No. 4/0 THW copper conductors
rated 230 A

Feeder

300MCM THW copper
conductors rated 285 A
to satisfy Sec. 220-10 (b)

Feeder

Feeder
overcurrent
device

CB with 225-A trip setting,
UL-listed for continuous operation
at 100% of its rating

Feeder
overcurrent
device

Circuit breaker not listed for
100% continuous duty—
minimum amp rating
= 1.25 × 224 A = 280 A.
300-A CB (400-A frame)
must be used.

CASE A   **CASE B**

FIG. 217 Sizing of feeder conductors and overcurrent protection affects the sizing of panelboards.

6. Although the letter of the **NEC** rules requires the panel busbar ratings shown in Fig. 217, there is no safety compromise in sizing the busbars at no more than 100 percent of continuous load, because application of the panelboard is independent of the conditions that dictate the 125 percent multiplier for feeder protection and feeder conductors. In Fig. 104 above, for instance, sizing of the panel busbars at 125 percent of 176 A calls for a panel of not less than 1.25 × 176 A or 220-A busbar capacity—dictating the use of a 225-A panel. But safety and the rule of **NEC** Section 220-3(d) would be fully served by using a panel with busbars rated for 100 percent of the minimum "computed" load of 176 A (although a 225-A panel is the next larger standard size above 176 A). The use of a panel rated at 125 A (on the basis that the actual load is only 116 A) would be contrary to the requirement of Section 220-3(d), calling for the "wiring system" (including the panelboard) to be sized as a minimum on a watts-per-square-foot basis.

IMPORTANT: In applications such as described above, it is extremely important to distinguish between a "lighting and appliance" panelboard (as defined in Section 384-14) and other panelboards. (See "Panel main protection" in Section 2 of this handbook.) Because a lighting and appliance panelboard requires main overcurrent protection (on the supply side of the panel) rated not over the panel busbar rating, when the rating of the feeder overcurrent device is increased to 125 percent of the continuous feeder load, a panel for which the feeder overcurrent device serves as main protection must have a busbar rating not lower than the current rating of the feeder protection—as shown in case B of Fig. 217. The feeder CB may not be rated less than 300 A, and that value of protection would not satisfy as "mains protection" for a 225-A panel at the load end of the feeder. The next standard size of panelboard above 225 A is 400 A, and that would be properly protected by the 300-A feeder CB (or fuses). Of course, a 225-A panel with its own main 225-A CB or fused switch could be used in case B, instead of the 400-A panel. This is a typical example of the need to interrelate and satisfy all **NEC** rules bearing on each design situation.

In case B of Fig. 217, if the panel fed by the feeder were not a lighting and appliance panel but a panel made up of multipole CBs feeding, say, continuous electric-heating loads connected phase to phase, the panel would not require main protection; and a panel rated at 225 A could be used for a total load of 224 A per phase busbar (although that would be contrary to the exact wording of the rule of Section 384-13).

Figure 218 shows how the rules of Sections 220-10(b) and 384-13 combine to dictate the minimum busbar rating of a branch-circuit or feeder panel that is not subject to the mains protection requirement for a lighting and appliance panel. The load on the panel is exactly the

Subfeeder distribution panel with
467 A of continuous load and
316 A of noncontinuous load

**1200-A PANEL
IS NEEDED.**

Sec. 384-13
[and Sec. 220-10(b)]
calls for a panel
rated not less than
467 A × 125%
+316 A or 900 A.

Minimum of
900-A feeder
conductors

800-A feeder
conductors

800-A PANEL IS OK!

Sec. 384-13 calls for
a panel rated at 100%
× 467 A +316 A
or 783 A.

Feeder protection *not* rated
for 100% continuous load—
must be rated at least 900 A

800-A feeder protection
UL-listed for 100%
continuous load

CASE A   **CASE B**

FIG. 218 NEC rules produce puzzling conflicts in sizing panels.

same in cases A and B. But, because the feeder protection in case A is not rated for 100 percent continuous loading, the sizes of the feeder protection, feeder conductors, and panel must be increased because of the continuous load. In case B, the 125 percent multiplier is not required. Because the panel in case A does not require a main protective device at its rating, there would be no problem in having an 800-A panel there instead of the 1200-A (or 1000-A) panel—even though the 900-A feeder protection exceeds the 800-A panel busbar rating. The size of the panel should not be required to be increased to 125 percent of the continuous load simply because of the unrelated heat characteristics of a remotely located feeder protective device. However, the wording of Section 384-13 does dictate such upsizing of the panel busbar rating.

Selecting Conductor Sizes

When the total design power (watts) or apparent power (voltamperes) has been determined and includes all the foregoing provisions, it and the voltage and other electrical characteristics of the feeder (or the total load in amperes) are used to obtain the required current-carrying capacity of the conductors. In addition to the circuit voltage, the feeder conductor size for a particular total load will depend upon the power factor and the voltage drop in the conductors.

In calculations of the feeder design load, care should be taken to distinguish between circuit load values in *watts* and those in *voltamperes*. When load values are in watts, the power factor must be taken into consideration.

The sizes of conductors are related to the total current, and they must provide the capacity to carry the current at the given value of power factor. Of course, at low power factor, the conductors must have more current-carrying capacity to supply a particular load wattage than at high power factor.

The required size of a feeder conductor is found by calculating the feeder load current at the feeder voltage:

$$\text{Feeder current rating} = \frac{\text{feeder load (kw or kva)}}{\text{circuit voltage}}$$

This general formula is expressed exactly in the following two forms:

$$I = \frac{\text{load watts}}{k \times E \times \text{pf}} = \frac{\text{load voltamperes}}{k \times E}$$

where I = current in any line wire which the feeder must be rated to carry (check tables of conductor current ratings), amps

k = 1.0 for 2-wire dc or single-phase ac; 1.73 for 3-wire, 3-phase ac; 2.0 for 3-wire dc or single-phase ac; 3.0 for 4-wire, 3-phase ac

E = voltage between ungrounded wire and neutral or, if no neutral exists, between any two ungrounded wires, volts

pf = power factor of load (decimal value)

NOTE: Whenever a voltage value is used in a formula to determine the current rating, the lowest voltage at which the current might operate should be used; this provides the highest current value which might flow in the circuit and will ensure the selection of conductors with fully adequate carrying capacity.

EXAMPLE

What is the minimum required feeder load current to be used in sizing a feeder for a 20,000-W lighting load that will operate continuously, fed by a 208/120-V, 3-phase, 4-wire feeder circuit and operating at 100 percent power factor?

SOLUTION (using the lower value of voltage, 115 V)

The formula gives

$$I = \frac{20{,}000}{3 \times 115 \times 1} = \frac{20{,}000}{345} = 58 \text{ A}$$

Then, as required by **NEC** Section 220-10(b), that value must be multiplied by 1.25 because the load is continuous (unless a 100 percent rated protective device is used for the circuit):

$$1.25 \times 58 \text{ A} = 73 \text{ A}$$

To meet this capacity requirement, any of the following feeder conductors (from **NEC** Table 310-16) could be used:

Copper: No. 3 TW, THW, RHW, XHHW, RHH, or THHN. (Because of the UL rule, the conductor must be sized for a 60°C ampere rating.)

Aluminum and copper-clad aluminum: No. 2 TW, THW, RHW, XHHW, RHH, or THHN.

EXAMPLE

What feeder current rating is needed in sizing the minimum 480-V, 3-phase, 3-wire circuit to supply a 100-kVA load?

SOLUTION

From the formula,

$$I = \frac{100{,}000}{1.73 \times 440} = \frac{100{,}000}{761} = 132 \text{ A}$$

To meet this capacity requirement, any of the following feeder conductors (from **NEC** Table 310-16) could be used:

Copper: No. 2/0 TW or No. 1/0 THW, RHW, RHH, XHHW, or THHN

Aluminum and copper-clad aluminum: No. 4/0 TW or No. 2/0 THW, RHH, XHHW, or THHN

Figure 219 shows the formulas that can be used to calculate power loss due to I^2R heating in feeder conductors. With today's electric-utility rates, it is often economical to shift to a larger size of conductor when the resulting reduction in energy costs over the first year or two covers the extra cost. (See Section 10 of this handbook for a table of formulas used for electrical calculations.)

Sizing the Feeder Neutral

NEC Section 220-22 covers requirements for sizing the neutral conductor in a feeder. It states that the neutral feeder load shall be the "maximum unbalance" of the feeder load and defines the maximum unbalance as "the maximum connected load between the neutral and any one ungrounded conductor." In a 3-wire, 240/120-V, single-phase feeder, the neutral must have a current-carrying capacity at least equal to the current drawn by the total 120-V load connected between the more heavily loaded hot leg and the neutral.

Because 240-V loads, connected between the two hot legs, do not place any load on the neutral, the neutral conductor of such a feeder must be sized to make up a 2-wire, 120-V circuit with the more heavily loaded hot leg. Actually, the 120-V circuit loads on such a feeder would

Assuming resistivity of conductor metal to be as follows:
K = 12 for circuits loaded to more than 50% (copper)
K = 11 for circuits loaded to less than 50% (copper)
K = 18 for aluminum conductors
For a 2-wire circuit (direct-current or single-phase):

$$P = \frac{2 \times K \times L \times I^2}{CM}$$

For a 3-wire, 3-phase circuit (assuming balanced load):

$$P = \frac{3 \times K \times L \times I^2}{CM}$$

in which, P = power lost in the circuit (watts)
      K = resistivity of conductor metal (circular mil-ohms/foot)
      I = current in each wire of the circuit (amps)
      L = one-way length of the circuit (ft)
  CM = cross-section area of each of the wires in circular mils
When the resistance of conductors is determined from a table, the power loss is calculated
from the relation

$$P = I^2 R$$

FIG. 219 These calculations show power loss in feeder conductors, which may
dictate the use of the next larger size of conductor.

be considered as balanced on both sides of the neutral. The neutral, then, would be the same
size as each of the hot legs if only 120-V loads also were supplied by the feeder. If 240-V loads
also were supplied, the hot legs would be sized for the total load, but the neutral would be sized
for only the total 120-V load connected between one hot leg and the neutral. Figure 220 shows
how phase-to-neutral loading affects the size of the neutral but phase-to-phase loads impose no
neutral load.

Section 220-22 contains three other provisions relative to the sizing of neutral conductors.

1. When a feeder supplies household electric ranges, the neutral conductor may be smaller than
   the hot conductors but must have a carrying capacity at least equal to 70 percent of the
   current capacity required in the ungrounded conductors to handle the load (i.e., 70 percent
   of the load on the ungrounded conductors). Table 220-19 gives the demand loads to be used
   in sizing feeders that supply electric ranges and similar cooking appliances. Figure 221 shows
   this application.

2. For feeders of three or more conductors (3-wire dc; 3-wire, single-phase; and 4-wire, 3-
   phase) a further demand factor of 70 percent may be applied to that portion of the unbalanced
   load in excess of 200 A. That is, in a feeder supplying only 120-V loads evenly divided
   between each ungrounded conductor and the neutral, the neutral must be sized at the same
   load as each ungrounded conductor up to 200 A capacity; however, its size may be reduced
   from the size of the ungrounded conductors for loads above 200 A by adding to the 200 A
   only 70 percent of the amount of load current above 200 A (Fig. 222). It should be noted
   that this 70 percent demand factor is applicable to the "unbalanced load" in excess of 200 A
   and not simply to the total load, which in many cases may include 240-V loads on 240/120-
   V, 3-wire, single-phase feeders or 3-phase loads or phase-to-phase–connected loads on 3-
   phase feeders.

3. This reduction of the neutral load to 200 A plus 70 percent of the current over 200 A does
   not apply to electric-discharge lighting. Figure 223 indicates the difference between feeder-
   neutral sizing for incandescent lighting and that for electric-discharge lighting (fluorescent,

Under unbalanced conditions, with one hot leg fully loaded to 60 amps and the other leg open, the neutral would carry 60 amps and must have the same rating as the loaded hot leg. Thus No. 4 TW hot legs would require No. 4 TW neutral (copper).

The neutral here must carry only the unbalance of the two 50-amp, hot-to-neutral loads and has nothing to do with the two straight 240-volt, 100-amp loads. Neutral must be sized for a maximum of 100 amps— the maximum unbalance from hot to neutral.

FIG. 220   Neutral sizing is determined only by neutral-connected loads.

DEMAND LOAD for 8 10-kw ranges = 23 kw
LOAD ON EACH UNGROUNDED LEG

$$= \frac{23{,}000 \text{ W}}{230 \text{ V}} = 100 \text{ amps}$$
(e.g. No. 3 THW)

$$\frac{\text{Required minimum}}{\text{Neutral capacity}} = 70\% \times 100 \text{ amps} = 70 \text{ amps}$$

FIG. 221   Because of diversity in the use of heater elements, a range neutral may be reduced.

metal-halide, etc.). In a feeder supplying ballasts for electric-discharge lamps, there must not be a reduction of the neutral capacity for the part of the load that consists of electric-discharge light sources. For feeders supplying only electric-discharge lighting, the neutral conductor load must be the same as the phase conductors, no matter how large the total load may be. And for feeders supplying motor loads as well as lighting loads, any loads that do not connect

Each phase leg must be rated at least 400 amps.
Neutral must be rated at least

200 amps + (70% × 200 amps) = 340 amps

FIG. 222   Feeder neutral size may be reduced for resistance loads over 200 A.

to the neutral are not involved in sizing of the neutral, as shown for the 3-phase motor loads in Fig. 224.

Full sizing of the neutrals of such feeders is required because, in a balanced circuit supplying ballasts, neutral current approximating the phase current is produced by third (and other odd-order) harmonics developed by the ballasts (Fig. 225). For large electric-discharge lighting loads, this factor affects the sizing of neutrals all the way back to the service. It also affects the rating of conductors in conduit because such a feeder circuit consists of four current-carrying wires, which requires application of an 80 percent wire-ampacity derating factor. [See Notes 8 and 10(c) to **NEC** Tables 310-16 through 310-19.]

It should be noted that the wording in **NEC** Section 220-22 prohibits reduction of the size of the neutral when electric-discharge lighting is used, even if the feeder supplying the electric-discharge lighting load over 200 A happens to be a 120/240-V, 3-wire, single-phase feeder. In such a feeder, however, the third-harmonic currents in the hot legs are 180° out of phase with each other and, therefore, would not be additive in the neutral as they are in a 3-phase, 4-wire circuit. In the latter circuit, the third-harmonic components of the phase currents are in phase with each other and add together in the neutral instead of canceling out. Figure 226 shows a 120/240-V circuit.

In the case of a feeder supplying, say, 200 A of fluorescent lighting and 200 A of incandescent, there can be no reduction of the neutral below the required 400-A capacity of the phase legs, because the 200 A of fluorescent lighting load cannot be used in any way to take advantage of the 70 percent demand factor on that part of the load in excess of 200 A.

Minimizing Voltage Drop

Voltage drop must be carefully considered in sizing feeder conductors, and calculations should be made for peak load conditions. The voltage drop must be calculated on the basis that full system voltage (for example, 120, 240, or 480 V) is available at the service entrance or transformer secondary supplying the feeder. The voltage drop is the amount of reduction from the nominal supply voltage. Feeder conductors should be sized such that the voltage drop up to the branch-circuit panelboards or the point at which the branch-circuits originate is not more than 1 percent for lighting loads or combined lighting, heating, and power loads, and not more than 2 percent for power or heating loads. Local codes may impose lower limits on voltage drop. Voltage-drop limitations are shown in Fig. 227, as follows:

1. For combinations of lighting and power loads on feeders and branch circuits, use the voltage drop percentages for lighting loads (at the left in Fig. 227).

Incandescent lighting only

Serving an incandescent load, each phase conductor must be rated for 1000 amps. But neutral only has to be rated for 200 amps plus (70% x 800 amps) or 200 + 560 = 760 amps.

Electric discharge lighting only

Because load is electric discharge lighting, there can be no reduction in the size of the neutral. Neutral must be rated for 1000 amps, the same as the phase conductors, because the third harmonic currents of the phase legs add together in the neutral. This applies also when the load is mercury-vapor or other metallic-vapor lighting.

Incandescent plus electric discharge lighting

Each phase leg carries a total of 400 amps to supply the incandescent load plus the fluorescent load. But because there can be no reduction of neutral capacity for the fluorescent and because the incandescent load is not over 200 amps, the neutral must be sized for the maximum possible unbalance, which is 400 amps.

FIG. 223 Type of lighting load affects neutral-conductor sizing.

Incandescent plus motor load

Although 1000 amps flow on each phase leg, only 500 amps is related to the neutral. Neutral, then, is sized for 200 amps plus (70% x 300 amps) or 200 + 210 = 410 amps. The amount of current taken for 3-phase motors cannot be "unbalanced load" and no capacity has to be provided for this in the neutral.

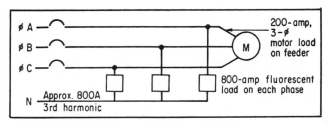

Electric discharge lighting plus motor load

Here, again, the only possible load that could flow on the neutral is the 800 amps flowing over each phase to the fluorescent lighting. But because it is fluorescent lighting there can be no reduction of neutral capacity below the 800-amp value on each phase. The 70% factor for that current above 200 amps DOES NOT APPLY in such cases.

FIG. 224 Phase-to-phase loads do not affect neutral sizing.

FIG. 225 Feeder to ballast loads must have a full-size neutral.

No reduction of neutral capacity even with zero neutral current?

As shown, both the fundamental and harmonic currents are 180° out of phase and both cancel in the neutral. This action is the same as it would be for an incandescent load. Under balanced conditions, the neutral current is zero. But the literal wording of Section 220-22 says there can be no reduction in neutral capacity when fluorescent lighting is supplied. As a result, there should be no use of the 70% factor for current over 200 amps as there would be for incandescent loading. Neutral here must be rated for 1000 amps.

FIG. 226 NEC rule does not permit reduction of neutral sizing in this case.

FIG. 227 Feeder voltage drop must be carefully calculated to be within the limits of good design.

2. The word "feeder" here refers to the overall run of conductors carrying power from the source (the service entrance, loadcenter substation, or local transformer at which the full value of system voltage is available) to the point of final branch-circuit distribution, including feeders, subfeeders, sub-subfeeders, etc.

3. The voltage-drop percentages are based on nominal circuit voltage being available at the load terminals of the source of each voltage level in the distribution system.

The voltage drop in any set of feeders can be calculated with the formulas given in Fig. 228. From this calculation, it can be determined if the conductor size initially selected to handle the load will be adequate to maintain the voltage drop within given limits. If it is not, the size of the conductors must be increased (or other steps taken where conductor reactance is not negligible) until the voltage drop is within the prescribed limits.

Two-Wire, Single-Phase Circuits (Inductance Negligible):

$$V = \frac{2k \times L \times I}{d^2} = 2R \times L \times I \qquad d^2 = \frac{2k \times I \times L}{V}$$

V = drop in circuit voltage (volts)
R = resistance per ft of conductor (ohms/ft)
I = current in conductor (amps)

Three-Wire, Single-Phase Circuits (Inductance Negligible):

$$V = \frac{2k \times L \times I}{d^2}$$

V = drop between outside conductors (volts)
I = current in more heavily loaded outside conductor (amperes)

Three-Wire, Three-Phase Circuits (Inductance Negligible):

$$V = \frac{2k \times I \times L}{d^2} \times 0.866$$

V = voltage drop of 3-phase circuit

Four-Wire, Three-Phase Balanced Circuits (Inductance Negligible):
For lighting loads: Voltage drop between one outside conductor and neutral equals one-half of drop calculated by formula for 2-wire circuits.
For motor loads: Voltage drop between any two outside conductors equals 0.866 times drop determined by formula for 2-wire circuits.

In above formulas

L = one-way length of circuit (ft)
d^2 = cross-section area of conductor (circular mils)
k = resistivity of conductor metal (cir mil-ohms/ft) (k = 12 for circuits loaded to more than 50% of allowable carrying capacity; k = 11 for circuits loaded less than 50%; k = 18 for aluminum conductor)

FIG. 228 These basic formulas indicate resistive voltage drop in feeders.

In application, the loadings and lengths of feeders can be adjusted to accommodate voltage-drop requirements, with additional advantages as shown in Fig. 229 for a modern office building. A total of nine risers supply lighting and receptacle circuits in all of the building's office space—floors 2 through 22. Each of these floors has three electric closets. At each feeder-riser location, one feeder supplies the bottom eight floors, the second feeder supplies seven floors in the middle of the building's height, and the third feeder supplies the top six floors. With the design demand load the same for each of the electric closets, the feeder serving the bottom eight closets has a heavier load than the one serving the middle seven closets, which in turn is more heavily loaded than the feeder to the top six closets. Because each feeder is designed to deliver its kilovoltampere capacity with only a 2 percent drop in voltage from the switchboard to any panel, the unbalanced loading of the feeders made possible the use of a single makeup—four 500-MCM THW conductors in 3½-in conduit—for all the feeders. This one size of feeder can supply a total load of eight closets with a 2 percent voltage drop for the distance from the switchboard to the lowest load, but it can only supply a total of six closets with a 2 percent drop when the feeder has to run to the top of the building.

The use of a single basic feeder makeup for all feeders provides two distinct advantages. First, four 500-MCMs in 3½-in conduit is a highly economical and efficient makeup, based on apparent power delivered, voltage drop, and cable power loss and measured against the installed cost of the feeder. Second, the use of a single feeder makeup minimizes construction costs. The use of only 500-MCMs in 3½-in conduit permitted a high degree of standardization and mechani-

FIG. 229 Feeder loads may be varied to balance voltage drop against circuit length and provide uniform circuit makeup.

zation in installing the conduit, pulling the conductors, coupling to panel enclosures, making taps, installing lugs, etc.

There are many cases in which the above-mentioned limits of voltage drop (1 percent for lighting feeders, etc.) should be relaxed in the interest of reducing the prohibitive costs of conductors and conduits that may be required by such low drops. In many installations, a 5 percent drop in feeders is not critical or unsafe. A recommended plan for realistic voltage drop in apartment houses in shown in Fig. 230, where the slightly greater drop is not considered detrimental to the lighting and appliance loads in the apartments.

Figure 231 shows how loadcenter distribution improves voltage level and regulation by deriving full-voltage feeders from transformers located at the centers of load groupings, thereby shortening the secondary feeders and reducing the voltage drop.

Voltage regulators

In general, when overcabling is needed to comply with **NEC** rules on voltage drop, the alternative of voltage regulators plus normal cabling should be considered, for it may turn out to be the most economical method.

Voltage regulators are used wherever it is necessary to eliminate objectionable variations in the voltage level of a supply circuit. Overvoltage and undervoltage have undesirable effects on the operation of a wide range of electrical and electronic load devices. In particular, electronic data-processing equipment and communication, instrumentation, and control devices require a fairly constant operating voltage.

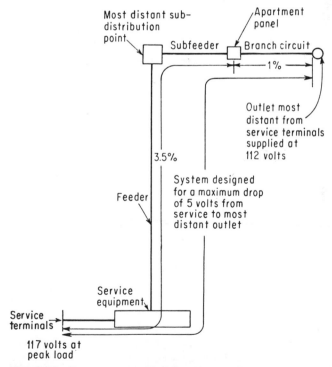

FIG. 230 Economy often indicates the maximum acceptable voltage drop for noncritical loads.

In the case of the load-center layout, shorter feeders cover the same distance over the plant as the long secondary feeders from the single switchboard, but the voltage drop is greatly reduced in the load-center system. Improved voltage regulation adds up to overall increase in operating efficiency of the system.

FIG. 231 Voltage drop can be greatly reduced by deriving feeders from loadcenter substations or transformer distribution centers.

Voltage regulators are particularly effective in high-ampere (thousands of amperes) distribution circuits for modern commercial and industrial properties, where long secondary-voltage runs (typically, 120/208 V) would require many conductors per phase leg to keep voltage drops within reasonable limits. They are extremely economical when used on long high-ampere circuits with constantly varying load demands (such as in apartment houses, schools, and hospitals), where they permit the use of only the amount of cable necessary for the load current, instead of that needed to maintain a low voltage drop.

Figure 232 shows the use of a voltage-regulator installation on the campus of a university. The utility company supplied secondary voltage from the transformer bank at 208Y/120 V nominal, at +5 percent and −2 percent from the 208-base. The local code limited the voltage drop between service-entrance equipment and the final distribution point to 2½ percent (or 5.2 V) on a 208-V base.

Calculations to determine the number and size of conductors that would be needed to serve the three buildings assigned approximately half the allowable voltage drop to the supply feeders, and the remaining half to the building distribution feeders. Since the utility transformer bank was close to the classroom building and the administration building, the supply feeders were relatively short, and the allowable voltage drop could be achieved by normal cabling. The library, however, required a 450-ft supply feeder from the transformer bank to the distribution center located in the basement. In addition, the library had an estimated load of 1900 A.

Note: Cost of building distribution feeders not included

FIG. 232 Voltage regulators also provide a solution to extreme
voltage-drop conditions.

Oversizing of the feeders to meet the voltage-drop requirement (half the allowable drop taking place in the supply feeder over a run of 450 ft at 1900 A) would have required twenty 500-MCM conductors per phase. The cost of materials and labor, the amount of space required, and the difficulty of installation made this method impractical.

Voltage regulators were located at the load end of the library feeder, to compensate for all the voltage drop in the supply feeder. With this design, only six 500-MCM conductors per phase were needed, since they could be selected to carry only the load current. Because the regulators would maintain nominal voltage at the distribution center in the library basement, the full 2½ percent voltage drop allowed by code could be assigned to the building's distribution feeders. The amount of cabling space was thus kept to a minimum.

The regulators maintain nominal voltage at the distribution center not only when the load is heavy and the utility supply voltage is low, but also when the load is light and the utility supply voltage is high. This results in the utilization equipment in the library building always operating at peak efficiency with maximum life.

The short-circuit current available from the utility for a 3-phase bolted fault was 116,000 rms symmetrical amperes. The impedance introduced by the all-cable method would have reduced the short-circuit current at the distribution equipment to approximately 40,000 A. However, with the regulator-cable installation, the additional impedance of the smaller cables reduces the short-circuit current to approximately 19,000 A—an added bonus.

Figures 233 and 234 show sets of curves correlating size of conductor, ampere load, and voltage drop for 3-phase circuits in steel conduit. The curves are for a one-V drop and 90 percent power factor, based on ampacities for types RHW and THW (75°C) 600-V insulation. Many such graphs and tabulated data on voltage drop are available in handbooks and from manufacturers.

Figures 235 and 236 are handy tables for determining maximum feeder-circuit lengths for 1 percent voltage drop on single-phase and 3-phase feeders.

The table in Fig. 235 gives the one-way lengths for 115/230-V circuits under balanced loading. With a balanced 3-wire load, the drop will be 1.15 V for each 115-V circuit—that is, for each side of the 115/230-V circuit. For a 230-V, 2-wire circuit, the given lengths produce a voltage drop of 2.3 V. An example is shown below the table.

The table in Fig. 236 gives the one-way lengths for 3-phase, 230-V, delta ac load circuits with 85 percent power factor. For balanced load conditions, the conductors and lengths shown produce voltage drops of 2.3 V (1 percent of 230 V). For other voltage-drop percents (2 percent, 3 percent, etc.) the lengths shown can be doubled, tripled, etc. Thus, if, as shown, a 90-A-loaded 3-phase circuit can be run 113 ft using No. 1/0 conductors for a 2.3-V drop, then the same load could be carried twice as far (226 ft) if double the voltage drop (4.6 V) were acceptable. The table may be used for calculating other feeder lengths for 1 percent voltage drop:

- For 208-V, 4-wire wye feeders, multiply the given lengths by 0.9.
- For 230-V, single-phase feeders, multiply the given lengths by 0.85.
- For 460-V, 3- or 4-wire feeders, multiply the given lengths by 2.
- For aluminum wire, multiply the given lengths by 0.7 or use the length of copper wire which is two sizes smaller than the aluminum size under consideration.

Multiple Conductors

In making up high-current-capacity feeders, there are distinct advantages to the use of multiple conductors in each leg. This may be accomplished in accordance with **NEC** Section 310-4, which reads as follows:

Conductors in Parallel. Aluminum, copper-clad aluminum, or copper conductors of size 1/0 and larger, comprising each phase or neutral, shall be permitted to be connected in parallel (electrically joined at both ends to form a single conductor) only if all of the following conditions are met: all of the parallel conductors shall be of the same length, of the same conductor material, same circular-mil area, same insulation type, and terminated in the same manner. Where run in separate raceways or cables, the raceways or cables shall have the same physical characteristics.

This specific **NEC** provision is intended to allow a practical means of installing large-capacity feeders and services. For instance, six conductors might be used per phase and neutral to obtain a 2000-A capacity per phase, which simply could not be accomplished without parallel conductors. To ensure equal division of the total current among all conductors involved, all the multiple

VOLTAGE DROP AND CAPACITIES

60 cycle—3-phase in steel conduit (49°C temp.)

GENERAL NOTES

Curves are for a 1-volt drop and 90% PF.
For any other voltage drop, ratio ordinates.
Capacities are for Type RHW 600-volt insulation.

TYPICAL EXAMPLE

Required: to carry 70 amps 400 ft with a maximum of 2 volts drop, what is the wire size?
a. 400 ft for 2 volts drop is the same as 200 ft for 1-volt drop.
b. Intersection of 70 amps and 200 ft is just below 400 MCM.
c. Use 400 MCM conductors to assure best operation of the feeder.

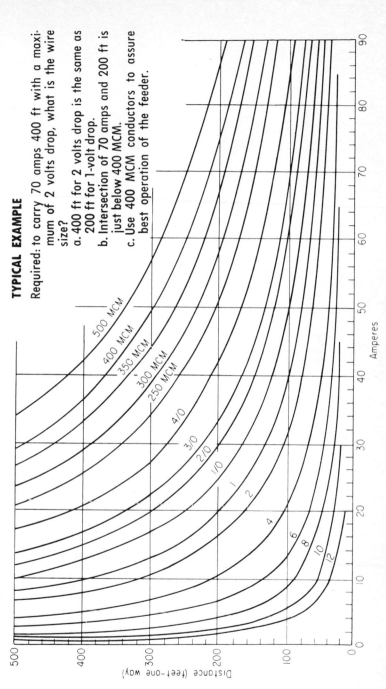

FIG. 233 Maximum lengths of 3-phase feeder circuits for a 1-V drop at loads up to 90 A.

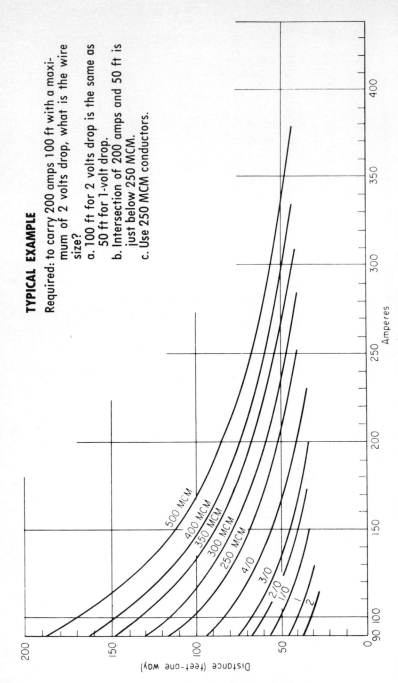

TYPICAL EXAMPLE

Required: to carry 200 amps 100 ft with a maximum of 2 volts drop, what is the wire size?

a. 100 ft for 2 volts drop is the same as 50 ft for 1-volt drop.

b. Intersection of 200 amps and 50 ft is just below 250 MCM.

c. Use 250 MCM conductors.

Distance (feet-one way)

500 MCM
400 MCM
350 MCM
300 MCM
250 MCM
4/0
3/0
2/0
1/0
1
2

Amperes

FIG. 234 Maximum lengths of 3-phase feeder circuits for a 1-V drop at loads from 90 to 380 A.

SINGLE-PHASE AC LOADS

115/230 volts, 60 cycles, 100% PF

Amp load	Wire size— circular mils					Wire size—B & S or AWG									
	500	400	350	300	250	4/0	3/0	2/0	1/0	1	2	3	4	6	8
40	1106	898	788	669	558	475	378	299	239	188	150	119	94	59	38
50	885	719	630	535	447	380	303	240	191	150	120	91	75	47	30
60	737	599	525	446	372	317	252	200	159	125	100	79	62	39	
70	632	513	450	382	319	271	216	171	136	107	86	68	53	34	
80	553	449	394	334	279	238	189	150	119	94	75	59	47		
90	491	399	350	297	248	211	168	133	106	83	67	53	42		
100	442	359	315	267	223	190	151	120	95	75	60	47			
110	402	327	286	243	203	173	138	109	87	68	55				
120	369	299	263	223	186	158	126	100	79	63					
130	340	276	242	206	172	146	116	92	73	58					
140	316	257	225	191	159	136	108	86	68						
150	295	240	210	178	149	127	101	80	64						
160	276	225	197	167	140	119	95	75	60						
170	260	211	185	157	131	112	89	70							
180	246	200	175	148	124	106	84	66							
190	233	189	166	140	117	100	80								
200	221	180	157	134	112	95	76								
210	211	171	150	127	106	90									
220	201	163	143	122	101	86									
230	192	156	137	116	97	83									
240	184	150	131	111	93										
250	177	144	126	107	89										
260	170	138	121	103	80										
270	164	133	117	99											
280	158	128	112	96											
290	152	124	109	92											
300	147	120	105												
310	143	116	102												
320	138	112													
330	134	109													
340	130	106													

Calculations based on copper resistance of 12.5 ohms per CM-ft at 50C (122F).

Reactance and impedance losses calculated for each wire.

Conductors closely grouped in metallic conduit.

REFER TO TABLE: For a load of 40 amps, it shows that No. 4 conductors must be used if the circuit length is between 60 and 94 ft. Under the condition shown in the diagram, there will be a 1.15-volt drop in each 115-volt circuit.

FIG. 235 Circuit lengths (in feet) for a 1 percent voltage drop in single-phase feeders.

3-PHASE DELTA AC LOADS

230 volts, 60 cycles, 85% PF

Amp load	Wire size—circular mils					Wire size—B & S or AWG									
	500	400	350	300	250	4/0	3/0	2/0	1/0	1	2	3	4	6	8
40	710	625	584	530	475	429	364	303	253	208	173	139	113	75	49
50	568	500	467	424	380	343	291	242	203	167	139	111	90	60	39
60	473	417	389	353	317	286	243	202	169	139	115	93	75	50	
70	406	357	333	303	271	245	208	173	145	119	99	79	64	43	
80	355	312	292	265	238	214	182	151	127	104	87	69	56		
90	316	278	259	235	211	191	162	134	113	93	77	62	45		
100	284	250	233	212	190	172	146	121	101	83	69	55			
110	258	227	212	193	173	156	132	110	92	76	63				
120	237	208	195	177	158	143	121	101	84	69	58				
130	218	192	180	163	146	132	112	93	78	64					
140	203	179	167	151	136	123	104	86	72						
150	189	168	156	141	127	114	97	81	67						
160	177	156	146	132	119	107	91	76							
170	167	147	137	125	112	101	86	71							
180	158	139	130	118	106	95	81	67							
190	149	132	123	112	100	90	77								
200	142	125	117	106	95	86	73								
210	135	119	111	101	90	82									
220	129	114	106	96	86	78									
230	123	109	101	92	83	75									
240	118	104	97	88	79										
250	114	100	93	85	76										
260	109	96	90	81	73										
270	105	93	86	78											
280	101	89	83	76											
290	98	86	80	73											
300	95	83	78												
310	92	81	75												
320	89	78													
330	86	76													
340	83	73													
350	81														
360	79														
370	77														
380	75														

Calculations based on copper resistance of 12.5 ohms per CM-ft at 50C (122F).

Reactance and impedance losses calculated for each wire.

Conductors closely grouped in metallic conduit.

FIG. 236 Circuit lengths (in feet) for a 1 percent voltage drop in 3-phase, 3-wire feeders.

conductors must be of the same length, the same material, the same circular-mil area, and the same insulation type; they must be terminated in the same manner; and, where multiple conductors are run in separate raceways or cables, the raceways or cables must have the same physical characteristics. The reason for the last condition is that the impedance of a circuit in a nonferrous raceway is different from the impedance of the same circuit in a ferrous raceway; the use of raceways or enclosures with different characteristics would unbalance currents.

From the **NEC** tables that give the current-carrying capacities of various sizes of conductors, it can be seen that small conductors carry more current per circular mil of cross section than large conductors. This results from the rating of conductor capacity according to temperature rise. The larger the cable, the smaller the radiating surface per circular mil of cross section. Losses due to "skin effect" (the apparent higher resistance of conductors to alternating current than to direct current) are also higher in the larger conductor sizes. And larger conductors cost more per ampere than smaller conductors.

All these factors point to the advisability of using a number of smaller conductors in multiple to obtain a particular carrying capacity, rather than using a single conductor of that capacity. In many cases, multiple conductors for a feeder provide distinct operating advantages and are more economical than the equivalent-capacity single-conductor makeup. But it should be noted that the reduced total cross section of conductor resulting from the use of multiple conductors per leg produces higher resistance and a greater voltage drop than in the same length of an equal-ampacity single conductor per leg. Voltage drop may therefore be a limitation.

Figure 237 shows a typical application of copper conductors in multiple, with the advantages of such use. Where more than three conductors are located in a single conduit, the ampacity (i.e., load-current rating) of each conductor must be derated from the ampacity value shown in **NEC** Table 310-16. The four circuit makeups show:

1. Without derating for conduit occupancy, circuit 2 would be equivalent to circuit 1.
2. A circuit of six 400-MCMs can be made equivalent in current-carrying capacity to a circuit of three 2000-MCMs by dividing the 400-MCMs between two conduits (three conductors per 3-in conduit). If three different phases are run in each of two 3-in conduits for this circuit, the multiple circuit would not require derating, and its 670-A rating would exceed the 665-A rating of circuit 1.
3. Circuit 2 is almost equivalent to circuit 3 in current rating.
4. Circuit 4 is equivalent to circuit 1 in current rating but uses less conductor copper and a smaller conduit. And these advantages are obtained even with occupancy derating.

Where large currents are involved, it is particularly important that the separate phase conductors be located close together to avoid excessive voltage drop and ensure equal division of current. It is also essential that all phases and the neutral (and grounding wires, if any) be run in each conduit, even where the conduit is of nonmetallic material.

Figure 238 shows a handy way of evaluating single-conductor feeder makeup versus multiple-conductor feeder makeup, based on achieving the maximum current rating (even with ampacity derating according to Note 8 to Table 310-16), lowest voltage drop and power loss, and maximum economy—all in one conduit.

Except where the conductor size is governed by voltage-drop considerations, it is seldom economical to use conductors of sizes larger than 1000 MCM, because above this size the increase in ampacity is very small in proportion to the increase in the size of the conductor. Thus, for a 50 percent increase in the conductor size (e.g., from 1000 to 1500 MCM), the ampacity of a Type THW conductor increases only 80 A, or less than 15 percent; for an increase in size from 1000 to 2000 MCM, a 100 percent size increase, the ampacity increases only 120 A, or about 20 percent. In any case where single conductors larger than 500 MCM would be required, it is worthwhile to compute the total installation cost using single conductors and the cost using two or more conductors in parallel.

It is important to note that when multiple conductors are used per circuit phase leg, more space may be required at equipment terminals to bend and install the conductors.

Exception No. 1 to Section 310-4 clearly indicates long-time **NEC** acceptance of the practice of paralleling conductors smaller than No. 1/0 for use in traveling cables of elevators, dumbwaiters, and similar equipment.

Exception No. 2 of Section 310-4 permits parallel circuit makeup using conductors smaller than No. 1/0, but all the given conditions must be observed. This exception permits the use of

1. 6″

A 3-phase circuit of three 2,000 MCM type THW conductors in a 6-in. conduit. Load current rating of each phase = 665 amps, the ampacity from Table 310-16. Cross-section area per phase = 2.9013 sq. in.

2. 4″

A 3-phase circuit of six 400 MCM type THW conductors (two per phase) in a 4-in. conduit.
Load current rating of each phase might appear to be = 2 X 335 = 670 amps.
But, because of the 80% load limitation required by Note 8 to Tables 310-16/19: Load current rating of each phase = 670 X 80% = 536 amps.
Cross-section area per phase = 1.3938 sq in. (two conductors).

3. 4″

A 3-phase circuit of three 1,000 MCM type THW conductors in a 4-in. conduit. Load current rating of each phase = 545 amps.
Cross-section area per phase = 1.5482 sq in.

4. 4-1/2″

A 3-phase circuit of six 600 MCM type THW conductors in a 5-in. conduit. Load current rating of each phase might appear to be = 2 X 420 = 840 amps.
But 80% load limitation must be applied because of the number of conductors in the conduit:
Load current rating of each phase = 840 X 80% = 672 amps.
Cross-section area per phase = 2.0522 sq in. (two conductors).

FIG. 237 Multiple-conductor feeder makeups must be related to conductor ampacity derating (Note 8, Table 310-16).

smaller conductors in parallel for circuit applications where it is necessary to reduce the conductor capacitance effect or to reduce the voltage drop over long circuit runs.

A common neutral for a feeder

A frequently discussed **NEC** requirement is that of Section 215-4, covering the use of a common neutral with more than one set of feeders. This section says that a common neutral feeder may be used for two or three sets of 3-wire feeders, or two sets of 4-wire feeders. It further requires that all conductors of feeder circuits employing a common neutral feeder must be within the same enclosure when the enclosure or raceway containing them is metal.

POWER FEEDER SPECIFICATION DATA

P	Cable Size	Cables Per Phase	Conduit Size	Cost in Dollars Per 100 Feet	Ampere Rating NE Code	Dollars Per 100 Feet Per Unit P	Economic Choice
2.0	4	1	1¼	$136.00	70	$68.00	
3.0	2	1	1¼	151.00	95	50.30	
3.7	1	1	1½	189.00	110	51.10	
4.6	0	1	2	237.00	125	51.50	
5.5	2/0	1	2	260.00	145	47.20	
6.6	3/0	1	2	286.00	165	43.40	
7.4	1	2	2½	342.00	176	46.30	B
7.7	4/0	1	2½	356.00	195	46.30	A
8.2	250	1	2½	421.00	255	51.40	B
9.2	0	2	2½	378.00	200	41.10	A
9.5	350	1	3	532.00	310	56.00	C
11.0	2/0	2	3	468.00	232	42.50	A
11.0	500	1	3	622.00	380	56.60	C
11.2	1	3	3	472.00	231	42.20	B
13.2	3/0	2	3	521.00	264	39.50	A
13.8	0	3	3	526.00	262	38.10	A
15.4	4/0	2	3	586.00	312	38.00	A
16.4	250	2	3½	755.00	408	46.00	B
16.5	2/0	3	3½	634.00	304	38.40	A
19.8	3/0	3	3½	713.00	346	36.00	A
23.1	4/0	3	4	848.00	410	36.70	A
24.5	250	3	4½	1085.00	535	44.30	A

Wiring in steel conduit for up to 600-volt three-phase 30 C ambient NE Code ratings.

Size 250 MCM and above — Heat and moisture resistant type cable insulation with 75C; rating.

Below 250 MCM — Thermoplastic cable insulation with 60 C rating.

A — Preferred   B — Second Choice   C — Uneconomical

ECONOMIC PERFORMANCE OF FEEDERS IN STEEL CONDUIT

A — 1 Cable per phase
B — 2 Cables per phase
C — 3 Cables per phase

A suitable unit for comparing the performance of cables is P, which is defined as (1)

$$P = LxI/dVx1000$$

where, L = feeder length in feet

I = load current in amperes

dV = feeder voltage drop

(1000 is included for brevity)

Based on the average value of load power factor, 0.8 to 0.95, it can be assumed that voltage drop in a feeder is equal to the impedance drop, so that dV in Equation (1) becomes $LIZ\sqrt{3}/1000$, Equation (1) then simplifies to $P = 1/(Z\sqrt{3})$

where, Z = ohms to neutral impedance per 1000 feet of feeder.

The higher the value of "P," the lower the voltage drop and power losses and the higher the extra capacity.

FIG. 238 "P factor" gives figure of merit for comparing single and multiple feeder-conductor makeup.

A common neutral is a single neutral conductor used as the neutral for more than one set of feeder conductors. It must have a current-carrying capacity equal to the sum of the neutral-conductor capacities that would be required if an individual neutral conductor were used with each feeder set. Figure 239 shows a typical example of a common neutral—here, used for three-feeder circuits.

FIG. 239 Single neutral conductor may be used with two or more sets of feeder phase legs.

A common neutral may be used only with feeders; it may never be used with branch circuits. A single neutral of a multiwire branch circuit is not a "common neutral." Rather, it is the neutral of only a single circuit, even though the circuit may consist of 3 or 4 wires. A feeder common neutral is used only with more than one feeder circuit.

OVERCURRENT PROTECTION

The most important task in modern feeder design is the selection of overcurrent protection. Certainly, the vital relation between effective overcurrent protection and the safety of personnel and property must be fully recognized. With electrical application continuing to grow at a rapid pace, experience has shown that carelessness or inadequacy in the selection of overcurrent protection can lead to shock and fire hazards. Of only slightly less importance is the need for full and thorough overcurrent protection for long, reliable life of equipment and systems.

In this total-electric age, all modern facilities are dependent upon electrical distribution systems for efficient, economical, and productive performance of activities within the facility. Office buildings, schools, stores, factories, hotels, hospitals—all modern buildings—have an almost total dependence on artificial internal environments created by air conditioning, heat, light, elevators, pumps, etc. Their electrical systems must be based on overcurrent protection that fully and surely protects against overloads up to the maximum possible short-circuit currents, that is carefully coordinated to minimize the extent of electrical outages, and that eliminates nuisance or unwanted opening by operating only when and as required for proper protection of life and property.

In the design of an electrical system, required current-carrying capacities are determined for the various feeders and subfeeders. Then these required capacities are translated into standard circuit conductors which have sufficient current-carrying capacities, based on the size of the conductors, the type of insulation on the conductors, the ambient temperature at the place of installation, the number of conductors in each conduit, the type and continuity of the load, and judicious inclusion of spare capacity to provide for future load growth. Or, if busway, armored cable, or other cable assemblies are to be used, similar considerations go into the selection of conductors with the required current-carrying capacities. In any case, then, the next step is to provide overcurrent protection for each and every circuit.

The overcurrent device for conductors or equipment must automatically open the circuit it protects if the current flowing in that circuit reaches a value which will cause an excessive or dangerous temperature in the conductors or conductor installation.

Overcurrent protection for conductors must also be rated for safe operation at the level of fault current obtainable at the point of application. Every fuse and circuit breaker for short-circuit protection must be applied in such a way that the fault current produced by a bolted short circuit on its load terminals will not damage or destroy the device. Specifically this requires that a short-circuit overcurrent device have a proven interrupting capacity at least equal to the current which the electrical system can deliver into a short on its load terminals.

But safe application of a protective device does not stop with adequate interrupting capacity to allow its use at the point of installation in the system. The speed of operation of the device must then be analyzed in relation to the thermal and magnetic energy which the device will permit to flow in the faulted circuit. A very important consideration is the provision of conductor sizing to meet the potential heating load of short-circuit currents in cables. With expanded use of circuit-breaker overcurrent protection, coordination of protection from loads back to the source has introduced time delays in the operation of overcurrent devices. Cables in CB-protected systems must be able to withstand any impressed short-circuit currents for the duration of the overcurrent delay. For example, a motor circuit to a 100-hp motor might be required to carry as much as 15,000 A for a number of seconds. To limit damage to the cable due to heating, a much larger size of conductor than necessary for the load current alone may be required.

A device may be able to break a given short-circuit current without damaging itself in the process; however, in the time it takes to open the faulted circuit, enough energy may get through to damage or destroy other equipment in series with the fault. This other equipment might be cable or busway or a switch or motor controller—any circuit component which simply cannot withstand the few cycles of high-amperage short-circuit current that flows in the period of time between initiation of the fault and interruption of the current flow.

Basic rules

Conductors in an electric circuit—whether it is a branch circuit or a feeder—must be protected against excessive current flow (Fig. 240). Such protection must be provided in accordance with

PROTECTIVE DEVICE IN SERIES

CT-RELAY RESPONDS TO LOAD CURRENT

FIG. 240 Protective device must respond automatically to excess feeder current.

the current-carrying capacities of the conductors, except where certain higher settings or ratings of protective devices are permitted or required—as in the case of motor branch circuits or motor feeders.

NEC Section 240-12, "Electrical System Coordination," is aimed at "industrial locations" where hazard to personnel would result from sudden disorderly shutdown of electrical equipment under fault conditions. For such systems, it is permissible to oversize protective devices so that conductors are not protected at their ampacities, and the circuit will not open on overload. But to satisfy the rule, the circuit protective devices must provide proper short-circuit protection (with selective coordination of the time-current characteristics of protective devices in series from the service to any load) so that, automatically, any fault will actuate only the short-circuit protective device closest to the fault on the line side of the fault, thereby minimizing the extent of an electrical outage due to a fault. In addition, each circuit must be provided with "overload indication based on monitoring systems or devices." In the event of an overload, although the circuit will not open, a visible and/or audible signal must be given to warn of the overload and indicate the need for proper attention to the matter. The rule says that "The monitoring system may cause the condition to go to alarm allowing corrective action or an orderly shutdown thereby minimizing personnel hazard and equipment damage." (See Fig. 241.)

An overcurrent device must not be used in a permanently grounded conductor, except where the device simultaneously opens all conductors of the circuit *or* where three fuses are used in a 3-phase, 3-wire circuit fed from a corner-grounded delta system and a fuse is required in the grounded leg by **NEC** Section 430-36 because the fuses are providing running overload protection (as well as short-circuit protection) for a motor.

Overcurrent protection for a conductor must be located at the point where the conductor receives its supply. This means that a conductor run of a particular current-carrying capacity must be protected at the point at which it is fed by a conductor of higher current-carrying capacity. In a feeder run, then, a change (reduction) in conductor size must be accompanied by pro-

Only in industrial systems
where needed to minimize hazard

Overload
indicator

A

B

C

On a fault in
this section of
the system,
device "A"
opens to
clear fault

On a fault
here, *only*
device "B"
opens—
device "A"
must not
operate

Short-circuit
devices rated
above circuit
conductor
ampacity at
A,B,C, and D

On fault here,
only "D" opens—
"A," "B" and "C"
remain closed

On fault here,
only "C" opens—
"A" and "B" stay
closed

D

Each circuit must have audible
or visual indication of overload

FIG. 241 If selectively coordinated, industrial systems may utilize short-circuit protection *only*.

tection for the smaller conductor (Fig. 242). Exceptions to this rule are made in the case of taps, which are discussed later in this section under "Feeder taps." Another exception is where the protection of the larger conductor meets the requirements for protection of the smaller conductor, that is, the rating or setting of the fuse or CB protecting the larger wires is not in excess of the ampacity of the smaller conductors or is suitable to protect the smaller tap wires in accordance with some specific **NEC** rule.

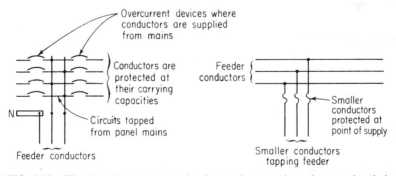

Overcurrent devices where
conductors are supplied
from mains

Conductors are
protected at
their carrying
capacities

Feeder
conductors

N

Circuits tapped
from panel mains

Smaller
conductors
protected at
point of supply

Feeder conductors

Smaller conductors
tapping feeder

FIG. 242 Effective design protects feeder conductors where they receive their supply.

For feeder circuits other than motor feeders, the overcurrent device that protects against short circuits and grounds can be sized as follows:

1. If the allowable current-carrying capacity of a conductor does not correspond to the rating of a standard-size fuse, the next larger rating of fuse may be used only where the rating of the fuse does not exceed 800 A.
2. Nonadjustable-trip breakers must be rated in accordance with the current-carrying capacity of the conductors they protect—except that a higher-rated circuit breaker may be used if the current-carrying capacity of the conductor does not correspond to a standard rating. In such a case, the next higher standard rating and setting may be used only where the rating is 800 A or less. An adjustable-trip breaker must be set to protect the feeder conductors against overload current.

The general rule, in both these cases, is that the device must be rated to protect conductors in accordance with their safe, allowable current-carrying capacities. The current rating of a protective device is commonly called the *continuous current rating*. The word "continuous" means an uninterrupted current flow for a period of 3 hours or more. It represents the maximum value of current that the device will carry without tripping open or blowing, under certain specified conditions. Of course, there will be cases where standard ampere ratings and settings of overcurrent devices will not correspond with conductor capacities. In such cases, the next larger standard size of overcurrent device may be used where the rating of the protective device is 800 A or less. A basic guide to effective selection of the ampere ratings of overcurrent devices for any feeder or service application is given in the following:

If a circuit conductor of, say, 500-MCM THW copper (not more than three in a conduit at not over 86°F ambient) satisfies design requirements and **NEC** rules for a particular load current not in excess of the conductor's table ampacity of 380 A, then the conductor may be protected by a fuse or CB rated at 400 A. **NEC** Section 240-6, which gives the "standard ampere ratings" of protective devices, shows devices rated at 350 and 400 A, but none at 380 A. In such a case, the **NEC** accepts 400 A as "the next higher standard device rating" above the conductor ampacity of 380 A.

But such a 400-A device would permit a load increase above the 380 A that is the safe maximum limit for the conductor. It would be more effective practice to use protection rated at 350 A and prevent such overload.

For the application of fuses and CBs:

1. If the ampacity of a conductor does not correspond to the rating of a standard-size fuse, the next higher rating of fuse may be used only where that rating is 800 A or less. Over 800 A, the next lower fuse rating must be used.
2. A nonadjustable-trip breaker (one without overload trip adjustment above its rating—although it may have adjustable short-circuit trip) must be rated in accordance with the current-carrying capacity of the conductors it protects, except that the next higher standard rating of CB may be used if the ampacity of the conductor does not correspond to a standard rating. However, in such a case, the next higher standard rating may be used only where the rating is 800 A or less.

An example of such an application is shown in Fig. 243, where a nonadjustable CB with a rating of 1200 A is used to protect the conductors of a feeder circuit, which are rated at 1140 A. As shown, the use of a 1200-A CB to protect a circuit rated at 1140 A (3×380 A = 1140 A) clearly violates Exception No. 1 of **NEC** Section 240-6, because the CB rating is the next higher rating above the ampacity of the conductors—on a circuit rated over 800 A. With a feeder circuit as shown (three 500-MCM THW conductors, each rated at 380 A), the CB must not be rated over 1140 A. A standard 1000-A CB would satisfy the **NEC** rule

1200 – amp nonadjustable
CB (next higher standard
rating to 1140 amps)

Feeder circuit to
large load

Feed circuit rated 1140 amps (3 parallel sets
of 500 MCM THW copper conductors, each
conductor rated 380 amps)

FIG. 243 This use of "the next higher standard
rated protective device" is *not* acceptable.

as being the next-lower-rated protective device from Section 240-6. Of course, if 500-MCM THHN or XHHW conductors were used instead of THW conductors, then each would be rated at 430 A; three per phase would give the circuit an ampacity of 1290 A (3 × 430 A); the 1200-A CB would satisfy the basic rule in the first sentence of Section 240-3; and Exception No. 1 would not be involved.

In the design of electrical systems for industrial manufacturing and process operations, special or unusual conditions often require suspension of the normal application of **NEC** rules on overcurrent protection. To meet such needs, **NEC** Article 685 allows the customized design and layout of overcurrent protection for "integrated electrical systems." In effect, the rules of this article exempt the interconnected and/or interrelated electrical components of manufacturing assembly or production lines and related industrial process operations from the conventional rules on overcurrent protection. Section 685-2 lists those **NEC** rules that allow special treatment for installations where orderly shutdown is required but would not be provided by the standard rules.

In the **NEC**, the scope of Article 685 is described as follows:

This article covers integrated electrical systems, other than unit equipment, in which orderly shutdown is necessary to ensure safe operation. An integrated electrical system as used in this article is a unitized segment of an industrial wiring system where all of the following conditions are met: (1) an orderly shutdown is required to minimize personnel hazard and equipment damage; (2) the conditions of maintenance and supervision assure that qualified persons will service the system; and (3) effective safeguards, acceptable to the authority having jurisdiction, are established and maintained.

Sizing Feeder Protection

As noted above under "Feeder sizing versus panelboard rating," **NEC** Section 220-10(b) must be fully satisfied when feeder overcurrent protective devices are sized. To repeat the basic concept, for emphasis: When a feeder supplies continuous loads or any combination of continuous and noncontinuous loads, the rating of overcurrent devices protecting the feeder and the ampacity of the feeder conductors shall not be less than the sum of the noncontinuous load plus 125 percent of the continuous load, unless the CB or fused switch protecting the feeder is UL-listed for continuous operation at 100 percent of its load rating. This is shown in Fig. 244.

To correspond to that rule, the UL provides clear information about available circuit breakers and fusible switches for use with continuous loading (maximum current flowing continuously for 3 hours or more).

The nominal or theoretical continuous current rating of a CB generally is taken to be the same as its trip setting—the value of current at which the breaker will open, either instanta-

THE RULE

FEEDER OVERCURRENT DEVICE must be rated not less than 125% of
the continuous load, *and* the feeder conductors must be sized on the
same basis to assure them effective protection in accordance with their
ampacity

EXAMPLE

For this feeder, with conductors rated at 380 amps, the maximum load permitted for a conventional fused switch is
80% of 380 amps, or 304 amps.

Rating of fuses and ampacity of feeder conductors are both at
least equal to 125% times the continuous load.

FIG. 244 Basic approach to the sizing of protection for a continuous-load feeder.

neously or after some intentional time delay. But the real continuous current rating of a CB—
the value of current that it can safely and properly carry for periods of 3 hours or more—
frequently is reduced to 80 percent of the nameplate value by rules of codes and standards.

The UL *Construction Materials Directory* contains a clear, simple rule in the instructions
under "Circuit Breakers, Molded-Case." It says: "Unless otherwise marked, circuit breakers
should not be loaded to exceed 80% of their current rating, where in normal operation the load
will continue for three or more hours."

A load that continues for three hours or more is a continuous load. If a breaker is marked
for continuous operation, it may be loaded to 100 percent of its rating and operated continuously.

Circuit breakers are available for continuous operation at 100 percent of their current rating,
but they must be used in the mounting and enclosure arrangements established by UL for 100
percent rating. Molded-case CBs of the 100 percent continuous type are made in frame ratings
from 600 A up, with trip settings from 225 A up. Information on the use of breakers with 100
percent ratings is given on their nameplates (Fig. 245). The nameplates clearly indicate that

EXAMPLE 1

Because of the relatively large volume in the switchboard structure, this is normally not a problem. Special UL tests must be performed if these specifications are not met.

The ventilation applies only to 100% rated applications.

WIRING DIAG.

S.O.#

VENTILATION 68 SQ. INCHES	SUITABLE FOR CONTINUOUS OPERATION AT 80% OF FRAME RATING IN AN ENCLOSURE WITHOUT VENTILATION.
	SUITABLE FOR CONTINUOUS OPERATION AT 100% OF FRAME RATING IF USED IN A MINIMUM ENCLOSURE 45" HIGH X 38" X 20" WITH MINIMUM VENTILATION PROVIDED AS SHOWN IN EITHER THE FRONT OR SIDE OF THE ENCLOSURE. **CAUTION** -TO PREVENT OVERHEATING WHEN CABLE CONNECTED USE COPPER CONDUCTORS ONLY.
VENTILATION 68 SQ. INCHES	

ENCLOSURE

Generally, bus bar extensions and lugs are connected to the circuit breaker within the switchboard. In that case, aluminum cable may be used.

EXAMPLE 2

WIRING DIAG.

S.O.

SUITABLE FOR CONTINUOUS OPERATION AT 80% OF FRAME RATING IN AN ENCLOSURE WITHOUT VENTILATION.

SUITABLE FOR CONTINUOUS OPERATION AT 100% OF FRAME RATING IF USED IN A MINIMUM ENCLOSURE 45" HIGH X 38" X 20". VENTILATION IS NOT REQUIRED. Caution—TO PREVENT OVERHEATING WHEN CABLE CONNECTED USE COPPER CONDUCTORS ONLY.

In this case, ventilation is not required.

FIG. 245 Circuit breakers that may be used for continuous loading to 100 percent of their rating are marked in detail.

ventilation may or may not be required. Because most switchboards have fairly large interior volumes, the "minimum enclosure" dimensions shown on these nameplates (45 by 38 by 20 in) usually are readily achieved. But, special UL tests must be performed if these dimensions are not achieved. Where busbar extensions and lugs are connected to the circuit breakers within the switchboard, the caution about copper conductors does not apply, and aluminum conductors may be used.

If a switchboard does not provide the ventilation pattern and the enclosure size specified on the nameplate, the circuit breaker must be applied at 80 percent of its rating. Switchboard manufacturers have UL tests conducted with a circuit breaker installed in a specific enclosure; the enclosure can receive a listing for operation at 100 percent of rating even though the ventilation pattern or overall enclosure size does not meet the specifications. In cases where the breaker nameplate specifications are not provided by the switchboard, the customer must request a letter from the manufacturer certifying that a 100 percent listing has been received. Otherwise, the breaker must be applied at 80 percent.

To realize savings with devices listed by UL at 100 percent of their continuous current rating, circuit-breaker manufacturers' data sheets must be used to determine the types and ampere ratings of available breakers that are rated at 100 percent, along with the frame sizes, approved enclosure sizes, and ventilation patterns required by UL, if any. According to UL Standard 489, Paragraph 33.37A, "A circuit breaker having a frame size less than 600 amperes shall not be marked suitable for continuous operation at 100% of rating." Of course, trip units with lower ampere ratings may be installed in the 600-A frame.

It is essential to check the instructions given in the UL listing to determine if and under what conditions a circuit breaker (or a fuse in a switch) is rated for continuous operation at 100 percent of its current rating.

Care must also be exercised in selecting fuse protection for feeders. In general, the rating of a fuse is taken as 100 percent of the rated nameplate current when the fuse is enclosed by a switch or panel housing. But, because of the heat generated by many fuses, the maximum continuous load permitted on a fused switch is restricted by a number of NEMA, UL, and **NEC** rules to 80 percent of its rating. The limitation of circuit load current to no more than 80 percent of the current rating of fuses in equipment protects the switch or other piece of equipment from the heat produced in the fuse element; it also protects attached circuit wires from excessive heating close to the terminals. The fuse itself can actually carry 100 percent of its current rating continuously without damage to itself, but its heat is conducted into the adjacent wiring and switch components.

NEMA standards require that a fused, enclosed switch be marked, as part of the electrical rating, "Continuous load current not to exceed 80 percent of the rating of fuses employed in other than motor circuits." That derating compensates for the extra heat produced by continuous operation. Motor circuits are excluded from that rule, but a motor circuit is required by the **NEC** to have conductors rated at least 125 percent of the motor full-load current; this, in effect, limits the load current to 80 percent of the conductor ampacity and limits the load on the fuses rated to protect those conductors. But, the UL *Construction Materials Directory* does recognize fused bolted-pressure switches and high-pressure butt-contact switches for use at 100 percent of their ratings on circuits with available fault currents up to 100,000, 150,000, or 200,000 rms symmetrical amperes—as marked. (See "Fused Power Circuit Devices" in that UL publication.) The ratings of fuses for feeder protection, therefore, are determined by the type of switch used and its suitability for continuous operation at 100 percent loading (Fig. 246).

Figure 247 shows how the sizing of feeder overcurrent protection and conductor size differs with the type of CB used. In practice, the ratings of the two breakers would be increased to the next highest standard trip ratings—2500 instead of 2255 A, and 2000 instead of 1804 A.

Figure 248 shows how the selection of a CB rated at 100 percent offers substantial economy in the size of the switchboard bus and in the assembly of related components to make up the switchboard. The startling difference in list price between using 80 percent and 100 percent breakers is $4952 ($13,968−$9016). Prices will, of course, vary among manufacturers, depending on available discounts and other variables of the marketplace. But, regardless of the variation, the savings with equipment rated at 100 percent will be about 30 percent. Put another way, failure to use CBs rated at 100 percent can add 50 percent to the cost of the CBs.

SWITCH NOT CONTINUOUS-RATED

Fusible switch must be selected to hold fuses rated at 125% of continuous-load current

100-A fuses in 100-A switch

80-A continuous load (3 or more hours) other than motor load

Circuit conductors must have 100-A rating (or may be sized so 100-A protection is next-higher standard rating of protective device above ampacity of conductors).

SWITCH CONTINUOUS-RATED

Bolted pressure switch or high-pressure, butt-contact switch assembly . . .

. . . with Class L fuses

To continuous load at 100% of fuse rating

FIG. 246 Rating of feeder fuse protection depends on the switch assembly.

Sizing feeder and CB using 80%-rated CB
[NE Code **Sec. 220-10(b)]**

CB
2500A frame
2255A trip

2255A feeder conductors

1804A load

1804 x 1.25 = 2255A

Sizing feeder and CB using 100%-rated CB
[NE Code **Sec. 220-10(b), Ex.]**

CB
2000A frame
1804A trip
100% rated

1804A feeder conductors

1804A load

1804 x 1.0 = 1804A

FIG. 247 Continuous-rated CB permits reduced sizing of feeder conductors and protection.

Cost of main switchboard components using 80%-rated main CB

3- 2500A structures — $4,704
1- 2500A tx. connection — 1,568
1- 2500A circuit breaker — 7,696

Total — $13,968

Cost of main switchboard components using 100%-rated main CB

3- 2000A structures — $3,864
1- 2000A tx. connection — 1,288
1- 2000A circuit breaker — 3,864

Total — $9,016

FIG. 248   Use of CBs rated 100 percent provides substantial overall economy.

The selection of a circuit breaker or fuse of the proper current rating must be carefully based on a determination of the real "continuous" rating—whether it is the nameplate value or some lesser value because of derating to supply loads that operate steadily for periods of 3 hours or more. In addition, that same selection task must be related to the need to ensure that the CB or fuse is correctly rated to automatically open the circuit if the current reaches a value that will cause an excessive or dangerous temperature in the conductor or conductor insulation or in other equipment being protected.

Fuses

Fuses comprise a simple and effective means of providing overcurrent protection. Containing only a fusible element and having no moving parts, a fuse requires no maintenance, although the periodic inspection of fuseholders is recommended to ensure good electrical contact. The fusible element of a fuse opens at a precisely determined time in proportion to the amount of current that flows through the fuse. This time-current characteristic varies with the size and type of fuse.

The cartridge fuses used for today's feeder-protection equipment are made in a wide range of types, sizes, and ratings. Various fuse types are designated by NEMA and UL standards. In broad terms, these fuses are classified as (1) one-time, (2) renewable, (3) dual-element, (4) current-limiting, and (5) high-interrupting-capacity.

By present NEMA and UL standards, the standard (so-called **NEC**-type) cartridge fuses of the one-time and renewable types are designated as Class H fuses. Such fuses have interrupting ratings of 10,000 A. Fuses in this category are used in applications where available short-circuit currents are relatively low.

The same standards classify cartridge fuses with interrupting-current ratings above 10,000 rms symmetrical amperes in Classes G, J, K, L, R, and T. These are high-interrupting-capacity or current-limiting fuses. The term "high-interrupting-capacity" indicates an interrupting rating between 10,000 and about 200,000 rms symmetrical amperes, depending upon the particular fuse. A "current-limiting" fuse is one that safely interrupts all available currents within its interrupting rating and limits the peak let-through current I_p and the total ampere-squared-seconds I^2t to specified values.

The term "current-limiting" indicates that the fuse, when tested on a circuit capable of delivering a specific short-circuit current (rms symmetrical amperes) at rated voltage, will start to melt within 90 electrical degrees and will clear the circuit within 180 electrical degrees (one-half cycle). Because the time required for a fuse to melt depends on the available current, a fuse that is current-limiting when subjected to a specific short-circuit current (rms symmetrical amperes) may not be current-limiting on a circuit with lower maximum available current.

Class J and L fuses: Class J (0 to 600 A, 600 V ac) and Class L (601 to 6000 A, 600 V ac) fuses are current-limiting, high-interrupting-capacity fuses. The interrupting ratings are 100,000 and 200,000 rms symmetrical amperes, and the designated rating is marked on the label of each Class J or L fuse. Class J and L fuses are also marked "Current limiting."

Class J fuse dimensions are different from those for standard Class H cartridge fuses of the same voltage rating and ampere classification. They require special fuseholders that do not accept non-current-limiting fuses. This arrangement complies with the last sentence of **NEC** Section 240-60(b), which reads, "Fuseholders for current-limiting fuses shall not permit insertion of fuses that are not current limiting."

Class K fuses: These are subdivided into Classes K-1, K-5, and K-9. Class K fuses have the same dimensions as Class H (standard **NEC**) fuses and are interchangeable with them. Class K-1, K-5, and K-9 fuses have different degrees of current limitation but are not permitted to be labeled "Current-limiting" because their physical characteristics allow them to be interchanged with non-current-limiting types. Classes R and T have been developed to provide current limitation and prevent interchangeability with non-current-limiting types.

Class R fuses: These fuses are made in two designations: RK1 and RK5. UL data are as follows: Fuses marked "Class RK1" or "Class RK5" are high-interrupting-capacity types and are marked "Current-limiting." Although these fuses will fit into standard fuseholders that take Class H and Class K fuses, special rejection-type fuseholders designed for Class RK1 and RK5 fuses will not accept Class H and Class K fuses. In that way, circuits and equipment protected in accordance with the characteristics of RK1 and RK5 fuses cannot have that protection reduced by the insertion of other fuses of a lower protective level.

Other fuse data: Other UL applications data that affect the selection of fuses are as follows:

■ Fuses designated as Class G (0 to 60 A, 300 V ac) are high-interrupting-capacity types and are marked "Current limiting." They are not interchangeable with other fuses mentioned above and below.

- Fuses designated as Class T (0 to 600 A, 250 and 600 V ac) are high-interrupting-capacity types and are marked "Current limiting." They are not interchangeable with other fuses mentioned above.

- Class K-1, K-5, and K-9 fuses are marked, in addition to their regular voltage and current ratings, with an interrupting rating of 200,000, 100,000 or 50,000 A (rms symmetrical).

- Class RK1, RK5, J, L, and T fuses are marked, in addition to their regular voltage and current ratings, with interrupting rating up to 200,000 A (rms symmetrical).

- Class G fuses are marked, in addition to their regular voltage and current ratings, with an interrupting rating of 100,000 A (rms symmetrical).

- Class L fuses are designed for use in equipment to which line and load connections are made by means of solid busbars. For this reason, temperature rises on Class L fuse blades may exceed those observed in connection with other cartridge-fuse designs. Terminal connections for wires in such equipment must be designed to avoid excessive temperatures on the wire insulation.

Some manufacturers provide fuses that are advertised and marked to indicate that they have "time-delay" characteristics. In the case of Class G, H, K, and RK fuses, their time-delay characteristics (minimum blowing time) have been investigated. Class G fuses, which can carry 200 percent of their rated current for 12 s or more, and Class H, K, and RK fuses, which can carry 500 percent of their rated current for 10 s or more, may be marked "D" or "Time delay," or with some equivalent designation. Class L fuses are permitted to be marked "Time delay" but have not been evaluated for such performance. Class J and T fuses are not permitted to be marked "Time delay."

Figure 249 shows the variation in basic operating characteristics of types of fuses commonly used for feeder protection in modern systems.

FIG. 249   Each type of fuse is suited to particular feeder protective functions.

Circuit breakers

Circuit breakers must be chosen to suit the application, whether they are molded-case breakers or low-voltage air CBs. Molded-case breakers may be of the thermal-magnetic or fully magnetic type; air CBs may be installed in conjunction with current-limiting fuses. Circuit breakers may incorporate electrical operation, adjustable time delays, and protective relay hookups; a variety of other features, such as trips for ground-fault protection, may be required and should be carefully selected from manufacturers' data.

The two basic types of breakers, according to general construction, are the molded-case CB and the air (ac power) CB. There is some overlap in the continuous-current ratings of these devices, but the air CB is a larger, heavier-duty device than the molded-case breaker, has generally higher short-circuit interrupting ratings, and is available with current ratings up to 6000 A, whereas the molded-case device is commonly used only up to 3000 A. The use of current-limiting fuses in molded-case or power CBs extends the application of both types with respect to short-circuit interrupting ratings.

The availability of molded-case CBs in sizes up to 3000 A has stimulated their use where air CBs would have been required in past years. Up to 3000 A, the selection of a molded-case CB instead of an air CB offers substantial economy in space and equipment costs. Required interrupting duty (symmetrical short-circuit current rating) is an important selection criterion that may often favor air CBs over molded-case CBs where either may be used. But the use of current-limiting fuses in conjunction with molded-case CBs can provide fully adequate interrupting capacity.

The selection of a molded-case CB often involves a choice of CB trip arrangment. For feeders up to 400 A, the most commonly used molded-case CB that fully meets the needs of circuit protection is the unit that has a noninterchangeable trip assembly made up of a fixed value of thermal overload pickup current with time delay, and a fixed magnetic trip element that provides instantaneous (no intentional delay) opening of the CB on values of current of 10 or more times its continuous current rating (its thermal-trip setting).

Where adjustable magnetic (instantaneous) trip setting is required, a molded-case CB can be selected with a noninterchangeable, built-in trip assembly that provides adjustment of the magnetic trip setting from, say, 3 to 10 times the continuous rating (the thermal-trip setting). And CB units with interchangeable trip assemblies allow the continuous current rating of installed CBs to be changed from one value to another, up to the CB-frame ampere rating. And each trip-assembly unit has an adjustable magnetic trip element. Such CBs offer flexibility of current rating to meet changes in circuit load conditions or conductor makeup.

Although economy indicates the use of molded-case CBs over air CBs for those circuit sizes where either might be used, a basic concern that affects selection is the presence of an instantaneous trip element in the molded-case CB. Because of this element, a series layout of such CBs can induce cascade operation of two or more breakers if all the CBs are subjected to a fault current above their instantaneous trip settings (Fig. 250). This potential for excessive system outage commonly dictates the use of an air CB as a main ahead of molded-case branch CBs; the main air CB has long-time-delay and short-time-delay trip characteristics but no instantaneous trip element. Thus, on any faults in the molded-case-CB feeders, the air CB will remain closed and allow only a molded-case CB to clear the fault. Selection of CBs for this highly desirable selective coordination depends on careful study of time-current characteristics, interrupting ratings, and available circuit fault current.

Although electromechanical tripping mechanisms have long been the most widely used in molded-case and power circuit breakers, solid-state (electronic) tripping circuitry has grown rapidly in both capability and popularity. Solid-state-trip molded-case circuit breakers are made with ratings of 100 A and higher. Offering adjustable pickup and time settings for continuous current rating, long time delay, short time delay, and the value of instantaneous operation, these solid-state breakers may also incorporate ground-fault protection. These "smart" CBs are fast becoming the standard type in ratings of 1200 A and above.

Another growing capability of both molded-case and power (air) CBs is that of current limitation. Short-circuit-duty ratings up to 200,000 A can provide the same fast clearing and current limitation as current-limiting fuses.

Hybrid-type molded-case CBs, called *encased* or *insulated-case* CBs, combine the characteristics of molded-case and air CBs. Such units, which very closely approach the capabilities of

This CB is either at the supply end of the feeder or is a main in the switchboard or panel

BRANCH: 200-amp CB with instantaneous trip setting of 2000 amps

Feeder

A fault here of over 5000 amps can open both main and branch CBs

MAIN: 1000-amp CB with instantaneous trip setting of 5000 amps (max. setting)

Subfeeders

NOTE: If the main is an air CB or encased CB without an instantaneous trip, the main can ride through the fault and remain closed while the instantaneous trip of the branch CB opens the faulted subfeeder.

FIG. 250 Lack of selective coordination knocks out all subfeeders when a fault develops on one subfeeder.

air CBs, provide for fully selective application with conventional load-side molded-case units. Such CBs include a solid-state trip assembly that practically eliminates the effect of instantaneous operation described above. With a short time delay of 18 cycles, such units without instantaneous trip (actually an instantaneous-trip override) afford selective coordination that has previously been available only from the larger, more expensive open-type air CBs.

Low-voltage, open-type air circuit breakers are heavy-duty switching and protective devices with high interrupting capacities. Such CBs have rugged contact assemblies, provisions for arc suppression, and either manually or electrically powered operation. They are available either as individually mounted units—in wall or floor mounting enclosures—or in switchboards or substations, as follows:

1. In ratings from 15 A up to 6000 A continuous current, with voltage ratings of 240, 480, and 600 V.

2. With interrupting ratings that vary with size and voltage rating from 15,000 to 150,000 rms asymmetrical amperes (and beyond 200,000 A in combination with suitable current-limiting fuses).

3. With fully magnetic tripping characteristics based on as many as three individual elements: long time delay, short time delay, and instantaneous trip. These separate elements may be used singly or in any combination to accomplish time-current curve shaping for any selective coordination scheme including omission of the instantaneous trip setting to assure full selectivity with load-side devices having instantaneous trip. Adjustments are provided for the long- and short-time-delay pickup settings. The instantaneous trip may be set at the factory or may be adjustable.

Low-voltage power circuit breakers provide for circuit-breaker application above the range of molded-case CBs (3000 A). They are suited to applications where high switching frequency demands a rugged unit to handle heavy current: for such applications, the air CB is the best choice even within the current ratings of molded-case CBs.

Short-Circuit Protection

The selection of circuit overcurrent devices must include careful consideration of the ability of the devices to operate properly and safely on short-circuit faults—phase-to-ground and phase-to-phase faults in grounded wiring systems, and phase-to-phase faults in ungrounded wiring systems. In all cases, the following points are important:

1. Under normal operation, a circuit draws current in proportion to the voltage applied and the impedance of the load. When a short circuit occurs, the source voltage no longer encounters the opposition to current flow that the normal load had presented. Instead, the voltage is applied across a load of much lower impedance, made up of the impedance of the conductors from the source of the voltage to the short-circuit fault, the impedance of the transformer from which the circuit is derived, and any other impedances due to equipment interposed in the circuit between the transformer and the fault.

   In the top part of Fig. 251, the amount of current which will flow through the short circuit is determined from the system voltage and the total impedance connected in the path of current flow from the source to the fault. This impedance includes resistance and reactance in the conductors, in one or more transformers (going from the fault back to the source), and in any equipment connected in the path of current flow. If, as shown, a source of infinite capacity is assumed on the primary side of the transformer—as if it were a source of voltage with no internal impedance—then the impedance that determines the amount of short-circuit current on the secondary of the transformer consists of the impedance of the transformer itself and the impedance of the secondary conductors up to the fault.

   In a transformer circuit, the "impedance" of the unit is involved in the ability of the transformer to supply short-circuit currents into faults on the load side of the transformer. The impedance of a transformer is the opposition which the transformer presents to the flow of short-circuit current through it.

   As shown at the bottom of Fig. 251, under normal conditions of operation (a), a transformer winding can be considered to be a source of voltage E_s. This source is made up of a generator producing open-circuit voltage E_{oc} and having internal resistance R'' and inductive reactance X_L''. Under normal conditions, load current I_L flows, determined by the transformer

A-Normal Operation   **B-fault condition**

FIG. 251 Elements of a short-circuit condition.

impedance, the impedance of the circuit conductors, and the impedance of the load. When a short occurs (*b*), the transformer open-circuit voltage E_{oc} is connected across a total load made up only of the transformer impedance and whatever part of the circuit-conductor impedance is in the short circuit. If the supply to the transformer primary can deliver the necessary primary current, the secondary short-circuit current is equal to E/Z. Thus the transformer impedance is the limiting factor for short-circuit current.

All transformers have impedance, which generally is expressed as a percentage of the normal rated primary voltage. That percentage of the primary voltage must be applied to the transformer to cause full-load rated current to flow in the short-circuited secondary. For instance, if a 480- to 120-V transformer has an impedance of 5 percent, then 5 percent of 480 V, or 24 V, applied to the primary will cause rated load current to flow in the short-circuited secondary. If 5 percent of the primary voltage will cause such current, then 100 percent of the primary voltage will cause 100/5 (or 20) times the full-load rated secondary current to flow through a solid short circuit on the secondary terminals.

From the foregoing, it can be seen that the lower the impedance of a transformer of given kilovoltampere rating, the higher will be the short-circuit current which it can deliver. Take two transformers, each rated at 500 kVA. Assume the rated secondary load current is the same for both transformers. If one transformer is rated at 10 percent impedance, it can supply 100/10 (or 10) times the rated secondary current into a short circuit on its secondary terminals. If the other transformer is rated at 2 percent impedance, it can supply 100/2 (or 50) times the rated secondary current into a short circuit on its secondary terminals. Thus, the second transformer can supply five times as much short-circuit current as the first, even though they have the same load-handling ability.

Figure 252 presents a simple calculation that yields the symmetrical, single-phase, short-circuit current. The short-circuit protective device must be capable of safely interrupting this

Assuming negligible line and other impedances between the transformer and the fault:

1. Transformer full-load secondary current $= \dfrac{100,000 \text{ va}}{240 \text{ v}} = 417 \text{ amps}$

2. Maximum short-circuit current based on transformer impedance $=$

$$\frac{100\%}{\%Z_T} \times \text{secondary current}$$

Max. $I_{sc} = \dfrac{100}{2.5} \times 417 = 16,680 \text{ amps symmetrical}$

FIG. 252   Basic idea behind short-circuit calculations.

value of current and the asymmetrical current value that is obtained by applying some multiplier (such as 1.25) to the symmetrical value. In a more rigorous analysis, this value would be reduced by all impedances in the circuit.

2. Every overcurrent protective device, in a circuit at any voltage level, must be capable of interrupting the maximum possible short-circuit current that might be delivered by the system into a solid short on the load terminals of the device—without destroying itself in the process. This is shown in Fig. 253, and the short-circuit protective device in Fig. 252 must be rated for at least the 16,680-A calculated current.

All short-circuit protective devices. . .

. . . must have an interrupting rating at least equal to the maximum fault current that the circuit could deliver into a short circuit on the *line side* of the device.

NOTE: That means that the fault current "avàilable" at the line terminals of all fuses and circuit breakers *must be known* in order to assure that the device has a rating sufficient for the level of fault current.

FIG. 253   All feeder fuses and circuit breakers must have adequate interrupting capacity.

3. Coordinated selective protection for modern circuits provides fast, effective isolation of any faulted section of a system but does not interrupt service to any other section. A careful study of the time-current characteristics of protective devices along with correct application can ensure clearing of a fault by the device nearest to the fault on its supply side, as shown in Fig. 254.

   Selective coordination of overcurrent protection, in which every fault is cleared by the closest protective device on the line side of the fault, is widely considered to be a particular capability of modern CB equipment—either CBs alone or CBs in conjunction with load-side fusible equipment. Selective coordination ensures fault-initiated opening at only one point in the system, such as on a branch circuit or a feeder; it eliminates cascade operation, where, say, both a branch device and a feeder device open on a branch-circuit fault.

   Curve-shaping of the time-current trip characteristics of both molded-case and air CBs by means of long-time-delay, short-time-delay, and/or instantaneous-trip elements is the technique by which sophisticated selective coordination can be achieved, even through a number of voltage transformations. Only the faulted circuit is opened, and the extent of the outage is minimized. But, again, such coordination must be carefully and precisely related to all other design, installation, operational, and economic factors.

4. Let-through current is another factor in the effective application of protective deices for short circuits. A given protective device may be able to interrupt the maximum short-circuit current at its point of installation safely, but it may take so much time to open the faulted circuit that damage will occur to equipment or devices connected in series with the fault. Unless the device operates quickly enough, the tremendous rupturing stresses created by short circuits can cause severe damage. The system components must be related to the let-through current,

NOTE: Any fault is cleared by nearest device on supply side of fault, and all other devices in series with the fault do not operate.

FIG. 254   Selection coordination of overcurrent devices minimizes the extent of an outage due to a fault.

the current which flows from the time the fault develops until the circuit is opened. That is required by **NEC** Section 110-10, as shown in Fig. 255. Current-limiting fuses, for instance, open a short circuit in much less than half a cycle, thereby squelching let-through current.

Because a short-circuit fault is an accidental condition that occurs at random with respect to the alternating voltage wave of a circuit, the fault can occur at any point in the wave form of a

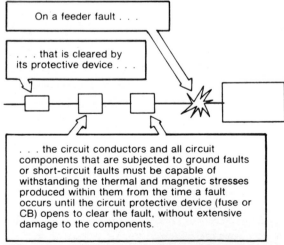

FIG. 255   Operating speed of short-circuit protection must be related to equipment "withstandability" ratings.

voltage cycle. The two extremes of the wave form are where the voltage wave passes through the "zero" axis and where the voltage is a maximum. Figure 256 shows these two extreme conditions at which short-circuit current can start and continue to flow if the circuit is not opened by fuses or breakers. Maximum asymmetry in the flow of short-circuit current is produced when the short circuit occurs at that point in the voltage wave for which the short-circuit current is exactly at its positive or negative peak value, as determined by the ratio of reactance to resistance in the short circuit; this ratio establishes the phase-angle difference between the voltage wave and the short-circuit current wave. In the case shown in Fig. 256a, with the short circuit occurring as shown on the voltage wave, the phase angle is such that the short-circuit current is exactly at its negative peak. Maximum asymmetry is thus produced. The instantaneous change in phase relation between current and voltage (from the normal condition to fault condition) produces an offset in the ac current wave which acts like a dc transient.

A–MAXIMUM ASYMMETRY

B–COMPLETE SYMMETRY

FIG. 256 How short-circuit currents could flow if they were not opened by overcurrent protection.

From the immediate condition of asymmetry, the dc component of the short-circuit current wave decays to zero at a rate determined by the ratio of reactance to resistance in the short circuit. For a purely resistive short circuit (zero ratio of reactance to resistance), which is a completely theoretical condition that can never be achieved in an ac circuit, the decay to zero of the dc component will be instantaneous. If the short circuit is purely reactive (infinite ratio of reactance to resistance), which is also a practical impossibility, the dc component never decays to zero, the current remains offset, and the asymmetrical rms current is equal to 1.73 times the symmetrical rms current. Of course, actual circuits have ratios of reactance to resistance up to about 20, although the vast majority of cases have ratios of less than 7. This means the dc offset current will decay to zero in several cycles.

In Fig. 256b, a short-circuit current wave of complete symmetry flows when the short occurs at that point in the voltage wave for which the instantaneously initiated short-circuit current is at its zero value. Again the exact point in the voltage wave which will correspond in time to a zero value for the short-circuit current wave is determined by the phase angle (or power factor, or ratio of reactance to resistance) of the short circuit.

Although short circuits can start at the two extremes shown in Fig. 256, chances are greatest that a fault will occur at some intermediate degree of asymmetry. But because design must account for the worst possibility, short-circuit calculations should be performed to determine the asymmetrical current which represents the most destructive current that the protective device might be called upon to break.

From the character of the asymmetry, it can be seen that the ratio of reactance to resistance is important in applying protective devices because it indicates how fast the dc will decay and how much current the device will be called upon to handle in an attempt to open the circuit in the first half cycle, or in the second cycle, or the third, etc.—since the current decreases with each succeeding cycle. A fast-acting fuse such as a current-limiting fuse, which opens the short circuit in the first half cycle, will break almost the maximum value of the asymmetrical wave and must be rated for that current value. On the other hand, a circuit breaker which will not act until the fourth cycle of the short-circuit current wave will have to interrupt only the level of current to which the asymmetrical wave has decayed.

On a fully asymmetrical short, the maximum instantaneous current—the peak asymmetrical current—occurs at the end of the first half cycle of the current wave. Owing to decay of the dc component, subsequent peaks are lower in value. Various values for this half-cycle asymmetrical current have been determined from study of actual circuits and are shown in Fig. 257.

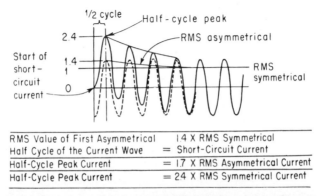

RMS Value of First Asymmetrical Half Cycle of the Current Wave	1.4 X RMS Symmetrical = Short-Circuit Current
Half-Cycle Peak Current	= 1.7 X RMS Asymmetrical Current
Half-Cycle Peak Current	= 2.4 X RMS Symmetrical Current

FIG. 257 Fast-acting fuses and CBs must be rated for the current value in the first half cycle.

The foregoing discussion explains the use of multipliers to determine the asymmetrical current from the calculated symmetrical value, such as that arrived at in Fig. 257. For fuses, which operate relatively fast, a multiplier of 1.4 is commonly applied to the rms symmetrical short-circuit current to get the rms asymmetrical value. For circuit breakers which operate almost as fast as some fuses, the same multiplier would be used in determining the possible asymmetrical current when the breaker opens the circuit. But for breakers which operate more slowly, say two or three cycles after the fault starts, a multiplier of only 1.25 or 1.1 will give the value to which the asymmetrical current has decayed when the circuit is opened.

The value of 1.4, mentioned above as the multiplier for determining the first-half-cycle value of the rms asymmetrical current, has been established for a ratio of reactance to resistance which is not exceeded in the majority of cases. Thus, this multiplier yields the value to which the asymmetrical current has decayed in the first half cycle. But in circuits with higher-than-normal ratio of reactance to resistance, the decay will not be as great in the first half cycle. A multiplier of, say, 1.5 or 1.6 would be required to compute the possible value of the first-half-cycle rms asymmetrical current in such circuits. In all applications, when multipliers are used, the X/R ratio to which the multiplier applies should be known, as well as the X/R ratio of the circuit being protected.

Figure 258 shows the application of current-limiting fuses to quickly clear ground and short-circuit faults and to limit the energy let-through on circuits which could deliver fault currents over 100,000 or 200,000 A. On a given short circuit, a current-limiting fuse may be called upon to interrupt either of the two extreme conditions shown here, or some intermediate condition. In the fully asymmetrical case, the fuse would limit both rms current and peak available current if the melting time of the fuse were anything less than one-half cycle. But within its current-limiting range, a true current-limiting fuse must clear the circuit in a time interval that is not greater than the duration of the first symmetrical current loop, and it must limit the peak let-through current to a value less than the peak available current. Thus, it is the action of a fuse in clearing a symmetrical short circuit that determines its current-limiting ability, as shown in the lower sketch. From the upper sketch, it can be seen that a fuse which has a melting time almost up to the peak of the asymmetrical first half cycle will limit current let-through on an asymmetrical fault but will not limit let-through if the short happens to occur as a completely symmetrical wave.

The rms value of the triangular wave of let-through current through the fuse is

$$I_{LT} = \frac{I_p}{1.7}$$

where I_{LT} = rms let-through current, A
$\quad\quad I_p$ = peak let-through current, A

FIG. 258 Current-limiting operation of fuse protection.

The application of standard thermal-magnetic or fully magnetic circuit breakers should be based on a correlation of their time-delay tripping curves and instantaneous trip setting with the requirements for overload protection, the current characteristics of the circuit being protected, and the need for coordination in the operation of overcurrent devices connected in series (such as a feeder device in series with a subfeeder device in series with a branch-circuit device). Breaker operation is readily analyzed in relation to a particular circuit by reference to the manufacturer's curves on the breaker.

Figure 259 shows a typical set of curves for a molded-case CB. The curves show trip time versus current for a 225-A breaker with interchangeable trip units. Each trip unit consists of a fixed thermal trip element and an independent adjustable magnetic trip element. The various trip units give the breaker its particular continuous current rating and are available in standard settings: 70, 100, 125, 150, 175, 200, and 225 A. Then the adjustable magnetic element associated with each thermal trip can be adjusted to from 3 to 10 times the current rating of the thermal trip. The curves are shown as bands to cover the tolerances to which the elements are manufactured and within which they will operate. A band can be taken to be a large number of individual curves laid one against another. The curves show time plotted against multiples of any current rating (i.e., multiples of 70 A, or 150 A, etc.). The two operating bands shown

FIG. 259   CB time-current curves reveal overload and short-circuit characteristics.

below 100 s represent the two extreme conditions of setting of the adjustable magnetic trip—3 times the CB rating at left, and 10 times the CB rating at right. Intermediate settings of the magnetic trip would produce bands between the two shown here.

Figure 260 shows the application of current-limiting fuses in combination with a power circuit breaker. Current-limiting fuses can be combined with both molded-case and power circuit breakers. A typical application is represented by the curves shown. The fuses extend the maximum interrupting capacity of the CB to values above 200,000 rms symmetrical amperes.

The three-pole CB is an electrically operated low-voltage power circuit breaker with a continuous rating of 3000 A and an interrupting capacity of 75,000 rms asymmetrical amperes at 480 V. The CB has three magnetic tripping characteristics in a 2000-A trip device. The first, the long-time-delay trip, is adjusted to pick up at 2000 A. The second, the short-time-delay trip, is adjusted to pick up at 6000 A. The third trip is the instantaneous trip, set for 12 times the coil rating, or 24,000 A. The tolerance on pickup settings is ±15 percent.

The time-delay trips are factory adjusted for required delays. The fuses used are rated at 3000 A continuous carrying capacity with interrupting capacity of 200,000 A. The coordination between the fuse operating band and the breaker tolerance band is shown in the graph. The

FIG. 260 Fuse-CB combinations depend upon coordinated operation.

combination provides CB operation on faults up to about 60,000 A. Both devices operate above that.

Fuse or CB?

In modern electrical systems, both fuses and circuit breakers are used to provide required overcurrent protection for circuits and equipment. The choice between fuses and a circuit breaker for any protective application involves careful evaluation of the cost, capabilities, and characteristics of the particular sizes and types to be used—all related to the characteristics of the load and the thermal and magnetic damage limits of equipment to be protected. Fuses and circuit breakers each have advantages and limitations, and different electrical users tend to prefer one device over the other, for various reasons. Although the many factors that are involved in a choice between fuses and CBs reveal almost as much art as science in the selection process, certain **NEC** rules and UL regulations dictate the use of only fuses in certain applications.

The basic concerns in choosing between fuses and breakers are obvious, but the choice should be carefully evaluated. A CB, for instance, is able to reclose immediately after a fault trip, because it usually is not damaged in the fault interruption and does not require maintenance or repair. Although there are exceptions to that ability, such immediate reclosure is often viewed as an important advantage over fuses, which must be replaced after blowing and which thereby entail labor and replacement costs as well as some loss of time in restoring the circuit. However, when a fault in a circuit opens a protective device, time and effort are usually required to repair the fault condition, so that immediate reclosing of the protective device is often not even possible. And the initial economy of fusible equipment relative to CBs, particularly in applications demanding high short-circuit capacity, commonly far outweighs the costs of labor and fuse replacements over the life of the overall system—even where the number of fault operations might be greater than average.

Certainly a major factor today in choosing between fusible protection and CBs is the matter of interrupting capacity (or short-circuit duty). With the continual beefing up of utility supply systems and the increasing use of transformers with higher kilovoltampere ratings to handle the growth in average electrical loads, overcurrent protection rated above 50,000 rms symmetrical amperes has become commonplace. For such applications, the ready availability of fuses rated at 100,000 and 200,000 A promotes the operating (as well as the economic) advantage of using fuses or fuse devices—either instead of CBs or in conjunction with CBs.

However, there are applications where the adjustable trip characteristics of circuit breakers—long time delay, short time delay, and instantaneous—provide the only means for precisely selective coordination in elaborate electrical systems. In such systems, the use of molded-case CBs with extremely versatile solid-state trip mechanisms (including ground-fault protection) can provide highly engineered protection. And the availability of current-limiting circuit breakers with 200,000 A short-circuit-duty ratings makes possible high-capacity total overcurrent protection using only circuit breakers, even for the largest and most sophisticated electrical systems.

Current-limiting circuit breakers of both the molded-case and power types can completely interrupt a fault current in less than one-quarter cycle. The action is similar to that shown for a current-limiting fuse in Fig. 258. Such breakers can sustain many such fault interruptions in circuits with up to 200,000 A available and may be restored to the closed position as soon as the fault is removed. A typical 100-A molded-case current-limiting CB, rated for 100,000 rms symmetrical amperes, can operate in a circuit with a prospective peak fault current (the current that could flow if the fuse were not in the circuit) of 230,000 A, and the CB will limit the current to not over 25,000 A. Such breakers are listed by UL for panelboard main protection where the branch CBs have 10,000-A interrupting capacity but the entire panelboard assembly is rated for 100,000-A short-circuit duty.

Locating Protection

All feeder overcurrent protective devices must satisfy **NEC** Section 240-24 with respect to installation location. The basic rule requires that fuses and circuit breakers must be readily accessible. In accordance with the definition of "readily accessible," they must be "capable of being reached quickly for operation, renewal, or inspections, without requiring those to whom ready access is requisite to climb over or remove obstacles or to resort to portable ladders, chairs, etc." (Fig. 261).

Although the **NEC** gives no maximum heights below which overcurrent protective devices are considered readily accessible, some guidance can be obtained from Section 380-8, which provides detailed requirements for the location of switches and CBs. This section states that switches and CBs must be installed so that the center of the grip of the operating handle, when in its highest position, is not more than 6½ ft above the floor or working platform.

An exception to the basic rule covers any case where an overcurrent device is used in a busway plug-in unit to tap a branch circuit from the busway. Section 364-12 requires that such devices consist of an externally operable CB or an externally operable fusible switch. These devices must be capable of being operated from the floor by means of ropes, chains, or sticks.

Section 240-24 clarifies the use of plug-in overcurrent protective devices on busway for the protection of circuits tapped from busway. After the general rule that overcurrent protective

Overcurrent protection is not "readily accessible" . . .

. . . if a portable ladder is needed to get at it.

6'6" max.

Handles of switches and CBs must be not more than 6½ ft above floor or platform (Sec. 380-8).

Overcurrent device in a panel, switch, CB, switchboard, MCC is not readily accessible . . .

. . . if crates or other obstacles block access to it.

FIG. 261   Feeder overcurrent devices should be installed so as to be "readily accessible."

devices must be readily accessible (capable of being reached without stepping on a chair or table or resorting to a portable ladder), an exception notes that it is not only permissible to use busway protective devices up on the busway but it is required by Section 364-12. Such devices on high-mounted busway are not "readily accessible" (not within reach of a person standing on the floor). The wording of Section 364-11 makes clear that this requirement for overcurrent protection in the device on the busway applies to subfeeders tapped from the busway as well as branch circuits tapped from the busway. Figures 262 and 263 cover the rules of **NEC** Sections 240-24, 364-12, and 380-8—all of which are related.

Overcurrent protection—either a fused switch or a circuit breaker—is usually required in each subfeeder tapping power from a busway feeder. This is necessary to protect the lower current-carrying capacity of the subfeeder, and the device should be placed at the point at which the subfeeder connects into the feeder. However, the **NEC** provides that overcurrent protection may be omitted "at points where busways are reduced in size, provided that the smaller busway does not extend more than 50 ft and has a current rating at least equal to one-third the rating or setting of the overcurrent device next back on the line" (Fig. 264).

Figure 265 summarizes requirements regarding the location of switches and circuit breakers.

Plug-in connection for tapping-off branch circuit shall contain overcurrent protection, which does not have to be within reach of person standing on floor.

Branch circuit tap from busway

Plug-in connection for tapping-off a feeder or subfeeder shall contain overcurrent protection, which does not have to be within reach of person standing on floor.

Feeder or subfeeder tap from busway

NOTE: To get at overcurrent protection in either case, personnel would have to use portable ladder or chair or some other climbing technique.

FIG. 262   These overcurrent devices are not required to be readily accessible.

Plug-in device must be externally operable circuit
breaker or fused switch that provides overcurrent
protection for the subfeeder or branch circuit
tapped from the busway

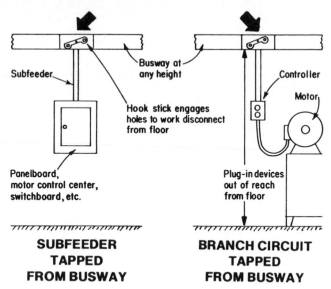

SUBFEEDER
TAPPED
FROM BUSWAY

BRANCH CIRCUIT
TAPPED
FROM BUSWAY

**IMPORTANT! The plug-in switch or CB unit must be able to be
operated by a hookstick or chain or rope operator if the unit is
mounted out of reach up on a busway and contains disconnect
means. Section 380-8 says *all* busway switches and CBs must
be operable from the floor.**

FIG. 263   Circuit tapped from busway must have protection on the
busway.

FIG. 264   Busway subfeeder not over 50 ft long does not *require*
protection.

6½ ft max. from floor or platform up to center of handle in its highest position

Floor

EXCEPTIONS

1. Fused switch or CB may be up on busway . . . .

Busway

Out of reach from floor

. . . . but means must *always* be provided to operate handle from the floor (i.e., hookstick, etc.)

2. Switch adjacent to motor, appliance, or other equipment it supplies, at high mounting, but accessible by portable ladder or similar means

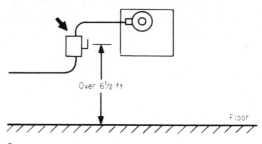

Over 6½ ft

Floor

3. Hookstick-operable isolating switches are permitted at heights over 6½ ft

FIG. 265 Switches and CBs must have ready access for effective operation.

Feeder Taps

Although basic **NEC** requirements dictate the use of an overcurrent device at the point at which a conductor receives its supply, as described above, exceptions to this rule are made in the case of taps to feeders. That is, to meet the practical demands of field application, certain limited lengths of unprotected conductors may be used to tap energy from protected feeder conductors.

Exceptions to the rule for protecting conductors at their points of supply are made in the case of 10-, 25-, and 100-ft taps from a feeder (as described in Exceptions No. 2, No., 3, and

No. 10 to Section 240-21). The tap exceptions should be applied carefully, to effectively minimize any sacrifice of safety. The exceptions permit smaller conductors to tap from larger conductors without overcurrent protective devices sized to protect the smaller conductors at their point of supply.

Unprotected taps not over 10 ft long may be made from feeders or transformer secondaries provided:

1. The smaller conductors have a current rating that is not lower than the combined computed loads of the circuits supplied by the tap conductors and have an ampacity of

Not less than the rating of the "device" supplied by the tap conductors, or

Not less than the rating of the overcurrent device (fuses or CB) that might be installed at the termination of the tap conductors.

2. The tap does not extend beyond the switchboard, panelboard, or control device which it supplies.

3. The tap conductors are enclosed in conduit, EMT, metal gutter, or other approved raceway when not a part of the switchboard or panelboard.

Figure 266 shows the basic concepts behind the use of 10-ft unprotected tap conductors. The **NEC** rule specifically recognizes that a 10-ft tap may be made from a transformer secondary in the same way it has always been permitted from a feeder. In either case, the tap conductors must not be over 10 ft long and must have ampacity not less than the ampere rating of the switchboard, panelboard, or control device. Or, the tap conductors may be terminated in an overcurrent protective device rated not more than the ampacity of the tap conductors.

A 10-ft feeder tap may terminate in main-lugs-only of a panelboard or other equipment supplied. It does not have to terminate at the terminals of a single circuit breaker or fused switch, as is required for the 25-ft taps described next.

Taps not over 25 ft long (Fig. 267) may be made from feeders or transformers without overcurrent protection for the tap conductors at their supply end, provided:

1. The smaller conductors have a current rating at least one-third that of the conductors from which they are tapped.

2. The tap conductors are suitably protected from mechanical damage by enclosure in a conduit or other raceway.

3. The tap is terminated in a single circuit breaker or set of fuses that will limit the load on the tap to the ampacity of the tap conductors as designated in **NEC** Table 310-16.

NOTE: The length of any unprotected feeder tap is the *cut length of the conductors,* and not the conduit length.

Examples of taps

Figure 268 shows the use of a 10-ft feeder tap to supply a single motor branch circuit. The conduit feeder may be a horizontal run or a vertical run, such as a riser. If the tap conductors are of such size that they have a current rating at least one-third that of the feeder conductors from which they are tapped, because they terminate in a single CB or set of fuses they satisfy Exception No. 3 of **NEC** Section 240-21 and could be run a distance of 25 ft without protection at the point of tapoff from the feeder or busway. But they must have an ampacity at least equal to the rating or setting of the fuses or CB used for motor branch-circuit protection. Because Section 364-12 requires that any busway used as a feeder must have overcurrent protection on the busway for any subfeeder or branch circuit tapped from the busway, the use of a cable tap

FIG. 266 A 10-ft unprotected tap may be made from either a feeder or a transformer secondary.

box on busway without overcurrent protection (which is acceptable for the conduit installation of Fig. 268) would be a clear violation of the **NEC**.

A common application of the 10-ft-tap exception is the supply of panelboards from conduit feeders or busways, as shown in Fig. 269. The figure shows an interesting requirement that arises from the rule of **NEC** Section 384-16, which requires that lighting and appliance panel-

FIG. 267 Feeder taps not over 25 ft (but over 10 ft) must terminate in individual overcurrent protection.

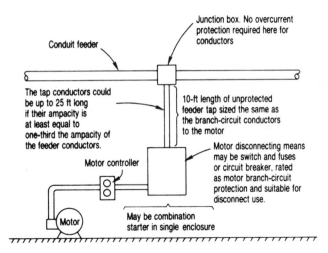

FIG. 268 Feeder tap to motor branch circuit may be 10 or 25 ft long.

boards be protected on their supply side by overcurrent protection rated not higher than the rating of the panelboard busbars. If the feeder is busway, the protection must be placed at the point of tapoff on the busway. In that case a 100-A CB or fused switch on the busway would provide the required protection of the panel, and a main would not be required in the panel. But if the feeder circuit is in conduit, the 100-A panel protection would have to be in the panel or just ahead of it. It could not be at the junction box on the conduit because that would make

Although not required for protection of the 10-foot tap, overcurrent protection is required for lighting and appliance panel to protect it in accordance with its main rating. Thus a 100-amp CB or 100-amp fuses must be installed AT THE PANEL. Such protection could not be installed at the point of tap-off because it would then not be readily accessible and would violate Section 240-24.

3-phase, 4-wire 600-amp feeder

600-amp feeder protection

① Tap length not over 10 feet does not require overcurrent protection at point of supply to tap conductors.

② Tap conductors must have current rating not less than panel main protection.

3-phase, 4-wire lighting and appliance panelboard with 100-amp mains.

FIG. 269 The 10-ft feeder tap is commonly made to supply a panelboard.

it not readily accessible and therefore a violation of Section 240-24. With a conduit feeder, a fused-switch or CB main in the panelboard could be rated up to the 100-A main rating.

In Fig. 269, the tap would be limited to 10 ft in length, even though it terminates in a single 100-A protective device, because the 100-A rating of the tap conductors does not satisfy the requirement that they have an ampacity at least one-third the 600-A rating of the feeder conductors being tapped. But, the 200-A rating of No. 3/0 THW copper conductors is one-third of 600 A, and such conductors could be used as tap conductors and run 25 ft to the panel's 100-A main CB or fused switch.

Figure 270 shows the basic conditions for the use of 10- and 25-ft unprotected taps from transformers. And, as shown in Fig. 271, more than one set of 10- or 25-ft tap conductors may be taken from a transformer secondary—provided all the requirements are satisfied. Where more than one 25-ft tap is taken from a single transformer, each set of tap conductors must have an ampacity at least equal to one-third the ampacity of the primary supply conductors as reflected into the secondary.

Figure 272 shows the application of Exception No. 8 of Section 240-21 in conjunction with the rule of Section 450-3(b)(2) covering transformer protection. As shown in Example 1, the 100-A main protection in the panel is sufficient protection for the transformer and the primary and secondary conductors when these conditions are met:

1. The tap conductors have ampacity at least equal to one-third that of the 125-A feeder conductors.
2. The secondary conductors are rated at least one-third the ampacity of the 125-A feeder conductors, based on the primary-to-secondary transformer ratio.
3. The total tap length is not over 25 ft, primary plus secondary.
4. All conductors are in conduit.
5. The secondary conductors terminate in the 100-A main protection that limits the secondary load to the ampacity of the secondary conductors, simultaneously provides the protection required by the lighting panel, and is not rated over 125 percent of the transformer secondary current.
6. The primary feeder protection is not over 250 percent of the transformer rated primary current, as recognized by Section 450-3(b)(2), and the 100-A main breaker in the panel serves

10-FT TAP

1. A 10-ft tap may be made from transformer secondary to a panel, switchboard, MCC, etc.

2. If this is a lighting panel that requires main protection, a fused switch or CB must be installed as a main protective device in the panel or just ahead of it, at the end of the 10-ft tap.

3. If the panel is *not* a lighting panel (such as a panel with 240-volt or 480-volt heating circuits or other makeup that does not make it a lighting panel as specified in Section 384-14), then main protection is not required at all, and the 10-ft tap conductors terminate in main lugs of the panel switchboard, or other equipment.

25-FT TAP

1. If transformer secondary feeds lighting panel **having a main CB or fused switch,** then . . .

2. . . . . secondary tap conductors from transformer may be 25 ft long, as permitted by Section 240-21, Exception No. 3, but *only* where the tap terminates in a single CB or set of fuses.

3. Or, a 25 ft tap may be made from a transformer to a CB or fused switch in an individual enclosure or serving as a main in a switchboard or MCC.

FIG. 270 Unprotected taps from a transformer secondary must satisfy the specific conditions of use.

as the required "overcurrent device on the secondary side rated or set at not more than 125 percent of the rated secondary current of the transformer."

In Example 2, the tap conductors from the primary feeder to each transformer may be the same size as the primary feeder conductors, or they may be smaller than the primary conductors if they are sized in accordance with Exception No. 8 of Section 240-21; this exception permits a 25-ft tap from a primary feeder to be made up of both primary and secondary tap conductors. The 25-ft tap may have any part of its length on the primary or secondary, but it must not be longer than 25 ft and must terminate in a single CB or set of fuses.

Figure 273 shows another application of Exception No. 8 and Section 450-3(b)(2). Because the primary wires tapped to each transformer from the main 100-A feeder are also rated 100 A and are therefore protected by the 100-A feeder protection, the entire primary circuit to each

Transformer with
400A-rated secondary

200A, 208/120V,
3 φ, 3W panels
with 200A main
CB or fused
switch in each

10-ft secondary
top conductors rated 200A

NOTE: From a single transformer secondary of adequate capacity, more than one set of 10-ft tap conductors may be run to more than one panel or other distribution equipment.

In the case shown here, because each of the two taps terminates in a fused switch or CB main in each panel, the condition for a 25-ft tap is satisfied and the taps may be run for 25 ft instead of 10 ft. The 200-A rating of each tap is greater than 1/3 of the 400-A transformer secondary rating to which the primary supply conductors are sized at the secondary-to-primary turns ratio.

FIG. 271 More than one set of 10- or 25-ft tap conductors may be used with a single transformer secondary.

EXAMPLE 1:

480V
2-wire
circuit

125A fuses

125A conductors

50A rated wire

480V

50A primary
full-load current

25 ft max
conductor
length

240/120V

100A rated wire

100A main
protection

240/120V 3-wire
panel for lighting

EXAMPLE 2:

Primary conductors
protected by fuses or CB

Top conductors from
primary feeder to
each transformer

Primary
protection,
not over 250 %
of primary
current of any
transformer

Each secondary
protection rated
not over 125 % of
that transformer's
rated secondary
current

FIG. 272 Two examples show a 25-ft tap of primary-plus-secondary conductors.

Any distance

100A fuses

100 A rated conductors

25 ft max conductor length

480 V 2-wire circuit

Each transformer has a rated primary current of 50 A

480V ⟩ 240/120V

100 A rated conductors

480V ⟩ 240/120V

Transformers

100 A rated secondary conductors are 1/3 ampacity of primary conductors (100 A x ratio of sec-to-pri voltage = 100 x 1/2 = 50 A, which is greater than 1/3 of 100 A).

Each set of 100 A secondary conductors is suitably protected by conduit and terminates in a 100A protective device which limits the load on the wires to their rated ampacity and also satisfies Sec. 450-3 (b) (2).

FIG. 273 This example ties tap rules into the need for transformer protection.

transformer is excluded from the allowable 25 ft of tap to the secondary main protective device. The 100-A protection in each panel is not over 125 percent of the rated transformer secondary current. It, therefore, provides the transformer protection required by Section 450-3(b)(2). The same device also protects each panel at its main busbar rating of 100 A.

Figure 274 compares the two different 25-ft-tap techniques covered by Exceptions No. 3 and No. 8.

As shown in Fig. 275, the **NEC** permits unprotected taps to be made from generator terminals to the first overcurrent device that the generator supplies—such as in the fusible switch or circuit breaker used for control and protection of the circuit supplied by the generator. No maximum length is specified for the generator tap conductors. But because the tap conductors terminate in a single circuit breaker or set of fuses rated or set for the tap-conductor ampacity, tap conductors up to 25 ft long would comply with the basic concept for 25-ft taps. The tap conductors must have an ampacity of at least 115 percent of the generator nameplate current rating.

Exception No. 10 of **NEC** Section 240-21 is another departure from the rule that conductors must be provided with overcurrent protection at their supply ends, where they receive current from larger conductors or from a transformer. Exception No. 10 permits a longer length than the 10-ft unprotected tap and the 25-ft tap. Under specified conditions that are similar to the requirements of the 25-ft-tap exception, an unprotected tap up to 100 ft in length may be used in "high-bay manufacturing buildings" that are over 35 ft high at the walls, but only "where conditions of maintenance and supervision assure that only qualified persons will service the system." The phrase "35 ft high at the walls" means that this exception cannot be applied where a triangular or curved roof section is over 35 ft high at the peak but less than 35 ft high at the walls.

The 100-ft tap must meet specific conditions:

1. From the point at which the tap to a larger feeder is made, not more than 25 ft of the tap may run horizontally, and the sum of horizontal run and vertical run must not exceed 100 ft. Figure 276 shows some of the almost limitless configurations of tap layout that would fall within these dimension limitations.

No. 1 Type THW tap,
not over 25 feet long

No. 0000
Type THW
feeder

100 amp
fuses

25-ft tap — EXCEPTION No. 3

480-V three wire
▽ (from 480 Y/277-V supply)

110 A

No. 6 TW(CU)-55 amp
FLPC = 45 amp

37.5-kVA
FLSC = 104 A

25 ft
max.

208Y/120 V

1/0 TW(CU)-125 A

125 A

125-A
lighting panelboard

Taps protected from physical damage.
Secondary-to-primary voltage ratio = 208:480 = 1:2.3

25-ft tap — EXCEPTION No. 8

FIG. 274   The two different 25-ft tap techniques that may be used.

2. The tap conductors must have an ampacity equal to at least one-third the rating of the overcurrent device protecting the larger feeder conductors from which the tap is made.

3. The tap conductors must terminate in a circuit breaker or fused switch, where the rating of the overcurrent protection is not greater than the tap-conductor ampacity.

4. The tap conductors must be protected from physical damage and must be installed in metal or nonmetallic raceway.

**GENERATOR TAP CONDUCTORS
WITHOUT PROTECTION AT SUPPLY END**

FIG. 275 Generator output conductors may be run (unprotected) to the generator controls.

FIG. 276 Unprotected feeder taps up to 100 ft long may be used in "high-bay manufacturing buildings."

5. There must be no splices in the total length of each of the conductors of the tap.

6. The tap conductors must not be smaller than No. 6 copper or No. 4 aluminum.

7. The tap conductors must not pass through walls, floors, or ceilings.

As shown in Fig. 276, the tap conductors from a feeder protected at 1200 A are rated at not less than one-third the protection rating, or 400 A. Although 500-MCM THW copper is rated at 380 A, that value does not satisfy the minimum requirement for 400 A. But if 500-MCM

THHN or XHHW copper, with an ampacity of 430 A, were used for the tap conductors, the rule would be satisfied. However, in such a case, those conductors would have to be used as if their ampacity were 380 A for the purpose of load calculation because of the general UL rule requiring 75°C conductor terminations for connecting to equipment rated over 100 A (such as the panelboard, switch, motor control center, or other equipment fed by the taps). And the conductors for the main feeder being tapped could be rated below the 1200 A shown in the sketch if the 1200-A protection on the feeder were selected in accordance with **NEC** Section 430-62 or 430-63 for supplying a motor load or motor and lighting load. In such a case, the overcurrent protection may be rated considerably higher than the feeder conductor ampacity. But the tap conductors must have ampacity at least equal to one-third the feeder protection rating.

Ground-Fault Protection

Fuses and circuit breakers, applied as described above, are sized to protect conductors in accordance with their current-carrying capacities. If the current exceeds the rating of the fuse or the trip setting of the CB, the device opens the circuit—either instantaneously or after some time delay, depending upon the actual level of the fault current. This excessive current might be caused by an operating overload, by a phase-to-ground fault, or by a short circuit between two circuit conductors. Thus, a 1000-A fuse will blow if current in excess of that value flows over the circuit. It will blow earlier on heavy overcurrent and later on lower overcurrent. But it will blow, and the circuit and equipment will be protected against damage due to the overcurrent.

Another type of fault condition is very common in grounded systems but will not be cleared by conventional overcurrent devices. That is the phase-to-ground fault (invariably by arcing contact between an energized circuit conductor and a grounded metal enclosure or conduit) having a current value below the rating of the overcurrent device.

Although the **NEC** does require GFP (ground-fault protection) on service entrances under certain conditions (Section 230-95), there is no general GFP requirement for feeders or branch circuits. (Section 517-14 does require GFP on feeders in health-care facilities.) However, the advantages of GFP do recommend its use for feeders, subfeeders, and even large branch circuits in many commercial and industrial buildings.

The vast majority of modern grounded electrical systems, for which huge sums of money are expended on meticulously designed systems of conventional overcurrent protection, are completely unprotected against the most common type of electrical failure—low-current ground faults (from one phase leg to a grounded equipment enclosure). The fuses and/or CBs might be selected to have the highest interrupting capacities required by potential short-circuit currents, and such factors as time delay and current limitation may be perfectly tailored to the needs of the particular system. But these vast sums of money are expended to protect very expensive electrical systems (and valuable industrial processes or commercial operations) against the type of electrical fault which almost never happens—the bolted 3-phase short circuit on the load terminals of a protective device. At the same time, in spite of these expensive overcurrent provisions, the grounded system is totally unprotected against the fantastically destructive effects of the very common phase-to-ground fault. (See Fig. 277.)

On a high-capacity feeder, a line-to-ground fault (i.e., a fault from a phase conductor to a conduit, to a junction box, or to some other metallic equipment enclosure) can and frequently does draw current whose value is below the rating or setting of the circuit protective device. For instance, a 500-A ground fault on a 2000-A protective device that has only a 1200-A load will not be cleared by the device. If such a fault is a "bolted" line-to-ground fault (a highly unlikely fault), a certain amount of heat will be generated by the I^2R effect of the current; but this will usually not be dangerous, and such fault current will merely register as additional operating load, with wasted energy (wattage) in the system. But, bolted phase-to-ground faults are very

FIG. 277 Fuses and circuit breakers will not respond to low-level arcing ground faults.

rare. The usual phase-to-ground fault exists as an intermittent or arcing fault, and an arcing fault of the same current rating as the essentially harmless bolted fault can be fantastically destructive because of the intense heat of the arc.

Of course, any ground-fault current (bolted or arcing) above the rating or setting of the circuit protective device will normally be cleared by the device. In such cases, the bolted fault current will be eliminated. But, even when the protective device eventually operates (in the case of a heavy ground-fault current that adds to the normal circuit load current to produce a total current in excess of the rating of the circuit protective device), the time delay of the device may be minutes or even hours; this is more than enough time for the arcing fault current to burn out conduit and enclosures, acting just like a torch, and even to propagate flames to create a fire hazard.

In spite of increasingly effective application of conventional overcurrent protective devices, the problem of ground faults continues to persist and even grows with expanding electric usage. In the interest of safety, definitive engineering design must include protection against such faults. Phase overcurrent protective devices are normally limited in their effectiveness because (1) they must have a time delay and a setting somewhat higher than full load to ride through normal inrushes, and (2) they are unable to distinguish between normal currents and low-magnitude fault currents which may be smaller than full-load currents.

Dangerous temperatures and magnetic forces are proportional to current during overloads and short circuits; therefore, overcurrent protective devices usually are adequate protection against such faults. However, the temperatures of arcing faults are, generally, independent of current magnitude; and arcs of great and extensive destructive capability can be sustained by currents that do not exceed the settings of overcurrent devices. Other means of protection are therefore necessary: A ground-detection device that responds only to ground-fault current can be coupled to an automatic switching device to open all three phases when a line-to-ground fault exists on the circuit.

With a ground-fault protection system of the type shown in Fig. 278 at the service entrance, a ground fault anywhere in the system is immediately sensed by the ground-relay system; however, action to open the circuit usually is delayed to allow a conventional overcurrent device near

FIG. 278 Ground-fault sensing may be performed on a bonding jumper from ground bus to neutral bus.

the point of fault to open if it can. As a practical measure, this time delay is designed to be only a few cycles or seconds, depending on the voltage of the circuit, the time-current characteristics of the overcurrent devices in the system, and the location of the ground-fault relay in the distribution system. If the conventional overcurrent protective devices fail to clear the circuit during the predetermined time delay, and if the fault continues, the ground-fault protective relays will open the circuit. This provides added overcurrent protection not available by any other means.

As shown in Fig. 278, the ground-fault protective relays are located in the main switchboard and will thus detect any ground faults that may develop at any point in the wiring system, from the furthest point back to—and including—the switchboard. If desired, another set of ground-fault protective relays can be located at any point downstream from the service equipment.

The ground-fault protection hookup in Fig. 278 makes use of ground-strap sensing; this scheme is applicable only to the service-entrance switchboard to disconnect the main service switch or CB on a ground fault somewhere in the system. Or, it may be applied on the secondary side of a transformer where the neutral and equipment grounding bus are bonded together. It includes a ground-sensing current transformer (CT) on the bonding strap that connects the frame or grounding bus of the switchboard to the neutral—all within the switchboard enclosure. The current transformer is a standard ratio-type CT and not a zero-sequence type CT. On any ground fault in the system, fault current will return to the neutral by flowing back over the conduit and other enclosures and then through the bonding strap to the neutral. Such current flow causes a current output from the CT, energizing a ground relay that, in turn, energizes the operating coil or shunt-trip coil of the main switch or CB, causing it to open.

This method of detecting ground-fault current in the neutral-frame bonding strap can only be used at service equipment, where such a connection is made. Moreover, selective coordination with downstream overcurrent devices is not easily accomplished. A ground fault on any branch circuit, feeder, or subfeeder—anywhere in the system—will cause current to flow on the bonding strap. To minimize nuisance tripping and the chance of disconnecting the entire service unnecessarily, the trip setting of the CT relay circuit must be high, and the time delay in its operation must be fairly long to give downstream protection a chance to clear the fault with minimal circuit outage. For instance, it could be necessary to set the CT response at 400 A, with a 5-s delay in the ground relay. On large feeders (say, 400 A or larger), conventional overcurrent devices would not clear a 400-A ground fault in the 5-s delay time, and the service main would open. Although this might not always be completely intolerable, it is generally not desirable to shut down the entire service because of a branch-circuit or subfeeder fault. Selective ground-fault protection is the obvious answer to the problem. But selective coordination between ground-fault protection and conventional protective devices (fuses and CBs) at main and feeder circuits is a very clear and specific task as a result of **NEC** Section 230-95(a), which calls for a maximum time delay of 1 s at any ground-fault current value of 3000 A or more.

Field experience has repeatedly shown that ground-fault interrupting equipment should be set as low as practical so that virtually no ground-fault current will remain unrecognized; introducing a time-delay action will permit the normal overcurrent protective device to operate. For a 4000-A circuit, the relay setting should usually be about 800 A. Ground-fault current of that magnitude would not be cleared by similarly sized overcurrent devices, but such current would be cleared automatically by the ground-fault protective system. On the other hand, downstream overcurrent devices with lower settings of, say, 200 A and less would open by themselves at such low ground-fault values.

Ground-fault-current values are highly unpredictable because of the many variable factors that are involved. As a consequence, the current settings of ground-fault protection equipment and the time delays to be allowed are matters of judgment.

The basic hookup offering widest application for ground-fault protection of individual circuits anywhere in a system is the zero-sequence sensing shown in Fig. 279; this hookup makes use of a current transformer and relay to actuate electrically operated switching devices. The CTs are "doughnut"-type or rectangular zero-sequence current transformers that readily fit over large groups of cables, switchboard busbars, or circuit-breakers studs. Such CTs are able

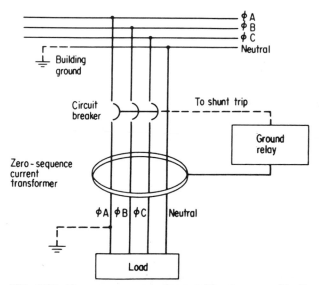

FIG. 279 Zero-sequence sensing provides for ground-fault protection on any circuit.

to detect milliamperes of current leaking out of systems with loads of many thousands of amperes.

This system allows selective protection for individual circuits. Each branch circuit, subfeeder, feeder, or main can be equipped with its own separate detection and operation hookup to deenergize only the circuit with the ground fault—with adjustable time delay in each relay to coordinate its operation with other lineside relays to ensure minimum outage on a ground fault. If the fault is on a branch circuit, the branch-circuit device will operate, and not the subfeeder or other device supplying the branch circuit further upstream.

For zero-sequence sensing, all conductors of a circuit being protected (including the neutral, if one is used) are run through the sensing current transformer. As long as the current associated with the circuit flows out and returns over those circuit conductors, the net flux in the CT is zero, and the CT has no output. Under such conditions, the CT indicates that no current is taking an unauthorized or fault path. That is, there is no current flowing to ground through grounded conduit or other metal enclosures. No current is generated in the CT winding, regardless of circuit balance, current magnitude, or any third or other odd-order harmonics (as from fluorescent ballasts or other magnetostrictive load devices). But, when a fault to ground does develop on one phase, the return of fault current over a path that is not enclosed by the CT (such as fault current on the conduit) causes an unbalance in the CT flux and generates an output signal from the CT to the relay, which then actuates the tripping mechanism of the associated electrically operated switching device—such as the shunt-trip of a CB.

The zero-sequence-CT method permits ground-fault protection to be as extensive or limited as dictated by the design and economics of any system. And, because separate GFP hookups can be used on individual circuits, with adjustable current trip settings and time-delay values, completely selective coordinated arrangements are possible to confine circuit outage to the faulted circuit.

Complete systems of zero-sequence CTs and related relays with adjustable time delay are available for application throughout a system. Figure 280 shows the idea behind a "total pro-

FIG. 280 For ground-fault protection system, time-delay variations can provide selectivity of operation.

tection" scheme that uses a zero-sequence CT and associated relay on each circuit to be provided with protection against low-level ground faults. Each CT-relay hookup will detect a ground fault on its load side and operate the shunt-trip of a circuit breaker, the shunt-trip of a bolted pressure switch, or the coil of an electrically operated switch (such as by breaking the holding-coil circuit of a contactor) to quickly clear the fault. The time delays of the relays, which are all responding to the same ground-fault current, can be used to coordinate operation so that any ground fault is cleared by the GFP device nearest to the fault on its line side. The individual GFP setups are unaffected by operating overloads and short-circuit currents, which cancel in the CT. The CT-relay control responds only to ground faults.

The system operates very quickly, to minimize the damage due to ground faults; the circuit opening speed is limited only by the need to coordinate the operating speeds of the relays to ensure that each ground fault will be cleared by the CT relay nearest to the fault on the line side of the fault. But the protection is completed with fuses or CBs that will open any circuit in which there is an operating overload or short-circuit fault above the continuous rating of these fuses or CBs.

Figure 281 shows series GFP layouts using a zone-selective system based on the use of a feedback lockout signal that enables each GFP device to be set for instantaneous operation for any fault on its load side and within its zone. The instantaneous trip at each step of the series-connected GFP hookups ensures the minimum amount of damage due to ground faults.

FIG. 281 Selectivity with maximum protection is provided by a zone-selective instantaneous system.

A less complicated way to include GFP on larger circuits is to use a molded-case CB with built-in ground-fault protection. Available in ratings from 400 to 3000 A, such CBs contain all the elements for GFP within the molded case; they do not require an external shunt-trip, zero-sequence CT, or relay. Air CBs and insulted-case CBs also are equipped with GFP.

Another sensing hookup that has been used for ground-fault detection in service switchboards is called the *residual method*. Four conventional CTs are used—one on each phase leg (with polarities carefully observed) and one on the neutral. The CT secondaries are hooked up to provide a signal output on ground fault and direct opening of the disconnect switch or CB.

In **NEC** Section 230-95, the rule requiring ground-fault protection for any service disconnect rated 1000 A or more (on 480/277-V services) specifies a maximum time delay of 1 s for ground-fault currents of 3000 A or more (Fig. 282). The maximum setting (the current at and above which the GFP will be actuated) of any service ground-fault protection hookup is 1200 A, but the time-current trip characteristic of the relay must ensure opening of the disconnect in not more than 1 s for any ground-fault current of 3000 A or more. That **NEC** rule is intended to establish a specific level of protection in GFP equipment by setting a maximum limit on I^2t for fault energy.

The reasoning behind the maximum time delay of 1 s on faults of 3000 A or more is as follows: The amount of damage done by an arcing fault is directly proportional to the time it is allowed to burn. Commercially available GFP systems can easily meet the 1-s limit. Some users, however, request time delays up to 60 s to allow downstream overcurrent devices plenty of time to trip thermally before the GFP on the main disconnect trips. But an arcing fault lasting 60 s can virtually destroy a service-equipment installation. Coordination with downstream overcurrent devices can and should be achieved by adding GFP on feeder circuits where needed, and the **NEC** should require a reasonable time limit for GFP. The value of 3000 A is 250 percent of 1200 A, and 250 percent of the setting is the calibrating point specified in ANSI 37.17. Specifying a maximum time delay starting at this current value will allow either flat or inverse time-delay characteristics for ground-fault relays with approximately the same level of protection.

Time-delay setting of GFP relay must *not* **exceed 1 second at 3000 amps.**

FIG. 282   Time delay is limited for GFP on a service disconnect.

GFP for health-care facilities

At least one additional level of ground-fault protection is required for hospitals and health-care facilities where ground-fault protection is used on service equipment (**NEC** Section 230-95). Where ground-fault protection is installed on the normal service disconnecting means, each main feeder must be provided with similar protective means. This requirement is intended to minimize the chance of extensive outage. By applying appropriate selectivity at each level, the ground fault can be limited to a single feeder, and service to the balance of the health-care facility may be maintained.

As shown in Fig. 283, if there is a GFP hookup on the service, then a GFP hookup must be put on each feeder derived from the service. The tripping time of the main GFP must be such that each feeder GFP will operate to open a ground fault on its feeder without opening the service GFP. And a time interval of not less than 0.1 s (i.e., the time equivalent of six cycles) must be provided between the feeder GFP trip and the service GFP trip. As shown, if the feeder GFP relays are set for instantaneous operation, the relay on the service GFP must have at least a 0.1-s time delay. A zone-selective GFP system with a feedback lockout signal to an instantaneous relay on the service could satisfy the rule on selectivity.

NOTE: For additional information on ground-fault protection at service entrances, as required by **NEC** Section 230-95, refer to Section 7 of this handbook.

FIG. 283 Hospital systems must have GFP on all main feeders fed by the service disconnect, as well as on the service disconnect itself.

SYSTEM AND EQUIPMENT GROUNDING

One of the most important, but least understood, considerations in the design of electrical systems is grounding. The name "grounding" comes from the fact that the technique involves making a low-resistance connection to the earth or ground. For any given piece of equipment or circuit, this connection may be a direct wire connection to a grounding electrode which is buried in the earth; or it may be a connection to some other conductive metallic element (such as a conduit or switchboard enclosure) that is connected to a grounding electrode. The purpose of grounding is to protect personnel, equipment, and circuits by eliminating the possibility of dangerous or excessive voltages.

There are two distinct considerations in the grounding of electrical systems: (1) the grounding of one of the conductors of the wiring system, and (2) the grounding of all metal enclosures that contain electrical wires or equipment to eliminate the possibility that an insulation failure in such an enclosure might place a potential on the enclosure and constitute a shock or fire hazard.

Wiring-system ground: This consists of the grounding of one of the wires of the electrical system to limit the voltage that might otherwise occur in the circuit through exposure to lightning or other voltages higher than that for which the circuit is designed (Fig. 284). Another purpose in grounding one of the wires of the system is to limit the maximum voltage to ground under normal operating conditions. Also, a system that operates with one of its conductors intentionally grounded will provide for automatic opening of the circuit if an accidental or fault ground occurs on one of its ungrounded conductors.

FIG. 284   "Grounded electrical system" operates with one of its circuit conductors connected to a grounding electrode.

The selection of the wiring-system conductor to be grounded depends upon the type of system. In 3-wire, single-phase systems, the midpoint of the transformer winding—the point from which the system neutral is derived—is grounded. In grounded 3-phase wiring systems, the neutral point of the wye-connected transformer (or transformers) or generator is usually the point connected to ground. In delta-connected transformer hookups, grounding can be effected by grounding one of the three phase legs, by grounding a center-tap point on one of the transformer windings (as in the 3-phase, 4-wire "red-leg" delta system), or by using a special grounding transformer that establishes a neutral point of a wye connection, which is grounded.

According to the **NEC**, all interior ac wiring systems must be solidly grounded (not through an intentional resistance or impedance) if they can be grounded in such a way that the maximum voltage to ground does not exceed 150 V. This rule makes it mandatory that, except for control circuits and isolated circuits in hospital operating suites (as covered in the exception to **NEC** Section 250-5), the following systems or circuits operate with one conductor grounded:

- A 120-V, 2-wire system or circuit must have one of its circuit wires grounded.
- A 240/120-V, 3-wire, single-phase system or circuit must have its neutral conductor grounded.
- A 208/120-V, 3-phase, 4-wire, wye-connected system or circuit must be operated with the neutral conductor grounded.

In all these systems or circuits, the neutrals must be grounded because the maximum voltage to ground from any other conductor of the system does not exceed 150 V when the neutral conductor is grounded.

In addition, **NEC** Section 250-5(b) requires that the neutral conductor of a 240/120-V, 3-phase, delta-connected, 4-wire system (with the neutral taken from the midpoint of one phase) be grounded at the transformer. It is also mandatory that a 480Y/277-V, 3-phase, 4-wire interior wiring system have its neutral grounded if the neutral is to be used as a circuit conductor—such as for 277-V lighting. And if 480-V autotransformer-type fluorescent or mercury-vapor ballasts are to be supplied from a 480/277-V system, then the neutral conductor must be grounded at the voltage source, even though the neutral is not used as a circuit conductor. Of course, it should be noted that 480/277-V systems are usually operated with the neutral grounded to obtain the automatic fault clearing of a grounded system. Figure 285 shows these rules.

According to these **NEC** rules, any 120-V, 2-wire circuit must normally have one of its conductors grounded. These requirements have often caused difficulty when applied to control circuits derived from the secondary of a control transformer that supplies power to the operating coils of motor starters, contactors, and relays. For instance, there are cases where a ground fault on the hot leg of a grounded control circuit can cause a hazard to personnel by actuating the control-circuit fuse of CB and shutting down an industrial process in a sudden, unexpected, nonorderly way. A metal-casting facility is an example of an installation where sudden shutdown due to a ground fault in the hot leg of a grounded control circuit could be objectionable. Because designers often wish to operate such 120-V control circuits ungrounded, Exception No. 3 of Section 250-5(b) permits ungrounded control circuits under certain specified conditions.

A 120-V control circuit may be operated ungrounded when all the following exist:

1. The circuit is derived from a transformer that has a primary rating below 1000 V.

2. Whether in a commercial, institutional, or industrial facility, supervision will ensure that only persons qualified in electrical work will maintain and service the control circuits.

3. There is a need to prevent circuit opening on a ground fault; i.e., continuity of power is required for safety or for operating reliability.

FIG. 285 Some systems or circuits must be used with one conductor grounded.

FOR SEPARATE OR BUILT-IN TRANSFORMER—

1. If transformer primary is rated less than 1000 volts ...

2 ... **and secondary** supplies only control circuits where loss of power to motors or other load would cause objectionable conditions . . .

120V

440V, 3∅

3. . . . it is **not** necessary to ground one leg of the 120-volt secondary . . .

Stop

Start

C

4. . . . **but,** some type of ground detector must be used to indicate when a ground fault occurs on the control system.

To other control circuits if separate transformer is used

M

FIG. 286 Ungrounded 120-V circuit may be used for control purposes.

4. Some type of ground detector is used on the ungrounded system to alert personnel to the presence of a ground fault, enabling them to clear the ground fault in the normal downtime of the system. (Refer to Fig. 286.)

Although no mention is made of secondary voltage in this rule, Exception No. 3 is significant primarily for 120-V control circuits. The **NEC** has long permitted 240- and 480-V control circuits to be operated ungrounded. This exception can be applied to any 120-V control circuit derived from a control transformer in an individual motor starter or magnetic contactor. Of course, the exception could also be used to permit ungrounded 277-V control circuits under the same conditions.

An application where ungrounded 120-V circuits are not only permitted but are specifically required is shown in Fig. 287. Any electrical circuit within an anesthetizing location must be supplied from an isolated ungrounded distribution system. (**NEC** Section 517-104). An anesthetizing location is any area in a hospital in which flammable or nonflammable anesthetics are or may be administered to patients—operating rooms, delivery rooms, anesthesia rooms, and any corridors, utility rooms, or other areas used for the administration of anesthetics.

The **NEC** does recognize the use of ungrounded circuits or systems operating at less than 50 V. But grounding of circuits under 50 V is required for the cases shown in Fig. 288.

Although the **NEC** does not require grounding for electrical systems in which the voltage to ground would exceed 150 V, it does recommend that ground-fault detectors be used with ungrounded systems which operate at more than 150 V and less than 1000 V. Such detectors indicate when an accidental ground fault develops on one of the phase legs of an ungrounded system. The indicated ground fault can be removed during the normal downtime of the industrial operation—i.e., when the production machinery is not running.

For industrial electrical systems (for manufacturing plants and process industries), many designers prefer ungrounded or resistance-grounded systems to solidly neutral-grounded systems such as the 480Y/277-V solidly grounded system. Ungrounded or resistance-grounded operation of an electrical system is preferred because the occurrence of a ground fault on only one phase of such a system will not cause the automatic operation of a fuse or CB and the interruption of power to important loads, as it does for solidly grounded systems. There is also much strong

Supply cannot be made directly from two phase legs of 480/277-v system. On such systems, two stages of transformation are needed.

COMPLETE PACKAGED UNGROUNDED DISTRIBUTION CENTER FOR HOSPITAL OPERATING ROOM: Main CB, isolating transformer, ground indicator and CB panel for ungrounded circuits. For in-wall or floor mounting close to operating suite.

FIG. 287 Any 120-V circuit *must* be operated ungrounded in an anesthetizing location within a hospital.

FIG. 288 Under certain conditions, circuits operating at less than 50 V are required to be grounded.

opinion that solidly grounded systems result in greater production losses, decreased reliability, and higher maintenance costs. And arcing ground faults are widely considered to be much less of a concern in ungrounded or resistance-grounded systems.

The use of resistance grounding is preferred where a ground fault is to be detected but the circuit is not to be automatically shut down. Although this condition can be (and frequently is) obtained by using an ungrounded system with suitable ground detectors, some designers believe there is greater advantage in the use of resistance grounding, which can very effectively be used with GFP equipment to give better protection against arcing-type faults. Resistance grounding is most often selected for industrial plants, where lighting may be no more than 10 percent of the total load, and the greater concern is for continuous power to production or process equipment—particularly where unplanned shutdowns of such equipment can cost hundreds or thousands of dollars per hour in lost production. In these systems, detected faults can be cleared during planned shutdowns of equipment.

The major operating difference between a grounded and an ungrounded system is that a single ground fault will automatically cause opening of the circuit in a grounded system but will not interrupt an ungrounded system. However, the presence of a single ground fault on an ungrounded system exposes the system to the very destructive possibility of a phase-to-phase short if another ground fault should simultaneously develop on a different phase leg of the system (Fig. 289).

UNGROUNDED SYSTEMS

Single accidental ground on any phase leg does not interrupt service

Simultaneous accidental grounds on two phases constitute a short circuit and open one or two protective devices

Ground detector devices may be used to signal presence of accidental ground on any phase

FIG. 289   Ungrounded system will not automatically operate overcurrent protection on a single ground fault.

Grounded-neutral systems are generally recommended for high-voltage (over 600 V) distribution. Although ungrounded systems do not undergo a power outage as the result of a one-phase ground fault, the time and money spent in tracing faults indicated by ground detectors and other disadvantages of ungrounded systems have favored the use of grounded-neutral systems. Grounded systems are more economical in operation and maintenance. In such systems, if a fault occurs, it is isolated immediately and automatically.

Grounded-neutral systems have many other advantages. The elimination of multiple faults caused by undetected restriking grounds greatly increases service and equipment reliability. The lower voltage to ground which results from grounding the neutral offers greater safety for personnel and requires lower equipment voltage ratings. And, on high-voltage (above 600 V) systems, residual relays can be used to detect ground faults before they become phase-to-phase faults, which have substantial destructive ability.

OSHA and system grounding: One of the most consistently troublesome and controversial areas of modern electrical design is application of the **NEC** rules and OSHA regulations concerned with system grounding. For both new electrical systems and existing systems, many questions are raised about the need to operate with the neutral or one of the phase legs intentionally grounded at the service or transformer secondary. When must one of the circuit conductors be a grounded conductor? (Note that this is a matter of system grounding and is separate from the question of equipment grounding, or the effective tying together and grounding of equipment enclosures, housings, and similar non-current-carrying exposed metal parts of electrical equipment.)

Several **NEC** sections relate to this very important subject:

- Section 210-6(a) regulates the use of lampholders, lighting fixtures, and receptacles on circuits with various voltages to ground. Although such load devices are basically limited to use on circuits rated not over 150 V to ground (240/120-V, 3-wire, single-phase and 208/120-V, 3-phase, 4-wire systems), incandescent and electric-discharge lighting may be used on circuits rated up to 300 V to ground in industrial, commercial, and institutional occupancies under the conditions given. The "voltage to ground" is the maximum voltage between an ungrounded circuit conductor and an intentionally grounded conductor (such as a grounded neutral or phase leg). For an ungrounded system, the **NEC** specifies "voltage to ground" as the maximum voltage between any two of the circuit conductors (which would occur if one of the circuit conductors became accidentally grounded).

- Section 210-6(b) permits electric-discharge lighting units on ungrounded circuits of not more than 600 V "between conductors"—but only for outdoor use or in tunnels.

- Section 410-78 requires that only a grounded system (one circuit conductor intentionally grounded) may be used to supply lighting equipment that contains an autotransformer-type ballast that raises the voltage to more than 300 V.

- Section 250-5 specifies when electrical systems must be operated with one of the circuit conductors grounded.

OSHA has made the rules of Section 250-5 retroactive, so that all existing installations that do not conform with that section must be altered to comply with its rules.

The effect of the foregoing rules on various types and voltages of electrical systems are as follows:

1. Any new or existing circuits rated 120 V must have one conductor grounded. And this applies to the secondary of even the smallest lighting transformer that might be used for a light at a machine or other local application. Exceptions permitting ungrounded 120-V circuits are made in Exception No. 3 of Section 250-5(b) for control circuits and in Section 517-104 for circuits in anesthetizing locations in hospitals and health-care facilities. Any new or existing 240/120-V, 3-wire, single-phase system or 208/120-V, 3-phase, 4-wire system must have its neutral grounded. For instance, an ungrounded 120-V, 3-wire delta system violates Section 250-5(b)(1) and must be converted to a corner-grounded delta or grounded wye (through a zigzag grounding autotransformer; see **NEC** Section 450-4).

2. Any new or existing 480/277-V wye system must have the neutral point of the wye grounded if a neutral conductor is derived from that point and used as a circuit conductor to supply any kind of load. But if a 480/277-V transformer secondary is used to supply only 480-V loads and no neutral conductor is used, the neutral point of the wye does not have to be grounded to satisfy the **NEC**. Utility-company rules may, however, require grounding of the wye, and such rules must be checked.

3. A 480-V ungrounded delta system may not be used to supply indoor lighting of any kind for a new system. In such a system, the voltage to ground is taken as 480 V, which would be a violation of Exceptions No. 1 and No. 2 of Section 210-6(a). A 480-V ungrounded delta

system may be used for outdoor and tunnel lighting as permitted by Section 210-6(b), but in such cases only two-winding transformer ballasts (not autotransformer-type ballasts) must be used to satisfy Section 410-78.

4. Existing 480-V ungrounded delta systems that supply 480-V indoor lighting fixtures (with either two-winding or autotransformer-type ballasts) are not in violation of retroactive Section 250-5(b) and therefore do not have to be converted to grounded operation. Such hookups are, however, contrary to the rules of Section 210-6(a) and may violate Section 410-78 if autotransformer ballasts are used. But OSHA does not require retroactive application of the rules contained in **NEC** Sections 210-6(a) and 410-78. Of course, it is always good practice to update existing systems and bring them into compliance, even if this is not required.

5. An existing 480-V, 3-phase, 3-wire ungrounded system derived from a 480/277-V wye transformer secondary and supplying 480-V indoor lighting connected phase to phase is not in violation of retroactive Section 250-5(b) and does not have to have its neutral point grounded. Again, such a hookup could not be used without the neutral grounded for new indoor lighting, because it violates Section 210-6(a). But it could be used for new outdoor and/or tunnel lighting as in Section 210-6(b). And, again, Section 410-78 must be observed for new installations.

6. New or existing 240-V, 3-phase, 3-wire ungrounded systems may supply incandescent or electric-discharge lighting in accordance with Section 210-6(a). In such a system, the voltage to ground is 240 V, which does not exceed the 300-V maximum in Exceptions No. 1 and No. 2 of Section 210-6(a). And a 240-V ungrounded delta system is not in violation of the retroactive rules of Section 250-5(b). But the use of an ungrounded system to supply auto-transformer ballasts that raise the voltage to more than 300 V violates Section 410-78.

Obviously, in all design work, it is imperative to observe all applicable regulations and care-fully relate them to each other.

Equipment ground: This is the permanent and continuous bonding together (i.e., connecting together) of all non-current-carrying metal parts of equipment enclosures—conduit, boxes, cab-inets, housings, frames of motors, and lighting fixtures—and the connection of this intercon-nected system of enclosures to the system grounding electrode, either at the service equipment of a premises or on the secondary side of a transformer within a system (Fig. 290). All metal enclosures must be interconnected to provide a low-impedance path for fault-current flow along the enclosures, to ensure the operation of overcurrent devices that will open a circuit in the event of a fault. By opening a faulted circuit, these devices prevent dangerous voltages from being present on equipment enclosures that could be touched by personnel, with consequent electric shock to such personnel.

Simply stated, the grounding of all metal enclosures containing electric wires and equipment prevents the occurrence of any potential above ground on the enclosures. Such bonding together and grounding of all metal enclosures is required for both grounded electrical systems (those systems in which one of the circuit conductors is intentionally grounded) and ungrounded elec-trical systems (systems with none of the circuit wires intentionally grounded.)

But effective equipment grounding is extremely important in grounded electrical systems, to provide the automatic fault clearing that is one of the important advantages of grounded systems. A low-impedance path for fault current is necessary to permit enough current to flow to operate the fuses or circuit breaker protecting the circuit, and low impedance is specifically required by **NEC** Section 250-51.

In a grounded electrical system with a high-impedance equipment ground-return path, if one of the phase conductors of the system (i.e., one of the ungrounded conductors of the wiring system) should accidentally come in contact with one of the metal enclosures in which the wires

FIG. 290 For either a grounded or ungrounded electrical system, all metallic enclosures of conductors and other operating components must be interconnected and grounded at the service-system source.

are run, it is possible that not enough fault current would flow to operate the overcurrent devices. In such a case, the faulted circuit would not automatically open, and a dangerous voltage would be present on the conduit and other metal enclosures. This voltage would present a shock hazard and a fire hazard through possible arcing or sparking from the energized conduit to some grounded pipe or other grounded metal.

In a grounded system with a high-impedance equipment ground-return system, a ground fault will not open the faulted circuit; a phase-to-phase fault must develop to operate the overcurrent device. Such a condition is quite hazardous.

For effective protection against common ground faults in grounded electrical systems, therefore, low impedance for the equipment bonding system is even more important than low impedance for the grounding-electrode connection to earth. And in long runs of magnetic-material conductor enclosures, the ground-circuit impedance should be taken as 10 times the dc resistance to allow for the many variables.

Ground Connections

When an electrical system is to be operated with one conductor grounded—either because it is required by the **NEC** (as in a 240/120-V, single-phase system) or because it is desired by the system designer (as in a 240-V, 3-phase, corner-grounded system)—a connection to a suitable grounding electrode must be made at the service entrance. That is, the neutral conductor or other conductor to be grounded must be connected at the service equipment to a conductor that runs to a grounding electrode or "grounding electrode system," as described in **NEC** Section 250-81. The conductor that runs to the grounding electrode is defined as the *grounding-electrode conductor* in the **NEC**.

The **NEC** rule says that the connection of the grounding-electrode conductor to the system conductor that is to be grounded must be made "at any accessible point from the load end of the service lateral" to the service disconnecting means. This means that the grounding-electrode conductor (which runs to building steel and/or water pipe or to a driven ground rod) must be connected to the system neutral or other system wire to be grounded either in the enclosure for the service disconnect or in some enclosure on the supply side of the service disconnect. Such connection may be made, for instance, in the main service switch or CB in a service panelboard or switchboard. Or, the grounding-electrode conductor may be connected to the system grounded conductor in a gutter, CT cabinet, or meter housing on the supply side of the service disconnect. The utility company should be checked on grounding connections in meter sockets or other metering equipment (Fig. 291).

FIG. 291 Grounding connection for a system grounded neutral must be made on the supply side of the service equipment.

In addition to the grounding connection for the grounded system conductor at the point of service entrance to the premises, it is further required that another grounding connection be made to the same grounded conductor at the transformer that supplies the system. This means, for example, that a grounded service to a building must have the neutral point of the supplying transformer connected to a grounded electrode at the utility transformer outside the building, as well as having the neutral grounded to a water pipe and/or other suitable electrode at the building, as shown in Fig. 292. And, in the case of a building served from an outdoor transformer

System "grounded" conductor must be connected
to electrode (S) at service entrance . . .

For instance, utility
transformer on pole,
with connection to a
ground rod

Outside

120/208-volt
service equipment
enclosure

Building
wall

. . . but must <u>also</u> be grounded outside
the building at the utility transformer
or customer's transformer.

FIG. 292   Grounding connection for a grounded service conductor must be made
at two points.

pad or mat installation, the conductor that is grounded in the building must also be grounded
at the transformer pad or mat.

The above grounding details refer to connections at the service-entrance equipment of systems fed by a utility service, as distinguished from grounding connections for a "separately derived" system, such as one fed from a local transformer or from an on-site generator. (The latter are covered under "Transformer grounding" in Section 6 of this handbook.)

These design rules are concerned with the grounding of a utility-fed ac electric supply circuit that has one of its conductors intentionally grounded—such as a 240/120-V single-phase system or 208/120-V, 3-phase, 4-wire system or 480/277-V, 3-phase, 4-wire system that is required to have its neutral conductor grounded.

The basic concepts are as follows:

1. For such grounding connection, a grounding-electrode conductor must be connected to the grounded conductor (the neutral conductor) anywhere from the "load end of the service drop or lateral" to the neutral block or bus within the enclosure for the service disconnect—which includes a meter socket, a CT cabinet, or an auxiliary gutter or other enclosure ahead of the service disconnect, panelboard, or switchboard.

2. The grounding-electrode conductor that is connected to the grounded neutral (or grounded phase leg) must be run to a "grounding electrode system," as described later in Section 7 on "Services."

3. Exception No. 5 of **NEC** Section 250-23 permits the grounding-electrode conductor to be connected to the equipment grounding bus in the service-disconnect enclosure—instead of to the neutral block or bus—where such connection is considered necessary to prevent the desensitization of a service GFP hookup that senses fault current through a CT-type sensor on the ground strap between the neutral bus and the ground bus (see the bottom of Fig. 291). However, in any particular installation, the choice between connecting to the neutral bus and connecting to the ground bus will depend on the number and types of grounding electrodes, the presence or absence of grounded building structural steel, bonding between electrical raceways and other metal piping on the load side of the service equipment, and the number and locations of bonding connections. The grounding-electrode conductor may be connected to either the neutral bus (or terminal lug) or to the equipment grounding block or bus in any system that has a conductor or a busbar bonding the neutral bus or terminal to the equipment grounding block or bus. But, where the neutral is bonded to the enclosure simply by a bond-

ing screw, the grounding-electrode conductor must not be connected to the equipment grounding block or bus. It must be connected only to the neutral-conductor terminal block or bus. This rule is based on the idea that a lightning strike on the neutral conductor outside would cause the lightning current to flow through the bonding screw in going to ground through a grounding-electrode conductor that is connected to the equipment ground terminal. The conductivity of a screw-type bond is not adequate to effectively pass the very high lightning currents to earth.

One of the most important and widely discussed considerations in the design of electrical distribution systems is the matter of making a grounding connection to the system grounded neutral or grounded phase wire. The **NEC** says, "Grounding connections shall not be made on the load side of the service disconnecting means." Once a neutral or other circuit conductor is connected to a grounding electrode at the service equipment, the general rule is that the neutral or other grounded leg must be insulated from all equipment enclosures and other grounded parts on the load side of the service. That is, bonding of subpanels (or any other connection between the neutral or other grounded conductor and equipment enclosures) is prohibited.

There are some exceptions to that rule, but they are few and are very specific:

1. In a system, even though it is on the load side of the service, when voltage is stepped down by a transformer, a grounding connection must be made to the secondary neutral, as described in Section 6 under "Transformer grounding."
2. When a circuit is run from one building to another, it may be necessary (or simply permissible) to connect the system "grounded" conductor to a grounding electrode at the other building, as covered below under "Grounding at separate buildings."
3. The frames of ranges, wall ovens, countertop cook units, and clothes dryers may be "grounded" by connection to the grounded neutral of their supply circuit (Section 250-60).

It is clear violation to bond the neutral block in a panelboard to the panel enclosure in other than a service panel. In a panelboard used as service equipment, the neutral block (terminal block) is bonded to the panel cabinet by the bonding screw provided. And such bonding is required to tie the grounded conductor to the interconnected system of metal enclosures for the system (i.e., service-equipment enclosures, conduits, busway, boxes, panel cabinets, etc.). It is this connection which provides for the flow of fault current and operation of the overcurrent device (fuse or breaker) when a ground fault occurs. But, there must not be any connection between the grounded system conductor and the grounded metal enclosure system at any point on the load side of the service equipment, because such connection would constitute connection of the grounded system conductor to a grounding electrode (through the enclosure and raceway system to the water pipe or driven ground rod). Such connections, like the bonding of subpanels, can be dangerous, as shown in Fig. 293.

This **NEC** rule prohibiting connection of the grounded system wire to a grounding electrode on the load side of the service disconnect must not be confused with the rule of Section 250-60 which permits the grounded system conductor to be used for grounding the frames of electric ranges, wall ovens, counter-mounted cooking units, and electric clothes dryers. The connection referred to in Section 250-60 is that of an ungrounded metal enclosure to the grounded conductor for the purpose of grounding the enclosure.

Grounding at Separate Buildings

As described above, bonding of a panel neutral block (or the neutral bus or terminal in a switchboard, switch, or circuit breaker) to the enclosure is required by the **NEC** in service equipment. But the **NEC** permits bonding of the neutral conductor on the load side of the service equipment

1. THIS CONDITION WILL EXIST... AND...

Typical load outlet

Typical subpanel

Neutral block with bonding screw installed — providing an objectionable connection between the neutral and the entire system of metal enclosures, through the metal panel.

Distribution panel

Service entrance panel or switchboard

Neutral is bonded to service equipment enclosure and thus to all interconnected metal raceways and enclosures.

N Ø Ø Ø

This block is not bonded.

Conductors in metal raceway or metal cable armor, connected to metal enclosures.

Ground wire to water pipe

Current flows over both the neutral conductor and the metal raceway or cable armor — which make up two parallel current paths. If neutral is opened at any point, the raceway or cable armor will be the only current return path.

2. THIS HAZARD COULD DEVELOP

PANEL ENCLOSURE AND OTHER METAL RACEWAY AND ENCLOSURES CONNECTED TO PANEL ARE <u>HOT</u>.

Subpanel

Neutral block bonded to panel enclosure

Restricted neutral current on raceway

No current on open neutral

Poor connection or open between raceway and panel

With the neutral wire opened at any point between the sub-panel and the service, and with a high impedance locknut connection or an open in the conductive raceway current path, a dangerous voltage could be placed on the enclosure of a bonded sub-panel.

FIG. 293 Bonding of subpanels is prohibited for the reason shown here.

in those cases where a panelboard (or switchboard, switch, etc.) is used to supply circuits in a building and the panel is fed from another building. Where two or more buildings are supplied from a single installation of service equipment in one of the buildings, a grounding electrode at each building must be connected to the ac system grounded conductor on the supply side of the building disconnecting means of a grounded system, as shown in Fig. 294, or connected to the metal enclosures of the building disconnecting means of an ungrounded system.

FIG. 294 Basic rule calls for grounding of a grounded conductor (usually the neutral) at each building fed from the main building.

That is the basic rule. But an exception to the rule notes that a grounding electrode at a separate building supplied by a feeder or branch circuit is not required where either of the following conditions is met: (1) Only one branch circuit is supplied, and there is no non-current-carrying equipment in the building that requires grounding. An example would be a small residential garage with a single lighting outlet or switch with no metal boxes, faceplates, or lighting fixtures within 8 ft vertically or 5 ft horizontally from any grounded object. (2) An "equipment grounding conductor" is run with the circuit conductors for the grounding of non-current-carrying equipment, water piping, or building metal frames in the separate building, and no livestock are housed in the building. (See Fig. 295). And, the need for a grounding

FIG. 295 Under these conditions, no grounding electrode or connection is required if an outbuilding does not house livestock.

electrode at the outbuilding is eliminated because the words "equipment grounding conductor" are understood to include metal "conduit," which the **NEC** recognizes as an "equipment grounding conductor . . . enclosing the circuit conductors." If the separate building has an approved grounding electrode and/or interior metallic piping system, the equipment grounding conductor must be bonded to the electrode and/or piping system.

A subpanel in a building fed from another building must always have its neutral block bonded and must be connected to a grounding electrode if livestock are housed in the building. (Animals have a much greater sensitivity to electric shock than humans, and the absence of grounding at, say, a dairy barn can cause shock to milk cows, through the milking machine, which would stop milk flow.) The necessity for bonding the neutral block in such a subpanel is based on Sections 250-24 and 250-54. The latter says, "Where an ac system is connected to a grounding electrode in or at a building as specified in Sections 250-23 and 250-24, the same electrode shall be used to ground conductor enclosures and equipment in or on that building." Although the **NEC** permits and even requires bonding of the neutral to equipment enclosures at both building "services," if the feeder circuit is in conduit, neutral current flows on the conduit because it is electrically in parallel with the neutral conductor, being bonded to it at both ends.

FIG. 296 Absence of an equipment grounding conductor in this case requires bonding and grounding at the outbuilding disconnect.

Figure 296 shows another situation in which a grounding-electrode connection must be made at the other building, as specified in the basic rule of Section 250-24(b). In the sketch shown, if the 3-phase, 3-wire, ungrounded feeder circuit to the outbuilding had been run with a separate equipment grounding conductor which effectively connected the metal enclosure of the disconnects in the outbuilding to the grounding-electrode conductor in the service entrance equipment of the main building, a connection to a grounding electrode would not be required at the outbuilding. But the equipment grounding conductor run to the outbuilding would have to be bonded to any interior metal piping, metal building frames, and grounding electrodes that might exist at the outbuilding. All grounding electrodes that exist at the outbuilding must be bonded to the ground bus or terminal in the disconnect at the outbuilding, whether or not an equipment grounding conductor is run with the circuit conductors from the main building. If, as shown in the sketch, an equipment grounding conductor is not run to the outbuilding, then a grounding-electrode conductor must be run from the ground bus or terminal in the outbuilding disconnect to a suitable grounding electrode which must be provided as specified in Sections 250-81 and 250-83. But, if an equipment grounding conductor is run to the outbuilding and connected to the enclosure of the building disconnect, there is no need to provide a grounding electrode if none exists at the outbuilding.

Figure 297 applies these rules to overhead feed from one building to another.

Three buildings served by a single service. The additional grounds are
required in most cases under the conditions described.

FIG. 297 Overhead feeders to outbuildings require bonding and grounding
at each building when an equipment grounding conductor is not carried with
the overhead circuits.

Equipment Grounding

The **NEC** requires that metal equipment enclosures, boxes, and cabinets be grounded; the
grounding connection may be made by metal cable armor, by the metal raceway that supplies
such enclosures (rigid metal conduit, intermediate metal conduit, EMT, flex, or liquid-tight flex
as permitted by Section 350-5 or 351-7), or by an equipment grounding conductor, such as
where the equipment is fed by rigid nonmetallic conduit. When a separate equipment grounding
conductor (i.e., other than the metal raceway or metal cable armor) is used for ac circuits, it
must be contained within the same raceway, cable, or cord or otherwise run with the circuit
conductors (Fig. 298).

**If equipment grounding conductor (other than raceway) is used
to ground motor, it must be run in raceway with circuit wires.**

Alternating–
current motor

FIG. 298 Equipment grounding conductors must be run with the
circuit wires *within* rigid conduit, IMC, or EMT and within flexible
metal conduit or liquid-tight flex when the length is over 6 ft.

External grounding of equipment enclosures, frames, or housings is a violation of the **NEC**
for ac equipment. It is not acceptable, for instance, to feed an ac motor with a nonmetallic
conduit or cable, without a grounding conductor in the conduit or cable, and then to provide
grounding of the metal frame through a grounding conductor connected to the metal frame and
run to building steel or to a grounding-grid conductor. An equipment grounding conductor must
be run with the circuit conductors, within the raceway, for all conduit runs over 6 ft long.

Keeping an equipment grounding conductor physically close to ac-circuit supply conductors
is absolutely essential to ensure minimum impedance in grounding current paths, to provide

most effective clearing of ground faults. When an equipment grounding conductor is kept phys-ically close to any circuit conductor that would be supplying the fault current (that is, the grounding conductor is kept in the "same raceway, cable, or cord or otherwise run with the circuit conductors"), the impedance of the fault circuit has minimum inductive reactance and minimum ac resistance because of mutual cancellation of the magnetic fields around the con-ductors and reduced skin effect. Under such a condition of "sufficiently low impedance," the voltage to ground is limited to the greatest possible extent, the fault current is higher because of the minimized impedance, and the circuit overcurrent device will operate at an earlier point on its time-current characteristic to ensure maximum fault-clearing speed; the overall effect is to facilitate the operation of the protective devices in the circuit.

Direct-current circuits do not present this need to keep the grounding conductor close to the circuit conductors. Because there are no alternating magnetic fields around dc conductors, there is no inductive reactance or skin effect in dc circuits. The only impedance to current flow in a dc circuit is resistance—which is the same for a dc ground-fault path whether or not the equip-ment grounding conductor is placed physically close to the circuit conductors which would sup-ply the fault current in the event of a ground fault. External equipment grounding by connection to grounded building steel or to an external ground grid is, therefore, acceptable for dc equip-ment, provided the external grounding path is effectively tied back to the grounded conductor of the dc system.

The arrangement shown in Fig. 299 violates the basic concepts of **NEC**-accepted grounding because the lighting fixture, which must be grounded, is not grounded by an equipment ground-ing conductor contained within the cord.

EXTERNAL EQUIPMENT GROUNDING IS A VIOLATION!

FIG. 299 Alternating-current equipment enclosures that are required to be grounded must be fed by a circuit that includes an equipment grounding conductor.

Except for dc circuits and for isolated, ungrounded power sources, an equipment grounding conductor of any type must not be run separately from the circuit conductors. The engineering reason for keeping the ground return path and the phase legs in close proximity (that is, in the same raceway) is to minimize the impedance of the fault circuit by placing conductors so their magnetic fields mutually cancel each other, keeping inductive reactance down and allowing suf-ficient current to flow to facilitate the operation of the circuit protective devices.

The hookup in Fig. 299 also violates the rule of Section 250-58(a), which prohibits the use of building steel as the equipment grounding conductor for ac equipment.

NOTE: Care must be taken to distinguish between an "equipment grounding conduc-tor" as covered by **NEC** Section 250-57 and an "equipment bonding jumper" as covered by Sec-tion 250-79(e). A "bonding jumper" may be external to equipment, but it must not be over 6 ft long.

Grounding Conductors

Equipment grounding conductors that are run within nonmetallic raceways (or even in metal raceways where the designer wants the benefit of a dedicated grounding conductor in addition to the ground path through the metal raceway) may be bare conductors, or they may be insulated conductors provided that the insulation is green or green with one or more yellow stripes. But there are exceptions to that rule for wires larger than No. 6.

A conductor of a color other than green may be used as an equipment grounding conductor if the conductor is stripped for its exposed length within an enclosure, so it appears bare, or if green coloring, green tape, or a green label is used on the conductor at its termination. As shown in Fig. 300, the phase legs may be any color other than white, grey, or green. The neutral must be white or gray (or any color other than green if it is larger than No. 6 and if white tape, marking, or paint is applied to the neutral near its terminations.) The grounding conductor, as noted, may be green or may be any color conductor if all insulation is stripped off for the exposed length. Alternatives to stripping the black insulated conductor used for equipment grounds include (1) coloring the exposed insulation green or (2) marking the exposed insulation with green tape or green adhesive labels.

EXAMPLE

Enclosure for switchboard, panel, motor control center, etc.

PVC conduit for feeder needs equipment grounding conductor

Five black insulated conductors in conduit; 3 phases, neutral and ground

To load

White tape on termination end of grounded leg (neutral)

Neutral bus

Ground bus

Black insulated conductor used as equipment grounding conductor has all insulation stripped from entire length exposed in enclosure

FIG. 300 Equipment grounding conductor larger than No. 6 does not have to be colored green under certain conditions.

On-the-job identification is permitted for an insulated conductor used as an equipment grounding conductor in a multiconductor cable. Such a conductor, regardless of size, may be identified in the manner described above for conductors larger than No. 6 used in raceway. The conductor may be stripped bare or colored green to indicate that it is a grounding conductor. But such usage is recognized only for industrial-type systems under conditions of competent maintenance.

Building-Frame Ground

The **NEC** clearly prohibits the use of structural building steel as an equipment grounding conductor for equipment mounted on or fastened to the building steel if the supply circuit to the equipment operates on alternating current. But structural building steel that is effectively grounded and bonded to the grounded circuit conductor of a dc supply system may be used as the equipment grounding conductor for the metal enclosures of dc-operated equipment that is conductively attached to the building steel.

As shown in Fig. 301, the use of building steel as a grounding conductor provides a long fault return path of very high impedance because the path is separated from the feeder-circuit hot legs. Ground-fault current returning over building steel to the point where the building steel is bonded to the ac system neutral (or other grounded) conductor is separated from the circuit conductor that is providing the fault current. The impedance is, therefore, elevated, and the conditions are less than optimal, so that the grounding cannot be counted on to facilitate operation of the fuse or CB protecting the faulted circuit. The current may not be high enough to provide fast and certain clearing of the fault.

Conduit terminations for metal raceways are important in ensuring a continuous low-impedance ground path through the metal raceway-and-enclosure system that forms the equipment grounding conductor for an electrical distribution system. Single locknut-and-bushing terminations are permitted for 120/240-V systems and 120/208-V systems. A 480/277-V grounded system, 480-V ungrounded system, or higher-voltage system must have double locknut-and-busing terminals on rigid metal conduit and IMC (Fig. 302). Where good electrical continuity is desired on installations of rigid metal conduit or IMC, two locknuts are always specified so that the metal of the box can be solidly clamped between the locknuts, one on the outside and one on the inside. The reason for not relying on the bushing (in place of the inside locknut) is that

FIG. 301 Grounded metal frame of a building is not an acceptable grounding conductor for ac equipment.

FIG. 302 Conduits carrying circuits operating above 250 V to ground must have double-locknut terminations.

both conduit and box may be secured in place, and if the conduit is placed so that it extends into the box a distance greater than the thickness of the bushing, the bushing will not make contact with the inside surface of the box. But that possible weakness in the single-locknut termination does not exclude it from use on systems up to 250 V to ground.

When an individual equipment grounding conductor is used in a raceway (either in a non-metallic raceway, as required, or in a metal raceway—where such a conductor is used for grounding reliability even though metal raceway is a suitable grounding conductor), the grounding conductor must have the minimum size shown in **NEC** Table 250-95. The minimum acceptable size for an equipment grounding conductor is based on the rating of the overcurrent device (fuse or CB) protecting the circuit that is run in the same raceway and for which the equipment grounding conductor is intended to provide a path of ground-fault current return, as shown in Fig. 303. Each size of grounding conductor in the table can carry enough current to blow a fuse or trip a CB of the corresponding rating (in the left-hand column of the table). In Fig. 303, if the fuses are rated at 60 A, Table 250-95 shows that the equipment grounding conductor must be at least No. 10 copper or No. 8 aluminum or copper-clad aluminum.

Whenever an equipment grounding conductor is used for a circuit that consists of only one conductor for each hot leg (or phase leg), the grounding conductor is sized simply and directly from Table 250-95, as described. When a circuit is made up of parallel conductors per phase, as in an 800-A circuit with two conductors per phase, an equipment grounding conductor is sized in the same way; in that case, at least No. 1/0 copper or No. 3/0 aluminum would have to be used, as indicated alongside 800 A in Table 250-95. But, if such a circuit is made up using

FIG. 303 Equipment grounding conductor must be of sufficient size to carry the current required to operate the circuit protective device under fault condition.

two conduits—that is, three phase legs and a neutral in each conduit—an individual grounding conductor must be run in each of the conduits, and *each* of the two grounding conductors would have to be at least No. 1/0 copper or No. 3/0 aluminum, as shown in Fig. 304.

Another example is shown in Fig. 305, where a 1200-A protective device on a parallel circuit calls for No. 3/0 copper or 250-MCM aluminum grounding conductor. [Note in that example that each 500-MCM XHHW circuit conductor has an ampacity of 430 A (Table 310-16), and three per phase gives a circuit ampacity of 3 × 430 A = 1290 A. Use of a 1200-A protective device satisfies the basic rule of Section 240-3, protecting each phase leg within its ampacity. Because the load on the circuit is continuous (over 3 hours), the circuit is loaded to not over 960 A, satisfying Section 220-10(b), which requires a continuous load to be limited to no more than 80 percent of the circuit conductor ampacity and no more than 80 percent of the protection rating. Each circuit conductor is actually made up of three 500-MCM XHHWs, with a total per-phase ampacity of 3 × 430 A = 1290 A. But the load is limited to 0.8 × 1200 A = 960 A per phase. Each 500-MCM is then carrying 960 A ÷ 3 = 320 A. Because that value is less than 380 A, which is the ampacity of 500-MCM THW copper, the use of XHHW conductors does comply with the UL requirement that, for equipment rated over 100 A, conductors connected to the equipment be rated at not over 75°C (such as THW) or, if 90°C conductors are used (such as XHHW), they must be used at no more than the ampacity of 75°C conductors of the same size. However, some authorities object to that usage on the grounds that 1200-A protection would not be acceptable according to Exception No. 1 of Section 240-3 if 75°C (THW)

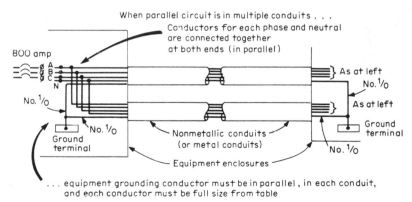

FIG. 304 Equipment grounding conductor must be installed in each conduit for multiconduit, parallel-conductor feeder makeup.

FIG. 305 Grounding conductor in each conduit of a multi-conduit feeder must be sized to correspond to the full value of the circuit protection.

conductors were used, with an ampacity of only 3 × 380 A = 1140 A per phase, because the rating is over the 800-A level, and it is not permitted to go to the next larger size of protective device. Therefore, they note, the XHHW conductors are not actually being used as 75°C conductors; the load current could later be increased above the 75°C ampacity; and the application does violate the letter and intent of the UL rule, thereby violating **NEC** Section 110-3(b).]

The same **NEC** rule points out that the equipment grounding conductor need not be larger than the circuit conductors. The main application of this exception is to motor circuits, where short-circuit protective devices are usually rated considerably higher than the motor branch-circuit conductor ampacity, as permitted in Section 430-52 (up to 400 percent), to allow starting of a motor without opening the circuit on inrush current. In such cases, the literal use of Table 250-95 could result in the use of grounding conductors that are larger than the circuit conductors.

Exception No. 3 points out that metal raceways and cable armor are recognized as equipment grounding conductors, and Table 250-95 does not apply to them.

Figure 306 shows details of a controversy that often arises. When two (or more) circuits are located in the same conduit, it is logical to conclude that a single equipment grounding conductor within the conduit may serve as the required grounding conductor for both circuits if it satisfies Table 250-95 for the circuit with the highest-rated overcurrent protection. The common contention is that if a single metal conduit is adequate as the equipment grounding conductor for all the contained circuits, a single grounding conductor can serve the same purpose when installed in a nonmetallic conduit that connects two metal enclosures (such as a panel and a homerun junction box) where both circuits are within both enclosures. As shown, a No. 12 copper conductor satisfies Table 250-95 as an equipment grounding conductor for the circuit protected at 20 A. The same No. 12 also may serve for the circuit protected at 15 A, for which a grounding conductor must not be smaller than No. 14 copper. Such use is specifically permitted by the **NEC**. Although this will have primary application for PVC conduit, where an equipment grounding conductor is required, it may also apply to circuits in EMT, IMC, or rigid metal conduit when an equipment grounding conductor is run with the circuit conductors to supplement the metal raceway as an equipment grounding return path.

FIG. 306 Single equipment grounding conductor may serve for two or more circuits run in a single raceway.

FIG. 307 Equipment bonding jumpers for multiple-conduit feeders provide better protection when run in parallel, as at bottom here.

Bonding Jumpers

Any bonding jumper on the load side of the service should conform to the requirements for equipment grounding conductors. As a result, the bonding of conduits for a parallel circuit makeup would have to comply with the **NEC** rule that requires equipment grounding conductors to be run in parallel "where conductors are run in parallel in multiple raceways." That rule would then be taken to require that bonding jumpers also must be run in parallel for multiple-conduit circuits. In the case of the 800-A circuit run in two conduits, as described above, instead of a single No. 1/0 copper jumper from one bushing lug to the other bushing lug and then to the ground bus, it would be better to use a separate No. 1/0 copper jumper from each bushing to the ground bus, so the jumpers are run in parallel, as required for equipment grounding conductors. Figure 307 shows the two possible arrangements. Bonding jumpers on the load side of service equipment are sized and routed in the same way as equipment grounding conductors, because such bonding jumpers and equipment grounding conductors serve identical functions. And Section 250-95 requires the equipment grounding conductor for each of the conduits of a parallel circuit to be the full size determined from the circuit rating. Here, No. 1/0 copper for each conduit would be based on the 800-A rating of the feeder protective device. [NOTE: Although it is not recommended practice, **NEC** Sec. 250-79(d) does permit the above described use of a single No. 1/0 jumper to bond both conduits to the equipment ground bus. In fact the **NEC** rule permits any number of metal conduits to be bonded to a switchboard or other equipment enclosure by a "single common continuous" jumper if it is sized from Table 250-95 for "the largest overcurrent device supplying" any of the circuits contained in all of the conduits.

section

Motor Feeders

When a load consists only of motors or of motors plus lighting, heating, and/or other types of loads, the feeder conductors and overcurrent protection must be carefully selected on the basis of all applicable design considerations.

The sizing of feeder conductors must satisfy applicable **NEC** rules and, among other considerations, ensure that voltage drop and copper loss are kept to reasonable values. The initial design procedure is based on the following requirements:

1. The current-carrying capacity of conductors supplying several motors must be at least equal to 125 percent of the full-load current of the highest rated motor plus the sum of the full-load current of the other motors supplied by the feeder (**NEC** Section 430-24).
2. The current-carrying capacity of feeder conductors supplying a single motor plus other loads must be at least equal to 125 percent of the full-load current of the motor plus other load current.
3. The current-carrying capacity of feeder conductors supplying a motor load and a lighting and/or appliance load must be sufficient to handle the lighting and/or appliance load as determined from the procedure for sizing lighting feeders, plus the motor load as determined from items 1 and 2 above.

The **NEC** permits inspectors to authorize the use of demand factors for motor feeders—based on reduced heating of conductors that supply motors operating intermittently, motors on duty cycles, or motors not operating together. Where necessary, the authority enforcing the **NEC** should be consulted to ensure that the conditions and operating characteristics are deemed suitable for reduced-capacity feeders.

Sizing Conductors

The **NEC** allows the sizing of motor feeders (and feeders supplying combination power and lighting loads) on the basis of maximum demand running current, calculated as follows:

$$\text{Running current} = (1.25 \times I_f) + (DF \times I_t)$$

where I_f = full-load current of largest motor, A
  DF = demand factor
  I_t = sum of full-load currents of all motors except largest, A

But modern design must include provisions for limiting the voltage drop under conditions of high motor-current inrush and therefore dictates the use of the maximum demand starting current in sizing conductors for improved voltage stability on the feeder. Thus current is calculated as follows:

$$\text{Starting current} = I_s + \text{DF} \times I_t$$

where I_s is the average starting current of the largest motor. (Use the percentage of motor full-load current given for fuses in Table 430-152.)

Voltage Drop and Power Factor

Voltage drop and I^2R loss must be carefully considered in sizing motor feeders. The design voltage-drop percentage may vary with the particular operating conditions and layout of motor loads; however, it should never exceed a 3 percent drop from the service entrance to the point of origin of motor branch circuits. And a maximum voltage drop of only 2 percent from the service to the motor branch-circuit protection is widely used for motor feeder design.

Both the reactance and the resistance of the feeder conductors must be included in voltage-drop calculations, as both contribute to the drop. Power factor must also be accounted for in these calculations. And voltage drop in a feeder must be analyzed in terms of the number of motors supplied and the size and operating duty of each motor.

When a number of motors may be starting simultaneously, or when several motors driving sluggish loads may be started at or near the same time, the voltage drop in the feeder could be large, unless it is sized to account for high load currents. Of course, such conditions, when analyzed, will often clearly indicate the need for further subdivision or adjustment of feeder loads, as well as the best types of motors and/or controllers to select for use. The initial value of the starting current—the locked-rotor current—must be used in studying the effect of motor loads on voltage drop.

The I^2R loss in a motor feeder—the power (watts) lost in the conductors due to heat developed by current flow through the conductors—is equal to the square of the total current drawn through the conductors times the total resistance of the conductors. This loss may frequently be substantial, even when the voltage drop in the feeder is within recommended limits. All voltage-drop studies should include the consideration of an increase in conductor size beyond that necessary to limit the voltage drop, for the purpose of limiting I^2R losses and the cost of the power they waste.

The power factor of a motor feeder is improved by placing a capacitor of the proper kilovar rating at the supply end of the feeder; this reduces both the required current flow to a given load and the voltage drop, by eliminating the reactive current from the feeder conductors.

The power factor of a circuit supplying several motors is determined as follows:

1. For each motor, multiply its horsepower by its power factor at 75 percent of rated load.
2. Add these products for all the motors.
3. Divide the sum obtained in step 2 by the total horsepower connected to the circuit, to obtain the approximate (but accurate enough for most calculations) power factor of the circuit.

The circuit can then be corrected to raise the power factor to a desired level by selecting a capacitor of the proper kilovar rating for the given load as directed by the literature of capacitor manufacturers.

Desired Power Factor in Percentage

Original Power Factor in Percentage

	80%	81	82	83	84	85	86	87	88	89	90	91	92	93	94	95	96	97	98	99	100
50%	.982	1.008	1.034	1.060	1.086	1.112	1.139	1.165	1.192	1.220	1.248	1.276	1.303	1.337	1.369	1.40?	1.441	1.481	1.529	1.590	1.732
51	.936	.962	.988	1.014	1.040	1.066	1.093	1.119	1.146	1.174	1.202	1.230	1.257	1.291	1.323	1.357	1.395	1.435	1.483	1.544	1.686
52	.894	.920	.946	.972	.998	1.024	1.051	1.077	1.104	1.132	1.160	1.188	1.215	1.249	1.281	1.315	1.353	1.393	1.441	1.502	1.644
53	.850	.876	.902	.928	.954	.980	1.007	1.033	1.060	1.088	1.116	1.144	1.171	1.205	1.237	1.271	1.309	1.349	1.397	1.458	1.600
54	.809	.835	.861	.887	.913	.939	.966	.992	1.019	1.047	1.075	1.103	1.130	1.164	1.196	1.230	1.268	1.308	1.356	1.417	1.559
55	.769	.795	.821	.847	.873	.899	.926	.952	.979	1.007	1.035	1.063	1.090	1.124	1.156	1.190	1.228	1.268	1.316	1.377	1.519
56	.730	.756	.782	.808	.834	.860	.887	.913	.940	.968	.996	1.024	1.051	1.085	1.117	1.151	1.189	1.229	1.277	1.338	1.480
57	.692	.718	.744	.770	.796	.822	.849	.875	.902	.930	.958	.986	1.013	1.047	1.079	1.113	1.151	1.191	1.239	1.300	1.442
58	.655	.681	.707	.733	.759	.785	.812	.838	.865	.893	.921	.949	.976	1.010	1.042	1.076	1.114	1.154	1.202	1.263	1.405
59	.618	.644	.670	.696	.722	.748	.775	.801	.828	.856	.884	.912	.939	.973	1.005	1.039	1.077	1.117	1.165	1.226	1.368
60	.584	.610	.636	.662	.688	.714	.741	.767	.794	.822	.849	.878	.905	.939	.971	1.005	1.043	1.083	1.131	1.192	1.334
61	.549	.575	.601	.627	.653	.679	.706	.732	.759	.787	.815	.843	.870	.904	.936	.970	1.008	1.048	1.096	1.157	1.299
61	.515	.541	.567	.593	.619	.645	.672	.698	.725	.753	.781	.809	.836	.870	.902	.936	.974	1.014	1.062	1.123	1.265
63	.483	.509	.535	.561	.587	.613	.640	.666	.693	.721	.749	.777	.804	.838	.870	.904	.942	.982	1.030	1.091	1.233
64	.450	.476	.502	.528	.554	.580	.607	.633	.660	.688	.716	.744	.771	.805	.837	.871	.909	.949	.997	1.058	1.200
65	.419	.445	.471	.497	.523	.549	.576	.602	.629	.657	.685	.713	.740	.774	.806	.840	.878	.918	.966	1.027	1.169
66	.388	.414	.440	.466	.492	.518	.545	.571	.598	.626	.654	.682	.709	.743	.775	.809	.847	.887	.935	.996	1.138
67	.358	.384	.410	.436	.462	488	.515	.541	.568	.596	.624	.652	.679	.713	.745	.779	.817	.857	.905	.966	1.108
68	.329	.355	.381	.407	.433	.459	.486	.512	.539	.567	.595	.623	.650	.684	.716	.750	.788	.828	.876	.937	1.079
69	.209	.325	.351	.377	.403	.429	.456	.482	.509	.537	.565	.593	.620	.654	.686	.720	.758	.798	.840	.907	1.049
70	.270	.296	.322	.348	.374	.400	.427	.453	.480	.508	.536	.564	.591	.625	.657	.691	.729	.769	.811	.878	1.020
71	.242	.268	.294	.320	.346	.372	.399	.425	.452	.480	.508	.536	.563	.597	.629	.663	.701	.741	.783	.850	.992
72	.213	.239	.265	.291	.317	.343	.370	.396	.423	.451	.479	.507	.534	.568	.600	.634	.672	.712	.754	.821	.963
73	.186	.212	.238	.264	.290	.316	.343	.369	.396	.424	.452	.480	.507	.541	.573	.607	.645	.685	.727	.794	.936
74	.159	.185	.211	.237	.263	.289	.316	.342	.369	.397	.425	.453	.480	.514	.546	.580	.618	.658	.700	.767	.909
75	.132	.158	.184	.210	.236	.262	.289	.315	.342	.370	.398	.426	.453	.487	.519	.553	.591	.631	.673	.740	.882
76	.105	.131	.157	.183	.209	.235	.262	.288	.315	.343	.371	.399	.426	.460	.492	.526	.564	.604	.652	.713	.855
77	.079	.105	.131	.157	.183	.209	.236	.262	.289	.317	.345	.373	.400	.434	.466	.500	.538	.578	.620	.687	.829
78	.053	.079	.105	.131	.157	.183	.210	.236	.263	.291	.319	.347	.374	.408	.440	.474	.512	.552	.594	.661	.803
79	.026	.052	.078	.104	.130	.156	.183	.209	.236	.264	.292	.320	.347	.381	.413	.447	.485	.525	.567	.634	.776
80	.000	.026	.052	.078	.104	.130	.157	.183	.210	.238	.266	.294	.321	.355	.387	.421	.459	.499	.541	.608	.750
81	—	.000	.026	.052	.078	.104	.131	.157	.184	.212	.240	.268	.295	.329	.361	.395	.433	.473	.515	.582	.724
82	—	—	.000	.026	.052	.078	.105	.131	.158	.186	.214	.242	.269	.303	.335	.369	.407	.447	.489	.556	.698
83	—	—	—	.000	.026	.052	.079	.105	.132	.160	.188	.216	.243	.277	.309	.343	.381	.421	.463	.530	.672
84	—	—	—	—	.000	.026	.053	.079	.106	.134	.162	.190	.217	.251	.283	.317	.355	.395	.437	.504	.645
85	—	—	—	—	—	.000	.027	.053	.080	.108	.136	.164	.191	.225	.257	.291	.329	.369	.417	.478	.620

Example: Total kw input of load from wattmeter reading 100 kw at a power factor of 60%. The leading reactive kva necessary to raise the power factor to 90% is found by multiplying the 100 kw by the factor found in the table, which is .849. Then 100 kw × 0.849 = 84.9 kva. Use 85 kva.

FIG. 308   Capacitor kilovars required to improve a load to a desired power factor.

Figure 308 is a handy table commonly used for computing the capacitor kilovars (reactive kilovoltamperes) required to raise the power factor from one value to another for a given kilowatt load.

Figure 309 presents data on the application of power-factor capacitors on plug-in busway supplying motor loads. The busways were used in a modern industrial plant to supply plug-in circuits to motor loads, and the data were used as follows: The total horsepower of the connected motor load was determined for each busway run. A demand factor was applied, based on past experience and metering, reducing the demand on the busway run. The horsepower ratings were then corrected for the efficiency of the motors, which was assumed to be 95 percent. This corrected horsepower was converted to kilowatts, and a corresponding kilovoltampere value was established for an 80 percent power factor, for each busway. Reactive kilovoltampere (kilovar) values were also determined. Then kilovoltampere and kilovar values were calculated for the given kilowatt values, based on an assumed power factor of 95 percent. The difference in kilovars was then noted for each busway, to establish the kilovars required to bring the power factor of each busway up to about 95 percent. The last column in Fig. 309 shows the amounts of capacitor equipment that were required, based on the use of 15-kvar unit capacitors rated at

Bus duct	Total hp	40% Demand	95% Eff.	Kw	Kva 80%	Kvar 80%	Kva 95%	Kvar 95%	Kvar diff.	Kvar use
1 A	364	146	153	109	136	82	115	35	47	45
1 B	374	150	157	117	142	85	124	38.4	46.6	45
1 C	461	185	194	145	181	109	153	47	62	60
2 A	735	(30%) 220	230	172	215	130	181	56	74	75
2 B	1300	(30%) 390	410	306	382	230	322	100	130	135
3 A	729	291	307	229	286	172	241	75	97	105
3 B	275	110	116	87	109	65	92	29	36	45

FORMULAS:

$$\text{Power Factor} = \frac{KW}{KVA}$$

$$KVA = \frac{1.73 \times E \times I}{1000}$$

$$KVA \ \text{input to induction motor} = \frac{HP \times .746}{P.F. \times Eff.}$$

$$\text{Power Factor} = \cos \emptyset$$

$$KW \ \text{output of motor} = HP \times .746$$

$$KW \ \text{input to motor} = \frac{HP \times .746}{Efficiency}$$

$\cos \emptyset_1 = .8 \quad \cos \emptyset_2 = .95$

$\sin \emptyset_1 = .6 \quad \sin \emptyset_2 = .312$

$$KVA = \frac{KW}{\cos \emptyset} \quad KVAR = KVA \times \sin \emptyset$$

Power factor can be increased with capacitors or with synchronous motors. The following shows how a 75 kVA synchronous motor load (which draws current that "leads" the voltage instead of current that "lags" the voltage, as with an induction motor) was used to provide 45 kvar (reactive kilovolt amperes) to offset part of the 135 kvar of the induction motor load. The kW components add up to 290 kW, and the lagging kvar is reduced to 90 kvar.

SUBSTATION

INCANDESCENT LIGHTS
50 KVA
1.0 PF

INDUCTION MOTOR LOADS
225 KVA
0.8 PF (LAG)

SYNCHRONOUS MOTOR LOADS
75 KVA
0.8 PF (LEAD)

50 KVA = KW

75 KVA
45 KVAR
60 KW

180 KW
135 KVAR
225 KVA

290 KW
90 KVAR
303 KVA
COMBINED LOADS

FIG. 309 In an industrial system, these design data were used to determine capacitor power-factor correction for plug-in busway to motor loads.

5.4

575 V for the 12 sections of 600-A, 480-V, 3-phase, 4-wire busway. The capacitor equipment was mounted on the busway sections.

Protecting Motor Feeders

Overcurrent protection for a feeder to several motors must have a rating or setting not greater than the largest rating or setting of the branch-circuit protective device for any motor of the group plus the sum of the full-load currents of the other motors supplied by the feeder. It is possible for motors of different horsepower ratings to have the same rating of branch-circuit protective device, depending upon the type of motor and the type of protective device. If two or more motors in the group are of different horsepower ratings but the rating or setting of the branch-circuit protective device is the same for both motors, then one of the protective devices should be considered as the largest for the calculation of feeder overcurrent protection.

Because **NEC** Table 430-152 recognizes many different ratings of branch-circuit protective devices (based on the use of fuses or circuit breakers and depending upon the particular type of motor), it is possible for two motors of equal horsepower rating to have vastly different ratings of branch-circuit protection. For instance, for a 25-hp motor protected by non-time-delay fuses, Table 430-152 gives 300 percent of the full-load motor current as the maximum rating or setting of the branch-circuit device. Thus, 250-A fuses would be used for a motor with a 78-A full-load rating. But for a motor of the same horsepower (and even of the same type), time-delay fuses must be rated at only 175 percent of 78 A, which would mean using 150-A fuses for branch-circuit protection (Fig. 310). If the two 25-hp motors were of different types—one being a wound-rotor motor, and the other a squirrel-cage induction motor—it would still be necessary to base the selection of the feeder protection on the largest rating or setting of a branch-circuit protective device, regardless of the horsepower rating of the motor.

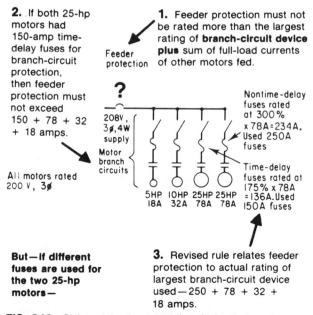

2. If both 25-hp motors had 150-amp time-delay fuses for branch-circuit protection, then feeder protection must not exceed 150 + 78 + 32 + 18 amps.

All motors rated 200 V, 3φ

1. Feeder protection must not be rated more than the largest rating of **branch-circuit device plus** sum of full-load currents of other motors fed.

Feeder protection

208V, 3φ,4W supply

Motor branch circuits

5HP IOHP 25HP 25HP
18A   32A   78A   78A

Nontime-delay fuses rated at 300% x 78A=234A. Used 250A fuses

Time-delay fuses rated at 175% x 78A =136A.Used 150A fuses

But—If different fuses are used for the two 25-hp motors—

3. Revised rule relates feeder protection to actual rating of largest branch-circuit device used—250 + 78 + 32 + 18 amps.

FIG. 310 Sizing of feeder protective device is based on the largest branch-circuit protective device, regardless of the actual motor horsepower ratings.

In large-capacity installations where extra feeder capacity is provided for load growth or future changes, the feeder overcurrent protection may be calculated on the basis of the rated current-carrying capacity of the feeder conductors. In some cases, such as where two or more motors on a feeder may be started simultaneously, feeder conductors may have to be larger than usually required for feeders to several motors.

The **NEC** calculation for selecting the size of a feeder overcurrent protective device is concerned with establishing the maximum setting or rating of the CB or fuse. If a lower value of protection is suitable, it may be used.

EXAMPLE

In Fig. 311, 250-A fuses are used to protect the feeder conductors to a group of four motors; the fuses were selected as follows:

The four motors supplied by the 3-phase, 440-V, 60-Hz feeder are not marked with a code letter (**NEC** Table 430-152) and are:

- One 50-hp squirrel-cage induction motor (full-voltage starting)
- One 30-hp wound-rotor induction motor
- Two 10-hp squirrel-cage induction motors (full-voltage starting)

Step 1: Branch-Circuit Loads

From **NEC** Table 430-150, the motors have full-load current ratings as follows:

50-hp motor: 65 A

30-hp motor: 40 A

10-hp motors: 14 A

Step 2: Conductors

The feeder conductors must have a carrying capacity that is computed as follows:

$$1.25 \times 65 \text{ A} = 81 \text{ A}$$

$$81 \text{ A} + 40 \text{ A} + (2 \times 14 \text{ A}) = 149 \text{ A}$$

The feeder conductors must be at least No. 3/0 TW, 1/0 THW, or 1/0 RHH, XHHW, or THHN (copper) wires.

FIG. 311   Rating of feeder protection and size of feeder conductors are based on motor full-load currents.

Step 3: Branch-Circuit Protection

The required overcurrent (branch-circuit) protection (from **NEC** Table 430-152 and Section 430-52) using nontime-delay fuses is found as follows:

1. The 50-hp motor must be protected at not more than 200 A (since 300 percent of 65 A is 3 × 65 A = 195 A, and the next larger standard size of fuse is 200 A).
2. The 30-hp motor must be protected at not more than 60 A (150 percent of 40 A).
3. Each 10-hp motor must be protected at not more than 45 A (300 percent of 14 A = 42 A).

Step 4: Feeder Protection

Based on step 3, the maximum rating or setting for the overcurrent device protecting such a feeder must not be greater than the largest rating or setting of branch-circuit protective device for one of the motors of the group plus the sum of the full-load currents of the other motors. From the above, then, the maximum allowable size of feeder fuses is 200 + 40 + 14 + 14 = 268 A.

This calls for a maximum rating of 250 A for the motor feeder fuses, since 250 A is the nearest standard fuse rating that does not exceed the maximum allowable value of 268 A.

NOTE: *There is no provision* in **NEC** Section 430-62 for the use of the next higher size, rating, or setting of protective device for a motor feeder when the calculated maximum rating does not correspond to a standard size of device.

EXAMPLE

Figure 312 shows a feeder that supplies five motors as follows:

- One 40-hp wound-rotor induction motor
- Two 25-hp squirrel-cage induction motors (full-voltage start)
- One 10-hp squirrel-cage induction motor (full-voltage start)
- One 5-hp squirrel-cage induction motor (full-voltage start)

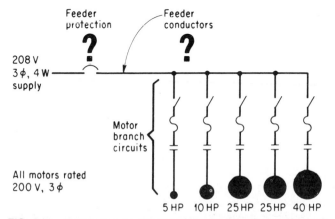

FIG. 312   Rating of a circuit breaker for motor feeder protection is determined by the largest rating of branch-circuit fuses in a case like this.

The design procedure involves these steps:

Step 1

The **NEC** full-load running currents of these motors are as follows:

40-hp motor: 119.6 A (use 120 A)

25-hp motors: 78.2 A (use 78 A)

10-hp motor: 32.2 A (use 32 A)

5-hp motor: 17.5 A (use 18 A)

These current values are derived from **NEC** Table 430-150, by using the multiplier of 1.15 to increase the values given in the table for 230-V motors to the correct current ratings for 200-V motors (as required in the footnote to the table).

Step 2

The minimum permissible ampacity of the feeder conductors supplying this group of continuously operating motors is the sum of the full-load running currents of all the motors plus 25 percent of the full-load current of the motor with the highest rated running current. (However, if some of the motors were operating on short-time, intermittent, periodic, or varying duty instead of all operating continuously, the sizing of the feeders would be modified as explained in **NEC** Section 430-24.)

Step 3

Based on the full-load running currents of these continuous-duty motors, the feeder conductors here must have an ampacity of not less than 356 A, that is, 120 A + 78 A + 78 A + 32 A + 18 A + (0.25 × 120 A).

Step 4

If THW aluminum conductors are used in conduit for the feeder, the smallest conductor size that would properly satisfy the need here is 700 MCM (375 A).

Step 5

The maximum permissible rating of a standard inverse-time CB used for the overcurrent protection of this feeder is 225 percent of the full-load current rating of the 40-hp motor plus the sum of the full-load currents of the other four motors of the group, when time-delay fuses are used for branch-circuit protection of the motors and a rating higher than 150 percent is needed to allow for inrush current. (See Exception No. 2*b*. of **NEC** Section 430-52.)

Step 6

Assuming that the fuses protecting the branch circuits to the motors are of the dual-element (time-delay) type, and the largest motor is fully capable of starting with its branch-circuit fuses sized according to Table 430-152, the maximum acceptable setting of the feeder CB is (1.50 × 120 A) + 78 A + 78 A + 32 A + 18 A, or 368 A.

Step 7

The maximum setting or rating of a standard, time-delay-type, molded-case CB that may be used for the protection of this feeder is 350 A, since Section 430-62 does not recognize the use of the next larger size of CB.

Power and Lighting Feeders

The conductors and protection for a feeder to both a motor load and a lighting and/or appliance load must be sized on the basis of both loads. The rating or setting of the overcurrent device must be sufficient to carry the lighting and/or appliance load plus the rating or setting of the motor branch-circuit protective device if only one motor is supplied, *or* plus the highest rating or setting of branch-circuit protective device for any one motor plus the sum of the full-load currents of the other motors, if more than one motor is supplied.

EXAMPLE

Figure 313 shows a feeder arrangement for a combination power and lighting load. As shown, the feeder supplies a 277-V lighting panel and four 3-phase motors. Such considerations as voltage drop, I^2R loss, spare capacity, and lamp dimming on motor starting would have to be accounted for in determining the actual wire size and overcurrent protection to use for the job. The circuiting shown would be safe, but perhaps not as efficient or effective as it could be.

The feeder conductors that will carry both the lighting-load current and the motor-load current must have sufficient capacity for the total load. The basic design steps are as follows:

Step 1

The total load is the sum of the motor load and the lighting load:

Motor load = 65 A + 40 A + 14 A + 14 A + (0.25 × 65 A) = 149 A per phase

Lighting load = 120 A × 1.25 = 150 A per phase

Total load = 149 A + 150 A = 299 A per phase

Note that, because the lighting load of 120 A per phase is a "continuous" load, **NEC** Section 220-10(b) requires that feeder conductors carrying that load must have ampacity at least equal to 125 percent of the load current. The multiplier 1.25 used above satisfies the requirement. But, only the phase conductors of the feeder are subject to that 125 percent factor; the neutral conductor of the feeder requires only an ampacity of 120 A. The 125 percent continuous-load factor is not applied to the feeder neutral because the neutral does not connect to the terminal of a switch, CB, or other device for which heating would be a problem under continuous load.

Step 2

NEC Table 310-16 shows that a combined power and lighting load of 299 A can be served by the following copper conductors:

500-MCM TW (320 A)

350-MCM THW (310 A)

WHAT IS MINIMUM SIZE OF THESE CONDUCTORS?

WHAT SIZE CB?

For motor feeder: 250A fuses

Conductors for 120A continuous lighting load must have ampacity of 125% of 120A [Sec. 220-10(b)]

277-volt lighting panel—continuous load

50 hp, 65 amps FL

30 hp, 40 amps FL

10 hp, 14 amps FL

10 hp, 14 amps FL

400/277 V
3-φ, 4-wire feeder

φA
φB
φC
N

FIG. 313 Conductors and protection for a power and lighting feeder must be sized to satisfy both loads.

5.10

Table 310-16 shows that this load can be served by the following aluminum or copper-clad aluminum conductors:

700-MCM TW (310 A)

500-MCM THW, RHH, or THHN (310 A)

Step 3

The protective device for a feeder supplying a combined motor and lighting load may not have a rating greater than the sum of the maximum rating of the motor feeder protective device and the lighting load; these are obtained as follows:

1. The maximum rating of the motor feeder protective device is the rating or setting of the largest branch-circuit device of any motor of the group being served plus the sum of the full-load currents of the other motors; this is

$$200 \text{ A } (50\text{-hp motor}) + 40 \text{ A} + 14 \text{ A} + 14 \text{ A} = 268 \text{ A}$$

This calls for a maximum standard rating of 250 A for the motor feeder fuses (the nearest standard fuse rating that does not exceed the maximum permitted value of 268 A).

2. For the lighting load, 120 A × 1.25 = 150 A.

A CB for the combined load must be rated at 268 A + 150 A = 418 A maximum. This calls for a 400-A CB (the nearest standard rating that does not exceed the 418-A maximum).

AGAIN: The **NEC** *does not* have a provision which permits the use of the next higher size, rating, or setting of protective device for a motor feeder when the calculated maximum rating does not correspond to a standard size of device.

EXAMPLE

Figure 314 shows a layout in which a main feeder is run to a point where it is tapped (such as at an auxiliary gutter) by two subfeeders—one for a lighting load and one for a motor load. The main switch for the combined feeder might be in a switchboard or panelboard, or it could be a main service switch (disconnect and protection at a service entrance). The conductors for the combined feeder and the fuses in the 200-A combined feeder switch are sized on the basis of the following loads: The motor load is four 3-phase, 230-V, squirrel-cage induction motors, designed for 40°C temperature rise, marked with code letter H, started across the line, and rated as follows:

- One 10-hp motor
- One 7½-hp motor
- Two 1½-hp motors

The lighting load is a 20-kw, single-phase, 115-V load.

Step 1

The average full-load motor currents (from **NEC** Table 430-150) are:

10-hp motor: 28 A

7½-hp motor: 22 A

1½-hp motors: 5.2 A each

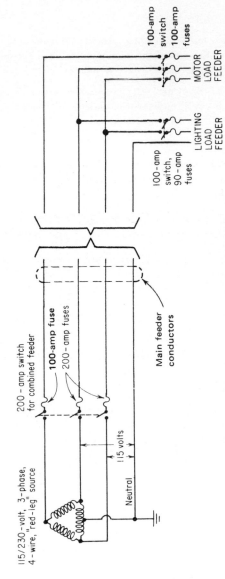

FIG. 314 Sizing of conductors and protection must be based on actual phase loadings.

115/230 - volt, 3 - phase,
4 - wire, "red - leg" source

200 - amp switch
for combined feeder

100-amp fuse

200 - amp fuses

Neutral

115 volts

Main feeder
conductors

100 - amp
switch,
90 - amp
fuses

LIGHTING
LOAD
FEEDER

100-amp
switch
100-amp
fuses

MOTOR
LOAD
FEEDER

Step 2

The conductors for the motor-load feeder must be selected to have an ampacity based on the following:

10 hp: 1.25 × 28 A	35 A	
7½ hp	22 A	
1½ hp: 2 × 5.2 A	10.4 A	
Total	67.4 A	

The required ampacity of 67.4 A could be provided by

Three No. 4 TW in 1-in conduit, or

Three No. 4 THW in 1-in conduit, or

Three No. 4 RHH (without covering) in 1-in conduit, or

Three No. 4 THHN in 1-in conduit

Step 3

The motor feeder protection (Section 430-62), using circuit breakers for the branch circuits but fuses in the feeder, is sized as follows:

10 hp: 28 A × 2.5	70 A (rating of branch CB)
7½ hp	22 A
1½ hp: 2 × 5.2 A	10.4 A
Total	102.4 A

The maximum size of the feeder fuses is 102.4 A. Therefore, a 100-A switch with 100-A non-time-delay fuses or smaller time-delay fuses would be used.

   NOTE: NEC Section 430-62 does not permit the motor-feeder fuses to be upsized to the next standard rating above 102.4 A, or 110 A. The fuses must be selected at 100 A, the next *lower* rating.

Step 4

For the lighting load, the full-load current would be

$$\frac{20 \times 1000\ W}{230\ V} = 87\ A$$

A 100-A switch with two 90-A fuses would be used for a lighting load that is not continuous—that is, one where the full load is not on for 3 hours or more. [If the 87-A load were continuous, a fused switch would be selected so that load was not over 80 percent of the fuse rating. This means that the minimum fuse size would be 87 A × 1.25 = 108.75 A, so a 110-A fuse would be used for each phase, in a 200-A switch (per NEMA standards).]

   Thus, three No. 2 TW, THW, RHH, or THHN in 1¼-in conduit, or in 1-in conduit (for THHN), would be used. (For a continuous 87-A load, where the fuse size would be increased to 110 A, the feeder conductors would have to be rated for at least 110 A as are No. 1 TW and No. 2 THW, RHH, or THHN.)

Step 5

Steps 1 to 4 cover the sizing of the individual feeders—one for the motor load and one for the lighting load. A single main feeder which supplies the combined motor and lighting load will have the 3-phase motor load on each of its three phase conductors and the lighting load on only two of its phase conductors and a neutral. A 3-phase, 4-wire, "red-leg" circuit would be designed as follows:

Step 6

The two phase conductors which supply both the motor load and a noncontinuous lighting load must have a minimum capacity of

Motor load	67.4 A
Lighting load	87 A
Total	154.4 A

Each of these two conductors must be at least No. 3/0 TW (165 A) or No. 2/0 THW (175 A) or RHH or THHN (185 A). [If the lighting load were continuous, a value of 108.75 A (125 percent of 87 A) would be used instead of 87 A in the previous calculation.]

Step 7

The third phase conductor of the main combined feeder serves only the motor load and must have a rating of at least 67.4 A, which calls for No. 4 TW, THW, RHH, or THHN as noted in step 2.

Step 8

Under the assumption that the 20-kW single-phase lighting load is all 115-V loads, equally balanced from hot legs to neutral, the neutral must be rated for the maximum unbalance; this is 87 A, the same as the load on each hot leg. This neutral conductor would then be a No. 2, TW, THW, RHH, or THHN, as shown in step 4. (If the lighting load were continuous, the neutral could still be sized at 87 A, although the phase legs would require an ampacity at least equal to 1.25 × 87 A in such a case.)

Step 9

Based on the loads on the three hot legs—154, 154, and 67.4 A—a three-pole, 200-A switch is needed for this combined feeder.

Step 10

Overcurrent protection for the two hot legs that carry the motor load and the lighting load is sized according to **NEC** Section 430-63:

Motor load	102.4 A setting
Lighting load	87 A
Combined load	189.4 A

Thus, the two fuses in the 200-A switch protecting the combined load must have a rating not greater than 200 A. (**NEC** Section 430-63 may be interpreted as setting a maximum fuse rating of 190 A—that is, 100 A for the rating of the motor-feeder fuses plus 90 A for the lighting-feeder fuses—thereby requiring the use of the next smaller size fuse, or 175 A.)

Step 11

The fuse which protects the third hot leg—the one serving only the motor load—must be rated not over 100 A.

   NOTE: This sizing calculation would not change at all if the feeder constituted the service-entrance conductors to a building with the given load, rather than the feeder to the combined load within the building.

Transformer Applications (to 600 volts)

The wide availability of economical, efficient, and highly effective dry-type transformers has strongly promoted the use of such transformers to carry electric power, at higher-than-utilization voltages, as close as possible to the center of a layout of utilization equipment—lighting, appliances, and receptacle loads.

Widespread acceptance and application of industrial medium-voltage systems (above 600 V and usually up to 35 kV) have stimulated the use of higher voltages in electrical systems for commercial and industrial buildings. Not too many years ago, 3-wire, single-phase, 120/240 V distribution systems and 3-phase, 4-wire, 120/208-V distribution systems were standards for nonindustrial buildings. Today, although these lower-voltage systems still find wide and efficient application, the 480Y/277-V, 3-phase, 4-wire system is established as the preferred secondary-voltage system for use where substantial motor loads and general area fluorescent or other electric-discharge lighting must be supplied.

Tremendous growth in the size of electrical loads in commercial buildings was responsible for the development of the 480Y system, which has many of the characteristics of industrial-type systems. Air-conditioning loads and business and other machine loads increased the ratio of power loads to lighting loads, calling for the type of loadcenter circuit treatments and layouts which are used in industrial plants to economically and efficiently serve heavy motor loads.

Higher-voltage feeders to motor loads and to step-down transformers for lighting and receptacle circuits proved to be the ideal solution. Less copper is needed to distribute the heavy power requirements, and voltage drop and other losses are effectively minimized. The wide availability of extensive lines of lighting equipment which operates on 277-V circuits has also contributed greatly to the success of this system.

Studies have shown that a typical 480Y system can provide savings of more than $40 per kilovoltampere demand over a 120/208-V system handling the same load. Many installations over the past years—in schools, office buildings, industrial plants, shopping centers, and sports arenas—have convincingly proved the economic and operating advantages of the 480/277-V system.

The 480Y system was developed to meet the requirements of commercial building load conditions. Usually, the general lighting is fluorescent and can be served by 277-V circuits. Motors for air-conditioning compressors, circulating fans, elevators, and pumps make up an average load of about 4 VA/ft². These motors can be more efficiently and economically supplied at 480 V than at lower voltages, and 480-V motors are less expensive than lower-voltage motors of the same horsepower ratings. The combined power and general lighting loads average between 5 and 15 W/ft². Receptacle and miscellaneous loads—desk lamps, local lights, business machines, appliances, water coolers, etc.—average only about 0.5 to 2 W/ft².

As can be seen from the above, about 80 percent of a building load may be served directly by 480/277-V feeders. The 120-V circuits may be provided either by using separate 120/208-V substations (fed by medium-voltage feeders of 4160 V, 13.2 kV, etc.) or by using dry-type transformers (480 to 120/240 V or 480-V, 3-phase to 120/208-V) installed locally at the center of each concentration of 120-V loads. The latter method offers greater economy, even when the 120-V loading is as high as 5 W/ft².

The design of electrical systems utilizing dry-type transformer applications must include careful consideration of a number of transformer characteristics. These are discussed in the next several paragraphs.

Voltage Rating

The primary and secondary voltage ratings of a transformer are determined very simply from the voltage rating of the supply to the primary winding and the required voltage level on the secondary. Both single- and 3-phase transformers are made with a wide variety of voltage ratings to afford ready application in standard system configurations and even in nonstandard arrangements.

The two basic selection factors are secondary voltage and secondary current. Together they determine the kilovoltampere capacity of the transformer, which is its load-handling ability. The two terms are defined as follows:

- The *rated secondary voltage* of a constant-potential transformer is the voltage at which the transformer secondary is designed to deliver the rated kilovoltampere capacity.

- The *rated secondary current* of a constant-potential transformer is the secondary current obtained by dividing the rated kilovoltampere capacity by the rated secondary voltage.

Transformer selection must first focus on those two basic ratings—depending on the load to be supplied. Then the primary voltage rating of the transformer must be matched to the voltage of the primary feeder circuit to the transformer.

To match actual system conditions, transformers are available that allow some variation from the nominal primary voltage but still produce the required secondary voltages. Figure 315 shows some typical transformer primary tap connections for varying input voltages. On general-purpose transformers the primary is commonly provided with a number of taps for input voltages that vary over a range of 15 percent; for instance, two 2½ percent taps above normal voltage and four 2½ percent taps below normal voltage may be provided. On a transformer with a 480-V primary, then, the lowest of these taps would be connected to a primary feeder which operated at 432 V to obtain the rated secondary output of 240/120 V (since 480 V − 4 × 2.5 percent of 480 V = 432 V). If the 432-V primary supply were connected to the 480-V transformer

SINGLE-PHASE

Primary with two
5% taps below normal

Common abbreviations:

FCBN — Full capacity
below normal

FCAN — Full capacity
above normal

THREE-PHASE

Primary with two 2½% taps
above and below normal

Taps:

No. 1   504 volts
 "   2   492   "
 "   3   480   "
 "   4   468   "
 "   5   456   "

FIG. 315 Transformer primary taps provide for matching
the transformer input-voltage rating to the voltage of the
supply circuit.

terminals or to any of the other taps, the secondary voltage would be lower than 240/120 V. The taps thus compensate for system voltage drops.

General-purpose transformers have no-load tap changers, requiring that the transformer be deenergized when tap changes are made. Invariably, suitable taps for the given primary voltage are set once and left alone, unless some change in load conditions alters the voltage level at the transformer primary. On large transformers serving fluctuating demand loads, automatic tap changers which operate under load can be used on the transformer to maintain a constant voltage level at the transformer secondary terminals or at a remote point on the secondary output feeder.

The selection of transformer taps should be based on the no-load or maximum-voltage conditions at the transformer. Taps above or below the nominal rated voltage should be selected to accommodate anticipated variations in supply voltage and, thereby, provide the required no-load secondary voltage. This requires that voltage drops in secondary circuits be taken into account to arrive at proper voltage levels at utilization devices.

It should be remembered that transformer taps change the voltage transformation by changing the turns ratio of the transformer. Because of this, the kilovoltampere rating of a transformer is not changed when taps are changed.

For a given voltage rating, a transformer winding has a certain current rating, which determines the kilovoltampere rating of the transformer. The current (ampere) rating is a maximum value determined by the impedance of the winding and the capability of the transformer to dissipate heat. The conditions of use must not place a higher current on either the primary or the secondary winding. If the voltage input to a set of primary transformer terminals of given voltage rating is reduced below that voltage rating, the maximum kilovoltampere rating is also reduced, because the current cannot be increased to maintain constant kilovoltamperes. And if an overvoltage is applied, overheating will result and will shorten the life of the transformer.

Figure 316 shows the basic current, voltage, and kilovoltampere relationships for single-phase and 3-phase transformers. These relationships are used in design calculations as shown in the following examples.

EXAMPLE

What are the primary and secondary current ratings of a 50-kVA transformer for stepping 480 V down to 240 V?

First use

$$I_p = \frac{\text{transformer kVA rating} \times 1000}{E_p}$$

$$= \frac{50 \times 1000}{480} = 104 \text{ A}$$

Then

$$I_s = I_p \frac{E_p}{E_s} = 104 \frac{480}{240} = 208 \text{ A}$$

Single-phase transformer fundamentals

CIRCUIT CONDITIONS

E_p = Primary voltage
I_p = Primary current
E_s = Secondary voltage
I_s = Secondary current
E_{vd} = Voltage drop in circuit conductors
E_L = Voltage across load
T_p = Number of turns on primary
T_s = Number of turns on secondary

BASIC TRANSFORMER

PRACTICAL RELATIONSHIPS

$$\frac{E_p}{E_s} = \frac{T_p}{T_s} \qquad T_p \times I_p = T_s \times I_s \qquad \frac{I_p}{I_s} = \frac{T_s}{T_p} \qquad \frac{I_p}{I_s} = \frac{E_s}{E_p} \qquad I_p \times E_p = I_s \times E_s \qquad \text{Transformer kva rating} = \frac{I_p \times E_p}{1000} = \frac{I_s \times E_s}{1000}$$

Three-phase transformers currents and voltages

For any delta connection
(pri. or sec.):

$$I_L = \frac{\text{Rated transformer kva} \times 1000}{1.73 \times E_L}$$

For any wye connection
(pri. or sec.):

$$I_L = \frac{\text{Rated transformer kva} \times 1000}{1.73 \times E_L}$$

Where, I_L = Line current (pri. or sec.) of transformer
I_w = Rated current of each transformer winding
E_L = Rated voltage (phase-to-phase) of transformer
E_w = Rated voltage of each transformer winding

NOTE: This data applied to 3-phase transformers and to single-phase transformers connected for 3-phase use.

FIG. 316   Transformer circuit design is based on these relationships.

EXAMPLE

What are the primary and secondary current ratings of a 500-kVA, 3-phase transformer stepping 480-V delta down to 120/208-V wye?

On the delta primary, with $E_L = 480$ V

$$I_L = \frac{500 \times 1000}{1.73 \times 480} = \frac{500,000}{830} = 603 \text{ A}$$

On the wye secondary, with $E = 208$ V

$$I_L = \frac{500 \times 1000}{1.73 \times 208} = \frac{500,000}{360} = 1388 \text{ A}$$

NOTE: The above formulas do not take transformer losses into consideration and, therefore, provide only approximate determinations. However, because transformers have extremely high efficiencies (e.g., 98 percent), the calculations shown are generally acceptable for practical applications of transformers.

One consideration in selecting a transformer is the "impedance" of the unit, which is indicative of its ability to supply short-circuit currents into faults on the load side of the transformer. The impedance of a transformer is the opposition which the transformer presents to the flow of short-circuit current through it.

Every transformer has an impedance, which is generally expressed as a percentage—the percentage of the normal rated primary voltage which must be applied to the transformer to cause full rated load current to flow in the short-circuited secondary. For instance, if a 480- to 120-V transformer has an impedance of 5 percent, then 5 percent of 480 V, or 24 V, applied to the primary will cause rated load current to flow in the short-circuited secondary. If 5 percent of the primary voltage will cause such a current, then 100 percent of the primary voltage will cause 100/5 (or 20) times the full rated load current to flow through a solid short circuit on the secondary terminals.

From the foregoing, it can be seen that the lower the impedance of a transformer of given kilovoltampere rating, the higher the short-circuit current which it can deliver. Consider two transformers, both rated at 500 kVA. Assume the rated secondary load current is the same for both transformers. If one transformer is rated at 10 percent impedance, it can supply 100/10 (or 10) times the rated secondary current into a short circuit on its secondary terminals. If the other is rated at 2 percent impedance, it can supply 100/2 (or 50) times the rated secondary current into a short circuit on its secondary terminals. Thus, the second transformer can supply five times as much short-circuit current as the first, even though they have the same kilovolt-ampere load-handling ability. Common impedance values for general-purpose transformers are between 3 and 6 percent.

Voltage regulation is another transformer consideration, and impedance is related to it. Voltage regulation is a measure of how the secondary voltage of a transformer varies as the load on the transformer varies from full load to zero, with the primary voltage held constant.

Voltage regulation is expressed as a percentage that is calculated as the no-load voltage minus the full-load voltage divided by the full-load voltage times 100 percent. If a transformer has a no-load secondary voltage of 240 V, and the voltage drops to 220 V when the transformer is fully loaded (supplying the rated current output), then the regulation percentage is

$$(240 - 220) \div 220 \times 100 \text{ percent} = 9 \text{ percent}$$

This indicates that 9 percent of the secondary voltage is being dropped across the internal impedance of the transformer. It is obvious, then, that the higher the impedance of a transformer, the greater the drop from no-load to full-load voltage and the higher the percentage regulation.

It is generally desirable to keep regulation as low as possible, to minimize variations in voltage as load-current demand varies. Typical regulation values are between 2 and 4 percent.

Low regulation becomes very important when the transformer supplies varying load demands for utilization equipment which is sensitive to voltage changes. In cases where the load current on the transformer is relatively constant, however, the transformer taps or the primary supply conditions can be adjusted to provide fixed compensation for the voltage drop in the transformer. The voltage level can be set as required at the utilization devices, and it will not change because the load current is not changing. In such cases, acceptable voltage conditions at the utilization devices can be achieved with transformers of either low or high regulation.

But it should be pointed out that the high-regulation transformer provides a high internal impedance, which reduces the level of possible short circuits on the secondary. As a result, fuses, circuit breakers, and other protective devices need not be rated for as high interrupting capacities as they would need if a low-regulation (and low-impedance) transformer were used.

Kilovoltampere Rating

The rated kilovoltampere output of a transformer is that output which it can deliver for a specified time, at the rated secondary voltage and rated frequency, without exceeding a specified temperature rise based on insulation life and ambient temperature. The output that a transformer can deliver without objectionable deterioration of the insulation may be more or less than the rated output, depending upon ambient temperature and load cycles.

Transformers may be loaded above their kilovoltampere ratings with no sacrifice of life expectancy only in accordance with prescribed temperature testing or short-time loading data from the manufacturer. In addition, the load that can be carried by self-cooled transformers may be increased considerably by the use of fans or other auxiliary cooling equipment. Such an application may be advantageous for short-time peak loading, although the design characteristics of the transformer, including increased voltage regulation, must be considered.

The basic loading conditions, for which the normal life expectancy of a transformer is determined, are:

1. The transformer is continuously loaded at its rated kilovoltamperes and rated voltage.
2. The average temperature of the cooling air during any 24-hour period is 30°C.
3. The temperature of the cooling air at no time exceeds 40°C.

The rated kilovoltamperes, rated voltage, and rated primary and secondary currents are considered to be maximum values. The kilovoltampere rating is based on drawing the rated current at the rated voltage; but, as noted previously, the kilovoltampere handling capacity cannot be kept constant by, say, operating at a lower voltage and a higher current. This capacity is determined by the maximum current the transformer can carry without reaching a dangerous temperature, and that current cannot be exceeded. Moreover, it should be noted that transformers are rated in *kilovoltamperes,* so that their load-handling ability is dependent only upon current and voltage, without regard for the effect of circuit power factor. The power-handling capacity (rated kilowatt load) of a transformer, therefore, is greater for high-power-factor loads than for low-power-factor loads.

A transformer selected for a particular application must have a kilovoltampere rating at least equal to the kilovoltampere rating of load to be supplied. But modern engineering practice, based on experience with load growth in electrical systems, demands the inclusion of spare capacity in every transformer installation. Of course, this spare capacity will be provided automatically if the feeders and branch circuits supplied by the transformer have spare capacity and the transformer is sized to accommodate that extra capacity.

In selecting transformers to supply motor loads, several general rules are used to determine the required capacity. One such rule says that a transformer should have at least 1 kVA of capacity for each horsepower of motor load. Another guide suggests at least 1¼ kVA per horsepower of motor load for motors rated 5 hp and above. Motors over 50 hp may be satisfactorily served by more or less than 1¼ kVA/hp, depending upon the torque characteristics (NEMA design) of the particular motor, the horsepower rating of the motor, the load characteristics, whether or not reduced-voltage starting is used, and what type of reduced-voltage starter is used. Of course, greater accuracy can be achieved in matching transformer capacity to a number of motors if the nameplate current ratings of the motors are added together and then multiplied by the rated voltage to obtain the required kilovoltamperes. Effective, economical, and reliable operation of large motors may often depend upon selection of a transformer with the correct kilovoltampere rating and internal impedance for the particular conditions of motor starting, duty cycle, and loading. Early engineering assistance from a transformer manufacturer can prevent many problems in motor applications.

Frequency Rating

By far the most common frequency of alternating current used for modern power and lighting electrical systems is 60 Hz. The transformers used on such systems are correspondingly rated for 60 Hz. Where other frequencies are used, care must be taken to ensure the suitability of the transformers.

Number of Phases

Single-phase transformers are used on single-phase systems and on single-phase circuits derived from 3-phase systems. To transform 3-phase voltage from one level to another, either three single-phase transformers or a 3-phase transformer can be used. A 3-phase transformer weighs less and requires less space than three single-phase transformers of the same type, construction, and total kilovoltampere capacity, and provides greater ease and economy in installation. But the use of three single-phase transformers has the advantage that failure of one of the units requires that only the one unit be replaced. Failure in a 3-phase transformer requires replacement of the entire device.

Noise Rating

A transformer that is used in a place with a low ambient noise level—school, office, library, church, etc.—requires a low decibel hum rating and/or careful installation to prevent objectionable noise.

Air-cooled, dry-type transformers are generally enclosed in sheet-metal housings. The enclosure may or may not be ventilated. Completely enclosed units are made for use in atmospheres where dust, moisture, chemicals, or other airborne matter might harm exposed coils in a ventilated case. Some units have their core laminations exposed to dissipate heat. Other small-size units have totally enclosed housings filled with epoxy compound, which seals the windings against atmospheric attack and greatly reduces operating hum.

Temperature Rating

Dry-type transformers are designed with various types of insulation on the windings; the rating and loading of a transformer are based on the temperature limits of the particular insulation system. The common systems are as follows:

Class 105 (formerly called "Class A") insulation systems, when properly loaded and applied in an ambient temperature of not over 40°C, will operate at no more than a 55°C average temperature rise on the winding conductors, with an added 10°C of hot-spot temperature gradient. (This gradient is the difference between the average conductor temperature and the temperature of the hottest spot—the point where the greatest possible insulation damage could occur.) The sum—40°C plus 55°C plus 10°C—results in the designation of 105°C for the system. Today's usage limits Class 105 insulation to the smallest single-phase transformers (e.g., up to 0.150 kVA).

Class 150 (formerly called "Class B") insulation systems reach a maximum operating temperature of 150°C—an 80°C rise on the winding, added to a maximum 40°C ambient, plus a 30°C hot-spot temperature gradient. Class 150 transformers are generally available in ratings up to 2 kVA, single phase, for 600, 480, or 240-V primary with 240/120-V secondaries.

Class 185 (formerly called "Class F") insulation systems reach a maximum operating temperature of 185°C—a 115°C rise on the windings, added to a maximum 40°C ambient, plus a 30°C hot-spot temperature gradient. Class 185 units are made in single-phase and 3-phase types from 3 through 30 kVA, for 600, 480, and 240-V primaries with 240/120-V or 208/120-V secondaries.

Class 220 (formerly called "Class H") insulation systems have a total temperature rise of 220°C—a 150°C temperature rise on the conductors plus a hot-spot temperature gradient of 30°C, plus 40°C ambient temperature. Single-phase and 3-phase transformers of the ventilated indoor type are available in ratings from 15 to 500 kVA for single-phase systems and up to 1000 kVA for 3-phase systems, for primaries up to 600 V. In those ratings, Class H insulation is the type most commonly used in dry-type transformers today.

Transformers must be carefully selected, with regard for all the factors that depend upon or are related to the temperature rating, as follows:

Relative loading: Transformers loaded at or very close to their kilovoltampere ratings will run hotter and be more disposed to thermal aging and damage than transformers that are lightly loaded. If, for instance, an 80-kVA demand load is to be fed by a 3-phase, 480- to 208/120-V transformer, a 100-kVA unit would be selected. But, based on this initial loading, the transformer would be only 80 percent loaded.

Duty cycle: A transformer may be operated fully loaded (100 percent of rating) for 24 hours a day, 7 days a week. Or, the same transformer may be loaded at less than 100 percent of its rating for all or part of each 24-hour period and for one or more days every week. The loading level and cycle can vary widely, and such variation should be known and factored into the selection.

Ambient temperature: The average and maximum ambient temperatures at the place where a transformer is to be installed must be determined or estimated and included in the analysis leading to the selection of the transformer.

Efficiency

With so much attention focused on spiraling energy costs today, the relative efficiency of a transformer—its core losses and conductor losses in relation to its kilovoltampere rating—has become an important element in evaluating its life cycle costs (first cost plus operating energy costs over the life of the equipment). Higher losses result in a higher transformer operating-temperature rise on the winding, for a given load.

Normally, transformer insulation is expected to remain effective for 20 to 25 years, if the transformer is loaded to 100 percent of its kilovoltampere rating and operated continuously at its rated voltage, with an average ambient temperature of not more than 30°C for any 24-hour period and a maximum ambient of 40°C, and at any altitude up to 3300 ft. Obviously, these are more severe conditions of operation than those found in the usual transformer application. Most transformers are not loaded to 100 percent of the rated load and operated 24 hours a day, every day, for their operating lives. And most installations are not subjected to a steady ambient temperature of 86°F (30°C) or a maximum of 104°F (40°C).

When transformers are applied at their maximum operating temperatures—that is, when their winding-temperature rise is at its highest value and the overall system is subjected to the maximum temperature—care should be taken to assure that conditions such as excessive ambient temperatures, 100 percent loading, and/or severe duty cycling do not represent a threat to the normal operating life of the equipment. Such conditions as frequent heavy-current motor-starting duty (that subjects a transformer to the exaggerated I^2R effects of repeated high motor inrush currents) and repeated thermal cycling due to off-on loading have been found to be more damaging to insulation than steady-state (continuous) operation.

There is no specific temperature that is critically related to transformer insulation life. That is, there is no single temperature above which insulation will deteriorate quickly, and below which deterioration will be slight or negligible. Thus, there is no fixed or correct operating temperature. Heat is always the enemy; insulation deteriorates at just about any temperature, and it is difficult to assign an absolute maximum operating temperature. Operating time is also involved in insulation failure, along with operating temperature—insulation destruction is a function of the product of time and heat. Equal insulation deterioration can result from short-time, high-temperature operation and from long-time, low-temperature operation. Continuous full-load operation and periods of overloading produce high heat that reduces transformer life below the life that would result from lower-temperature operation.

Transformers with Class 220 insulation systems—the most commonly used insulation for the widest range of transformer ratings—are available in three distinct designs, based on temperature of operation. Each design has specific characteristics, and must be carefully selected by relating its rating and capabilities to all the requirements and conditions of the particular application. The three types of Class 220 transformers are as follows:

1. *Class 220, 150°C-rise transformer.* A unit of this type operates with the full 150°C temperature rise on the winding conductors. Such a transformer is used at the maximum temperature capability of the overall insulation system. When fully loaded to its kilovoltampere rating and operated continuously in an average ambient temperature not over 30°C (86°F), with the maximum ambient never over 40°C (104°F), such a transformer can be expected to offer what is described as "full economic life." This design of transformer is commonly applied with an 80 percent load limitation under continuous loading, as a conservative measure to afford some degree of spare capacity for future load growth. Of course, such 80 percent loading reduces the production of operating heat and retards thermal aging and ultimate insulation failure. But Class 220, 150°C-rise units are not designed with any continuous overload capability.

2. *Class 220, 115°C-rise transformer.* A unit of this design has an insulation system that is rated for the full 220°C temperature limit and is capable of withstanding a 150°C rise on the windings; however, the actual operating temperature of the windings is limited by design to only 115°C (instead of 150°C) when the transformer is fully and continuously loaded to its

kilovoltampere rating. The reduced temperature rise is achieved by using larger conductors, with lower I^2R losses, in the windings; this results in a larger coil, core, and case. Such a unit is larger, weighs more, and costs more than a 150°C-rise unit of the same kilovoltampere rating. Such a transformer runs cooler than a Class 220, 150°C-rise unit, and for that reason may be considered to have a built-in "temperature reserve" somewhat similar in effect to the 80 percent load limitation that designers use on Class 220, 150°C-rise units. The reduced operating temperature provides the unit with slower thermal aging—and, therefore, greater life expectancy—than 150°C-rise units under full, continuous loading. The Class 220, 115°C-rise unit, because of its ultimate capability of 150°C rise on the windings, has a continuous overload capability of 15 percent above its nameplate kilovoltampere rating. And the lower losses of the unit (for a given capacity rating) result in energy savings and reduced operating costs.

3. *Class 220, 80°C-rise transformer.* This unit has an insulation system rated for a maximum operating temperature of 220°C but is designed to provide full, continuous kilovoltampere loading with a temperature rise of only 80°C on the windings. As the next step in reduced operating temperature below the level designed into Class 220, 115°C units, the 80°C-rise unit has even larger winding conductors to further reduce I^2R heat losses. Class 220, 80°C-rise units therefore have an even greater tolerance for the extreme heating effects of full, continuous loading, with greater life expectancy than hotter units under the same operating conditions. Class 220, 80°C units cost more and are generally larger and heavier than either 150°C- or 115°C-rise units. However, such units have a designed continuous overload capability of 30 percent without exceeding the safe operating temperature of the insulation, and they offer even greater reductions in energy losses and operating costs than Class 220, 115°C-rise units.

Class 220, 115°C-rise and 80°C-rise transformers are available as UL-listed devices for their capacity ratings at their specific temperature rise—either 115 or 80°C. But it should be noted that such units achieve reduced temperature ratings through their core and coil design, which reduces the temperature rise by using larger conductors with lower resistance (I^2R) losses. They are specifically designed to deliver rated kilovoltamperes at the reduced temperatures; they are not simply 150°C-rise units operated at reduced current loadings. Using an oversized 150°C-rise unit to obtain a lower temperature rise on the conductors will reduce conductor I^2R losses but result in unnecessarily high core losses (heat losses due to hysteresis—"magnetic friction"—in the lamination structure of the core and losses due to eddy currents).

Electric users select 115°C- or 80°C-rise units where they judge that conservative transformer loading is dictated by heavy, continuous loadings in areas with elevated ambients, so that excessive temperature is a concern. Or, these higher-priced transformers are selected where extremely high reliability or extra long life is needed.

In applications where transformers are fully and continuously loaded, the higher efficiency (lower losses) of 115°C- or 80°C-rise units may justify their higher first cost, since savings in electric-energy charges can recoup their added cost in a few years. Cases have been cited in which energy-cost savings over the life of a lower-rise transformer far exceeded its original cost. Of course, detailed analysis and calculations should relate transformer cost, local utility energy rates, required transformer loading and duty cycle, and the ratings and characteristics of the transformer in determining the economic advantages of low-rise units.

Comparisons can be developed between 150°C-rise units and either 115°C- or 80°C-rise units, as regards first costs and operating costs over the life of the transformers. Data on transformer losses are available and can be combined with utility energy rates and operating data to evaluate the potential economy of using either 115°C- or 80°C-rise units. And the overload capabilities of the low-rise Class 220 transformers may strongly suggest the use of such units—

for the overload capabilities alone or in conjunction with energy savings and greater life expectancy.

UL listing

The *Electrical Construction Materials List* of the UL lists "Transformers—Power." To satisfy **NEC** and OSHA regulations, as well as local code rules on the acceptability of equipment, any transformers of the types and sizes covered by UL listing must be so listed. The use of an unlisted transformer of a type and size covered by UL listing would certainly be considered a violation of the spirit of **NEC** Section 110-2.

UL listing covers "air-cooled" types rated up to 333 kVA for single-phase transformers and up to 1000 kVA for three-phase units (all up to 600 V rating).

Enclosure

The housing of a transformer must be suitable for use where the transformer is installed. Units used outdoors must be approved for such use, and any conditions that prevail at the installed location must be taken into account, as required by **NEC** Section 110-3.

Mounting: Every transformer must be selected so that it is suitable for the manner in which it is intended to be mounted—such as wall-hung, floor-mounted, or ceiling-suspended.

Grounding provisions

Careful evaluation of bonding and grounding provisions should be an important element of the transformer selection task. Either a built-in jumper conductor or provision for a bonding jumper must be included to afford ready compliance with **NEC** Section 250-26, which covers bonding and grounding of the secondary neutral and transformer case and is discussed below. The choice between internal grounding connection and connection to an external grounding stud on the case should be made with reference to local code rules and inspectors' preferences.

Winding connections

The winding connections in 3-phase transformers must match primary and secondary electrical requirements on voltage values.

Wye-wye transformer connections should generally be avoided because of the objectionable characteristics of such connections. First, a connection of this type generally requires that the primary neutral point of the transformer windings be connected to the neutral of the source. Without this connection of the primary neutral, unbalanced loading from phases to neutral on the secondary side produces a "floating neutral" on the primary side, with a serious unbalancing of secondary wye voltages and reduced power-handling capacity. And because the feeder neutral to the transformer primary is absolutely necessary to properly supply wye-connected loads on the wye secondary, the source from which the primary feeder originates must be provided with a neutral point, and the extra neutral conductor must be run with the supply circuit. Of course, this objection to wye-wye connection does not apply with balanced wye or delta loading on the secondary.

The second, and major, objection to the general use of wye-wye–connected transformers arises from the character of the harmonic currents generated in such a hookup. Under balanced conditions, a 3-phase system operates with a 120° phase difference between the phases of the fundamental current frequency and the phases of each harmonic frequency of current, except for the third harmonic and multiples of the third harmonic which alternate in phase with each other. With sinusoidal applied voltages, magnetizing currents in iron-core transformers contain appreciable third-harmonic current. But since the vector sum of the currents at the neutral junction in a wye hookup must equal zero, third-harmonic currents and multiples of the third harmonic cannot exist in the primary of a wye-wye connected transformer if the primary neutral point does not have a connection to the neutral of the source supplying the transformer. Without the neutral connection, the third harmonic and its multiples are eliminated.

When the third harmonic of the primary magnetizing current is eliminated by not connecting the primary neutral, the flux wave is distorted so that appreciable third-harmonic voltage is induced in the secondary. In a wye-wye connection the third-harmonic voltages cancel each other between line terminals, but they show up in voltages between line terminals and neutral. These harmonic voltage waves assume a phase relationship with the fundamental voltage wave such that the windings can be subjected to severe overvoltages. Voltage peaks of nearly twice normal amplitude can subject the transformer insulation to destructive magnetic stresses, and even destroy the transformer.

There are basically two ways to eliminate these effects of wye-wye transformer connection. The simplest way is to provide a primary neutral to permit third harmonics in the transformer exciting currents, thereby producing sinusoidal flux and sinusoidal secondary voltages. However, with this solution, the primary harmonic current flow may possibly produce interference on telephone lines. The second method of eliminating unsafe voltage stresses due to harmonic voltages is to use a tertiary winding in each transformer, in a delta connection. The delta tertiary windings short-circuit the third-harmonic voltages, producing the third harmonic of the magnetizing current for sinusoidal flux. Thus, voltage spikes due to wave distortion are virtually eliminated.

The problem of third harmonics occurs in 3-phase, wye-wye hookups of single-phase transformers and in 3-phase transformers with shell-type core and winding construction. However, wye-wye 3-phase transformers with core-type construction produce almost no third-harmonic flux and, therefore, do not produce third-harmonic currents or the overvoltage operating condition.

For given conditions of application, a transformer manufacturer can provide the proper equipment for wye-wye transformation. However, careful attention to voltage levels and the circuit design configuration—whether to use wye or delta connections—can often afford real economy by eliminating the need for special provisions, such as tertiary windings, in transformers.

Transformer Protection

An important element in the design of transformer layouts for electrical distribution systems is the provision of effective overcurrent protection for each transformer. With regard to transformer protection, careful observance of the rules of **NEC** Section 430-3(b) will ensure both compliance with the **Code** and complete design adequacy for transformers rated up to 600 V on their primaries.

Figure 317 shows the basic rule as applied to protection of the dry-type transformers used for power stepdown to 120/240 V, single-phase or to 120/208 V, 3-phase, 4-wire: Overcurrent protection for a dry-type transformer is provided by fuses or CBs rated at not more than 125 percent of the transformer primary full-load current (TPFLC) to protect the circuit conductors that supply the transformer primary. These circuit conductors must have an ampacity that is properly protected by the rating of the overcurrent protection.

In the layout in Fig. 317, a CB or a set of fuses rated not over 125 percent of the transformer rated primary full-load current provides all the overcurrent protection required by the **NEC** for

Primary circuit protection...

...does not qualify as suitable protection
for secondary conductors

Conductors also rated
125% TPFLC

Max rating or
setting of CB
or fuses is
125% TPFLC

Δ   Y
480V | 208/120V

Panel,
switchboard,
or MCC

←——— Any distance ———→|←———(A)———→|

Distance "A" from transformer to panel is limited to 10 or 25 ft,
subject to the requirements of Exceptions 2 and 3 to Section 240-
21. If overcurrent protection is placed at the transformer secondary connection to protect secondary conductors, the circuit can
run any distance to the panel. If the panel fed by the transformer
secondary is a lighting panel and requires main protection, the
protection must be on the secondary [Section 384-16(d)].

FIG. 317 Most common layout for protection of dry-type 600-V
transformers.

the transformer. This overcurrent protection is in the feeder circuit to the transformer, and it is
logically placed at the supply end of the feeder so it may also provide the overcurrent protection
required for the primary feeder conductors. There is no limit on the distance between the primary protection and the transformer. When the correct maximum rating of transformer protection is selected and installed at any point on the supply side of the transformer (either near to
or far from the transformer), the feeder circuit conductors must be sized so that the CB or fuses
selected will provide the proper protection as required for the conductors. The ampacity of the
feeder conductors must be at least equal to the ampere rating of the CB or fuses unless Exception
No. 1 of Section 240-3 is satisfied. That is, when the rating of the overcurrent protection selected
is not more than 125 percent of the rated primary current, the primary feeder conductor may
have an ampacity such that the overcurrent device is the next higher standard rating.

The rules for protecting a 600-V transformer with a CB or set of fuses in its primary circuit
are given in Fig. 318. Note that, for transformers with rated primary current of 9 A or more,
"the next higher standard" rating of protection may be used, if needed. Figure 318 shows the
absolute maximum values of protection for smaller transformers. When the factor 1.67 (167
percent) or 3 (300 percent) is used, if the resultant current value is not exactly equal to a standard rating of fuse or CB, then the next *lower* standard rated fuse or CB must be selected.

It is important to note that transformer primary-circuit protection is not acceptable as suitable protection for the secondary-circuit conductors—even if the secondary conductors have an
ampacity equal to the ampacity of the primary conductors times the primary-to-secondary voltage ratio. On 3- and 4-wire transformer secondaries, it is possible that an unbalanced load may
greatly exceed the secondary-conductor ampacity that is selected assuming balanced conditions.
Because of this, the **NEC** does not permit the protection of secondary conductors with overcurrent
devices operating from the primary through a transformer having a 3-wire or 4-wire secondary.
For other than 2-wire–to–2-wire transformers, protection for secondary conductors has to be

A transformer with rated primary current of *9 amps or more* . . .

. . . must have an overcurrent device rated or set at *not more than 1.25 times (125%)* rated primary current

NOTE: Where 1.25 times primary current does not correspond to a standard rating of protective device, the next higher standard rating from Section 240-6 is permitted.

A transformer with rated primary current of *less than 9 amps* . . .

. . . may be protected by a primary overcurrent device rated or set at *not more than 1.67 times (167%)* rated primary current

A transformer with rated primary current of *less than 2 amps* . . .

. . . may be protected by a primary overcurrent device rated or set at *not more than 3 times (300%)* rated primary current

FIG. 318 Basic rules on protecting a transformer with an overcurrent device on the primary.

provided completely separately from any primary-side protection. Section 384-16(d) states that required main protection for a lighting panel on the secondary side of a transformer must be located on the secondary side. However, Exception No. 5 of Section 240-3 permits the secondary circuit from a transformer to be protected by means of fuses or a CB in the primary circuit to the transformer—if the transformer has no more than a 2-wire primary circuit and a 2-wire secondary. As shown in Fig. 319, the 2-to-1 primary-to-secondary turns ratio of the transformer allows 20-A primary protection to protect against any secondary current in excess of 40 A, thereby protecting, say, secondary No. 8 TW wires rated at 40 A. The protection on the primary (here, 20 A) must not exceed the product of the secondary-conductor ampacity (40 A) multiplied by the secondary-to-primary transformer voltage ratio (120 ÷ 240 = 0.5). Here, 40 A × 0.5 = 20 A, and the rule is satisfied.

FIG. 319   This kind of protection of secondary conductors is permitted.

When the foregoing overcurrent-protection rules are observed, the transformer itself is properly protected, and the primary feeder conductors, sized to correspond, are provided with the required protection. But the secondary side must be evaluated separately and independently for any transformer with a 3-wire or 4-wire secondary. When a transformer is provided with primary-side overcurrent protection, a whole range of design and installation possibilities are available for a secondary arrangement that satisfies the **NEC**. The basic approach is to provide the required overcurrent protection for the secondary conductors right at the transformer—as with a fused switch or CB attached to the transformer enclosure, as shown in Fig. 320. Or, 10- or 25-ft taps may be made, as covered under "Feeder taps" in Section 4 of this handbook.

In designing transformer circuits, the rules of **NEC** Section 450-3 on overcurrent protection for transformers can be coordinated with Exception No. 8 of Section 240-21, which provides special rules for tap conductors used with transformers. This exception would be used mainly where the primary overcurrent devices are rated at 250 percent of rated transformer primary current, according to Section 450-3(b)(2), or where a secondary tap would exceed 10 ft. (Refer to the discussion of "Feeder taps.")

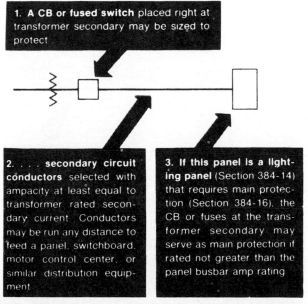

1. A CB or fused switch placed right at transformer secondary may be sized to protect

2. . . . secondary circuit conductors selected with ampacity at least equal to transformer rated secondary current. Conductors may be run any distance to feed a panel, switchboard, motor control center, or similar distribution equipment.

3. If this panel is a lighting panel (Section 384-14) that requires main protection (Section 384-16), the CB or fuses at the transformer secondary may serve as main protection if rated not greater than the panel busbar amp rating

FIG. 320   Protection of secondary-circuit conductors must be independent of primary protection.

Where secondary feeder taps do not exceed 10 ft in length, the requirements of Exception No. 2 of Section 240-21 could apply as in the case of any other feeder tap, with no restriction on the size of the feeder overcurrent device ahead of the tap. In applying the tap rules in Exceptions No. 2 and No. 8, the requirements of Section 450-3 on transformer overcurrent protection must always be satisfied.

Another acceptable way to protect a 600-V transformer is described in **NEC** Section 450-3(b)(2). In this method, the transformer primary may be fed from a circuit which has overcurrent protection (and circuit conductors) rated up to 250 percent (instead of 125 percent, as above) of the rated primary current; but, in such cases, there must be a protective device on the secondary side of the transformer, and that device must be rated or set at not more than 125 percent of the transformer's rated secondary current, as shown in Fig. 321. This secondary protective device must be located right at the transformer secondary terminals or not more than the length of a 10- or 25-ft tap away from the transformer, and the rules on tap conductors must be fully satisfied.

The secondary protective device rated at not over 125 percent of the rated secondary current may readily be incorporated as part of other required provisions on the secondary side of the transformer, such as protection for a secondary feeder from the transformer to a panel or switchboard or a motor control center fed from the switchboard. And a single secondary protective device rated not over 125 percent of the rated secondary current may serve as required panelboard main protection as well as the required transformer secondary protection, as shown at the bottom of Fig. 322.

The use of a transformer circuit with primary protection rated up to 250 percent of the rated primary current offers the opportunity to avoid situations in which a set of primary fuses or CB rated at only 125 percent would cause nuisance tripping or opening of the circuit on transformer inrush current. But the use of 250 percent rated primary protection has a more widely applicable advantage: It makes possible the feeding of two or more transformers from the same primary feeder. The number of transformers that might be used in any case would depend on the amount of continuous load on all the transformers. But, in all such cases, the primary protection must

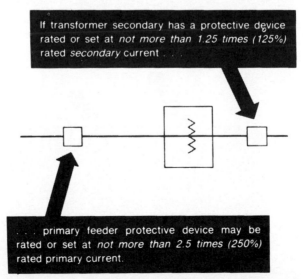

FIG. 321 Use of secondary protection permits higher-rated primary protection and circuit.

① Fuses or CB
not over 250%
rated primary current

Secondary feeder
conductors

Panel or
switchboard
or MCC

Fuses or CB, right at transformer secondary, rated
not over 125% of rated secondary current satisfies
Section 450-3(b)(2) as transformer protection, **and
the same fuse or CB** may serve as protection for
secondary feeder conductors **and also** as main
protection that might be required for a lighting panel
fed from the transformer.

② Secondary pro-
tection, not over
125% of second-
ary current

|← 25 ft max →|

This may be an individual
CB or fused switch or may
be a panel or switchboard
main protection.

NOTE: In both cases, primary and secondary feeder
conductors must be sized to be properly protected by
the fuses or CB in both the primary and secondary
circuit, which will give them more than adequate
ampacity for the transformer full-load current.

FIG. 322 Secondary protection rated at 125 percent of the
secondary current may be placed at one of several points.

be rated not more than 250 percent of any one transformer, if they are all the same size, or 250 percent of the smallest transformer, if they are of different sizes. And for each transformer fed, there must be a set of fuses or CB on the secondary side rated at not more than 125 percent of the rated secondary current, as shown in Fig. 323.

Figure 324 shows an application of 250 percent primary protection to a feeder supplying three transformers (such as at the bottom of Fig. 323). This example indicates how the rules of **NEC** Section 450-3(b) must be carefully related to Section 240-21 and to other **NEC** rules:

Exception No. 8 of Section 240-21 covers the use of a 25-ft unprotected tap from feeder conductors, with a transformer inserted in the 25-ft tap. This rule does not eliminate the need for secondary protection; it sets a special condition for placement of the secondary protective

NOTE: Each set of tap conductors from primary feeder to each transformer may be same size as primary feeder conductors **OR** may be smaller than primary conductors if sized in accordance with Section 240-21, Exception No. 8—which permits a 25-ft tap from a primary feeder to be made up of both primary and secondary tap conductors. The 25-ft tap may have any part of its length on the primary or secondary but must not be longer than 25 ft and must terminate in a single CB or set of fuses.

FIG. 323   Transformer primary protection rated at 250 percent protects one or more transformers.

device. It is a restatement of Exception No. 3, as applied to a tap containing a transformer, and applies to both single-phase and 3-phase transformer feeder taps.

Figure 324 shows a feeder supplying three 45-kVA transformers; each transformer is fed through a 25-ft feeder tap that conforms to Exception No. 8 of Section 240-21. Although each transformer has a rated primary current of 54 A at full load, the demand load on each transformer primary was calculated to be 41 A, based on secondary loading. Thus, No. 1 THW copper feeder conductors were considered adequate for the total noncontinuous demand load of the three transformer primaries, which is equal to 3 × 41 A, or 123 A. In the following step-by-step analysis of this system, the letters refer to the circled letters on the sketch:

A. The primary circuit tap conductors are No. 6 TW, rated at 55 A, which gives them "an ampacity at least ⅓ that of the conductors or overcurrent protection from which they are

THIS LAYOUT...

...IS HOOKED UP LIKE THIS:

FIG. 324   This layout of feeder and ransformer taps provides economical application with **NEC** compliance.

tapped," because these conductors are tapped from the feeder conductors protected at 125 A. The use of No. 6 TW conductors is permissible for the 41-A primary current.

B. The 125-A fuses in the feeder switch properly protect the No. 1 THW feeder conductors, which are rated at 130 A and feed the taps for the three transformers.

C. According to Exception No. 8*b* of Section 240-21, the conductors supplied by the transformer secondary must have "an ampacity that, when multiplied by the ratio of the secondary-to-primary voltage, is at least ⅓ the ampacity of the conductors or overcurrent protection from which the primary conductors are tapped." The secondary-to-primary voltage ratio of the transformer is

$$\frac{208 \text{ V}}{480 \text{ V}} = 0.433$$

(Note that phase-to-phase voltages must be used to determine this ratio.) Then, for the secondary conductors, according to Exception No. 8*b*,

$$\text{Minimum ampacity} \times 0.433 = \frac{1}{3} \times 125 \text{ A} = 41.67 \text{ A}$$

and

$$\text{Minimum ampacity} = \frac{41.67 \text{ A}}{0.433} = 96 \text{ A}$$

The No. 1 TW secondary conductors, rated at 110 A, are above the 96-A minimum and are, therefore, satisfactory.

D. The total length of the unprotected tap—i.e., the primary conductor length plus the secondary conductor length $(x + y)$ for any circuit leg—must not be greater than 25 ft.

E. The secondary tap conductors from the transformer must terminate in a single CB or set of fuses that will limit the load on those conductors to their rated ampacity from Table 310-16. Note that there is no exception to that requirement, and the next higher standard device rating *may not be used* if the conductor ampacity does not correspond to the rating of a standard device.

The overcurrent protection required at E, the load end of the 25-ft tap conductors, must not be rated higher than the ampacity of the No. 1 TW conductors:

$$\text{Maximum rating of fuses or CB at E} = 110 \text{ A}$$

But a 100-A main would satisfactorily protect the 96-A secondary load.

NOTE: The overcurrent protective device required at E could be the main protective device required for a lighting and appliance panel fed from the transformer.

Watch out for this trap!!

The foregoing calculation shows how unprotected taps may be made from feeder conductors by satisfying Exception No. 8 of Section 240-21. However, the rules of Section 240-21 are concerned with the *protection of conductors only.* Consideration must now be given to transformer protection, as follows:

1. Note that Section 240-21 makes no reference to transformer protection. But Section 450-3 calls for the protection of transformers, and no exception is made for the conditions of Exception No. 8 to Section 240-21.

2. It is clear that each transformer shown in Fig. 324 is *not* protected by a primary-side overcurrent device rated at not more than 125 percent of the rated primary current (54 A), as required by Section 450-3(b)(1), because 1.25 × 54 A = 68 A.

3. But Section 450-3(b)(2) does offer a way to provide the required protection. The 110-A protection at E is secondary protection rated not over 125 percent of the rated secondary current (1.25 × 125 A secondary current = 156 A). With that secondary protection, a primary feeder overcurrent device rated not more than 250 percent of the rated primary current would satisfy Section 450-3(b)(2). That would call for fuses in the feeder switch (or a CB), at B in the diagram, rated at not over

$$2.5 \times 54 \text{ A (primary current)} = 135 \text{ A}$$

Because the fuses in the feeder switch are rated at 125 A (which is not in excess of 250 percent of the transformer primary rated current), those fuses satisfy Section 450-3(b) (2).

In addition to the two basic methods described above for protecting transformers, Section 450-3(b)(2) provides for protection with built-in thermal overload protection, as shown in Fig. 325.

    <u>NOTE:</u>   Refer back to Figs. 272 to 274 under "Feeder taps" (in Section 4) for additional discussion of the relationships among tap conductors, conductor protection, and transformer protection.

If transformer is equipped with coordinated thermal overload protection by manufacturer and arranged to interrupt primary current . . .

. . . primary feeder overcurrent device may be rated or set at
   a. not more than **.6 times** rated primary current for transformers with **not more than 6%** impedance. or
   b. not more than **4 times** rated primary current for transformers with **over 6% but less than 10%** impedance.

FIG. 325 "Built-in" overload protection is an alternative to the other methods shown for transformer protection.

Typical applications

The following examples are based on the foregoing design considerations and procedures.

EXAMPLE

In Fig. 326, the feeder to the 3-phase dry-type transformer supplies 480-V power for step-down to 120/208 V for a lighting panel with a continuous load of 30 A on each phase. What are the required minimum sizes of conductors and protective devices?

FIG. 326 Sizing of conductors and overcurrent protection must satisfy all transformer-circuit conditions.

NOTE: Knowing these minimums under conditions of continuous use, the designer can effectively provide spare capacity without depriving the conductors and the transformer of the **NEC**-required levels of protection.

SOLUTION

Step 1

The kilovoltampere rating of the 30-A, 120/208-V load on the panel is

$$30 \times 208 \times 1.73 \div 1000 = 10.8 \text{ kVA}$$

That load will require at least a 15-kVA transformer, assuming a larger one is not needed for even greater possible load growth.

Step 2

Assuming eventual use of the entire transformer capacity, the rated primary current is computed as

$$\frac{15,000}{480 \times 1.732} = \frac{15,000}{831} = 18 \text{ A}$$

That value of primary current would require conductors with a minimum current-carrying capacity of 20 A: No. 12 copper or No. 10 aluminum. Because the footnote to **NEC** Table 310-16 says that No. 14 copper wire must not have a load-current rating over 15 A, the 18-A current prohibits the use of No. 14 conductors.

Step 3

NEC Section 450-3(b)(1) says that primary overcurrent protection for the transformer must be rated or set at not more than 125 percent of the rated primary current. In this example,

$$1.25 \times 18 \text{ A} = 22.5 \text{ A}$$

Because Exception No. 1 of Section 450-3(b)(1) says that the next higher rated device may be used here, this would call for a 25-A, 3-pole CB, or 25-A fuses. This transformer overcurrent protection may be installed at the supply end of the primary circuit feeder.

Step 4

Section 220-10(b) says that the feeder overcurrent device must be rated at least 125 percent of the continuous load. Since

$$1.25 \times 18 \text{ A} = 22.5 \text{ A}$$

the 25-A CB protection computed in step 3 above for an 18-A full-load transformer current would not violate the limitation set by Section 220-10(b).

Step 5

The primary-circuit conductors to the transformer are also required by Section 220-10(b) to have an ampacity of not less than 125 percent of 18 A, or 22.5 A. This would call for No. 12 copper or No. 10 aluminum circuit conductors. From **NEC** Table 310-16, No. 12 TW copper wire has an ampacity of 25 A, and No. 10 aluminum has that same ampacity. And No. 12 copper has a load rating of 20 A (from the footnote to Table 310-16), which exceeds the 18-A load current.

Step 6

The rated secondary current of a 15-kVA transformer would be

$$\frac{15,000}{208 \times 1.732} = 41.7 \text{ A}$$

Section 220-10(b) requires that the secondary-circuit conductors from the transformer to the panel have an ampacity of at least 125 percent of that value, or 52 A. To carry this load, the following secondary conductors could be used in conduit:

Copper: No. 6 TW (55 A), THW (65 A), or THHN (75 A)

Aluminum: No. 4 TW (55 A), THW (65 A), or THHN (75 A)

Step 7

But the secondary conductors cannot be protected "through" the transformer. Exception No. 5 of Section 240-3 permits the protection of transformer secondary conductors by overcurrent devices placed on the primary side of the transformer only for single-phase transformers with 2-wire primary and 2-wire secondary circuits. That is not the case here.

Therefore, protection must be provided for the secondary feeder conductors which run 50 ft from the transformer to the panel. Such protection must be located right at the transformer—that is, "at the point where the conductor to be protected receives its supply," as covered in Section 240-21. Of course, secondary conductors from the transformer can be used as 10-ft or 25-ft tap conductors provided they fully comply with Exception No. 2, No. 3, or No. 8 of Section 240-21. But where the length from the transformer to a panel is over 25 ft, as in this case, separate overcurrent protection must be inserted to protect the secondary conductors at their prescribed capacity.

A 60-A CB or 60-A fuses would protect any of the secondary conductors listed in step 6 above in accordance with its ampacity. An enclosure containing this 60-A protection could be placed right on the transformer case. Or, it could be placed so that 10-ft tap conductors or 25-ft tap conductors could reach it.

NOTE: The protection for the secondary conductors must never be placed further away from the transformer than the distance at which 25-ft tap conductors could be used. Overcurrent protection must be placed at *A* in Fig. 326, and the protection for the secondary conductors may also provide required protection for the lighting panel supplied—if the rating of the protection is not higher than the mains rating of the panel.

Thus, in this example, a 60-A device, for instance, could protect, say, both No. 6 THW secondary conductors and a 60-A lighting panel fed from the transformer.

EXAMPLE

The circuit components in Fig. 327 are sized as follows:

Step 1

The lighting load will be served by 15-A, 2-wire circuits. Because this lighting load will be operating continuously to supply fixtures in a store, each circuit must not be loaded to more than 80 percent of its rating. At 15 A and 120 V, each circuit has a maximum capacity of 15 × 120 or 1800 W. The 80 percent limit gives 1800 W × 0.8 = 1440 W. The required number of circuits is determined as follows:

$$\text{Number of circuits} = \frac{\text{total load in watts}}{\text{watts per circuit}} = \frac{17,000}{1440} = 12 \text{ circuits}$$

FIG. 327 Transformer feeder circuits may be sized for the load capacity or for the full capacity of the transformer selected.

Step 2

The feeder to the 120/240-V panel must have sufficient capacity for the 17-kW load, which operates continuously. The required feeder conductor ampere loading is found as follows:

$$I = \frac{\text{load in watts}}{2 \times \text{phase-to-neutral voltage}} = \frac{17,000}{240} = 71 \text{ A}$$

Although the feeder circuit is described as being rated at 120/240 V, these are high values of voltage that usually will not occur during normal operation. It is better to use 115/230 V as more likely and typical of the voltage conditions that will exist. And, since the current is determined by dividing wattage by voltage, it is safer to use the lower voltage value, which yields a higher current value to be used in sizing the feeder conductors. This will provide a higher-rated conductor and reduce the chance of the circuit being overloaded. Thus,

$$I = \frac{17,000}{230} = 74 \text{ A}$$

The demand factors which may be applied to an overall feeder load, as covered in **NEC** Section 220-11, are not applicable because this is a store building and all the lights will be operating simultaneously and constantly.

Step 3

But, because the load on the feeder is continuous (operates for longer than 3 hours), the value of 74 A must be multiplied by 125 percent to satisfy **NEC** Section 220-10(b) for continuous loading. The feeder must be sized for

$$1.25 \times 74 \text{ A} = 93 \text{ A}$$

To meet that ampacity requirement, the secondary feeder could be made up of three No. 2 TW copper conductors (rated at 95 A); or, No. 2 THW, XHHW, or THHN copper conductors could be used.

Step 4

According to **NEC** Section 384-13, the busbars of a panel must have a current rating not lower than the minimum feeder capacity required for the load. This means that, in this case, the

120/240-V panel must have a rating of at least 93 A—the load on the feeder. A 100-A panel would satisfy this requirement.

Step 5

Based on a total load of 17 kW, the smallest standard size of transformer which could be used to continuously supply the load would be rated at 25 kVA. If the load were not continuously at the maximum load value, it might be possible to use a 15-kVA transformer and allow it to operate overloaded during peak load times. But a 25-kVA transformer is properly required for a continuous 17-kW load.

Step 6

To obtain data for sizing the transformer's overcurrent protection, its primary and secondary full-load currents should be calculated using the formula for a single-phase transformer:

$$\text{Primary current} = \frac{\text{transformer kVA rating} \times 1000}{\text{primary voltage}} = \frac{25 \times 1000}{480} = 52 \text{ A}$$

Its full-load secondary current is found with the same formula, applied to the secondary:

$$\text{Secondary current} = \frac{\text{transformer kVA rating} \times 1000}{\text{secondary voltage}} = \frac{25,000}{240} = 104 \text{ A}$$

Step 7

Under the load conditions given, the load current on the transformer secondary will be the 74 A of load current drawn by the lighting panel. The load current drawn by the transformer primary is found from the formula which relates the currents and voltages on the primary and secondary sides of the transformer:

$$\frac{I_p}{I_s} = \frac{E_s}{E_p}$$

where I_p = primary current, A
I_s = secondary current, A
E_s = secondary voltage, V
E_p = primary voltage, V

For a secondary current of 74 A, the primary current is calculated as

$$I_p = I_s \frac{E_s}{E_p} = 74 \frac{240}{480} = 37 \text{ A}$$

To handle a continuous load of 37 A, the primary circuit conductors must have an ampacity of

$$1.25 \times 37 \text{ A} = 46 \text{ A}$$

Two No. 6 TW copper conductors (rated at 55 A) could be used, as could No. 6 THW, THHN, XHHW, or RHH conductors.

But, because a 25-kVA transformer has been selected, good design practice would call for sizing the entire feeder for the capacity of the transformer—even though the present loading is lower than that capacity.

Step 8

In an application of this sort, the transformer must be protected by a primary CB or fuses in a switch, rated or set at not more than 125 percent of the transformer rated primary current. The rated primary current is computed as 25,000 VA ÷ 480 V = 52 A, and 1.25 × 52 A = 65 A. That rating of protection would not properly protect No. 6 TW conductors (55 A) but would protect No. 6 THW (65 A) or No. 6 THHN or XHHW (75 A) conductors. However, to obtain the full output of the transformer for future use, primary conductors rated at 1.25 × 52 A or 65 A would have to be used. And because equipment rated up to 100 A is required by UL to be wired with 60°C conductors (TW)—or with higher-temperature conductors that are sized as if they were TW conductors—the primary circuit would require at least No. 4 TW (70 A) or THW, THHN, or XHHW conductors.

Step 9

If the primary feeder is sized to allow the full 52 A of rated primary current for a 25-kVA transformer, the secondary conductors must be resized at 25,000 VA ÷ 240 V × 1.25 or 130 A to allow for future load increase up to the transformer's rated 25-kVA capacity. That calls for No. 1 THW copper conductors (130 A) or No. 2/0 TW copper conductors (145 A). If THHN or XHHW conductors are used, they must be sized as if they were 75°C conductors (THW)—which means No. 1 copper must be used.

Step 10

The 100-A lighting panel must then be equipped with 100-A main protection. That protection could be placed right at the transformer secondary, or it could be placed at the load end of unprotected tap conductors that run no more than 25 ft from the transformer secondary to the main 100-A CB or fused switch that serves as main protection in the panel.

EXAMPLE

Figure 328 shows a single-phase transformer with a 2-wire primary circuit and a 2-wire secondary, that is used to feed ungrounded circuits for a hospital operating room.

FIG. 328 Primary protection will protect feeder-circuit conductors.

Step 1

In such a circuit, the secondary conductors may be protected by the overcurrent device protecting the primary circuit to the transformer if the size of the conductors and the rating of the protective device are calculated on the basis of the transformer turns ratio.

Step 2

In the hookup shown, the 30-A, two-pole CB protecting the primary circuit to the transformer is acceptable as protection for the transformer if the transformer rated primary current is more than 20 A [since 1.25 × 21 A = 26 A, and a 30-A CB is the next higher standard rating of protective device, as permitted by **NEC** Section 450-3(b)(1)]. The minimum size of copper conductors acceptable for use as the primary circuit to the transformer is No. 10 TW. The minimum acceptable size of secondary-circuit conductor is No. 6 TW copper (55 A), for which the 30-A primary CB acts as a 60-A protective device through the 2-to-1 current stepup from primary to secondary. And such 60-A protection is the "next higher standard rating" of CB above 55 A, as permitted by Exception No. 1 of Section 240-3.

Step 3

If the transformer shown in the sketch were rated at 5 kVA, it would have a rated primary current (at 240 V) of 5000 VA ÷ 240 V, or 20.8 A. The maximum rating of overcurrent device for such a transformer would be 125 percent of 20.8 A, or 26 A. Because Exception No. 1 of Section 480-3(b)(1) permits the next higher standard rating of protective device from Section 240-6 when 1.25 times the rated primary current does not correspond to a standard rating, the maximum standard rating of CB acceptable as protection for this transformer is 30 A, which is the next higher standard size of CB above 26 A.

EXAMPLE

Calculations of the various loads supplied by the 208/120-V switchboard shown in Fig. 329 reveal that the switchboard would have a noncontinuous demand load of 762 A per phase.

Step 1

The minimum transformer kilovoltampere rating that can handle that load is determined with the formula

$$\text{kVA rating} = \text{line-to-line voltage} \times \text{current per phase} \times 1.732 \div 1000$$

$$= 208 \text{ V} \times 762 \text{ A} \times 1.732 \div 1000 = 274 \text{ kVA}$$

Step 2

The nearest standard rating of transformer, which must be used to handle the load, is 300 kVA.

Step 3

A transformer with a 300-kVA rating has a rated primary current of $300 \times 1000 \div (480 \times 1.732)$, or 361 A.

Step 4

The CB or fuses used to protect the primary feeder circuit to the transformer must have a rating or setting of not more than 125 percent of the transformer rated primary current if the primary-circuit protection is to qualify by itself as the required overcurrent protection for the transformer.

FIG. 329   Switchboard loading dictates transformer size, which then dictates conductor and protection ratings.

Step 5

Therefore, the primary overcurrent device must have a rating or setting not over 1.25 × 361 A or 451 A.

Step 6

The maximum standard rating of protective device permitted by Exception No. 1 to **NEC** Section 450-3(b) is 450 A. (Actually, Exception No. 1 would permit the use of 500-A protection as the next higher rating above 451 A, as calculated in step 6 above.)

Step 7

Based on the use of primary overcurrent protection of the rating established in step 6 above, the primary-circuit conductors may be of the same ampacity or of such ampacity that the primary overcurrent device in step 6 is of the "next higher standard rating or setting" above the ampacity rating of the conductors. For this circuit, that condition is achieved if each phase leg of the primary circuit has an ampacity of over 400 A; either a single conductor per phase leg of that ampacity may be used, or two conductors per leg may be used, with each conductor rated over 200 A.

Step 8

If the primary circuit to the transformer is made up of two parallel sets of three conductors, with all six conductors in the same conduit, each conductor could be a 300-MCM THHN or XHHW aluminum conductor (in a dry location), which has a normal rating of 255 A (from **NEC** Table 310-16); but these conductors must be derated to 80 percent of their ampacity (0.8 × 255 A, or 204 A), as required by Note 8 to Table 316-16 because there are more than three current-carrying conductors in the conduit.

Step 9

With two parallel sets of conductors as described in step 8 above, the ampacity of each phase leg of the primary circuit is 2 × 204 A or 408 A, and the overcurrent protection sized in step 6 is the next higher rating above the ampacity of the circuit phase legs.

Step 10

The minimum size of conduit required for the six 300-MCM THHN or XHHW circuit conductors of step 8 above is 3 in.
   NOTE: Although 90°C conductors are being used (THHN or XHHW), they will be operating at only 361 A when the transformer is loaded to its capacity. Because 361 A is lower than the 368-A rating of two 300-MCM THWs in parallel (2 × 230 A × 0.8 = 368 A), the equipment terminal lugs will be within the 75°C operating limit that UL places on conductor terminal operations.

Step 11

If the load on the switchboard fed from the transformer secondary were continuous instead of noncontinuous, the calculations given above for sizing the primary-circuit overcurrent protection and the primary conductors would have to be evaluated in light of **NEC** Section 220-10(b), which requires feeder conductors and overcurrent protection to have an ampacity at least equal to 125 percent of the load current. The 274-kVA load on the transformer draws a primary current equal to 274 × 1000 ÷ (480 × 1.732), or 330 A of primary current. Then 1.25 × 330 A = 412 A. The primary-circuit conductors selected in step 9 above have a derated capacity of 408 A per phase, which is not quite up to the required 412-A level. The 300-MCM conductors are, therefore, not fully adequate for the continuous load. If 350-MCM THHN aluminum conductors (280 A) are used instead of the 300-MCM conductors, their derated ampacity of $2 \times 280 \text{ A} \times 0.8 = 448$ A will satisfy the need for an ampacity of 412 A. And the 450-A rating of the primary protection is acceptable because it is greater than 125 percent of the continuous load of 330 A.

Step 12

The above calculations cover the primary protection, primary-circuit makeup, and transformer protection. The conductors and other components on the secondary side of the transformer must be sized separately, on the basis of other, independent calculations; the primary protection does not affect the required secondary protection or the sizing of the secondary feeder conductors.

Step 13

The secondary conductors from the transformer to the switchboard must be sized on the basis of the demand load on the panel. For the noncontinuous load indicated, the secondary conductors must have an ampacity of not less than 762 A.

Step 14

The secondary-circuit conductors are selected on the basis of the load on the conductors, the distance of the switchboard from the transformer, and consideration of the need for overcurrent protection for the conductors.

Step 15

If the transformer is located adjacent to the switchboard, as shown at the top of Fig. 330, so that secondary conductors only 9 ft long can be used in conduit to connect the transformer secondary terminals to the main lugs in the switchboard, those conductors need only have an ampacity that is not less than the main current rating of the switchboard; in addition, they qualify as 10-ft tap conductors and may be used without overcurrent protection.

Step 16

If the secondary conductors are run in conduit from the transformer to the switchboard located 13 ft away, as at the bottom of Fig. 330, they may be run without overcurrent protection at the transformer secondary to a single CB or set of fuses in the switchboard, provided that the conductors have an ampacity of at least 762 A (for a noncontinuous load), they are not more than 25 ft long, and the protective device in which they terminate protects them at their rated ampacity.

Step 17

If the secondary conductors are run in conduit to the switchboard located 50 ft away from the transformer, they must be supplied with overcurrent protection at the transformer (their supply end); this overcurrent protection must protect the secondary conductors at their rated ampacity, which must be not less than 762 A.

Step 18

If the load on the switchboard consisted of 500 A of continuous load and 262 A of noncontinuous load, the rating of the overcurrent protection at the transformer in step 17 above

FIG. 330 Distance between a transformer and the equipment it feeds can alter calculations.

would have to be not less than 125 percent of 500 A plus 262 A, or 887 A. And the current rating of the switchboard would then have to be at least 1000 A.

Step 19

The condition described in step 18 would call for a standard secondary overcurrent device (CB or set of fuses) rated not below 1000 A. Then the secondary-circuit conductors would have to have at least that ampacity and not a lower ampacity, because the overcurrent device would be rated over 800 A.

Transformer Grounding

With the great popularity of distribution systems using transformer loadcenters, particularly 480/277-V systems, effective grounding of transformer secondaries has become an important consideration in modern electrical-system design. In a 480/277-V system, 120-V circuits for incandescent lighting and receptacle outlets may be provided either by using single-phase transformers to step 480 V down to 240/120-V, single-phase secondaries or by using 3-phase transformers to step the 480-V, 3-phase primary supply down to 208/120-V, 3-phase, 4-wire secondaries. The ac systems supplied from the secondaries of such transformers are considered to be "separately derived ac wiring systems" and are subject to the grounding requirements of **NEC** Section 250-26. The same rules on grounding and bonding apply to the output circuit of an engine generator used for emergency or standby power supply and to that of a generator used for on-site supply of the normal power to a building or facility.

Details on the grounding of transformer secondaries are as follows:

1. Any system that operates at over 50 V but not more than 150 V to ground must be grounded; that is, it must be operated with one of its circuit conductors connected to a grounding electrode.

2. This requires the grounding of secondaries of dry-type transformers serving 208/120-V, 3-phase or 240/120-V, single-phase circuits for lighting and appliance outlets and receptacles at loadcenters throughout a building, as shown in Fig. 331. And 480/277-V, wye-connected secondaries must also be grounded if indoor lighting is to be supplied—with loads connected from line to neutral and/or from line to line. Electric-discharge lighting that is supplied by 480-V ballasts indoors *must* be fed from a 480/277-V transformer secondary with a grounding connection to the neutral point to provide a voltage that is not over 300 V to ground, as required by Exceptions No. 1 and No. 2 of **NEC** Section 210-6(a).

3. All rules applying to both system and equipment grounding must be satisfied in such installations.

FIG. 331 Transformer secondary neutral must be both grounded and bonded to the case.

The wiring-system conductor to be grounded in compliance with items 1 and 2 above depends upon the type of system. In 3-wire, single-phase systems, the midpoint of the transformer winding—the point from which the system neutral is derived—is grounded. In grounded 3-phase wiring systems (either 3-wire or 4-wire systems), the neutral point of the wye-connected transformer or generator is the point connected to ground. In delta-connected transformer hookups, grounding can be effected by grounding one of the three phase legs, by grounding a center-tap point on one of the transformer windings (as in the 3-phase, 4-wire "red-leg" delta system), or by using a special grounding transformer which establishes a neutral point of a wye connection which is grounded.

The steps involved in designing transformer secondary connections are as follows (see Fig. 332):

Step 1

A bonding jumper must be installed between the transformer secondary neutral terminal and the metal case of the transformer. The size of this bonding conductor is obtained from **NEC** Table 250-94, based on the size of the transformer secondary phase conductors (as if they were service-entrance conductors), and is the same as the size of the required grounding-electrode conductor.

STEP 1 – BONDING JUMPER

STEP 2 – GROUNDING ELECTRODE CONDUCTOR

STEP 3 – GROUNDING ELECTRODE

FIG. 332 These design steps provide for transformer secondary connections.

For cases where the transformer secondary-circuit conductors are larger than 1100-MCM copper or 1750-MCM aluminum per phase leg, the size of the bonding jumper must not be less than 12½ percent of the cross-sectional area of the secondary phase leg.

EXAMPLE

A 75-kVA transformer has a 208/120-V, 3-phase, 4-wire secondary, with current of

$$75,000 \div (208 \times 1.732) = 209 \text{ A}$$

If No. 4/0 THW copper conductors (with a 230-A rating) were used for the secondary phase legs, then the size of the required bonding jumper would be obtained from Table 250-94 as if 4/0 service conductors were being used. The table shows that 4/0 copper service conductors require a minimum of No. 2 copper or No. 1/0 aluminum for a grounding-electrode conductor. The bonding jumper would have to be of either of those two sizes.

If the transformer were a 500-kVA unit with a 120/208-V secondary, its rated secondary current would be

$$\frac{500 \times 1000}{1.732 \times 208} = 1388 \text{ A}$$

If say, THW aluminum conductors were used, then each secondary phase leg would be made up of four 700-MCM aluminum conductors in parallel (each 700-MCM THW aluminum is rated at 375 A, and four are rated at 4 × 375 A or 1500 A, which suits the 1388-A load). Then, because 4 × 700 MCM equals 2800 MCM per phase leg, which is in excess of 1750 MCM, Section 250-79(c) would require the bonding jumper from the case to the neutral terminal to be at least equal to 12½ percent of 2800 MCM (0.125 × 2800 MCM), or 350-MCM aluminum.

Step 2

A grounding-electrode conductor must be installed from the transformer secondary neutral terminal to a suitable grounding electrode. This grounding conductor is sized the same as the required bonding jumper in the above example. That is, this grounding-electrode conductor is sized according to Table 250-94 as if it were a grounding-electrode conductor for a service with service-entrance conductors equal in size to the phase conductors used on the transformer secondary side. But, this grounding-electrode conductor does not have to be larger than 3/0 copper or 250-MCM aluminum when the transformer secondary phase legs are over 1100-MCM copper or 1750-MCM aluminum.

EXAMPLE

For the 75-kVA transformer in step 1, the grounding-electrode conductor must be not smaller than the required minimum size shown in Table 250-94 for 4/0 phase legs, which makes it the same size as the bonding jumper—that is, No. 2 copper or No. 1/0 aluminum. For the 500-kVA transformer, the grounding-electrode conductor is sized directly from Table 250-94, which requires 3/0 copper or 250-MCM aluminum where the phase legs are over 1100-MCM copper or 1750-MCM aluminum. Note that the grounding-electrode conductor for this 500-kVA transformer is smaller than the bonding jumper sized in step 1.

In the 1978 **NEC**, this rule called for the bonding and grounding of a grounded transformer secondary or generator output (for example, 208/120-V wye) "at the source" of the separately derived system. That phrase was frequently interpreted very rigidly to mean "only at the transformer itself." In the present **NEC**, the last sentence of Sections 250-26(a) and (b) permits the bonding and grounding connections to be made either right at the transformer or generator or at the first disconnect or overcurrent device fed from the transformer or generator, as in Fig. 333.

The last sentence of Sections 250-26(a) and (b) does, however, say that the bonding and grounding "shall be made at the source"—which appears to mean right at the transformer or generator, and not any other point—where the transformer supplies a system that "has no disconnecting means or overcurrent device." A local transformer that has no disconnect or overcurrent devices on its secondary is one that supplies only a single circuit and has overcurrent protection on its primary, such as a control transformer to supply motor starter coils. But in

FIG. 333 Secondary neutral grounding and bonding may be accomplished at either of two places.

such applications, the transformer does not supply a separately derived "system" in the usual sense that a "system" consists of more than one circuit. However, for a transformer supplying only one circuit, any required bonding and grounding of a secondary grounded conductor—such as the grounded leg of a 2-wire, 120-V control circuit—would normally have to be done right at the transformer.

Another interpretation that might be put on that last phrase is that it refers to a hookup in which the transformer secondary feeds main lugs only in a panelboard, switchboard, or motor control center. In such cases, the absence of a main CB or fused switch would mean that there is no disconnect means or overcurrent device for the overall "system" fed by the transformer, even though there are disconnects and overcurrent protection for the individual "circuits" that make up the "system." That interpretation would require bonding and grounding of the secondary right at the source (the transformer itself).

Step 3

The grounding-electrode conductor, installed and sized as in step 2 above, must be properly connected to a grounding electrode that must be "as near as practicable to and preferably in the same area as the grounding conductor connection to the system." That is, the grounding electrode must be as near as possible to the transformer itself. In order of preference, the grounding electrode must be one of the following:

1. The nearest available structural steel of the building, provided it is established that such building steel is effectively grounded
2. The nearest available water pipe, provided it is effectively grounded

Section 250-112 clarifies the term "effectively grounded" by noting that the grounding connection to a grounding electrode must "assure a permanent and effective ground." Thus, when a nearby cold-water pipe is used as a grounding electrode, it would appear to be necessary to bond around any unions or valves that might be opened and thereby might break the piping connection to the earth [Fig. 334]. And Section 250-81(a) requires a bonding jumper to be used

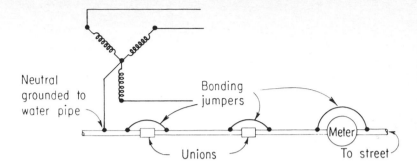

Neutral grounded to water pipe

Bonding jumpers

Unions

Meter

To street

NOTE: The bonding shown here may or may not be required by individual inspectors. The NE Code does not specifically clarify this point, although literal wording appears to require bonding.

FIG. 334   Reliability and conductivity of grounding-electrode connections must be assured.

around all indoor water meters to ensure continuity to earth or through the interior water-pipe system. Such bonding jumpers must be at least the same size as the grounding conductor from the transformer to the water pipe and other electrodes.

At least 10 ft of the metal water pipe must be buried in earth outside the building for the water-pipe system to qualify as a grounding electrode for the transformer secondary neutral. However, even where there is not 10 ft of metal pipe in the earth, there must always be a connection between an interior metal water-pipe system and the service entrance grounded conductor (the neutral of the system that feeds the primary of transformers in the building). The grounding connection for the service neutral or other system grounded conductor must be made at the service. Where a metallic water-pipe system in a building is fed from a nonmetallic underground piping system or has less than 10 ft of metal pipe underground, the water pipe is not a grounding electrode, and the service neutral or other service grounded conductor must have a connection to a ground rod or other electrode in addition to the connection to the interior metal water-pipe system.

Note that Section 250-26 accepts an "effectively grounded" metal water pipe as a suitable grounding electrode for a "separately derived system" such as a transformer secondary. But, Section 250-81 *does not* permit the use of a water-pipe electrode (with at least 10 ft in the ground) as the *sole* grounding electrode for the service to a building. Any such water-pipe electrode must be supplemented by bonding to at least one other grounding electrode—such as grounded building steel, reinforcing bars in the footing or foundation, or a driven ground rod. (See "Grounding Electrodes" in Section 7 of this handbook.)

Figure 335 shows the various elements involved in grounding and bonding at a transformer. Transformer housings must be grounded by connection to grounded cable armor or metal raceway, by use of a grounding conductor run with the primary circuit conductors (either a bare conductor or a conductor with green-colored covering), or by a separate equipment grounding conductor. Such an effective equipment grounding conductor (metal raceway or a separate conductor) is necessary to provide for fault-current return in case an energized conductor of a grounded primary supply circuit (such as a 3-phase, 480-V primary circuit derived from a grounded-wye, 480/277-V system) faults to the metal transformer housing. Even though the secondary neutral is bonded to the case and grounded to building steel, a low-impedance ground-return path must be provided for primary circuit faults in the metal housing to clear a primary protective device—as required by **NEC** Section 250-57(b). But the clearing of ground faults on

FIG. 335 All the elements of transformer grounding and bonding.

the *secondary* side of the transformer is facilitated by the bonding that Section 250-26 requires between the secondary neutral and the metal case of the transformer, as shown in Fig. 336.

Figure 337 shows an important detail for effective grounding of a transformer case and secondary neutral. A common technique for protecting a bare or insulated system grounding conductor (one which grounds the wiring system and equipment cases) and for protecting equipment-only grounding conductors makes use of a metal conduit sleeve, run open or installed in concrete. In all such cases, the **NEC** says that "metallic enclosures for grounding conductors shall be electrically continuous from the point of attachment to cabinets or equipment to the grounding electrode, and shall be securely fastened to the ground clamp or fitting." This means that the grounding conductor must be connected to its protective conduit at both ends so that any current which might flow over the conductor will also have the conduit as a parallel path.

The necessity for making a grounding conductor electrically parallel with its protective conduit applies to any grounding conductor. If the protective conduit were arranged so that the conductor and conduit were not acting as parallel conductors—such as would be the case in Fig. 337 if there were no bonding jumper from the conduit bushing to the conductor lug—the presence of magnetic metal conduit (steel) would serve to greatly increase the inductive reactance of the grounding conductor in limiting the flow of current to ground. The steel conduit would act as the core of a "choke" to restrict current flow.

FIG. 336 Secondary neutral bonding provides for fault clearing on secondary faults, but not on primary faults.

Dry-type transformer

Inside case:
Neutral point of
secondary wye
point connected
to stud on case

Insulated common grounding conductor
with lug on end, bolted to case

Bonding jumper to assure continuity
from case to conduit, sized per **NEC** Sec. 250-79

Grounding bushing with
terminal lug, on end of
conduit

Steel conduit
in concrete

Conduit and conductor run to
water pipe electrode. Both
conduit and conductor are
connected to electrode.

FIG. 337 Protective metallic conduit sleeving for a grounding-electrode conductor *must* be
bonded to the conductor at both ends.

Where building steel or a metal water pipe is not available for the grounding of in-building
transformers (or generators), other electrodes may be used, based on **NEC** Section 250-81 or 250-83.

Figure 338 shows techniques of transformer grounding that have been used in the past but
are no longer acceptable.

As shown in Fig. 339, the Exception to Section 250-26(b) exempts small control and signaling transformers from the basic requirement for a grounding-electrode conductor run from the
bonded secondary grounded conductor (such as a neutral) to a grounding electrode (nearby
building steel or a water pipe). A Class 1 remote-control or signaling transformer that is rated
not over 1000 VA simply has to have a grounded secondary conductor bonded to the metal case
of the transformer; no grounding-electrode conductor is needed—provided that the metal transformer case itself is properly grounded by grounded metal raceway supplying its primary or by
means of a suitable equipment grounding conductor that ties the case back to the grounding
electrode for the primary system. The Exception to Section 250-26(a) permits the use of a No.
14 copper conductor to bond the grounded leg of the transformer secondary to the transformer
frame, leaving the supply conduit to the transformer to provide the path to ground back to the
main service ground, but depending on the connection between neutral and frame to provide an
effective return for clearing faults, as shown. Transformer housings must be grounded by connection to grounded cable armor or metal raceway or by use of a grounding conductor run with
circuit conductors (either a bare conductor or a conductor with green-colored covering).

Because the rule on bonding jumpers for the secondary neutral point of a transformer refers
to Section 250-79(c) and therefore ties into Table 250-94, the smallest size that may be used is
No. 8 copper, as shown in that table. But for small transformers (such as those used for Class
1 remote-control or signaling circuits) a bonding jumper that large is not necessary and is not
suited to termination provisions. For that reason, the Exception to Section 250-26(a) permits
the bonding jumper for such transformers rated not over 1000 VA to be smaller than No. 8 wire.
The jumper simply has to be at least the same size as the secondary phase legs, but in no case
smaller than No. 14 copper or No. 12 aluminum.

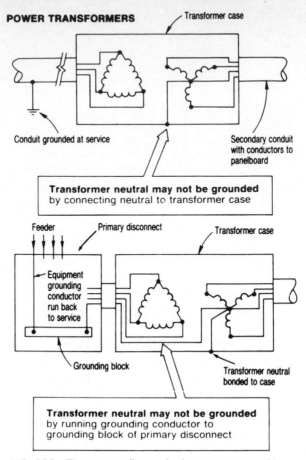

POWER TRANSFORMERS

Transformer case

Conduit grounded at service

Secondary conduit with conductors to panelboard

Transformer neutral may not be grounded by connecting neutral to transformer case

Feeder   Primary disconnect   Transformer case

Equipment grounding conductor run back to service

Grounding block

Transformer neutral bonded to case

Transformer neutral may not be grounded by running grounding conductor to grounding block of primary disconnect

FIG. 338   These grounding methods are *not* acceptable.

Other Design Examples

In Fig. 340, for a feeder of a 480/277-V, 3-phase, 4-wire system, a 140-kVA demand load of incandescent lighting and receptacle circuits must be fed from a transformer which will feed two 120/208-V, 3-phase, 4-wire panelboards for the required branch circuits to the loads:

1. What size 3-phase, dry-type transformer must be used for the total 120/208-V load?
2. What size primary and secondary conductors must be used to carry the transformer current?
3. What size bonding jumper and grounding-electrode conductors must be used at the transformer?
4. What ratings and arrangements of primary and secondary overcurrent protection are required?

These questions are answered in the next several paragraphs, for both noncontinuous and continuous loading.

**CONTROL OR SIGNALING TRANSFORMERS
NOT OVER 1000 VA**

Transformer may be grounded
by a No. 14 Cu bonding jumper
connecting one leg of 120-V
secondary to transformer case,
which is grounded through
primary conduit.

Fuse or CB anywhere
on secondary side
of control transformer
(typically protection
within a motor starter)
will operate on a ground fault

**Grounding electrode conductor
is not required**

FIG. 339 Grounding-electrode connection is *not* required
here.

FIG. 340 Sizing of all circuit components must be carefully coordinated for all conditions.

1. Transformer

The total demand load to be supplied from the 120/208-V, 3-phase, 4-wire secondary is 140 kVA, and the next highest standard size of dry-type transformer is 150 kVA. If the extra 10 kVA of capacity is sufficient to accommodate anticipated load growth, a 150-kVA transformer would do the job.

2. Conductors

The full-load primary current for a 150-kVA transformer with a 480-V, 3-phase, 3-wire delta primary hookup is determined using the standard formula:

$$\text{Rated primary current} = \frac{\text{rated transformer kVA} \times 1000}{1.73 \times \text{rated primary phase-to-phase voltage}}$$

$$= \frac{150 \times 1000}{1.73 \times 480} = \frac{150{,}000}{831} = 180 \text{ A}$$

This means that the ampacity of the primary-circuit conductors from the CB in the 480-V switchboard to the transformer must be at least 180 A to provide full use of the 150-kVA rating of the transformer.

NOTE: If the load fed by the circuit is continuous, a 125 percent multiplier must be used in sizing all circuit conductors, as explained later.

Conductors are selected with **NEC** Table 310-16. If the primary conductors are to be copper and in raceway, then No. 4/0 Type TW conductors (rated 195 A) or No. 3/0 Type THW (rated 200 A) or No. 3/0 Type THHN or XIIIIW (rated 225 A) appear to be acceptable. All these conductors would be operating within the ampacity and heating limitation for 75°C (THW) conductors. If aluminum conductors are to be used, Table 310-16 shows that three No. 4/0 TW in 2-in conduit, or three No. 3/0 THW in 2-in conduit, or three No. 3/0 THHN or XHHW in 1½-in conduit would be acceptable.

As shown in Fig. 340, the primary circuit run to the transformer from the 480-V switchboard is long, but the two panels fed by the transformer are located immediately adjacent to the transformer, and each is fed by tap conductors not over 10 ft long from the transformer to each panel.

We know that the rated secondary current of the transformer is

$$\text{Rated secondary current} = \frac{\text{rated transformer kVA} \times 1000}{1.73 \times \text{rated secondary phase-to-phase voltage}}$$

$$= \frac{150 \times 1000}{1.73 \times 208} = \frac{150,000}{360} = 417 \text{ A}$$

But, in this case, the transformer capacity will be divided equally between the two 200-A panels; i.e., a set of secondary conductors rated 200 A will be used for the 10-ft tap to each panel (assuming a separate raceway to each panel).

From **NEC** Table 310-16, each 10-ft feeder tap could be made up of four of any of the following copper conductors:

250-MCM Type TW (rated 215 A)

No. 3/0 Type THW (rated 200 A)

No. 3/0 Type THHN or XHHW (rated 225 A)

NOTE: The 225-A rating of the THHN or XHHW and the 215-A rating of the 250-MCM TW make the full secondary capacity of the transformer available, so that two conductors of any of these types, when paralleled, would have a current-carrying capacity greater than 417 A, the full secondary current rating. If a single set of secondary conductors had been taken from the transformer to supply, say, a single panel, the conductors would have had to be rated at least 417 A (as are 600-MCM Type THW, rated 420 A). However, in this example the secondary capacity is divided between two panels.

Taps from the transformer, if not more than 10 ft long, are acceptable, just as taps from a feeder with no change in voltage level would be. It is only necessary that all design factors be accounted for:

- The primary-circuit conductors to the transformer must be sized and protected properly.
- The transformer must be protected.
- The tap conductors must be sized and installed in accordance with Exception No. 2 of Section 240-21:

1. The tap conductors must not be over 10 ft long.
2. The ampacity of the tap conductors must not be less than the combined computed loads on

the panel fed, and not less than the rating of the panel mains or not less than the rating of fuses or a CB in which they terminate.

3. The tap conductors must be enclosed in raceway from the transformer case to each panel.

Protection of the panel mains must also be considered. If either of these panels is a "lighting and appliance" panel (more than 10 percent of its CB poles or fuses rated 30 A or less, with a neutral terminal block provided in the panel), each such panel must have main protection as described in Section 384-16. Basically, this means that a lighting and appliance panel must be protected by a main CB or fused switch with a rating not greater than the rating of the panel mains (i.e., not over 200 A).

3. Grounding

The bonding jumper and grounding-electrode conductor required by **NEC** Section 250-26 for the transformer would be sized as follows:

The bonding jumper used to connect the system neutral of the transformer to the metal housing of the transformer must be sized in accordance with Section 250-79(c), which says that Table 250-94 must be used just as in sizing the grounding-electrode conductor that runs to a water pipe or building steel.

The required size of the bonding jumper, either copper or aluminum, is selected by referring to Table 250-94 and using the size of the transformer secondary conductors as if they were service-entrance conductors. In this example, with the secondary circuit made up of two sets of, say, No. 3/0 copper THW, it is necessary to convert the two No. 3/0 conductors to an equivalent single copper conductor with a cross section equal to that of the two No. 3/0 THWs.

NOTE: In **NEC** Table 250-94, the phrase "or Equivalent for Parallel Conductors" in the heading of the left-hand column must be taken as meaning equivalent in cross-sectional area, not in ampacity. And the type of insulation on the service-entrance conductors is of no concern in this calculation. Physical size of conductor material—effective cross section per phase—is the sole determinant of the required size of grounding-electrode conductor (and bonding jumper).

The cross-sectional area of a single No. 3/0 stranded conductor is 167,800 cmil (from Table 8 of **NEC** Chapter 9). The single conductor "equivalent" to two No. 3/0 conductors would have a cross-sectional area of 2 × 167,800 or 335,600 cmils. From Table 8, the single conductor having cross-sectional area closest to this figure is the 350-MCM, which has a cross-sectional area of 350,000 cmils.

Table 250-94 shows that, for 350-MCM service conductors, a minimum of No. 2 copper or No. 1/0 aluminum must be used for a grounding-electrode conductor. For the transformer in this example, then, the bonding jumper must not be smaller than No. 2 copper or No. 1/0 aluminum.

The grounding-electrode conductor used to connect the transformer neutral (and the bonded equipment enclosure) to the grounding electrode must be sized according to Section 250-26, which says in effect that the grounding-electrode conductor must be sized the same as the bonding jumper, using Table 250-94 as described above. Thus, for this transformer application, the grounding-electrode conductor must not be smaller than No. 2 copper or No. 1/0 aluminum (again, from Table 250-94).

IMPORTANT: The size of a grounding-electrode conductor and bonding jumper for a dry-type transformer application is based on the conductor material and cross-sectional area of the secondary phase conductors. Thus, the use of higher-temperature-rated secondary phase conductors to obtain the required capacity in a smaller conductor (say THW instead of TW) will result in a smaller required grounding-electrode conductor and bonding jumper.

In this example, if TW conductors had been used for the secondary phase conductors rather than THW, they would have had to be at least 250-MCM conductors. Two per phase would have been equivalent in size to one 500-MCM, which would require a minimum of No. 1/0 copper or No. 3/0 aluminum for the grounding-electrode conductor and transformer bonding jumper (Table 250-94).

When transformer secondary conductors are selected at 125 percent of the "continuous" load current, they are still larger than those required for noncontinuous operation—and thereby often require larger grounding-electrode conductors and bonding jumpers.

If the 150-kVA transformer of this example had been used to supply, say, one 400-A panel with a single set of conductors, each secondary phase conductor might have been a single 600-MCM THW, rated at 420 A. The minimum required size of grounding-electrode conductor and bonding jumper would then have been No. 1/0 copper or No. 3/0 aluminum.

4. Overcurrent protection

Overcurrent protection for this 150-kVA dry-type transformer must conform to **NEC** Section 450-3(b). According to subparagraph (1) of that section, a dry-type transformer must be protected by an overcurrent device rated or set at not more than 125 percent of the rated primary current. This device may be right at the transformer primary or at the supply end of the circuit that feeds the transformer primary.

In this example, say, three No. 3/0 THW conductors are used for the primary supply circuit to carry the 180 A of rated primary current (assuming future full use of the transformer). These conductors must then be protected at not more than their 200-A ampacity. A 200-A CB or 200-A fuses would provide such protection.

Because Section 450-3(b)(1) sets a basic limit of 125 percent of the primary current on the rating of the transformer protective device, a CB or fuses used to protect this transformer must be rated at not more than 1.25×180 A or 225 A. Any rating or setting below this value would be acceptable according to Section 450-3(b)(1), so that a 200-A CB or 200-A fuse could be used.

Could these No. 3/0 THW primary conductors be used, then, for the circuit to the transformer, with a 200-A CB or 200-A fuses in a 200-A switch for both transformer and circuit protection? Before this question can be answered, other rules must be accounted for. If the 120/208-V load fed by the transformer is a 180-A "continuous" load (the full load operates continuously for 3 hours or more), the following must be considered.

1. When a feeder supplies a continuous load, the rating of the overcurrent device protecting the feeder and the ampacity of the primary feeder conductors must be not less than 125 percent of the continuous load [**NEC** Section 220-10(b)].

2. If the CB or fused switch feeding the transformer is located in a panelboard, the load on either one of them (180 A, under full-load conditions) must not exceed 80 percent of the rating of the CB or fuses [**NEC** Section 384-16(c)].

3. A fused switch used for other than a motor circuit must not be loaded in excess of 80 percent of the rating of the fuses in the switch (UL requirements). And a general UL rule limits the continuous loading on any molded-case circuit to not over 80 percent of the CB rating.

If the full load on the transformer is, say, lighting that will operate for 3-hour periods or longer, or if it is assumed that the load will be continuous, the foregoing rules would prohibit the use of a 200-A CB or fuses for this transformer circuit. Certainly the circuit to the transformer is a feeder which must have protection complying with item 1 above. Then, the use of a 200-A CB or 200-A fuses is not acceptable.

If the transformer's rated primary current of 180 A must not exceed 80 percent of the rating of the overcurrent protection (that is, the rating of the device is not less than 125 percent of 180

A), then the minimum acceptable rating of overcurrent protection is 180 A × 1.25 = 225 A. Thus, two conditions must be met:

- The primary circuit protection device (CB or fuses) must not be rated below 225 A [Section 220-10(b)].

- The same protective device, if it is also to provide protection for the transformer, must not be rated above 225 A.

But, when 225-A overcurrent protection is substituted for the initial selection of a 200-A CB or 200-A fuses, the size of the primary circuit conductors must be increased from No. 3/0 THW, because the 225-A protection does not protect these 200-A conductors within their allowable ampacity. It is important to note that Section 220-10(b) also requires the circuit conductors to be rated not less than 180 A × 1.25, or 225 A to correspond with standard fuse or CB ratings.

Primary-circuit conductors which are acceptable for use with 225-A protection must be selected from Table 310-16. The conductors could be as follows:

Copper: 300-MCM TW (rated 240 A), No. 4/0 THW (rated 230 A), or No. 4/0 THHN or XHHW (rated 260 A)

Aluminum: 400-MCM TW (rated 225 A), 300-MCM THW (rated 230 A), or 300-MCM THHN or XHHW (rated 255 A)

NOTE: For other than continuous loads, where conductors have a current rating that does not correspond to a standard fuse or CB rating, the next larger size of fuse or CB may be used. However, for continuous loading, Section 220-10(b) does not provide that type of exception to the rule that the conductors be rated at 125 percent of the load.

But, for transformer protection, it must be understood that exceptions to Section 450-3(b) do give permission to use the next larger standard overcurrent device rating when the transformer full-load current does not correspond to a standard rating of CB or fuse. However, in some cases, the next smaller size must be used to avoid exceeding the maximum value. (Permission for use of the "next higher size" is also given in Section 430-52, covering motor branch-circuit protection.)

EXAMPLE

In Fig. 341, a 500-kVA transformer is used to step 480-V, 3-phase power down to 208/120 V for incandescent lighting, electric heating and receptacle outlets. No motor loads are fed

FIG. 341 Loading, conductor sizing, and the relative locations of equipment influence design calculations.

by the secondary. The total load on the transformer is 422 kVA, of which 70 percent is continuous.

Step 1

The secondary current rating of the transformer is determined from the formula

$$I = \text{kVA rating} \times 1000 \div (1.732 \times 208)$$

Step 2

The rated secondary current is

$$\frac{500 \times 1000}{1.732 \times 208} = 1388 \text{ A}$$

Step 3

To use the full secondary capacity of the transformer, a circuit makeup of four 700-MCM THW aluminum conductors per phase leg (each rated at 375 A) would be acceptable for the load (4 × 375 A = 1500 A). Of course, to maintain that load ampacity, each set of three phases and a neutral must be installed in a separate conduit.

Step 4

Based on the use of four 700-MCM aluminum conductors per phase, the bonding jumper must be sized for a phase-leg size of 4 × 700 MCM, or 2800 MCM. The minimum acceptable size of aluminum bonding jumper would be 12½ percent of 2800 MCM, or 350 MCM, which would require the use of a standard 350-MCM jumper. If a copper bonding jumper were used, it would have to be sized at 12½ percent of 2000 MCM [four 500-MCM THW copper conductors, which have a total ampacity of 4 × 380 A, or 1520 A, and comprise the copper equivalent of four 700-MCM THW aluminum conductors per phase, as required by **NEC** Section 250-79(c)]. That would call for a 250-MCM copper jumper.

Step 5

The minimum acceptable size of grounding-electrode conductor to connect this transformer neutral to a nearby grounded structural-steel column is 3/0 copper or 250-MCM aluminum.

Step 6

The secondary conductors from the transformer to the switchboard are provided with overcurrent protection right at the transformer, and the protective device must have a rating or setting that is at least equal to 125 percent of the continuous load plus the noncontinuous load. The total load on the secondary is equal to 422 kVA ÷ (208 V × 1.732), or 1171 A. The continuous part of that load is 70 percent of 1171 A, or 820 A. The secondary conductors and their protection must be rated for at least 1.25 × 820 A + 351 A, or 1376 A.

Step 7

With such loading, the circuit breaker protecting the secondary would have to be at least a 1400-A standard size (the next rating above 1376 A). The 1500-A rating of the secondary circuit of four 700-MCM THW aluminum conductors fully satisfies the need for an ampacity of at least 1376 A. And the 1400-A CB is acceptable protection for conductors that have an ampacity of 1500 A.

NOTE: Under those conditions the entire circuit and transformer are loaded to within 24 A of the maximum permissible load on a 1400-A circuit breaker that is *not* UL-listed for continuous operation at full-load rating (1400 A − 1376 A). Up to 24 A of non-continuous load may be added. If it is added, the load on the transformer will then be 1171 A + 24 A, or 1195 A. But the additional transformer capacity of 193 A (1388 A rated less 1195 A load) could not be utilized. This example makes a strong case for use of 100 percent continuous-rated circuit breakers or 100 percent fusible switching equipment, in which case a CB or fused switch rated at 1200 A could be used for the 1171-A secondary load, with conductors having an ampacity of at least 1200 A.

Step 8

If the switchboard were located 5 ft closer to the transformer, there would be no need for overcurrent protection at the transformer, provided the secondary conductors were tap conductors not over 25 ft long, terminating in a single main CB or fused switch in the switchboard with protection rated at not more than the ampacity of the conductors.

Autotransformers

Most buildings with 480/277-V electrical systems obtain 208/120-V power by means of two-winding (or insulating) transformers. The same function can be obtained from autotransformers. Figure 342 shows a comparison of the 2-winding hookup vs. the autotransformer hookup.

FIG. 342   Autotransformer may be used to derive a 120/208-V, 3-phase, 4-wire system from a 480/277-V system.

In a large, modern commercial, institutional, or industrial building, the usual practice is to locate small step-down transformers—generally in the 15- to 45-kVA range, with 480-V delta primary and 208/120-V wye secondary windings—at regular intervals to provide power for convenience outlets and miscellaneous small equipment. And, quite often, relatively large amounts of 208/120-V power may be required for larger load concentrations such as computer equipment and kitchen loads. In such cases, transformers rated up to 500 kVA are commonly used.

The following comparison of one manufacturer's 225-kVA dual-winding transformers and autotransformers indicates the relative efficiency, weight, and sound-level advantages of the autotransformer for the common 480/277- to 208/120-V transformation.

	Dual-winding transformer	Autotransformer
Full-load loss, W	5800	3200 (55%)
Sound level, db	47	44 (94%)
Weight, lb	1750	1050 (60%)

NEC Section 210-9, headed "Circuits Derived from Autotransformers," says "branch circuits shall not be supplied by autotransformers" (transformers "in which a part of the winding is common to both primary and secondary circuits") unless the system supplied has an identified grounded conductor which is solidly connected to a similar identified grounded conductor of the system supplying the autotransformer. In effect, this section requires that a grounded neutral conductor be included in the supply circuit to the primary side of a 480/277- to 208/120-V autotransformer. And the secondary side of the autotransformer must also have a grounded neutral which is connected to the primary grounded neutral.

The addition of this fourth conductor for the primary supply to an autotransformer is not a significant disadvantage in the usual layout of 480/277-V systems. Dry-type transformers for stepping down to 208/120 V are located very close to (commonly within a few feet of) the 480/277-V panel. The neutral feeder conductor must be brought to this panel to provide for the 277-V grounded circuits for fluorescent lighting. Then the only additional requirement for supplying an autotransformer instead of a two-winding transformer is the short run of neutral conductor from the 480/277-V panel to the autotransformer.

Although the 480-V primary of a two-winding transformer (normally a delta-wye connection) requires only connections of the three phase conductors (no neutral is needed), an **NEC** rule requires a service-type grounding connection to be made at each two-winding transformer (as described previously for two-winding transformers, which are "separately derived systems" as covered by **NEC** Section 250-26). The definite grounding connection required by the **NEC** for all two-winding units adds labor and material costs when two-winding transformers are used. These added grounding costs for two-winding units are generally greater than the cost of neutral connection for an autotransformer primary.

In specifying autotransformers to derive 4-wire, 208/120-V systems from 480-V feeders, it is very desirable to require three-legged core-type units. This construction leads to reduced third-harmonic currents and is more tolerant of unbalanced phase loads. Autotransformers are also lighter and, therefore, easier and less expensive to install. Since they are more efficient, they require less ventilation. They are also significantly quieter in the larger sizes.

Figure 343 shows two other applications for autotransformers. The top diagram shows how a 110-V system for lighting may be derived from a 220-V system by means of an autotransformer. The 220-V system may be either a single-phase system or one leg of a 3-phase system. In the case illustrated, the "supplied" system has the required grounded wire solidly connected to a grounded wire of the "supplying" system: a 220-V single-phase system with one conductor grounded.

Exception No. 2 of **NEC** Section 210-9 permits the use of an autotransformer in existing installations for an individual branch circuit without connection to a similar identified grounded conductor where it is used for transforming from 208 to 240 V or vice versa, as shown at the bottom of Fig. 343. Typical applications are concerned with cooking equipment, heaters, motors,

Autotransformer used to derive a two-wire 110-
V system for lighting from a 220-V power system.

Autotransformers without grounded conductors are recognized.

FIG. 343 Two common uses of autotransformers, mainly on branch circuits.

and air-conditioning equipment. For such applications, transformers are commonly used. This is a long-established practice for voltage ranges where a hazard is not considered to exist.

Buck or *boost* transformers are designed for use on single- or 3-phase circuits to supply 12/24- or 16/32-V secondaries with a 120/240-V primary. When connected as autotransformers, they will handle kilovoltampere loads that are large in comparison with their physical size and relative cost.

Grounding autotransformers

In recent years, autotransformers have been used increasingly to convert existing ungrounded 480-V industrial distribution systems to grounded operation. There are two basic ways to convert an ungrounded system to a grounded type:

First, one of the three phase legs of the 480-V delta can be intentionally connected to a grounding-electrode conductor that is then run to a suitable grounding electrode. Such grounding gives the two ungrounded phases (A and B) a voltage of 480 V to ground. The system then operates as a grounded system, so that a ground fault (phase to conduit or other enclosure) on the secondary can cause fault-current flow that opens a circuit protective device to clear the faulted circuit.

But corner grounding of a delta system does not give the lowest possible phase-to-ground voltage. In fact, the voltage to ground of a corner-grounded delta system is the same as it is for an ungrounded delta system, because the voltage to ground for ungrounded circuits is defined as the greatest voltage between the given conductor and any other conductor of the circuit. Thus, the voltage to ground for an ungrounded delta system is the maximum voltage between any two conductors, on the assumption that an accidental ground on any one phase puts the other two phases at full line-to-line voltage above ground.

In recognition of increasing emphasis on the safety of grounded systems over ungrounded systems, **NEC** Section 450-4 covers the use of zig-zag grounding autotransformers to convert 3-phase, 3-wire, ungrounded delta systems to grounded wye systems. Such grounding of a 480-V delta system, therefore, lowers the voltage to ground from 480 V (when ungrounded) to 277 V (the phase-to-grounded-neutral voltage) when the system is converted to a wye system as shown in Fig. 344.

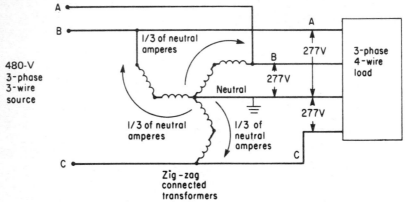

FIG. 344 Zig-zag grounding autotransformer can be used to change voltage to ground from 480 V to 277 V.

The *zig-zag* grounding autotransformer gets its name from the angular phase differences among the six windings that are divided among the three legs of the transformer's laminated magnetic-core assembly. The actual hookup of the six windings is an interconnection of two wye configurations, with specific polarities and locations for each winding. Just as a wye or delta transformer hookup has a graphic representation that looks like the letter Y or the Greek letter Δ, so a zig-zag grounding autotransformer is represented as two wye hookups with pairs of windings in series but phase-displaced, as in Fig. 345.

With no ground fault on any leg of the 3-phase system, current flow in the transformer windings is balanced, because equal impedances are connected across each pair of phase legs. The net impedance of the transformer under balanced conditions is very high, so that only a low level of magnetizing current flows through the windings. But when a ground fault develops on one leg of the 3-phase system, the transformer windings assume a very low impedance in the fault path, permitting a large fault current to flow and operate the circuit protective device— just as it would on a conventional grounded-neutral wye system, as shown in Fig. 346.

Because the kilovoltampere rating of a grounding autotransformer is based on short-time fault current, the selection of such a transformer is much different from the sizing of a conventional two-winding transformer to supply a load. Careful consultation with a manufacturer's sales engineer should precede any decisions about the use of these transformers.

NEC Section 450-4 points out that a grounding autotransformer may be used to provide a neutral reference for grounding purposes or for the purpose of converting a 3-phase, 3-wire delta system to a 3-phase, 4-wire, grounded wye system. In the latter case, a neutral conductor

ZIG-ZAG
REPRESENTATION

WINDINGS
ON CORE

FIG. 345 In zig-zag connection, a path is provided for the flow of fault current or neutral current.

FIG. 346 Zig-zag autotransformer provides the same automatic fault clearing as obtained on a solidly grounded wye system.

can be taken from the transformer to supply loads connected phase to neutral—such as 277-V loads on a 480-V delta system that is converted to a 480Y/277-V system.

Such transformers must have a continuous rating and a continuous neutral current rating. The phase current in a grounding autotransformer is one-third the neutral current, as shown in Fig. 344.

The same **NEC** rule requires the use of a three-pole CB rated at 125 percent of the transformer phase current. The requirement for "common-trip" in the overcurrent device excludes the conventional use of fuses in a switch as overcurrent protection. A three-pole CB prevents single-phase opening of the circuit.

7

Services

The service for any building consists of the conductors and equipment used to deliver electric energy from the utility supply lines (or from an on-site generator) to the distribution system of the building (or other premises being supplied). Service may be brought to a building either overhead or underground—from a utility pole line or from an underground transformer vault. Figure 347 shows the various component parts of services.

The word "service" includes all the materials and equipment involved with the transfer of electric power from the utility distribution line to the electrical wiring system of the premises being supplied. Although service layouts vary widely, depending upon the voltage and ampere rating, the type of premises being served, and the type of equipment selected to do the job, every service generally consists of *service-drop* conductors (for overload service from a utility pole line) or *service-lateral* conductors (for an underground service from either an overhead or underground utility system), along with metering equipment, some type of switch or circuit-breaker control, overcurrent protection, and related enclosures and hardware. A typical layout of "service" for a one-family house breaks down as shown in Fig. 347.

That part of the electrical system which directly connects to the utility supply line is referred to as the *service entrance*. Depending upon the type of utility line serving the building, there are two basic types of service entrances—overhead and underground.

The overhead service has been the most commonly used type. In a typical example, the utility supply line is run on wood poles along the street property line or back-lot line of the building, and a cable connection is made overhead from the utility line to a bracket installed high up on the building (Fig. 348). This line of wood poles also carries the telephone lines, and the poles are generally called "telephone" poles.

The aerial cable that runs from the overhead utility lines to the bracket on the outside wall of the building is called the *service drop*. This cable is installed by the utility line worker. At the bracket which terminates the service drop, conductors are then spliced to the drop cable conductors to carry power down to the electric meter and into the building.

The service-
entrance
conductors
extend from the
connection of
the service drop
down to the line-
side terminals of
the service-
entrance
equipment

Service-entrance head

Insulator

Service
mast
assembly

Splices

Roof seal
and flashing

Service drop
from pole

(Twisted drop
includes a
steel cable as
neutral and
as a support
for the two
"hot"
conductors)

Conduit

Meter socket

Watthour
meter

Entrance
ell

Service panel inside house

FIG. 347   Typical service arrangement for a residential building.

Service drop
conductors

Utility
pole-line

Meter

Service-entrance
conductors in cable
or raceway

FIG. 348   Basic overhead service.

In the underground service, the conductors that run from the utility line to the building are carried underground. Such an underground run to a building may be tapped from either an overhead utility pole line or an underground utility distribution system. Figure 349 compares overhead with underground services. Although underground utility services tapped from a pole line at the property line have been used for many years to eliminate the unsightliness of overhead wires coming to a building, the use of underground service tapped from an underground utility system has only recently started to gain widespread usage in residential areas. This latter technique is called *URD,* for underground residential distribution.

As noted above, when a building is supplied by an overhead drop, an installation of conductors must be made on the outside of the building to pick up power from the drop conductors and carry it into the meter enclosure and service entrance equipment (switch, CB, panelboard, or switchboard) for the building. On underground services, the supply conductors are also brought into the meter enclosure on the building and then run into the service equipment installed, usually, within the building. "Service conductors" is a general term that covers all the conductors used to connect the utility supply circuit or transformer to the service equipment of the premises served. This term includes *service drop* conductors, *service lateral* (underground service) con-

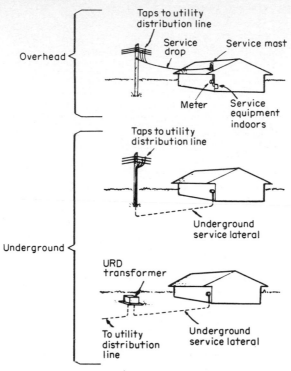

FIG. 349 Service may be made overhead *or* underground.

ductors, and *service entrance* conductors. In an overhead distribution system, the service conductors begin at the line pole where connection is made. If a primary line is extended to transformers installed outdoors on private property, the service conductors to the building proper begin at the secondary terminals of the transformers. Where the supply is from an underground distribution system, the service conductors begin at the point of connection to the underground street mains. In every case the service conductors terminate at the service equipment.

Design Procedure

The design of the service entrance for any premises can be reduced to a procedure involving five basic steps:

1. Calculate the required current rating and power capacity of the service feeder.
2. Lay out the elements of the service connection to the source of power supply.
3. Provide a disconnect means and effective overcurrent protection.
4. Consider an emergency supply source.
5. Consult with the supplying electric utility company and carefully observe all the detailed regulations set out in their published standards. Figure 350 shows part of the table of contents of a typical utility company booklet covering such requirements.

   NEC Article 230 covers services.
   Although the **NEC** states that a building or other premises must be supplied by "only one service"—that is, by only one set of service drop or service lateral (underground) conductors—

INTRODUCTION

This Booklet is published for the benefit of our customers, architects, engineers and contractors to provide a convenient reference. However, design or construction should not be undertaken until complete information is obtained from us. Such information and assistance is available from our Energy Management Services or Electric Operations Departments. See pages 6 thru 17 for location of our offices.

We supply electricity subject to Rules and Regulations, Terms and Conditions, policies and procedures, rate schedules, and industry standards —all of which are made a part of these requirements. Because these change from time to time, they are not included in this booklet, but are available upon request from our offices.

Legal restrictions, changes in the art, judgment and safety require this booklet to be revised from time to time, and we reserve the right to make such revisions. Our present schedule calls for publishing any such revisions in 1983.

We endeavor to supply electricity adequately and reliably. We do not guarantee a continuous supply and do not assume liability for direct or consequential loss or damage to persons or property due to the supply delivered, or as a result of any interruption or variation in the supply.

Failure to comply with our requirements, applicable codes, or orders of an enforcement authority can result in our refusal to energize the service or in the disconnection of an existing service.

FIG. 350 Excerpts from a utility regulations booklet.

1. . . . .when the total de-mand load of all feeders is greater than 3000 amps (up to 600 volts), or

2. . . .when the load de-mand of a single-phase installation is higher than the utility's normal max-imum for a single service, or

Large commercial, industrial or institutional building

3. . . . .when special permission is obtained from the inspection authority.

NOTE: "Two or more services" means two or more service drops or service laterals—not sets of service-entrance conductors tapped from one drop or lateral.

FIG. 351 Under certain conditions, a building may be supplied by more than a single service.

exceptions are made. If a separate service is required for fire pumps or for emergency lighting and power purposes, more than one set of service drop or lateral conductors may be used to supply the building. In the case of apartment houses, shopping centers, and other multiple-occupancy buildings, two or more sets of service entrance conductors may be tapped from a single set of service drop or lateral conductors, or two or more subsets of service entrance conductors may be tapped from a single set of main service entrance conductors.

Two or more services (drops or laterals) to one building are permitted when the total demand load of all the feeders is more than 3000 amps (600 V) or is more than the capacity of one drop (because the premises' load requirements are greater than the capacity of the utility's largest drop) or by special permission (Fig. 351). Cases in which separate light and power services are run to a single building, or separate services are provided to water heaters or electric space heaters to allow the use of different rate schedules, are also exceptions to the general rule requir-ing a single service (Fig. 352). And if a single building is so large, special permission can be obtained for additional services.

An important **NEC** rule says that for any installation where more than one service is permitted to supply one building, a "perma-nent plaque or directory" must be mounted or placed "at each service drop or lateral or at each service-equipment location" to advise personnel that there are other services to the premises and to tell where they are and what building parts they supply.

As discussed thoroughly later, when a sin-gle service is brought to a building or struc-ture, the usual requirement of electrical codes

230-volt, 3-phase 3-wire service

115/230-volt, 1-phase, 3-wire service

FIG. 352 A building may be supplied by two services when the services are for different pur-poses—e.g., a 3-phase power service for motors and a single-phase lighting service.

is that the service disconnect means must consist of not more than six normal disconnect switches or circuit breakers. Some local codes mandate the use of a single disconnect for a service. There are many **NEC** rules that modify the basic arrangement of service disconnects—as described below.

In its simplest form, a service to a building would consist of a single service drop from a pole line or a single underground service (a lateral) supplying a single set of service entrance conductors, which are connected to a single service disconnect (a circuit breaker or a fused switch). A minimal variation on that form is a single drop or lateral supplying a single set of service entrance conductors, which are run to a location just outside or inside the building, where two or more service disconnects (up to a maximum of six switches or CBs) are tapped from the SE (service entrance) conductors, with the taps made within a panelboard or switchboard housing the switches or CBs or with taps made in an auxiliary gutter and nippled into individual enclosures for the two to six service disconnects.

However, the **NEC** permits more than one set of SE conductors to be tapped from a single drop or lateral. Where a single service drop supplies a building which uses two to six service disconnecting means, each in a separate enclosure and each feeding a separate load, a separate set of service-entrance conductors may supply each of the disconnects (see **NEC** Section 230-45). Thus, the two to six sets of service entrance conductors are tapped from the single drop. Figure 353 shows how two to six separate sets of service-entrance conductors may be supplied by a single service drop for either single- or multiple-occupancy buildings. Disconnects can be of same or different ratings, and each set of service entrance conductors can be installed using any approved wiring method. This sketch shows some possible combinations that may be used. And it should be noted that this **NEC** permission to use up to six sets of service entrance conductors applies to single-occupancy buildings as well as to multiple-occupancy buildings.

NEC Section 230-2 provides the basic rule requiring that any "building or other structure" be supplied by "only one service" (that is, by a single drop or a single lateral). Exception No. 7 adds an important qualification of that rule but applies only to Section 230-45, covering service entrance layouts where two to six service disconnects are to be fed by underground service (lateral) conductors and are installed in separate individual enclosures at one location, with each disconnect supplying a separate load. As described in Section 230-45, such a service equipment layout may have a separate set of service entrance conductors run to "each or several" of the two to six enclosures. Exception No. 7 notes that where a separate set of underground conductors of size 1/0 or larger is run to each or several of the two to six service disconnects, the several sets of underground conductors are considered to be one service (that is, one service lateral) even though they are run as separate circuits, that is, connected together at their supply end (at the transformer on the pole or in the pad-mount enclosure or vault) but not connected together at their load ends. The several sets of conductors are taken to be "one service" in the meaning of Section 230-2, although they actually function as separate circuits (Fig. 354).

Although Section 230-2 applies to "service-entrance conductors" and service equipment layouts fed by either a "service drop" (overhead service) or a "service lateral" (underground service), Exception No. 7 is addressed specifically and only to service lateral conductors (as indicated by the word "underground") because of the need for clarification based on the **NEC** definitions of "service drop," "service lateral," "service-entrance conductors, overhead system," and "service-entrance conductors, underground system." (Refer to these definitions in the **NEC** to clearly understand the intent of Exception No. 7 and its relation to Section 230-45.)

The matter involves these separate but related considerations:

1. Because a "service lateral" may (and usually does) run directly from a transformer on a pole or in a pad-mount enclosure to gutter taps where short tap conductors feed the terminals of the service disconnects, most layouts of that type literally do not have any "service-entrance

**For apartment houses and other
multiple-occupancy buildings**

THIS IS OK BY THE PRESENT **NEC**

**For single-occupancy buildings
such as factories, schools and stores**

FIG. 353   More than one set of service-entrance conductors (up to six sets) may be tapped from a single drop.

conductors" that would be subject to the application permitted by Section 230-45—other than the short lengths of tap conductors in the gutter or box where splices are made to the lateral conductors.

2. Because Section 230-45 refers only to sets of "service-entrance conductors" as being acceptable for individual supply circuits tapped from one drop or lateral to feed the separate service disconnects, that rule clearly does not apply to "service lateral" conductors which, by definition, are not "service-entrance conductors." So there is no permission in Section 230-45 to split up "service lateral" capacity. And the basic rule of Section 230-2 has the clear, direct requirement that a building or structure be supplied through only one lateral for any underground service. That is, either a service lateral must be a single circuit of one set of conductors, or, if circuit capacity requires multiple conductors per phase leg, then the lateral must be made up of sets of conductors in parallel—connected together at both the supply and load ends—in order to constitute a single circuit (that is, one lateral).

FIG. 354 In both these cases, the building is supplied by a "single" service lateral.

3. Exception No. 7 permits "laterals" to be subdivided into separate, nonparallel sets of conductors in the way that Section 230-45 permits such use for "service-entrance conductors"—but only for conductors of size No. 1/0 and larger and only where each separate set of lateral conductors (each separate lateral circuit) supplies one or several of the two to six service disconnects.

Exception No. 7 recognizes the importance of subdividing the total service capacity among a number of sets of smaller conductors rather than as a single parallel circuit (that is, a number of sets of conductors connected together at both their supply and load ends). The single parallel circuit would have much lower impedance and would, therefore, require a higher short-circuit interrupting rating in the service equipment. The higher impedance of each separate set of lateral conductors (not connected together at their load ends) would limit short-circuit current and reduce short-circuit duty at the service equipment, permitting the use of lower interrupting-capacity–rated equipment reducing the destructive capability of any faults at the service equipment.

Current Rating of Service Feeders

The basic requirement of the **NEC**, dictated by reason and logic, is that the service conductors have adequate capacity for the total load current that might possibly be drawn simultaneously.

Unless tests and experience (such as demand-meter studies) indicate that something less than 100 percent of all load currents will flow simultaneously or unless control provisions (such as sequence switches or interlocked contactors) assure a maximum demand of less than the total connected load, it is imperative that the conductors have adequate current rating for the entire system load.

The sizing of service entrance conductors usually involves the same type of step-by-step procedure as set forth previously for sizing feeders. A set of service entrance conductors is considered to be a feeder and must be sized just as if it were a feeder (Fig. 355). In general, the service entrance feeder conductors must have a minimum current-carrying capacity sufficient to handle the total lighting and power load served. Where the **NEC** gives demand factors that may be used (Tables 220-11, 220-13, 220-18, 220-19, and 220-20) or where it allows the use of other acceptable demand factors based on sound engineering determination of a less-than-100 percent demand requirement (as in Exception No. 1 of Section 220-15 and Section 220-17), the minimum required ampacity for service entrance conductors may be reduced from the total of the connected loads being supplied.

From the analysis and calculations given in the section of this handbook on feeder circuits, a total power and lighting demand load can be developed for use in sizing service entrance conductors. Of course, where two or more sets of service entrance conductors are used (as shown in Fig. 353), the basic conductor sizing procedure must be applied to each separate set of service entrance feeder conductors.

When a total load (consisting of both initial and spare capacity for the future) has been established for the service entrance conductors, the required current-carrying capacity is easily determined by dividing the total load in kilovoltamperes (or in kilowatts with proper correction for the power factor of the load) by the voltage of the service. But, it should be carefully noted that **NEC** Section 230-41 establishes certain minimum capacities for service entrance conductors. These are for small-size service entrances, and the minimum sizes of conductors must be used regardless of the actual loads supplied. A 100-A service conductor ampacity is the **NEC** manda-

SERVICE-ENTRANCE FEEDER
   1. Overhead from utility pole line or
from transformer on pole
   2. Underground from pole-mounted,
pad-mounted, or underground transformer

Point of power supply to the service-entrance conductors

SE conductors are protected by a single service CB or fused switch — or up to six service CBs or fused switches

One or more feeders on the load side of the service disconnect and overcurrent protection

OVERCURRENT PROTECTION (?)
is at the load end, not the
supply end, of the service-
entrance feeder conductors.

FIG. 355 Evaluation and application of **NEC** rules must be based on an understanding that service entrance conductors constitute a "feeder" but depart from the usual feeder requirements.

tory minimum if the system supplied is a one-family dwelling with more than five 2-wire branch circuits (or the equivalent of that for multiwire circuits) or a one-family dwelling with an initial computed load of 10,000 W. Now that three 20-A small-appliance branch circuits are required in a single-family dwelling, the average new home will need a 100-A, 3-wire service, because even without electric cooking, heating, drying, or water-heating appliances, more than five 2-wire branch circuits will be installed. And **NEC** Sections 220-30 through 220-40 offer "optional calculations" for sizing service entrance conductors for dwelling units, schools, and farm buildings—where experience dictates the use of reasonable demand factors.

From the required current rating of the conductors, the required size of the service entrance phase conductors is determined. Sizing of the service neutral is the same as for feeders. Although suitably insulated conductors must be used for the phase conductors of service entrance feeders, the **NEC** does permit the use of bare grounded conductors (such as neutrals).

For service-lateral conductors (underground service), an individual grounded conductor (such as a grounded neutral) of aluminum without insulation or covering may not be used in underground raceway. A bare *copper* neutral may be used in raceway, in a cable assembly, or even directly buried in soil where local experience establishes that soil conditions do not attack copper. An aluminum grounded conductor of an underground service lateral may be without individual insulation "when part of a cable assembly identified for underground use," where the cable is directly buried or run in a raceway. Of course, a lateral made up of individual insulated phase legs and an insulated neutral is acceptable in underground conduit or raceway.

Except for the use of a bare neutral (as permitted), all service entrance conductors must be insulated. Again, a bare individual aluminum grounded conductor (grounded neutral or grounded phase leg) may not be used in raceway or for direct burial. Any such bare conductor is acceptable only when enclosed in an "identified" jacketed cable assembly—whether installed directly on the side of a building, structure, or pole or installed within a raceway on a building structure.

NOTE: Any nonmetallic cable used for underground installation—either directly buried or in raceway—must be "Type USE" cable. "Type SE" cable is recognized by UL only for above-ground installation.

An extremely important element of service design is that of fault consideration. Service busway and other service conductor arrangements must be sized and designed to ensure their safe application with the service disconnect and protection. That is, service conductors must be capable of withstanding the let-through thermal and magnetic stresses on a fault, as determined by the service overcurrent device.

Just as the sizing of a feeder involves adding the continuous and noncontinuous current loads for lighting, heating, motors, and/or other power equipment and then applying any applicable demand factors, the service entrance conductors are sized for the entire connected load in a building or other premises. The following simple example shows the procedure.

EXAMPLE

Assume a 120,000-ft^2 store building has the following loads:

252 kVA of continuous lighting

73 kVA demand load for receptacles

116 kVA of electric heating

346 kVA of motor loads

38 kVA of appliances

The general approach to the sizing of service entrance conductors might be as follows:

Step 1

Although the conductors must have sufficient capacity for the 252 kVA of continuous lighting, the **NEC** requires a minimum feeder capacity of 3 W/ft^2 for general lighting—for a total of 120,000 × 3 = 360,000 W. Assuming the use of high-power-factor lighting equipment, which can be taken as 360 kVA, **NEC** Table 220-11 requires it be taken at 100 percent demand.

Step 2

The 73-kVA demand load for receptacles was calculated from **NEC** Section 220-13. Refer to "Receptacle loads" in Section 2 of this handbook.

Step 3

The 116 kVA of electric space heating is a 100 percent demand figure, based on connected load and control provisions.

Step 4

The 346-kVA motor load is the sum of 125 percent of the load of the largest motor in the building and the demand loads of all the other motors (in accordance with **NEC** Sections 430-24, 430-26, and 430-33, second paragraph).

Step 5

The 38 kVA of appliance load is the total load of all appliances that could be operating simultaneously.

The service entrance conductors would thus seem to be required to have sufficient capacity for a total of

$$360 + 73 + 116 + 346 + 38 = 933 \text{ kVA}$$

On a 480Y/277-V system, the minimum required current capacity is most effectively calculated using a voltage value of 460 V, which is a multiple of 115 and 230 V. Using the lower value of voltage yields a higher value of current, with greater assurance of adequacy in the conductor size. This gives

$$\frac{933,000}{460 \times 1.732} = 1171 \text{ A}$$

This result indicates that service entrance conductors with 1171-A usable capacity would be adequate for the initial demand load.

But immediately, the critical need to add even minimal spare capacity would dictate the use of conductors rated for at least 1200 continuous amperes (which is the lowest standard **NEC** rating of overcurrent protection above 1171 A). And it is always necessary to focus on "usable" ampacity, because **NEC** Section 220-10(b) requires "feeder" conductors (and, therefore, service entrance "feeder" conductors) to have ampacity at least equal to "the noncontinuous load plus 125 percent of the continuous load" if the service protective device is not a circuit breaker listed for continuous loading at 100 percent of its rating or if the device is not a switch-and-fuse assembly listed for 100 percent continuous load. And Note 8 to **NEC** Table 310-16, if applicable here, requires ampacity derating for continuous or noncontinuous loads where more than three current-carrying conductors are installed in a raceway or in a cable, or directly buried in the earth.

In this example, where service conductors are being sized for a store, the minimum calculated lighting load of 360 kVA must be taken at 125 percent of its value because it is continuous (operating steadily for over 3 hours). And any parts of the other loads that might operate continuously must also be taken at 125 percent of their values. Ventilation fans or air conditioning that would run for 3 hours or more would be included. It might be determined, for instance, that 20 percent of the nonlighting load is continuous. Then, of the 573 kVA of nonlighting load, 114.6 kVA (0.2 × 573 kVA) would be taken at 125 percent and the balance at 100 percent.

Still another, typical factor that would have to be considered is the effect of "noncoincident" loads on the required service conductor capacity. If it can be determined that none of the 116-kVA heating load would operate at the same time as the air-conditioning load, then either the heating load or the air-conditioning load, whichever is smaller, may be omitted from the total demand load, because capacity available for the larger of the two loads would more than satisfy the current draw of the smaller load. Of course, all conditions must be accounted for, such as simultaneous operation of the air conditioning and electric-resistance reheat coils in air-conditioning ducts.

If the air-conditioning load is greater than the 116-kVA electric space-heating load in the store building, then the nonheating part of the nonlighting load is 73 kVA + 346 kVA + 38 kVA, or a total of 457 kVA, of which, say, 20 percent is continuous. Then, the sizing of the service conductors would proceed as follows:

Lighting load: 360 kVA × 1.25	450 kVA
Nonlighting load: 457 kVA × 0.2 × 1.25	114 kVA
Nonlighting load: 457 kVA × 0.8 × 1.00	366 kVA
Total demand	930 kVA

To handle that demand load, the conductor ampacity would have to be

$$\frac{930,000}{460 \times 1.732} = 1167 \text{ A}$$

which indicates that service conductors with an ampacity of 1167 A would be adequate for the initial design load; but, again, there would be no spare capacity for load growth. And note that 1167 A of ampacity from Table 310-16 would be adequate only if the makeup of the circuit did not require derating because more than three current-carrying conductors were installed in a single raceway or in a cable, or directly buried.

If the 1167-A circuit is made up of conductors installed in conduit, multiple conductors will be used for the phase legs and the neutral, with three phases and a neutral in each of the multiple conduits. If the major part of the load on a service feeder is electric-discharge lighting, such as fluorescent, metal-halide, or high-pressure sodium, then the neutral conductor of the feeder must be counted as a current-carrying conductor. In such a case, the three-phase conductors and the neutral conductor would have to be taken as four current-carrying conductors in each of the conduits making up the multiple 1167-A circuit, requiring the application of the rule of Note 8 to **NEC** Table 310-16, which calls for an ampacity-derating factor of 80 percent to be applied to all of the conductors in the service conduits.

If that 80 percent ampacity-reduction factor must be applied to the conductors making up the phase legs for the above 1167-A service feeder, then the conductor selected from Table 310-16 to be used in multiple for each phase leg must have an ampacity value such that its derated ampacity, multiplied by the number of conductors making up the phase leg, equals the required ampacity of 1167 A. For instance, if the service is made up of four conduits, with each one containing a neutral conductor plus three 500-MCM THW copper conductors for phases A, B, and C in that conduit, the derating factor of 80 percent would have to be applied to the ampacity of 380 A shown in Table 310-16 for a 500-MCM THW copper conductor. With each service feeder phase leg made up of four 500-MCM THW copper conductors (one in each of the four conduits), then the ampacity of each phase leg would be

$$4 \text{ (conductors)} \times 380 \text{ A (per conductor)} \times 0.8$$

$$\text{(derating factor)} = 1216 \text{ A}$$

This derated ampacity of 1216 A is the basic ampacity of the conductors under those conditions of application. It fully satisfies the previous calculated requirement for an ampacity of at least 1167 A in the service feeder. In addition, protection of that feeder by a 1200-A single main circuit breaker or 1200-A fuses in a 1200-A switch would also satisfy the rule of Section 220-10(b) that calls for overcurrent protection to have a rating not less than the value of 1167 A.

On a 480Y/277-V, 3-phase, 4-wire service to the example store building, a neutral conductor will be brought in to provide for 277-V lighting and other 277-V loads, so that each set of conductors of the parallel circuit will consist of four conductors: three phase legs and a neutral. But that neutral will be providing for current flow to a load that is only 252 kVA, out of a much larger total. Although the electric-discharge lighting load will be carrying significant third-harmonic current, even under balanced phase-loading conditions, the lighting is not a major portion of the load and the neutral is not required by Note 10(c) of **NEC** Table 210-16 to be counted as a current-carrying conductor. As a consequence, if the circuit is made up of multiple sets of four conductors in each conduit, Note 8 of Table 310-16 does not require derating of the ampacity of each of the four conductors, and conductors for the above calculated 1167 A may be selected directly from Table 310-16.

The value of 1167 A of ampacity for the service conductor is, of course, based solely on providing adequate rating for the initial load. But spare capacity must be added. Although economics always constrains the decision to provide spare capacity for unforeseen future loads, experience consistently confirms that such provision is not only prudent but economical in the long run. Loads start getting added the day the premises is occupied, and they keep piling up with the passage of time. Many designers argue that spare capacity must always be at least 25 percent of the connected demand load in commerical and institutional buildings, and at least 10 percent in industrial buildings, where loads are better accounted for and future changes are more readily anticipated.

In the store building, the original connected load is 825 kVA (252 + 73 + 116 + 346 + 38); adding a full 25 percent would give 825 × 1.25 or 1031 kVA. Dividing 1,031,000 VA by 460 × 1.732 yields an ampacity of 1294 A in the service conductors. The 1294-A value is the minimum load rating for conductors whose ampacity is required by Section 220-10(b) to be equal to 125 percent of the load current for continuous operation. That would call for a minimum ampacity of 1.25 × 1294 A or 1617 A. And even the minimum value of 1167 A for service conductors to the initial load is based on a spare capacity of 108 kVA (360 kVA of calculated general lighting minus the actual connected load of 252 kVA).

Service conductors with an ampacity of 1600 A would constitute an economically realistic basis of effective design for present and future loads, to ensure the full capacity of 1294 A, with 1600 A being the next larger standard rating of protective device above 1200 A. But it also becomes important to make sure that the conductors selected actually have that ampacity.

Assuming use of aluminum service entrance conductors in conduit, each phase leg of the circuit could be made up of four 700-MCM THHN conductors, each with an ampacity of 420 A (Table 310-16), for a total of 1680 A per phase leg, with each of four conduits containing a phase-A conductor, a phase-B conductor, a phase-C conductor, and a neutral conductor (which would be sized differently than the phase legs and would be smaller). Because electric-discharge lighting does not constitute the *major* portion of the load on the service conductors (only 252 kVA out of the total), Note 10(c) to Table 310-16 does not require that the neutral be counted as a current-carrying conductor in each conduit, and derating for conduit fill (Note 8) is not required. Dividing the 1294-A design load among the four conductors per phase leg of the service feeder shows that each 700-MCM THHN aluminum conductor is loaded to 1294 ÷ 4, or 324 A. Because that value is less than the 375-A ampacity of 700-MCM THW (75°C) aluminum conductors, the THHN (90°C) conductors are not being used in excess of the 75°C ampacity of 700-MCM aluminum conductors. They would, therefore, satisfy the UL (and **NEC**) requirement that conductor connections to equipment terminals (the one or more service CBs or switches) must not operate above the ampacities of 75°C conductors for equipment rated over 100 A (or above 60°C ampacities up to a 100-A equipment rating) unless the equipment is marked otherwise. To satisfy that UL rule, the maximum load must not exceed 4 × 375 A, or 1500 A.

As will be discussed later, **NEC** Section 220-10(b) is also a factor in the selection of the one or more circuit breakers or fused switches that are required for disconnect and overcurrent protection of service entrance conductors. Section 220-10(b) requires a rating of 125 percent of the load current for conductor ampacities and overcurrent protection, applied to that part of the load current that is continuous (operates steadily for 3 hours or longer). But the Exception to that rule permits continuous operation with conductors and overcurrent protection rated at 100 percent (instead of 125 percent) *if* the CB or fused switch is UL-listed for full-load continuous operation. In the foregoing analysis, the use of service conductors rated at 1200 A and protected by a CB or fused switch rated for continuous loading to 100 percent of its rating could be considered an effective and economical alternative in supplying spare capacity.

Although the foregoing might seem to be an involved and almost laborious evaluation of required service conductor ampacity, there is really no alternative to that type of procedure if all the related factors are to be effectively included. Any other approach quickly becomes a guess and includes the risk of being inadequate, dangerous, and costly.

Figure 356 shows a service and distribution panel for a small building, where sizing of the service conductors and overcurrent protection would follow a typical feeder-sizing approach:

1. Size of motor branch-circuit conductors: 125 percent of 28 A = 35 A. This requires No. 8 TW, THW, RHH, or THHN copper; or, No. 6 TW, THW, RHH, or THHN aluminum. Conductors are sized on the basis of the ampacity of 60°C conductor insulation.

FIG. 356   Sizing of a small service follows the numbered steps to obtain the required ampacity and protection.

2. Size of motor branch-circuit fuses (using one-time, non-time-delay fuses): 300 percent of 28 A = 84 A. This requires a maximum fuse size of 90 A. Smaller fuses, such as the time-delay type, may be used.

3. Size of service entrance conductors: 125 percent of 28 A + 80 A (lighting load) = 115 A. Conductors for the hot legs must be No. 1/0 TW or No. 2 THW, RHH, or THHN copper. (Note that if the lighting load is continuous, the SE conductors must be sized for 125 percent of 28 A + 125 percent of 80 A.)

4. Size of main fuses: 90 A (from item 2 above) + 80 A = 170 A. This requires a maximum fuse size of 175 A, in a 200-A switch. Again, smaller fuses may and should be used where possible to improve the overload protection on the circuit conductors.

Residential service entrance conductors

The sizing of service entrance conductors for a residential building—an apartment house, condominium, or 2- or 4-family house—would follow the feeder-sizing procedures that apply to other types of buildings. And that same general procedure may be followed in sizing service entrance conductors for a one-family house or an apartment in a multifamily dwelling.

EXAMPLE

A house has two floors with outside dimensions 30 by 25 ft. For both floors, that gives a total living area of 30 ft × 25 ft × 2 = 1500 ft². The house has oil-fired heating and hot water and only one large electric appliance, a 12-kW electric range. What is the minimum required size of circuit conductors?

SOLUTION

Using the standard **NEC** method for feeder sizing (as covered in **NEC** Sections 220-10 through 220-21), the loads are totaled as follows:

General lighting:	
1500 ft² × 3 W/ft²	4500 W
Two kitchen-appliance circuits: 1500 W per circuit	3000 W
One laundry circuit	1500 W
Total	9000 W

Then the demand factors are applied:

3000 W at 100%	3000 W
6000 (9000 − 3000) W at 35%	2100 W
Basic service entrance load	5100 W

The feeder demand load for the 12-kW range must be added to the 5100 W. As stated in **NEC** Section 220-19, the range feeder demand load is selected from Table 220-19. In this case, it is 8 kW (column A, one appliance):

Basic load	5100 W
Range feeder capacity	8000 W
Total feeder load	13,100 W

The minimum required ampacity for the ungrounded service entrance conductors of a 230/115-V, 3-wire, single-phase service is readily found by dividing the total feeder load by 230 V:

$$\frac{13,000 \text{ W}}{230 \text{ V}} = 56.96 \text{ or } 57 \text{ A}$$

Although a load current of that rating could be readily supplied by No. 4 copper TW, THW, THHN, or XHHW conductors, an important provision of **NEC** Section 230-41(b)(2) comes into play here. Where the initial computed total feeder load is 10 kW or more (and it is 13.1 kW here) for a single-family dwelling, the ungrounded conductors of a service feeder must be rated at least 100 A; that is, the minimum capacity of the service entrance hot legs is 100 A.

That would call for a minimum of one of the following (using 60°C ampacity):

- No. 1 TW copper conductors (110 A)
- No. 1 THW copper conductors (130 A)
- No. 1 THHN copper conductors (150 A)
- No. 1 XHHW copper conductors (150 A)
- No. 1/0 TW, THW, THHN, or XHHW aluminum conductors (100 to 125 A)

The above example represents the **NEC** minimum sizing of service entrance conductors in accordance with the steps indicated in Sections 220-10 through 220-21. The calculation accounts only for the initial loads and includes no capacity for load growth. A simplified calculation procedure for residential services can be based on the same **NEC** capacity requirements but can include a measure of extra capacity to cover the virtually inevitable growth in connected loads. Such a design procedure is as follows:

1. Calculate the general lighting load and the general-purpose receptacle outlet load—for those outlets served by 20-A, 115-V, general-purpose circuits. Although each circuit has a full capacity of 20 × 115 or 2300 W, the circuit will be considered as being rated at only 2000 W for purposes of this calculation, thereby allowing some spare capacity.

   For each 500 ft^2 of floor area in the house (excluding porches, garages, unused spaces, unfinished areas, etc.), allow one circuit. Allow an extra circuit for any part of 500 ft^2 left over.

   Multiply the required number of circuits by 2000 W to get the total load in watts. (The same result can be obtained by multiplying the total floor area in square feet by 4 W/ft^2.)

2. Find the total circuit capacity—in watts—to be allowed for the appliance load in the kitchen, dining room, pantry, laundry, and utility area that is to be served by 20-A, 115-V small-appliance circuits. This total capacity can be obtained by multiplying the number of such circuits by 2000 W. If the exact number of such circuits is not known, a 6000-W load can be assumed (two kitchen appliance circuits and one laundry circuit).

3. Take 3000 W of the sum of the amounts of steps 1 and 2 at 100 percent demand.

4. Add to this figure 35 percent (demand) of the remainder of the sum of the amounts of steps 1 and 2. The result is the capacity that must be provided in the service entrance conductors to supply the general lighting and receptacle loads.

5. Add 8000 W for an electric range (rated not over 12 kW). (Refer to **NEC** Table 220-19.) If the electric cooking appliances consist of a built-in oven and range top, the **NEC** must be consulted to get the proper demand load. The **NEC** allows the nameplate ratings of a counter-mounted cooking unit and not more than two wall-mounted ovens to be added together and treated as a single range of that total rating.

6. Add together the wattage ratings of all appliances to be served by individual circuits not previously accounted for in the calculation. A demand factor of 75 percent may be applied to the sum of the nameplate ratings of these appliances—excluding cooking units, clothes dryers, air-conditioning units, and any electric space-heating load.

   If both electric heating and air conditioning are to be used in the house, the rating (in watts) of only the larger of these two connected loads need be used in this total—since the two loads will not be used simultaneously.

7. Add together the following: the general lighting and general-purpose receptacle load (from step 4), the electric-range demand load (from step 5), and the appliance/air-conditioning/heating loads (from step 6).

8. Divide the total number of watts obtained in step 7 by 230 V (for 230/115-V, 3-wire, single-phase service) to get the required ampere rating of the service conductors.

Optional Service Entrance Calculation

The **NEC** provides an optional method for sizing residential service entrance conductors in Section 220-30. That optional calculation may be used:

1. Only for a one-family residence or an apartment in a multifamily dwelling, or other "dwelling unit," that is

2. Served by a 115/230- or 120/208-V, 3-wire, 100-A or larger service, and

3. Where the total load is supplied by one set of service-entrance conductors.

The method recognizes the greater diversity attainable in large-capacity installations. It therefore permits a smaller size of service entrance conductors for such installations than would be permitted by using the load calculations of Sections 220-10 through 220-21.

A typical example of the optional calculation of Section 220-30, based on Table 220-30 (shown in Fig. 357), is as follows.

Load (in kW or kVA)	Demand Factor Percent
Largest of [see Section 220-30(c)]	
Air conditioning and cooling, including heat pump compressors	100
Central electric space heating including integral supplemental heating in heat pumps	65
Less than four separately controlled electric space heating units.	65
Plus:	
First 10 kW of all other load	100
Remainder of other load	40

The loads identified in Table 220-30 as "other load" and as "remainder of other load" shall include the following:

(1) 1500 watts for each 2-wire, 20-ampere small appliance branch circuit and each laundry branch circuit specified in Section 220-16.

(2) 3 watts per square foot (0.093 sq m) for general lighting and general-use receptacles.

(3) The nameplate rating of all fixed appliances, ranges, wall-mounted ovens, counter-mounted cooking units, and including four or more separately controlled space heating units.

(4) The nameplate ampere or kVA rating of all motors and of all low-power-factor loads.

FIG. 357 Optional calculation for residential service conductors is based on this table.

EXAMPLE

A 1500-ft^2 house (excluding unoccupied basement, unfinished attic, and open porches) contains the following specific electric appliances:

12-kW range

2.5-kW water heater

1.2-kW dishwasher

9 kW of electric heat (in five rooms)

5-kW clothes dryer

6-A, 230-V air-conditioning unit

There is recognition in Section 220-21 that if "it is unlikely that two dissimilar loads will be in use simultaneously," then it is permissible to omit the smaller of the two in calculating the required capacity of feeder or service entrance conductors. In Section 220-30, that concept is spelled out to require adding only the larger of either an air-conditioning load or the connected load of four or more separately controlled electric space-heating units. For the residence considered here, these loads would be as follows:

Air conditioning: 6 A $\times$ 230 V = 1.38 kW

Heating (five separate units): 9.00 kW

It is permissible, therefore, to disregard the air-conditioning load in using the optional method.

The heating and other loads (except air conditioning) must be totaled in accordance with Section 220-30:

1. 1500 W for each of two small-appliance circuits (2-wire, 20-A) required by Section 220-3(b)(1): 3000 W

2. Laundry branch circuit (3-wire, 20-A): 1500 W

3. 3 W/ft^2 of floor area for general lighting and general-use receptacles (3 $\times$ 1500 ft^2): 4500 W

4. Nameplate rating of fixed appliances:

   Range: 12,000 W

   "4 or more" separately controlled heating units: 9000 W

   Water heater: 2500 W

   Dishwasher: 1200 W

   Clothes dryer: 5000 W

5. Total: 38,700 W

The first three load categories given in Table 220-30 are not applicable here. "Air conditioning" has already been excluded as a load because the heating load is greater. There is no "central" electric space heating; and there are not "less than four" separately controlled electric space-heating units.

The total load of 38,700 W, as summed above, is classified as "all other load," as referred to on line 6 of **NEC** Table 220-30. The last two references in Table 220-30 constitute all the "optional" calculations in this particular example:

1. Take 10 kW of "all other load" at 100 percent demand to obtain 10,000 W.

2. Take the "remainder of other load" at 40 percent demand factor to obtain 0.4 (38,700 − 10,000) = 11,480 W.

3. Add the results of steps 1 and 2 to obtain 10,000 W + 11,480 W = 21,480 W.

At 230 V, single phase, the minimum ampacity of each service hot leg would then have to be

$$\frac{21,480 \text{ W}}{230 \text{ V}} = 93.39 \text{ or } 93 \text{ A}$$

But Section 230-41(b)(2) requires a minimum conductor rating when the demand load is 10 kW or more. Thus,

Minimum service conductor rating = 100 A

The neutral service entrance conductor size is calculated in accordance with Section 220-22, based on Section 220-10. The 230-V loads have no relation to the required neutral capacity. The water heater, clothes dryer, and electric space-heating units operate at 230 V, 2-wire and have no neutrals. By considering only those loads served by a circuit with a neutral conductor and determining their maximum unbalance, the minimum required size of neutral conductor can be determined.

When a 3-wire, 230/115-V circuit serves a total load that is balanced from each hot leg to neutral—that is, half the total load is connected from one hot leg to neutral, and the other half of the total load from the other hot leg to neutral—the condition of maximum unbalance occurs when all the load fed by one hot leg is operating and all the load fed by the other hot leg is deenergized. Under that condition, the neutral current and hot-leg current are equal to half the total load wattage divided by 115 V (half the voltage between hot legs). But that current is exactly the same as the current that results from dividing the total load (connected hot leg to hot leg) by 230 V (which is twice the voltage from hot leg to neutral). Because of this relationship, it is easy to determine the neutral-current load by simply calculating the hot-leg current load—the total load from hot leg to hot leg divided by 230 V.

In this example, calculation of the neutral current load begins with the following steps, in which the components of the neutral load are summed:

1. Multiply 1500 ft^2 by 3 W/ft^2 to obtain 4500 W.

2. Add three small-appliance circuits (two kitchen, one laundry) at 1500 W each to obtain 4500 W.

3. Add the results of steps 1 and 2 to obtain the total lighting and small-appliance load as 9000 W.

4. Take 3000 W of that value at 100 percent demand factor to obtain 3000 W.

5. Take the balance of the load at 35 percent demand factor to obtain 0.35 (9000 − 3000) = 2100 W.

6. Add the results of steps 4 and 5 to obtain 5100 W.

The remainder of the calculation is performed with reference to Fig. 358.

Assuming an even balance of this 5100-W load on the two hot legs, under maximum unbalance the corresponding neutral current will be the same as the total load (5100 W) divided by 230 V:

$$\frac{5100 \text{ W}}{230 \text{ V}} = 22.17 \text{ A}$$

Neutral current for lighting and receptacles.

Neutral for lighting, receptacles, and range.

Neutral load = 22.17 + 24.34 + 10.43 = 56.94 amps

Neutral current for total load.

FIG. 358 These loads determine the required neutral-conductor ampacity.

The neutral unbalanced current for the range load can be taken as equal to the 8000-W range demand load multiplied by the 70 percent demand factor permitted by Section 220-22 and then divided by 230 V:

$$\frac{8000 \times 0.7}{230} = \frac{5600}{230} = 24.34 \text{ A}$$

The neutral current load that is added by the 115-V, 1200-W dishwasher must also be included:

$$\frac{1200 \text{ W}}{115 \text{ V}} = 10.43 \text{ A}$$

The minimum required neutral capacity is, therefore, the sum 22.17 A + 24.34 A + 10.43 A = 56.94 A, or, after rounding up, 57 A. From **NEC** Table 310-16, the minimum neutral conductors are

Copper: No. 4 TW, THW, THHN, or XHHW

Aluminum: No. 3 TW, THW, THHN, or XHHW

And the 75 or 90°C conductors must be used at the ampacity of 60°C conductors, as required by the UL *Electrical Construction Materials Directory.*

NOTE: The above calculation of the minimum required capacity of the neutral conductor differs from the calculation and results shown in Example No. 2(a) in **NEC** Chapter 9. There, the 1200-W dishwasher load is added as a 230-V load to the range load and general lighting and receptacle load. To include this 115-V load as a 230-V load (and then to divide the total by 230 V, as shown) does not accurately represent the neutral load that the 115-V, 1200-W dishwasher will produce. In fact, it yields exactly half the neutral load that the dishwasher represents. The optional calculation method of Section 220-30 does indicate, in part (3), that fixed appliances be added at nameplate load, and it does not differentiate between 115-V devices and 230-V devices. It simply totals all loads and then applies the 100 percent and 40 percent demand factors as indicated. That method clearly is based on well-founded data about load diversity and is aimed at obtaining a reasonable size of service hot legs. But, calculation of the feeder neutral in accordance with Section 220-22 is aimed at determining the maximum unbalanced current to which the service neutral might be subjected.

Although there is only a small difference between the **NEC** value of 51.7 A and the value of 56.9 A determined here, precise calculation should be made to ensure the adequacy of conductor ampacities. The difference of 5.2 A actually changes the required minimum size of neutral conductor from No. 6 up to No. 4 for copper, and from No. 4 up to No. 3 for aluminum. A load like a dishwasher, which draws current for a considerable period of time and is not used for just a few minutes like a toaster, should be factored into the calculation with an eye toward ensuring adequate capacity for conductors.

Another example of the application of Section 220-30 to the sizing of residential service conductors shows the very heavy loading that is permitted on a 100-A service.

EXAMPLE

The following loads apply to a 1500-ft^2 house with all-electric utilization:

- 1500 W for laundry receptacle circuit: 1500 W
- 1500 W for each of two (minimum of two) required kitchen-appliance circuits: 3000 W
- 1500 ft^2 at 3 W/ft^2 for general lighting and receptacles: 4500 W

- 14 kW of electric space heating from more than four separately controlled units: 14,000 W
- 12-kW electric range: 12,000 W
- 3-kW water heater: 3000 W
- 5-kW clothes dryer: 4500 W
- 3-kW load of unit air conditioners (because this load is smaller than the space-heating load and will not be operated simultaneously with it, no load need be added): 0 W
- Total: 42,500 W

SOLUTION

First 10 kW at 100 percent	10,000 W
Remainder at 40 percent = 32,500 W × 0.4	13,000 W
Total demand load	23,000 W

Size of service = 23,000 W ÷ 230 V = 100-A service

Under certain load conditions, the calculated required service capacity may be substantially less than 100 A. In such cases, however, 100 A is the minimum size service which can be used.

Service Entrance Wiring Methods

A number of specific design requirements apply to the layout and installation of raceways containing service entrance conductors:

1. Service entrance conductors may be run in rigid metal conduit, intermediate metal conduit, EMT, rigid nonmetallic conduit, or wireway. Busway or cable bus may also be used; and service entrance conductors may be run as service entrance cable, Type MC cable (metal-clad cable of the interlocked-armor or corrugated type), or Type MI cable. Cables that are suitable for use as service entrance conductors may be run in cable tray.

   The list of acceptable wiring methods for running service entrance conductors does not include flexible metallic conduit (Greenfield) or liquid-tight flexible metallic conduit. Although such raceways were permitted in previous **NEC** editions if a bonding jumper was used to assure ground continuity between the ends of the flex, they may no longer be used to enclose service entrance conductors, such as to provide flexible routing around a pipe or other obstruction (see Fig. 359).

2. A very important design requirement for service entrances calls for the service overcurrent protection (and the service entrance disconnect) to be located right at the point where service entrance conductors enter a building, either inside or outside the building. But for those cases where it may be difficult or impossible to satisfy that requirement directly, a standard technique is available. Conductors in conduit or duct enclosed by concrete or brick not less than 2 in thick are considered to be outside the building, even though they are actually run within the building. Figure 360 shows how a service conduit was encased within a building so that the conductors are considered as entering the building right at the service protection and disconnect, where the conductors emerge from the concrete, to satisfy the **NEC** rule requiring the service disconnect to be as close as possible to the point where the service entrance conductors enter the building. In the typical application, forms are hung around the service conduit and then filled with concrete to form the required case.

FIG. 359 Flexible conduit may not be used as service raceway.

FIG. 360 Conduit encased in concrete is considered to be outside the building.

3. The need to prevent moisture from entering energized equipment applies to underground service conduits (Fig. 361). Where service raceways are required to be sealed—as where they enter a building from underground—the sealing compound used must be marked on its container or elsewhere as suitable for safe and effective use with the particular cable insulation, with the type of shielding used, and with any other components it contacts. Some sealants attack certain insulations, semiconducting shielding layers, etc.

Service raceways must be sealed or plugged at either or both ends if moisture could contact live parts.

FIG. 361 Underground service conduit may have to be sealed to keep out water.

4. Service conductors—whether directly buried cables, conductors in metal conduit, conductors in nonmetallic conduit, or conductors in EMT—must satisfy all the rules of **NEC** Section 300-5 for protection against physical damage.

5. Service-entrance conductors in EMT or rigid conduit must be made rain-tight, using rain-tight raceway fittings, and must be equipped with a drain hole in the service ell at the bottom of the run or must be otherwise provided with a means for draining off condensation.

6. When rigid metal conduit, IMC, or EMT is used for a service, the raceway must be provided with a service head (or weather head). Figure 362 shows details of service-head installation.

Service heads must be located above the service drop attachment. Although this arrangement alone will not always prevent water from entering service raceways and equipment, such an arrangement will solve most water-entrance problems. But it is permissible for a service head to be located not more than 24 in from the service drop termination where it is impractical for the service head to be located above the service-drop termination. In such cases a mechanical connector is required at the lowest point in the drip loop to prevent siphoning.

FIG. 362 Service head position must exclude water from the raceway.

Connections or conductor arrangements, both at the pole and at the service, must be made so that water will not enter connections and siphon under head pressure into service raceways or equipment. Where the pole connection is higher than the connection of the service entrance conductors at the service head, the connections at the pole should be made in such a manner that moisture will not enter the conductor. In many cases the connection at the utility pole is higher than the connection at the building. Any stranded service-drop conductors act as a hose, and, at the service head, moisture is forced up through the service entrance conductors by this head pressure and down into the meter or service equipment.

7. Figure 363 shows the difference between copper and aluminum when used for a bare, grounded, underground service entrance conductor. For service-lateral conductors (underground service), an individual grounded conductor (such as a grounded neutral) of aluminum without insulation or covering may not be used either directly buried or in raceway underground. A bare copper neutral may be used in raceway, in a cable assembly, or even directly buried in soil where local experience establishes that soil conditions do not attack copper.

As the rule covering this matter was worded in the 1975 **NEC**, an individual bare alumi-

INSULATED PHASE CONDUCTORS and a bare copper neutral for an underground service lateral in buried raceway. Note: A bare aluminum or copper-clad aluminum neutral could be used here when part of a moisture- and fungus-resistant cable.

BARE COPPER NEUTRAL in a direct-buried cable assembly with moisture- and fungus-resistant outer covering. Note: A bare aluminum or copper-clad aluminum could be used like this, but it must be within the same type of cable assembly.

TYPE USE PHASE CONDUCTORS and a bare copper neutral directly buried where soil conditions are suitable for the bare copper.

FIG. 363 Underground, a bare aluminum conductor may not be directly buried or located in raceway.

num neutral (along with individual insulated aluminum phase legs) of an underground service lateral appeared to be permitted if the circuit was installed in conduit or other raceway, and that interpretation was common. But now, an aluminum grounded conductor of an underground service lateral may be used without individual insulation or covering "when part of a cable assembly identified for underground use" (such as USE cable) where the cable is directly buried or run in a raceway. Of course, a lateral made up of individual insulated phase legs and an insulated neutral is acceptable in underground conduit or raceway, or it may be directly buried if it is made up of all single-conductor type USE cables. Figure 364 shows these details.

Service entrance conductors must be insulated, except for a bare grounded conductor, such as a grounded neutral. But, again, a bare individual aluminum grounded conductor (grounded neutral or grounded phase leg) may not be used in raceway or for direct burial. Any such bare conductor is acceptable only when enclosed in an "identified," jacketed cable assembly such as Type SE (above ground), which may be used outdoors, with or without conduit enclosure. Underground, Type USE must be used.

NOTE: At upper right, the cable assembly must be UL-listed. Type USE (underground service entrance) cable and may be directly buried without the conduit shown. At bottom, if the four insulated aluminum conductors are single-conductor Type USE conductors, they may also be directly buried.

FIG. 364 Bare aluminum service conductor may be used underground only when part of a jacketed, listed cable assembly.

Service Disconnect

Service entrance conductors must be equipped with a readily accessible means of disconnecting the conductors from their source of supply.

An extremely important **NEC** rule requires that switches and circuit breakers used as service entrance disconnecting means must be approved for use as service equipment. The switch or CB must be listed and labeled by the UL as suitable for service entrance use. Manufacturers' catalogs should be checked on this point.

The disconnect means for service entrance conductors may consist of not more than six switches or six circuit breakers, in a common enclosure or individual enclosures, located either inside or outside the building wall, as close as possible to the point at which the conductors enter the building (Fig. 365). The intent of this rule [**NEC** Section 230-72(a)] is to limit to six operations of the hand the disconnection of all conductors of the service. The limitation is applicable to cases where all the service disconnecting means are located in one place or "grouped" as required in Section 230-72.

The first sentence of Section 230-72(a), which refers to "two to six disconnects" for *each* service, ties directly to the requirements of Section 230-2. It is the intent of this basic rule that where a multiple-occupancy building (such as an apartment building, store building, or office building that is not over two floors high) is provided with more than one set of service entrance

FIG. 365 Service disconnect arrangement may take any one of these three forms.

conductors tapped from a single drop or lateral, each of those sets of service entrance conductors may have up to six switches or circuit breakers to serve as the service disconnect means for that set. The rule does recognize that six disconnects for each set of service entrance conductors at a building with, say, 10 sets of SE conductors tapped from a drop or lateral does result in a total of 60 (6 × 10), disconnect devices for completely isolating the building's electrical system from the utility supply. And the same permission for use of up to six disconnect devices applies to "separate" services for fire pumps, emergency lighting, etc., which are recognized in Section 230-2 as being separate services for specific purposes.

For any type of occupancy, a power panel (not a lighting and appliance panel, as described in Section 384-14) containing up to six switches or circuit breakers may be used as the service disconnect. A lighting and appliance panel used as service equipment for the renovation of an existing service in an individual residential occupancy (but not for new installations) may have up to six main breakers or fused switches. However, a "lighting and appliance" panel used as service equipment for a new building of any type must have not more than two main devices—with the sum of their ratings not greater than the panel bus rating.

The first sentence of Section 230-71(a) and that of Section 230-72(a) note that from one to six switches (or circuit breakers) may serve as the service disconnecting means for each service to a building. For example, if a single-occupancy building has a 3-phase power service and a separate single-phase lighting service, each such service may have up to six disconnects (Fig. 366). Where the two sets of service equipment are not located adjacent to each other, a clear, conspicuous, and legible plaque or directory must be installed at each service equipment location indicating where the other service equipment is—as required by the second paragraph of Section 230-2. This is a very important safety consideration, to ensure that the building will be disconnected from *all* supply conductors in the case of an emergency such as a fire. Design of the service must always fully account for conditions of emergency, panic, or chaos.

Single-pole switches or circuit breakers equipped with handle ties may be used in groups as "single" disconnects for multiwire circuits, with fuses in the switch poles or the CB trip mechanism simultaneously providing overcurrent protection for the service (Fig. 367). Of course, multipole switches and circuit breakers may also be used as single disconnects. The requirements of the **NEC** are satisfied if all service entrance conductors can be disconnected with no more than six operations of the hand—regardless of whether each hand motion operates a single-pole unit, a multipole unit, or a group of single-pole units with "handle ties" or a "master handle" controlled by a single hand motion. Of course, a single main device for service disconnect and overcurrent protection—such as a single main CB or fused switch—gives better protection to the service conductors, is often required by local codes, and is a preferred design concept.

AS PERMITTED BY EXCEPTIONS TO SEC. 230-2:

Lighting service → 120/240 V

Power service → 440 V 3-phase

Each class of service may consist of 1 to 6 fused switches or CBs in a common enclosure, or a group of separate enclosures, grouped together at a common location.

FIG. 366 Each separate service may have up to six disconnect devices.

Single-pole CBs with handle tie make up a single disconnect

All poles rated over 30 amps, so this is not a lighting and appliance panel

Typical multiwire circuit

N

Twelve 1-pole CBs with handle ties

FIG. 367 This arrangement constitutes *six* disconnect devices.

Section 384-16(a) requires a higher degree of overcurrent protection for lighting and appliance branch-circuit panelboards when used as service equipment or on the load side of service equipment. Each such panelboard must be protected on the supply side by not more than two main circuit breakers or two sets of fuses having a combined rating not greater than that of the panelboard. But, as noted before, up to six main protective devices may be used in a lighting and appliance branch-circuit panelboard where such a panelboard is used as service entrance equipment for the renovation of an existing installation (but not for new jobs) in an individual residential occupancy. (Refer to the discussions of Figs. 114 and 115 in Section 2 of this handbook.) It should be noted that these rules concern only a lighting and appliance branch-circuit panelboard, which is defined as a panelboard having more than 10 percent of its overcurrent devices rated 30 A or less, for which neutral connections are provided. Other types of panelboards used as service equipment may still utilize the concept of six main disconnect devices fed by the supply conductors to the panelboard.

Grouping of disconnects

In a service disconnect arrangement of more than one disconnect—such as where two to six disconnect switches or CBs are used—all the disconnects making up the service equipment "for each service" must be grouped and not spread out at different locations, as required by **NEC** Section 230-72. The basic idea is that anyone operating the two to six disconnects must be able to do so while standing at one location. Service conductors must be readily disconnected from all loads at one place. And each of the individual disconnects must have lettering or a sign to specify what load it supplies, as shown in Fig. 368. The two to six service disconnects that are permitted for each service, or for each set of service entrance conductors in a multiple-occupancy building, must be grouped. But the individual groups of two to six breakers or switches do not have to be located together, and each grouping of two to six disconnects may be within a unit occupancy—such as an apartment—of the building.

Special or emergency service equipment does not have to be grouped with the normal service equipment. It should also be noted that **NEC** Section 700-12(d) requires emergency services to be widely separated from other services, to prevent failure of both due to a single fault. And Section 230-72(b) makes it mandatory to install emergency disconnect devices where they will not be disabled or affected by any fault or violent electrical failure in the normal service equipment. If a service disconnect for emergency and exit lighting is installed very close to a normal

120/208 – VOLT FEEDER TO TOOL ROOM LTG. PANEL

Sign on switch tells the load fed

FIG. 368 Each service disconnect must be "marked" to indicate the load served.

service switchboard, an equipment burndown or fire near the main switchboard might knock out the emergency circuit.

Each separate service permitted for fire pumps or for emergency service may be equipped with up to six disconnects, just as the normal service (or any service) may have up to six service entrance disconnects. But, again, the disconnect or disconnects for a fire-pump or emergency service must be remote from the normal service disconnects, as shown in Fig. 369.

Another exception to the general **NEC** rule on the grouping of service disconnects is made for those cases where one of the two to six disconnects is for a circuit to a water pump that is counted on for firefighting purposes. This exception permits (but *does not require*) one of the two to six

THIS DISTANCE SHOULD ISOLATE EMERGENCY DISCONNECT(S) FROM FAULTS IN NORMAL SERVICE EQUIPMENT

This tap should be made away from normal service equipment

Service-entrance conductors

Disconnect for fire pump, exit lights, or emergency service may consist of one to six fused switches or CBs

One to six disconnects for normal service in switchboard, panel, or separate enclosures

FIG. 369 Emergency disconnects must be isolated from the effects of destructive faults in normal service entrance disconnects.

service disconnects to be located remote from the other disconnecting means that are grouped in accordance with the basic rule—provided that the remote disconnect is used only to supply a water pump that is also intended to provide fire protection. In a residence or other building that gets its water supply from a well, spring, or lake, a remote disconnect for the water pump affords improved reliability of the water supply for fire suppression in the event that fire or other faults disable the normal service equipment. Remote location, and marking on the disconnect, will help to distinguish the water-pump disconnect from the other normal service disconnects, minimizing the chance that firefighters will unknowingly open the pump circuit when they routinely open service disconnects during a fire. This exception ties into the rule of Section 230-72(b), which requires (not simply permits) remote installation of a fire-pump disconnect switch that is permitted to be tapped ahead of the one to six switches or CBs that constitute the normal service disconnecting means (see Exception No. 5 to Section 230-82). The exception provides remote installation of a normal service disconnect when it is used for the same purpose (water pump used for firefighting) as the emergency service disconnect (fire pump) covered in Section 230-72(b). In both cases, remote installation of the pump disconnect isolates the criৄally important pump circuit from interruption or shutdown due to fire, arcing-fault burndown, or any other fault that might knock out the main (normal) service disconnects.

A wide variety of layouts can be made to satisfy the **NEC** permission for remote installation of a disconnect switch or CB serving as a normal service disconnect (one of a maximum of six) supplying a water pump. Figure 370 shows three typical arrangements that would provide the isolated fire-pump disconnect.

Disconnect location

The disconnecting means required for every set of service entrance conductors must be located at a readily accessible point nearest to the point at which the service conductors enter the building, on either the inside or the outside of the building (Fig. 371). The service disconnect switch (or circuit breaker) is generally placed on the inside of the building as near as possible to the point at which the conductors enter.

Although the **NEC** does not set maximum distance from the point of conductor entry to the service disconnect, various inspection agencies do set limits on this distance. For instance, service cable may not run within the building more than 18 in from its point of entry to the point at which it enters the disconnect. Or, service conductors in conduit must enter the disconnect within 10 ft of the point of entry. Or, as one agency requires, the disconnect must be within 10 ft of the point of entry, but overcurrent protection must be provided for the conductors right at the point at which they emerge from the wall into the building. The concern is to minimize the very real and proven potential hazard of having unprotected service conductors within the building. Faults in such unprotected service conductors must burn themselves clear, and such application has caused fires and fatalities.

Special design considerations apply to service disconnects for multiple-occupancy buildings, such as apartment houses, condominiums, town houses, office buildings, and shopping centers. The disconnect means for each unit in a multiple-occupancy building must be accessible to ocupants of that unit. For instance, the disconnect means for deenergizing the circuits in an apartment of an apartment building must be in the apartment (such as in a panel), in an accessible place in the hall, or in a place in the basement or outdoors where it can be reached.

Multiple-occupancy buildings having individual occupancy above the second floor must have service equipment grouped in a common accessible place, and the disconnect means may consist of not more than six switches (or six CBs) as in Fig. 372. Although specific provisions on that application have been deleted from the present **NEC**, such practice was recognized and required in previous editions, is not objectionable under the present **NEC**, and must be understood to be

① Utility pole

Total of not more than six disconnects

Water pump that also serves as fire-protection water supply

Meter →

One to five normal service disconnects **grouped** at main service location or in common enclosure

Single normal service disconnect installed remote from main location

② Metering from point of drop connection

Utility pole

Meter →

Two to five service disconnects, each fed by a separate set of SE conductors

③ One or more conduit risers for one or more laterals on pole

C. OR, remote disconnect could be fed by lateral circuit from main location

Utility pole

One to five service disconnects

A. One lateral to grouped disconnects or separate lateral to each disconnect

B. One lateral run directly to remote disconnect for water pump

FIG. 370 Remote installation is permitted for the service disconnect on a fire-pump circuit.

GENERAL CONCEPT

TYPICAL COMPLIANCE

FIG. 371 Service disconnect must be placed at the point where the service entrance conductors enter the building.

required by the present **NEC** because no proposal was made to delete the requirement. That there was no intention to remove the rule can be determined from the following sequence:

1. The above-described provision for multiple-occupancy buildings with individual occupancy above the second floor has been deleted from Section 230-72(d) and Exception No. 4 of Section 230-90. In the discussions and comments published by the **NEC** code-making panel on the change in these sections, a clear explanation is not given of the intent of the deletion. The original proposal that resulted in the deletion of all but the first sentence of Section 230-72(d) sought a new exception as follows:

   Exception: Subject to serving agency approval and where Section 230-2, Exception No. 5, conditions are met, more than one service location shall be permitted. The area served by each service shall be clearly defined.

2. The substantiation presented in favor of the addition of that rule made clear that the intent was to give an exception to the rule of the second sentence of the section, which said, "A multiple occupancy building having individual occupancy above the second floor shall have service equipment grouped in a common accessible location." Note that the use of the word "a" indicates *one* location. The exception sought recognition of grouping of service equipment in two or more common accessible locations for those multiple-occupancy buildings that have individual occupancy above the second floor but cover such a large property area that the utility company must provide two or more service drops or laterals spaced along the length

Such a building may have no more than six service disconnect switches-but may have any number of disconnect switches for individual occupancies depending upon the number of occupancies.

Apt. 3A 3B 3C
Apt. 2A 2B 2C — Feeders to occupancies
Apt. 1A 1B 1C — A single service entrance layout is used to feed one or more meter banks as required by size and layout of building.

Service equipment grouped in basement or outdoors

A. GROUPED WITH SINGLE MAIN DISCONNECT

One to each apt.

CBs or fused switches

Meters

One service switch or CB

Meter bank of nine meters and apartment disconnects

B. GROUPED WITH NOT MORE THAN SIX DISCONNECT DEVICES

Meter bank
No. 1
No. 2   Each bank
No. 3   has any
No. 4   number of apartment
No. 5   feeders
No. 6

Service conductors

Maximum {Six service switches (1-pole, 2-pole or 3-pole)}

FIG. 372   These basic rules must guide the design of services for multiple-occupancy buildings with individual occupancy above the second floor.

of the building, as provided by Exception No. 5 of Section 230-2. Thus, the use of a common location for the one to six disconnects fed by each drop or lateral required an exception to the rule requiring "a" (that is, one) common location for "not more than six switches or six circuit breakers" for any multiple-occupancy building with individual occupancy above the second floor.

3. Because the code-making panel voted to accept the idea, such a service layout is now acceptable to the **NEC**, as shown in Fig. 373. But deletion of the last three sentences that appear in Section 230-72(d) of the 1978 **NEC** cannot be taken to mean that there has been a change in the intent of the rules for locating service disconnects for multiple-occupancy buildings.

The phrase "individual occupancy" means any space such as an office or apartment that is independent of any other occupancy in the building. Generally, each such space is supplied through a separate meter. Each apartment intended for use as living quarters by one family is

Three utility services, overhead or underground,
as permitted by Section 230-2, Ex. No. 5
for "Buildings of Large Area "

FIG. 373 Buildings covering large areas may have more than
one service but must have service equipment grouped for each
service.

an individual occupancy. Each apartment might be supplied through a separate meter, or all
might be supplied through one meter.

Multiple-occupancy buildings that do not have individual occupancy above the second floor
may have service conductors run to each occupancy. The service disconnecting means in each
occupancy may then consist of not more than six switches (or six CBs) as shown in Fig. 374.
Example B in that illustration could be a hotel with stores on the lower two floors. The hotel
owner would represent an "occupant" who has occupancy on the lower floors and is the only
occupant of the upper floors. Such a building would not have "individual" occupancy above the
second floor.

Although specific permission for the use of separate sets of service conductors to each occupancy of a multiple-occupancy building that does not have individual occupancy above the second floor was deleted from Section 230-72 in the 1981 **NEC**, there is no indication that such
practice is objectionable. Some members of the **NEC** code-making panel warned that the word
"deletions" must not be construed as revoking permission to run separate sets of service entrance
conductors to an individual occupancy above the second floor—a practice that has long been
allowed by this **NEC** section. It was stated that:

> Separate service to individual occupants of multiple occupancy buildings has been permitted since
> 1933 with a good record of safety. The multiple services serve to reduce available fault current at
> the service location, and provide the consumer's ready access to the service equipment.

Three panel members objected to the deletion as unclear and confusing—and possibly inter-
pretable as permitting a separate set of service entrance conductors to be run to individual unit
occupancies in buildings with individual occupancy above the second floor.

Service drop is carried along top of building with service entrance conductors for each occupancy tapping the drop through a service head fitting. Or service can be made underground to a splice box or gutter.

All service conductors run on outside of building.

Could be a one-story building

Single common metering shown. Individual metering could be used.

Service entrance equipment in each occupancy may consist of up to 6 switches or CBs.

Occupant of unit 8 also occupies all floors above. Therefore building does not have individual occupancy above second floor.

Main set of underground service conductors.

Any number of meters grouped in outdoor bank.

Etc. floor
4th floor
3rd floor
2nd floor
1st floor

Service equipment in each unit may consist of up to 6 switches or 6 CBs.

Sub-sets of service conductors tapped from main service conductors, per Sec. 230-2, Excep. 3(b), run on outside of building.

FIG. 374 Buildings without individual occupancy above the second floor may have any number of sets of service entrance conductors and disconnects.

Because no objection or complaint was ever raised about the rules of Section 230-72(d) as they apply to buildings without individual occupancy above the second floor, it must be assumed that running service entrance conductors to individual occupancies in such buildings—which has long been permitted—is still completely acceptable.

In buildings with occupancy above the second floor, it was never acceptable to the **NEC** to run separate sets of service entrance conductors to individual occupancies, and such practice was not the letter or intent of any proposals for the 1981 **NEC**. In addition, such practice would be extremely poor design and completely grotesque in, say, a 10-story apartment house.

It is clear that the only intended change is that described in Fig. 373 to permit grouping of disconnects at more than "a" common location for multiple-occupancy buildings over two floors high. Where there is individual occupancy above the second floor, the disconnecting means must be located in one or more commonly accessible places and must not consist of more than six normal service disconnects at each location, as explained in Figs. 372 and 373.

It should be noted that the requirement for access to the disconnect means for each occupant would be modified where the building was under the management of a building superintendent or the equivalent and where electrical service and maintenance were furnished. [See the Exception to Section 240-24(b).]

To comply with the requirement that the disconnect be located where conductors enter a building, the conductors should either be run on the outside of the building to each occupancy or, if run inside the building, be encased in 2 in of concrete or masonry in accordance with Section 230-44. In either case the service equipment should be located "nearest to the entrance of the conductors inside the buiding," and each occupant should have "access to his disconnecting means," which is satisfied by CBs or switches in the occupant's premises.

Any desired number of sets of service entrance conductors may be tapped from the service drop or lateral, or two or more subsets of service entrance conductors may be tapped from a single set of main service conductors.

Type of disconnect

Any switch or CB used for service disconnect must be manually operable. In addition to manual operation, a switch may have provision for electrical operation—such as for remote control of the switch—provided it can be manually operated to the OPEN or OFF position (Fig. 375). But an electrically operated switch does not have to provide manual closing. An electrically operated breaker with a mechanical trip button that will open the breaker even if the supply power is dead is suitable for use as a service disconnect. The manually operated trip button assures that the breaker "can be opened by hand." **NEC** Section 240-80 requires circuit breakers to be capable of being closed and opened by hand. To provide manual closing of electrically operated circuit breakers, manufacturers provide emergency manual handles as standard accessories. Thus, such breaker mechanisms can be both closed and opened manually if operating power is not available.

Local requirements on the use of electrically operated service disconnects should be carefully included in service entrance design involving that type of equipment. If a switch can be opened and closed without exposing the operator to contact with live parts, it is an externally operable switch, even though access to the switch handle requires opening the door of a cabinet. Electrically operated switches and circuit breakers are required to be externally operable only to the open position; they are not required to be externally operable by hand to the closed position.

Switch for service disconnect must be manually operable by a direct-connected handle external to the enclosure.

To operating coil in switch or CB

Pushbutton station-in or outside service entrance room

Remote operation through an electro-magnetic control circuit may be used in addition to the required manual operation to the open position.

Note: Some local codes require both manual and electrical means of operation.

FIG. 375 Service disconnect *must* be manually operable, but may also be electric-power operable.

Disconnect rating

In general, the service disconnecting means must have a rating that is not below 60A; this rule is applicable to both fusible switches and CBs. However, 100 A is the minimum rating for a single switch or CB service disconnect for any "one-family dwelling" with an initial load of 10 kW or more, or where the initial installation contains more than five 2-wire branch circuits. It should be noted that, because the definition of a one-family dwelling is given in **NEC** Article 100 as "a building consisting solely of one dwelling unit," that rule applies to one-family houses only; it does not apply to apartments or similar dwelling units that are in two-family or multi-family dwellings.

If the demand of a total connected load, as calculated according to **NEC** Article 220, is 10 kW or more, a 100-A service disconnect, as well as 100-A service entrance conductors, must be used. Any one-family house with an electric range rated 8¾ kW or more must have a disconnect (or service equipment) rated at 100 A; such a range provides a demand load 8 kW, the two required 20-A kitchen-appliance circuits provide a demand load of 3 kW at 100 percent demand (from Table 220-11), and the sum of the two loads exceeds the 10-kW level at which the minimum 100-A service is required.

If 100-A service is used, the demand load may be as high as 23 kW. By using the optional service calculations of Table 220-30, a 23-kW demand load is obtained from a connected load of 42.5 kW. This shows the effect of diversity on large-capacity installations.

Where two to six disconnects are used for a service instead of a single main service disconnect switch or CB, another basic **NEC** rule applies. Figure 376 shows that the total rating of the multiple service disconnects (two to six) must be at least equal to the rating of the single service disconnect that would be required if a single disconnect were used instead of multiple disconnects. It should be noted that the sum of the CB ratings in Fig. 376 is well above 400 A, but it does comply with the rule of **NEC** Section 230-80, even though the 400-A service entrance conductors could be heavily overloaded. Exception No. 3 of Section 230-90(a) exempts this type of layout from the need to protect the conductors at their rated ampacity, as required in the basic

From Art. 220 (Secs. 220-10 through 220-21), calculation of demand load indicated that a single disconnect for this service must be rated at least 400 amps. The rating of multiple disconnects must total at least that value.

FIG. 376   Sum of the ratings of multiple service disconnects must *at least* be equal to the minimum required rating of a single disconnect.

rule. The 400-A rating of the service entrance conductors must be based on the minimum calculations set forth in Article 220 and is considered by the **NEC** to be safe because it is adequate
for the maximum sum of the demand loads fed by the five disconnects shown in the layout.

The continuous current rating of a service (or another) disconnect must be sufficient for full
load current, and its momentary current rating must be such that the device can withstand possible short-circuit currents for the length of time it takes for the circuit to be cleared. The interrupting capacity of a disconnect must be sufficient for any possible conditions of operation.
Requirements for interrupting loads and overloads must be related to the time-current characteristics of service protection devices. On primary circuits, the disconnect should usually be able
to break full load current up to its own rating. On secondary circuits, a switch may need an
interrupting capability ranging anywhere from its own current rating up to more than 10 times
its current rating.

Multibuilding premises

In a group of buildings or other structures under single management, disconnect means must be
provided for each building. Each building or other structure in the group must always be provided with a readily accessible means, installed either inside or outside the building or other
structure, at a location nearest the point of entrance of the conductors, for disconnecting all
ungrounded conductors from the supply (Fig. 377). The disconnect means required for each
outbuilding must be suitable for use as service equipment. The **NEC** rule on the provision of
disconnect means for other buildings is usually interpreted to permit the disconnect for each
building or structure to be the same kind as permitted for a service disconnect—that is, up to
six switches or CBs.

The wording of that **NEC** rule (in Section 230-84) has often caused confusion in applying the
rule. Some authorities had allowed use of the feeder switch in the main building as the only
disconnect for each feeder to the outlying buildings, provided the switches in the main building
were accessible to the occupants of the outlying buildings. But in the present **NEC**, the effect of
the last sentence of Section 230-84(a) [requiring location of the disconnect in accordance with

FIG. 377 Each separate building must have its own disconnect for the supply conductors to it.

Sections 230-72(c) and (d)] is to clearly require that the disconnect for any outbuilding be located within or just outside each building "nearest the point of entrance of the service-entrance conductors."

Overcurrent Protection

Each ungrounded service entrance conductor must be protected by a short-circuit overcurrent device in series with the conductor. Although there are many situations where, of necessity, the rating or setting of the overcurrent device will be higher than the service-conductor ampacity, the basic rule starts with the concept that the overcurrent device must have a rating or setting that is not higher than the allowable current capacity of the conductor. The idea is to ensure that the overcurrent protection required in the service entrance equipment protects the service entrance conductors from "overload." Because the service overcurrent protection devices are in the service disconnect equipment, which is installed at the load end of the service entrance conductors, these devices cannot protect the service entrance conductors from a "fault" that occurs in the service entrance conductors on the line side of the protection; but these devices can protect them from overload if they are of the proper rating. Conductors on the load side of the service equipment are considered as feeders or branch circuits and are required by the **NEC** to be protected as described in Articles 210, 215, and 240.

Each ungrounded service entrance conductor must be protected by an overcurrent device in series with the conductor. Figure 378 shows that such protection is always required where a single circuit breaker or fused switch is used for service disconnect and protection. The overcurrent device must have a rating or setting not higher than the allowable current capacity of the conductor, with the exceptions noted in **NEC** Section 230-90:

Exception No. 1: If the service supplies one motor in addition to some other load (such as lighting and heating), the overcurrent device may be rated or set in accordance with the protection required for a branch circuit supplying the one motor (Section 430-52) plus the other load, as shown in Fig. 379 and calculated as follows:

1. Size of motor branch-circuit fuses: 300 percent of 28 A or 84 A. This requires a maximum fuse size of 90 A. Smaller fuses, such as the time-delay type, may be used.

No. 1 TW
110 amps
115/230 v

Meter

100 amp

60A 60A 30A 30A 30A

Range   Dryer   Ironer
12 kW   47 kW   1600 watts

Five 20 amp circuits –
two for lighting,
three for small appliances

Water
heater
45 kW

Service equipment with one overall 100-amp main disconnect and fuse. Current through service conductors limited to 100 amperes. Without a main disconnect and overcurrent device, current is not limited and current over 110 amps could flow. Sum of protective devices is 210 amps per hot leg.

NOTE: Service-entrance conductors must be selected with adequate ampacity for the calculated service demand load, from Secs. 220-10 through 220-21.

FIG. 378 Single main service protective device must normally protect service entrance conductors at their ampacity.

Service conductors:
No. 2 THW copper

③

Fuses rated
175 amps

200 amp serv sw for
115/230V service

④

Lighting load
10 amps/
circuit =
80 amps
at 115 volts

①

Disconnect
switch 5 H.P.

Motor control
and OL
protection

②

No. 8

M

5 H.P. 230 V
1 φ 28 amps

FIG. 379 Motor load permits the sizing of service entrance
protection above the conductor ampacity.

2. Size of service-entrance conductors: Must be adequate for a load of 125 percent of 28 A plus 80 A (lighting load), or 115 A.

3. Size of main fuses: 90 A (from item 1 above) plus 80 A, or 170 A. This requires a maximum fuse size of 175 A. Again, smaller fuses may and should be used where possible to improve the overload protection on the circuit conductors.

The use of 175-A fuses where the calculation calls for 170 A conforms to Exception No. 2 of Section 230-90—the next higher standard rating of fuse (Section 240-6). For motor branch circuits and feeders, Articles 220 and 430 permit the use of overcurrent devices having ratings or settings higher than the capacities of the conductors. Article 230 makes similar provisions for services where the service supplies a motor load or a combination of both motors and other loads.

If the service supplies two or more motors as well as some other load, then the overcurrent protection must be rated in accordance with the required protection for a feeder supplying several motors plus the other load (Section 430-63). Or, if the service supplies only a multimotor load (with no other load fed), then Section 430-62 sets the maximum permitted rating of service overcurrent protection.

Exception No. 3:   Not more than six CBs or six sets of fuses may serve as overcurrent protection for the service entrance conductors, even though the sum of the ratings of the overcurrent devices is in excess of the ampacity of the service conductors supplying the devices—as illustrated in Fig. 380. The grouping of single-pole CBs as multipole devices, as permitted for disconnect means, may also apply to overcurrent protection. And a "set" of fuses is all the fuses required to protect the ungrounded service entrance conductors.

This exception ties into Section 230-80. Service conductors may be sized for the total maximum demand load, with permitted demand factors from Table 220-11 applied. Then each of the two to six feeders fed by the service entrance conductors is also sized from Article 220, based on the load fed by each feeder. When those feeders are given overcurrent protection in accordance with their ampacities, it is frequently found that the sum of the ratings of the overcurrent devices is greater than the ampacity of the service entrance conductors, which were sized by applying the applicable demand factors to the total connected load of all the feeders. Exception No. 3 recognizes that possibility as acceptable, even though it departs from the rule in the first

For a demand load of 125 amps, SE conductors could be No. 1 THW copper (130 amps).

In this case, service conductors could be overloaded (up to 240 amps, if CB's here are 2-pole). If main overcurrent protection, rated at 125 amps, were installed at point "A", service conductors would be protected against any load in excess of the calculated demand.

Meter   Point "A"

Current-carrying capacity of service entrance conductors determined by demand load, calculated as described in Secs. 220–10 through 220–21.

6Ɵ) 6Ɵ) 3Ɵ) 3Ɵ) 3Ɵ) 3Ɵ)

This may be
① Group of six single-pole CB's or switches, or
② Group of six multi-pole CB's or switches, or
③ Group of more than six single pole CB's or switches serving multi-wire circuits and arranged as multi-pole devices by "handle ties" or other means to provide disconnect of all ungrounded conductors with no more than six operations of the hand

Code permits use of up to six circuit breakers or fused switches as service disconnect means and service ovecurrent protection. Or one unfused main switch at point "A" and six sets of fuses (for multi-wire circuits) may also satisfy code requirements on disconnect and protection

FIG. 380 With two to six service entrance protective devices, conductors could be overloaded.

sentence of Section 230-90(a). The assumption is that if the demand load for the service entrance conductors is correctly calculated, there will be no overloading of those conductors because the diversity of feeder loads (some loads ON and some OFF) will be adequate to limit the load on the service entrance conductors.

Suppose the load of a building, computed in accordance with Article 220, is 255 A, and conductors are selected at that ampacity (250-MCM THW copper is rated at 255 A). Under Exception No. 1 of Section 240-3, 300-A fuses or a 300-A CB may be considered as overcurrent protection of the proper size for service conductors rated between 255 and 300 A if a single service disconnect is used. If the load were separated in such a manner that six 70-A CBs could be used instead of a single service disconnect means, then the total rating of the CBs would be 6×70 A, or 420 A, which is greater than the ampacity of the service entrance conductors. And that would be acceptable to the **NEC**, although heavy overload is possible.

NEC Section 230-90 covers service overcurrent protection; the requirement on subdivision of the protection for multiple-occupancy buildings is covered in Exception No. 4, which must be correlated to the previously discussed rules on service disconnect as set forth in Section 230-72(d). As was thoroughly discussed with relation to service disconnects above, the deletion of all but the first sentence of Exception No. 4 in the 1981 **NEC** was not a response to objections to the conditions of that rule. A proposal to rephrase Exception No. 3 of Section 230-90(a)—an entirely different consideration—somehow led to the radical deletion from Exception No. 4 of the rules on overcurrent protection for service layouts at multiple-occupancy buildings. The only logical assumption is that what was permitted by the wording of Exception No. 4 of Section 230-90(a) in the 1978 and previous editions of the **NEC** is still a completely acceptable concept, one that has been used widely and safely for over 40 years.

Figure 381 shows a very important design consideration. The **NEC** makes it mandatory to oversize the protection for the fire-pump circuit. That rule is intended to prevent opening of the fire-pump circuit on any overload up to and including stalling or even seizing of the pump motor. Because the conductors are "outside the building," operating overload is no hazard; and, under

If the service conductors to the fire-pump
room enter the fire-pump service equipment
directly from the outside or if they are
encased in 2-in.-thick concrete . . .

. . . they are judged to be
"outside of the building,"
and . . .

Fire pump
equipment

Fire pump
service
equipment

. . . the overcurrent protective device (fuses
or CB) must be rated or set to carry the
motor locked-rotor current indefinitely.

FIG. 381   Overcurrent protection for a fire pump must permit overload up to
locked-rotor current.

fire conditions, the pump must have no prohibition on its operation. It is better to lose the motor
than to attempt to protect it against overload when it is needed.

The basic design considerations on overcurrent protection for conductors feeding from one
building to other buildings are covered in Fig. 377, which also illustrates the requirements on
disconnect means for each feeder to a separate building. The rule of **NEC** Section 230-90(c)
permits the overcurrent protection for a feeder from one building to another to be at the supply
end of the circuit, in the building where the circuit originates, provided it is accessible to the
occupants of the building being fed. Because a switch or CB must be used as the disconnect in
each building being fed, a CB would constitute protection, and the use of fuses in a fusible switch
would also eliminate concern about access to protection in the building where the feeder originates.

Ground-Fault Protection

NEC Section 230-95 requires ground-fault protection equipment to be provided for each service
disconnecting means rated 1000 A or more in a solidly grounded-wye electrical service that
operates with its ungrounded legs at more than 150 V to ground (Fig. 382). This applies to the
rating of the disconnect, and not to the rating of the overcurrent devices or to the capacity of the
service entrance conductors.

The wording of the **NEC** rules makes clear that service ground-fault protection (GFP) is
required only for grounded-wye systems that have voltages over 150 V to ground. In effect, that
means the rule applies only to 480/277-V grounded-wye systems and not to 208/120-V systems
or any other commonly used systems. And GFP is not required on any systems operating over
600 V, although it is generally good design practice to include such protection.

A detailed discussion of ground-fault protection is given in Section 4 of this handbook. As
noted there, the basic rules on ground-fault protection at a service entrance are as follows:

1. In a typical service GFP hookup, a ground-fault current of 1200 A or more must cause the
   disconnect to open all ungrounded conductors. Thus the maximum permitted GFP pickup
   setting is 1200 A, although the device may be set lower.

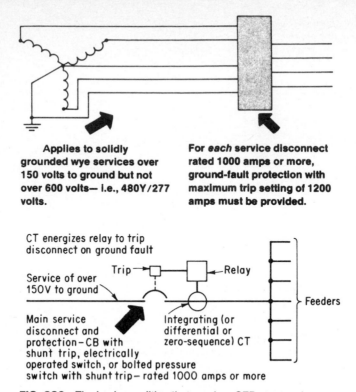

Applies to solidly grounded wye services over 150 volts to ground but not over 600 volts— i.e., 480Y/277 volts.

For *each* service disconnect rated 1000 amps or more, ground-fault protection with maximum trip setting of 1200 amps must be provided.

CT energizes relay to trip disconnect on ground fault

Trip → Relay

Service of over 150V to ground

Main service disconnect and protection – CB with shunt trip, electrically operated switch, or bolted pressure switch with shunt trip – rated 1000 amps or more

Integrating (or differential or zero-sequence) CT

Feeders

FIG. 382   The basic condition that requires GFP at a service.

2. The rule requiring GFP for any service disconnect rated 1000 A or more (on 480/277-V services) specifies a maximum time delay of 1 s for ground-fault currents of 3000 A or more. The maximum permitted setting of a service GFP hookup is 1200 A, but the time-current trip characteristic of the relay must ensure opening of the disconnect in not more than 1 s for any ground-fault current of 3000 A or more. This is intended to establish a specific level of protection in GFP equipment by setting a maximum limit of I^2t for fault energy.

3. Selective coordination between GFP and conventional protective devices (fuses and CBs) on service and feeder circuits is a very clear and specific task as a result of the wording of Section 230-95(a) that calls for a maximum time delay of 1 s at any ground-fault current of 3000 A or more.

Additional considerations for service GFP are as follows:

1. In applying the rule of Section 230-95, the rating of any service disconnect means is to be determined as shown in Fig. 383.

2. Because the rule on required service GFP applies to the rating of each service disconnect, in many instances GFP would be required if a single service main disconnect were used but not if the service subdivision option of using up to six service entrance disconnects were taken, as shown in Fig. 384.

3. Continuous industrial process operations are exempted from the GFP rules of Section 230-95(a) where the electrical system is under the supervision of qualified persons who will effect orderly shutdown of the system and thereby avoid the hazards, greater than ground fault itself, that would result from the nonorderly, automatic interruption that GFP would produce

FUSED SWITCH (bolted pressure switch, service protector, etc.)

Rating of switch is taken as...   ...the amp rating of the largest
                                       fuse that can be installed in the
                                       switch fuseholders.

EXAMPLE

If 900-amp fuses are used in this service switch, ground-fault protection would be required, because the switch can take fuses rated 1200 amps—which is above the 1000-amp level at which GFP becomes mandatory.

CIRCUIT BREAKER

Rating of breaker   ...the maximum continuous current rating
is taken as...   (pickup of long time-delay) for which the
                               trip device in the breaker is set or can
                               be adjusted.

Example: GFP would be required for a service CB with, say, an 800-amp trip setting if the CB had a trip device that can be adjusted to 1000 amps or more.

FIG. 383   Service-disconnect rating must be carefully determined.

480/277V
1200A service

Single 1200A disconnect
requires GF protection

480/277V
1200A service

Three 400A disconnects
do not require GF protection

FIG. 384   Subdividing the service disconnect to eliminate any 1000-A (or higher) disconnects evades mandatory GFP at the service.

in the supply to such critical continuous operations. The GFP requirements are not applicable where a nonorderly shutdown will introduce additional or increased hazards. The idea behind that is to provide maximum protection against service outage for industrial processes. With highly trained personnel at such locations, design and maintenance of the electrical system can often accomplish safety objectives more readily without GFP on the service. Electrical design can include provisions to minimize any danger to personnel resulting from loss of process power while also protecting electrical equipment.

4. Another exception in Section 230-95 excludes fire-pump service disconnects from the basic rule that requires ground-fault protection on any service disconnect rated 1000 A or more on a grounded-wye 480/277-V system. Because fire pumps are required by Exception No. 5 of Section 230-90 to have overcurrent protection devices large enough to permit locked-rotor current of the pump motor to flow without interruption, larger fire pumps (100 hp and more) would have disconnects rated 1000 A or more. Without the exception, those fire-pump disconnects would be subject to the basic rule and would have to be equipped with ground-fault protection. GFP protection on any fire pump is objectionable for the same reason that leads to Exception No. 5 of Section 230-90, which permits locked-rotor current to flow continuously. The intent is to give the pump motor every chance to operate during a fire—to prevent opening of the motor circuit on an overload up to and including stalling or seizing of the shaft or bearings. For the same reason, Section 430-31 exempts fire pumps from the need for overload protection, and Exception No. 4 to Section 430-72 requires overcurrent protection to be omitted from the control circuit of a starter for a fire pump. And it should be noted that Exception No. 2 in Section 230-95 says that ground-fault protection "shall not apply" to fire-pump motors—which appears to make omission of GFP mandatory, as shown in Fig. 385.

5. Obviously, the selection of ground-fault equipment for a given installation merits detailed study. The option of subdividing services, so that up to six service entrance disconnects are fed by one service drop or lateral, should be evaluated. A 4000-A service, for example, could be divided using five 800-A disconnecting means, and in such cases GFP would not be required.

6. One very important consideration that should be studied is the potential desensitizing of ground-fault sensing hookups when an emergency generator and transfer switch are provided in conjunction with the normal service to a building. The problem arises in those cases where a solid neutral connection from the normal service is made to the neutral of the generator through a three-pole transfer switch. With the neutral both grounded at the normal service and bonded to the generator frame, ground-fault current on the load side of the transfer switch can return over two paths, one of which will escape detection by the GFP sensor, as shown in Fig. 386. Such a hookup may also cause nuisance tripping of the GFP due to

FIG. 385 Fire-pump service disconnects of 1000 A or more must not be equipped with ground-fault protection.

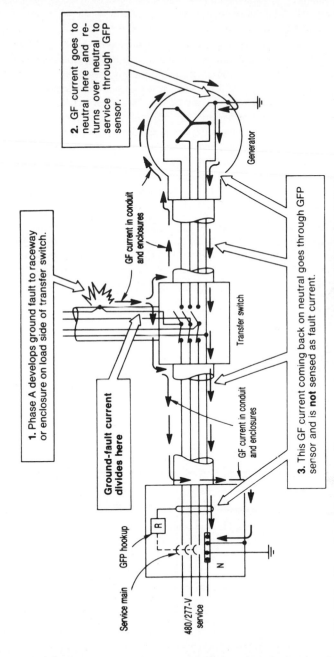

2. GF current goes to neutral here and returns over neutral to service through GFP sensor.

1. Phase A develops ground fault to raceway or enclosure on load side of transfer switch.

GF current in conduit and enclosures

Ground-fault current divides here

GF current in conduit and enclosures

Transfer switch

Generator

3. This GF current coming back on neutral goes through GFP sensor and is **not** sensed as fault current.

Service main

GFP hookup

R

N

480/277-V service

FIG. 386 Alternative fault-current paths can cause improper operation of GFP sensing devices.

normal neutral current. Under normal (nonfault) conditions, the neutral current due to normal load unbalance on the phase legs can divide at the common neutral connection in the transfer switch, with some current flowing toward the generator and returning to the service main on the conduit—and indicating falsely that a ground fault exists and causing nuisance tripping of the GFP. A four-pole neutral-switched transfer switch will eliminate this problem and ensure proper and effective operation of the GFP hookup, as shown in Fig. 387.

7. Although item 6 above is concerned with the desensitizing of service GFP where connection is made from an emergency generator through an automatic transfer switch, the whole question of using a GFP hookup on the generator-output disconnect means must be considered. Because a disconnect for the generator output constitutes a "service disconnect," such a unit rated 1000 A or more would be required by the wording of Section 230-95 to have GFP. But, recognizing that the reliability and continuity of an emergency supply is a design consideration that should be left to the judgment of the designer, **NEC** Section 700-26 states that the emergency generator disconnect does not require ground-fault protection. This rule clarifies the relationship between Section 230-95 and disconnect means for emergency generators. To afford the highest reliability and continuity for an emergency power supply, the **NEC** does not require GFP on any disconnect for an emergency generator, although it may be used if desired (Fig. 388).

  <u>VERY IMPORTANT:</u>   Because there have been many reports of improper and/or unsafe operation (or failure to operate) of ground-fault protective hookups, Section 230-95(c) requires (a mandatory rule) that every GFP hookup be "performance tested when first installed." This rule requires that such testing be done according to "approved instructions . . . provided with the equipment." A written record must be made of the test and made available to the inspection authority.

Service Grounding

The basic grounding concepts are presented under "System and equipment grounding" in Section 4 of this handbook. Every service entrance must have a grounding-electrode connection—that is, grounding of an intentionally grounded circuit conductor (such as a neutral conductor)

FIG. 387 Transfer switch with a switched pole for the neutral prevents the occurrence of two return paths for neutral current.

GFP Trip   GFP Relay

Transfer switch

Emergency generator

GFP Sensor

LOAD

NORMAL SUPPLY

NOTE: On the output of "optional standby" generator, a disconnect rated 1000 A or more on a 480/277-V grounded wye system *must* be equipped with GFP.

FIG. 388 Ground-fault protection is optional for the disconnect means on the output of an emergency or legally required standby generator.

and interconnection and grounding of the metal enclosures of the electrical system, or simply grounding of the metal enclosures where the system does not include an intentionally grounded circuit conductor. In all such cases, a major design consideration is the selection and layout of conductors and electrodes to provide the grounding connection to earth.

A grounding-electrode arrangement is required at the service entrance of a premises or in a building or other structure fed from a service in another building or other structure, as described under "Grounding at separate buildings" in Section 4. But the methods and requirements involved here do not apply to the grounding of a separately derived system, such as a local step-down transformer, which is covered under "Transformer grounding" in Section 6.

To satisfy the **NEC**, it is now necessary to provide a suitable "grounding electrode system," and not simply a "grounding electrode" as required by previous **NEC** editions. Several considerations are involved in this design task.

Until the 1978 **NEC**, the "water-pipe" electrode was the premier electrode for service grounding, and "other electrodes" or "made electrodes" were acceptable only "where a water system (electrode) . . . is not available." If a metal water pipe to a building had at least 10 ft of its length buried in the ground, that pipe had to be used as the grounding electrode, and no other electrode was required. The underground water pipe was the preferred electrode and the best electrode.

Of all the electrodes previously and still recognized by the **NEC**, the water pipe is the least acceptable to the present **NEC** and the only one that may never be used by itself as the sole electrode. It must always be supplemented by at least one "additional" grounding electrode. Any of the other grounding electrodes recognized by the **NEC** is acceptable as the sole grounding electrode, by itself.

Consider a typical water supply of 12-in-diameter metal pipe running, say, 400 ft underground to a building with a 4000-A service. According to Section 250-81(a), that water pipe, connected by a 3/0 copper conductor to the bonded service-equipment neutral, must be used as a grounding electrode; but it may not serve as the only grounding electrode. It must be supplemented by one of the other electrodes from Section 250-81 or 250-83. So the installation can be made acceptable by, say, running a No. 6 copper grounding-electrode conductor from the bonded service neutral to an 8-ft, ½-in-diameter ground rod. Although that seems like using a mouse to help an elephant pull a load, it is the literal requirement. And if the same building did not have 10 ft of metal water pipe in the ground, the 8-ft ground rod would be entirely acceptable as the only electrode.

The basic **NEC** rule on grounding-electrode systems requires that all or any of the electrodes specified in Sections 250-81(a) to (d), if available on the premises, must be bonded together to form a "grounding electrode system":

1. If there is at least a 10-ft length of underground metal water pipe, a grounding-electrode conductor must be connected to the water pipe.
2. If, in addition, the building has a metal frame that is "effectively grounded" (although that phrase is not defined or explained), the frame must be bonded to the water pipe—or vice versa, because the rules do not spell out where actual connections are to be made for grounding-electrode conductors and bonding jumpers.
3. Then, if there is at least a total of 20 ft of one or more ½-in-diameter steel reinforcing bars or rods embedded in the concrete footing or foundation, a bonding connection must be made from one of the other electrodes to one of the rebars—and, obviously, that has to be done before concrete is poured for the footing or foundation.

When two or more grounding electrodes are to be combined into a "grounding electrode system," the bonding jumper between pairs of electrodes must not be smaller than the size of grounding-electrode conductor indicated in Table 250-94 for the particular size of the largest phase leg of the service feeder, up to No. 3/0 copper, which is shown as the maximum required size of copper grounding-electrode conductor in Table 250-94.

The "unspliced" grounding-electrode conductor at the service may be connected to whichever one of the interbonded electrodes provides the most convenient and effective point of connection.

The grounding-electrode conductor must be sized as if it were the grounding-electrode conductor for whichever one of the interbonded electrodes requires the largest grounding-electrode conductor. If, for instance, a grounding-electrode system consists of a metal undergound water-pipe electrode bonded to a driven ground rod, the grounding-electrode conductor must be sized from Table 250-94; and on, say, a 2000-A service, it would have to be No. 3/0 copper or 250-MCM aluminum, whether it is connected to the water-pipe electrode (which would require that size of grounding-electrode conductor) or to the driven ground rod (which requires only a No. 6 copper or No. 4 aluminum grounding-electrode conductor if it is used by itself as a grounding electrode). And the bonding jumper between the water-pipe electrode and the ground rod would also have to be of that size.

The "20 feet of bare copper conductor" referred to in Sections 250-81(c) and (d) is going to be "available" at a building only if the 20-ft arrangement has been specified by the electrical designer, because it is clearly and only a grounding electrode; the other electrodes mentioned in Sections 250-81(a) to (c) (including the rebars) are specified by persons other than the electrical designer, and they may be "available." If a building has all or some of the electrodes described, they must all be bonded together and used as a single electrode system. If it has none, than any one of the electrodes described in Section 250-83 may be used for service grounding.

An electrode (such as a driven ground rod) that supplements an underground water-pipe electrode may be "bonded" to any one of several points in the service arrangement. It may be "bonded" (1) to the grounding-electrode conductor; or (2) to the grounded service conductor (grounded neutral), such as by connection to the neutral block or bus in the service panel or switchboard or in a current-transformer cabinet, meter socket, or other enclosure on the supply side of the service disconnect; and (3) to grounded metal service raceway; or (4) to interior metal water pipe at any convenient point. A bonding jumper must always be used around any water meter within a building and any place where piping on both sides of the meter is to be grounded.

The several requirements set by **NEC** rules and the conditions established for application of the rules can best be understood through a step-by-step approach relative to a typical installation:

Consider a building fed by an underground water-pipe system. At least 10 ft of metal water pipe is buried in the earth ahead of the point at which the metal pipe enters the building. As shown in Fig. 389, such a buried pipe is a grounding electrode, and connection must be made to the underground pipe with a grounding-electrode conductor sized from Table 250-94 and run from the grounding point in the service equipment. But now, a number of other factors must be accounted for, as follows:

1. Even though the water pipe is a suitable grounding electrode, at least one more grounding electrode must be provided and must be bonded to the water-pipe electrode. A water pipe, by itself, is not an adequate grounding electrode and must be supplemented by at least one other electrode to provide a "grounding electrode system."

2. The additional electrode may be:

   - The metal frame of the building, provided the frame is effectively grounded (embedded in earth and/or in buried concrete).

   - Or, a concrete-encased electrode within, and near the bottom of, a concrete foundation or footing in direct contact with the earth. The electrode must consist of at least 20 ft of one or more steel reinforcing bars or rods of not less than ½-in diameter, or it must consist of at least 20 ft of bare solid copper conductor not smaller than No. 4 AWG. Because the

GROUNDING ELECTRODE SYSTEM: BONDING JUMPERS, SIZED FROM TABLE 250-94, TIE WATER PIPE, BUILDING STEEL, AND ½-in.-DIA STEEL REINFORCING BARS TOGETHER.

Grounded metal frame of building

Metal water piping system in building

Metal water pipe extends at least 10 ft in earth

Water meter (A)

Service disconnect

Rebars in footing

Neutral bus in SE enclosure is bonded to enclosure

Water pipe **is** a grounding electrode but must **always** be supplemented by another electrode

Unspliced grounding-electrode conductor, sized from Table 250-94, must connect bonded service neutral to any **one** of the electrodes making up the grounding-electrode system

NOTE: At point "A", a bonding jumper **must** be used around the water meter and must be not smaller than the grounding-electrode conductor.

FIG. 389 Water pipe, metal building frame, and reinforcing in footing must be bonded into an "electrode system."

building footing or foundation in Fig. 389 contains such steel reinforcing, connection of a bonding conductor has to be made to the steel, and the conductor must be brought out for connection to the water pipe, the building steel, or the bonded service neutral.

■ Or, a "ground ring encircling the building or structure," buried directly in the earth at least 2½ ft down. The ground ring must be "at least 20 feet" of bare No. 2 or larger copper conductor. (In most cases, the conductor will have to be considerably longer than 20 ft to "encircle" the building or structure.)

■ Or, underground bare metal gas piping or other metal underground piping or tanks.

■ Or, a buried 8-ft ground rod or pipe or a plate electrode. An example of the bonding of a supplemental grounding electrode is shown in Fig. 390. And an interesting point about the bonding of the ground rod to the neutral bus in the service entrance enclosure relates to the size of the bonding jumper. Because the first paragraph of Section 250-81 requires that all the electrodes that make up a grounding-electrode system be "bonded together" by a jumper sized from Table 250-94, and because the last sentence of Section 250-81(a) requires the supplemental electrode to be "bonded" to the service neutral or other specified point, it seems clear that the conductor shown connecting the ground rod to the neutral bus in the sketch is a "bonding jumper" and must be the same size as the grounding-electrode

Grounding electrode conductor sized from Table 250-94 must connect bonded service neutral to clamp on water-pipe electrode, on either side of water meter, with same size bonding jumper around water meter.

Water pipe **is** an electrode but must be supplemented by another electrode

ELECTRODE THAT SUPPLEMENTS WATER-PIPE-ELECTRODE MAY BE BONDED TO SERVICE NEUTRAL.

Meter with jumper

More than 10 ft of metal pipe in earth

Metal water piping in building

Neutral bus in SE enclosure bonded to enclosure

Because building does not have grounded metal frame or either of the electrodes described in Sections 250-81(c) and (d), the next-to-last sentence of Section 250-81(a) requires that one of the electrodes of Section 250-83 be used to supplement the water-pipe electrode—such as a driven ground rod.

FIG. 390 An additional electrode must always be used to supplement a water-pipe electrode.

conductor (from Table 250-94). Because that conductor is not described as a "grounding electrode conductor," use of No. 6 copper or No. 4 aluminum—as recognized by Exception No. 1 of Section 250-94—would appear to be a violation. That concept is supported by the last sentence of the first paragraph of Section 250-81.

Section 250-81(c) states that the encased grounding electrode may be 20 ft of bare No. 4 copper—and it may be stranded. It was required to be "solid" copper conductor in the 1978 **NEC**.

3. According to the wording of Section 250-81, any of the four types of grounding electrodes mentioned in parts (a) to (d) of that section, if present, must be bonded together. Note that the rule does not state that any of those electrodes must be provided. But if any or all of them *are* present, they must be bonded together to form a "grounding electrode system." And, where a water-pipe electrode, as described, is present, any one of the other three electrodes in Section 250-81 may be used as the required "additional electrode."

In the case of a building fed by a nonmetallic underground piping system or one that does not have 10 ft of metal pipe underground, the water-pipe system is not a grounding electrode; however, the interior metal water-pipe system must be bonded to the service grounding. But when the water pipe is not a grounding electrode, another type of electrode must be provided to accomplish the service grounding. Any one of the other three electrodes of Section 250-81 may be used as the required grounding electrode. For instance, if the metal frame of the building is effectively grounded, a grounding-electrode conductor, sized from Table 250-94 and run from the bonded service neutral or ground terminal to the building frame, may satisfy the **NEC**, as shown in Fig. 391. Where none of the four electrodes described in Section 250-81 is present, one of the electrodes from Section 250-83 must be used (e.g., a ground rod, as shown in Fig. 392). Note, again, that any type of electrode other than a water-pipe electrode may be used by itself as the sole electrode.

As shown in Fig. 392, the wording of Exception No. 1 of Section 250-94 would permit the use of a No. 6 copper or No. 4 aluminum grounding-electrode conductor. Because the ground rod is the only grounding electrode, there is no relationship to the rules given in the last sentence of Section 250-81 and the last sentence of Section 250-81(a), which seem to require a larger "bonding jumper" in cases where a ground rod is a supplemental electrode, as shown in Fig. 390. Actually, it is more consistent with long-time **NEC** practice, based on tests, that the ground rod shown in Fig. 390 be connected to the neutral terminal or to the water pipe by a conductor that is not required to be larger than No. 6 copper or No. 4 aluminum.

The four illustrations shown in this discussion are only typical examples of the many specific ways in which **NEC** rules on grounding electrodes may be applied.

As noted previously, "continuity of the grounding path or the bonding connection to interior piping shall not rely on water meters." A bonding jumper always must be used around a water meter. This is required because of the chance of loss of grounding if the water meter is removed or replaced with a nonmetallic water meter. The bonding jumper around a meter must be at least the same size as the grounding-electrode conductor required for the given size of service conductors.

The concrete-encased electrode that is described in Section 250-81(c), known as the "Ufer system," has particular merit in new construction, where the bare copper conductor or steel reinforcing bar or rod can be readily installed in a foundation or footing form before concrete is poured. Electrodes of this type using a bare copper conductor have been installed as far back as 1940, and tests have proved the system to be highly effective (Fig. 393).

The intent of the words "bottom of a concrete foundation" is to completely encase the electrode within the concrete, in the footing near the bottom. The footing must be in direct contact

No. 3/0 bare standard copper ground cable buried in concrete footing and slab

Concrete footing

Steel piling

Ground rod 3/4" Copperweld 10'-0" long

Ground connector bolted to web of column

Ground connector

Elevation

Structural column

Plan

Exterior wall in building

Ground rod

Copper ground cables connected to adjacent columns in structure

ONE METHOD of grounding building structural members to ground cable system.

Section 250-81(b) recognizes the grounded metal building frame as a suitable grounding electrode by itself, connected to bonded service neutral by conductor sized from Table 250-94.

Service disconnect

Bonded neutral bus

Required bond around water meter

Metal piping

Grounded building metal

Nonmetallic water pipe

Metal water pipe extends **less** than 10 feet in earth and, therefore, it is **not** a grounding electrode.

Section 250-80(a) requires interior metal water piping to be bonded to service grounding, such as by bonding to the grounding electrode (the building frame in this case), with jumper sized from Table 250-94.

NOTE: Rebars in foundation could also serve as the only electrode if building did not have metal frame.

FIG. 391 Building's metal frame may be the sole grounding electrode.

Section 250-80(a) requires interior metal water piping to be bonded to service grounding, such as by bonding to grounded neutral bus.

Neutral bus in SE enclosure bonded to enclosure.

Interior metal water pipe

Nonmetallic water pipe

Meter with jumper

Metal water pipe is less than 10 feet in ground and does not qualify as electrode

Grounding electrode conductor connected to a made electrode need not be larger than No.6 copper or its equivalent ampacity

Because building does not have a grounded metal frame or either of the electrodes described in Section 250-81(c) and (d), Section 250-83 requires that at least one of the electrodes described in that Section be used — such as a ground rod.

Service panel

Neutral bar

No. 2 aluminum grounding conductor

No. 6 copper grounding conductor

Ground rod

Grade

Metal water piping

Plastic water piping

FIG. 392 For many systems, a driven ground rod may be the sole grounding electrode serving a building without other available electrodes.

BARE COPPER CONDUCTOR ... OR ...

... 1/2 IN. DIAMETER REBARS OR RODS

FIG. 393 The so-called "Ufer" system grounding electrode is especially effective.

with the earth, which means that dry gravel or polyethylene sheets between the footing and the earth are not permitted.

It may be advisable to provide additional corrosion protection in the form of plastic tubing or sheath at the point where the grounding electrode leaves the concrete foundation.

For concrete-encased steel reinforcing bar or rod systems used as grounding electrodes in underground footings or foundations, welded-type connections (metal-fusing methods) may be used for connections that are encased in concrete. Compression or other types of mechanical connectors may also be used.

"Made" electrodes

As a general rule, if a water-pipe system or other approved electrode is not available, a driven rod or pipe is used as the electrode. A rod or pipe driven into the ground does not always provide as low a ground resistance as is desirable, particularly where the soil is very dry. In some cases where several buildings are supplied, grounding at each building reduces the ground resistance.

Where it is necessary to bury more than one pipe or rod to lower the resistance to ground, they must be placed at least 6 ft apart. If they were placed closer together, there would be little improvement.

Where two driven or buried electrodes are used to ground two different systems that should be kept entirely separate from one another (such as a grounding electrode for a wiring system for light and power and a grounding electrode for a lightning rod), care must be taken to guard against the conditions of low resistance between the two electrodes and high resistance from each electrode to ground. If two driven rods or pipes are located 6 ft apart, the resistance between the two is sufficiently high and cannot be greatly increased by increasing the space. Thus, at least 6 ft of spacing is required between electrodes serving different systems.

Whenever a ground rod is used, it must be driven straight down into the earth, with at least 8 ft of its length in the ground (in contact with soil). If rock bottom is hit before the rod is 8 ft into the earth, it is permissible to drive it into the ground at an angle—not over 45° from the vertical—to ensure that at least 8 ft of its length is in the ground. However, if rock bottom is so shallow that it is not possible to get 8 ft of the rod in the earth at a 45° angle, then it is necessary to dig a 2½-ft-deep trench and lay the rod horizontally in the trench. Figure 394 shows these rules.

Another **NEC** requirement calls for the upper end of the rod to be flush with or below the ground level—unless the above-ground end and the conductor clamp are protected either by

FIG. 394 In a ground-rod installation, there must be at least 8 ft of rod in contact with soil.

locating them in a place where damage is unlikely or by placing a metal, wood, or plastic box or enclosure over the end.

This two-part rule was added to the **NEC** because it had become common practice to use an 8-ft ground rod driven, say, 6½ ft into the ground with the grounding-electrode conductor clamped to the top of the rod and run to the building. Not only is the connection subject to damage or disconnection by lawnmowers or vehicles, but also the length of unprotected, unsupported conductor from the rod to the building is a tripping hazard. The rule says, in effect, *bury everything or protect it!*

Of course, a buried conductor-clamp assembly that is flush with or below grade must be resistant to rusting or corrosion that might affect its integrity, and it must be UL-listed as suitable for direct earth burial. Such clamps are available.

In addition to ground rods, plate electrodes are a form of "made" (manufactured) electrodes. Such electrodes are listed in the UL *Electrical Construction Materials Directory* under the heading "Grounding and Bonding Equipment," which also covers bonding devices, ground clamps, grounding and bonding bushings, ground rods, armored grounding wire, protector grounding wire, grounding wedges, ground clips for securing the ground wire to an outlet box, water-meter shunts, and similar equipment. Only listed devices are acceptable for use. And listed equipment is suitable only for use with copper, unless it is marked "AL" and "CU."

When a made electrode is used for grounding, care must be taken to ensure that it provides an effective low-resistance path to the earth (or ground). A basic **NEC** rule calls for a made electrode to have a resistance to ground of 25 Ω or less. In any case where a single made electrode (rod, pipe, or plate) shows a resistance to ground over 25 Ω, the **NEC** simply requires that one additional made electrode be used in parallel; but there is then no need to make any measurement or add more electrodes or to be further concerned about the resistance to ground. In previous **NEC** editions, this rule implied that additional electrodes had to be used in parallel with the first one until a resistance of 25 Ω or less was obtained. Now, once the second electrode is added, it does not matter what the resistance to ground is, and there is no need for more electrodes, as shown in Fig. 395. Of course, good design practice and the specifications of many engineering organizations generally require a maximum resistance of 25 Ω.

When more than one made electrode is used for grounding, there must be at least 6-ft spacing between any pair of made electrodes. That minimum spacing is essential where more than one ground rod, pipe, or plate is connected to a single grounding-electrode conductor and the resistance of a single grounding electrode is over 25 Ω to ground. Even greater spacing is better for rods longer than 8 ft. Separation of the rods reduces their combined resistance to ground.

Insofar as made electrodes are concerned, wide variation in resistance is to be expected; the present requirements of the **NEC** concerning the use of such electrodes do not provide for a system that is in any way comparable to that which can be expected where good underground metallic piping can be utilized.

It is recognized that some types of soil may cause a high rate of corrosion, resulting in a need for periodic replacement of grounding electrodes. Also, intimate contact between two dissimilar metals such as iron and copper, under wet conditions, can lead to electrolytic corrosion.

When a cross occurs between a high-tension conductor and one of the low-tension secondary conductors, the electrode may be called upon to conduct a heavy current into the earth. The voltage drop in the ground connection, including the conductor leading to the electrode and the earth immediately around the electrode, will be equal to the current multiplied by the resistance. This results in a difference in potential between the grounded conductor of the wiring system and the ground. It is therefore important that the resistance be as low as practicable.

Where made electrodes are used for the grounding of interior wiring systems, resistance tests should be conducted on a sufficient number of electrodes to determine the conditions prevailing in each locality. The tests should be repeated several times a year to determine whether conditions have changed because of corrosion of the electrodes or drying out of the soil.

THE TWO RODS MUST BE
AT LEAST 6 FT APART

Driven rod electrodes
or other made electrodes

ONE ROD MEASURES LESS THAN 25 OHMS	ONE ROD MEASURES MORE THAN 25 OHMS	ONE ROD MEASURES OVER 25 OHMS. SECOND ROD IS USED; NO RESISTANCE READING TAKEN
OK	**NOT OK**	**OK**

FIG. 395   Resistance to earth of a "made" electrode must be measured.

Grounding connections

(Refer to the discussion of "Ground connections" in Section 4 of this handbook, and to Figs. 291 and 292.)

In addition to connection of the grounding-electrode conductor to the neutral terminal or the equipment grounding terminal as described in Section 4 above, another concern is bonding of the neutral terminal to the equipment grounding terminal at the service entrance.

The equipment grounding conductor at a service (such as the ground bus or terminal in the service equipment enclosure, or the enclosure itself) must be connected to the system grounded conductor (the neutral or grounded phase leg). The equipment ground and the neutral or other ground leg must be bonded together *on the supply side* of the service disconnecting means—which means either within or ahead of the enclosure for the service equipment, as shown in Fig. 396.

The ground bus or the enclosure must be simply bonded to the grounding-electrode conductor within or ahead of the service disconnect for an ungrounded system.

As shown in Fig. 397, some switchboard sections or interiors include neutral busbars that are factory-bonded to the switchboard enclosure and marked "Suitable for use only as service equipment." They may not be used as subdistribution switchboards; i.e., they may not be used on the load side of the service, except (1) where they are used, with the inspector's permission, as the first disconnecting means fed by a transformer secondary or a generator and where the bonded neutral satisfies Section 250-26(a) for a separately derived system, or (2) where they are used as the main disconnect for a building that is fed from another building (as described under "Grounding at separate buildings" in Section 4 of this handbook).

After service enclosures are bonded to equipment grounding buses and to the grounded neutral (or grounded phase leg) of a grounded system, a common connection to ground must be made for the interbonded components. All the bonded components—the service equipment enclosure, the grounded neutral or grounded phase leg, and any equipment grounding conductors that come into the service enclosure—must be connected to a common grounding electrode

Bonding is the insertion of a bonding screw into the panel neutral block to connect the block to the panel enclosure, or it is use of a bonding jumper from the neutral block to an equipment grounding block that is connected to the enclosure.

NOTE: Bonding—the connection of the neutral terminal to the enclosure or to the ground terminal that is, itself, connected to the enclosure—might also be done in an individual switch or CB enclosure.

Ground bus is and always *must be* bonded to the metal switchboard enclosure.

Bonding of the neutral is the connection between the neutral bus and the equipment grounding bus or between the neutral bus and the metal enclosure itself.

FIG. 396   Equipment ground terminal must be "bonded" to the grounded conductor at the service equipment.

with a single grounding-electrode conductor. A common grounding-electrode conductor must be run from the common point so obtained to the grounding electrode, as shown in Fig. 398.

Connection of the system neutral to the switchboard frame or ground bus within the switchboard provides the lowest impedance for the equipment ground return to the neutral. The main bonding jumper that bonds the service enclosure and equipment grounding conductors (which may be either conductors or conduit, EMT, etc.) to the grounded conductor of the system must be installed within the service equipment or within a service conductor enclosure on the line side of the service, as shown in Fig. 399. And it should be noted that in a service panel, equipment grounding conductors for load-side circuits may be connected to the neutral block, and there is no need for an equipment grounding terminal bar or block.

FIG. 397 Bonding of the service neutral may be accomplished by the equipment manufacturer.

FIG. 398 "Common" grounding-electrode conductor must be used for the service *grounded* conductor (neutral) and the equipment ground terminals.

Main bonding jumper (wire, bus, screw or similar conductor) must be *within* service equipment or service conductor enclosure

FIG. 399   Main service bonding jumper must be within the service equipment.

If one grounding conductor is used to ground the neutral to a water pipe or other grounding electrode, and a separate grounding conductor is used to ground the switchboard frame and housing to the water pipe or other electrode—without the neutral and the frame being connected together in the switchboard—then the length and impedance of the ground path are increased. The proven hazard is that the impedance of the fault-current path may then limit the fault current to a level too low to operate the overcurrent devices "protecting" the faulted circuit.

Note that a number of grounding electrodes that are bonded together are considered to be one grounding electrode.

The electrode or electrodes used to ground the neutral or other grounded conductor of an ac system must also be used to ground the entire system of interconnected raceways, boxes, and enclosures. The single common grounding-electrode conductor then connects to the single grounding electrode or electrode system and thereby grounds the bonded point of the system and equipment grounds.

In any building housing livestock, all piping systems, metal stanchions, drinking troughs, and other metalwork with which animals might come in contact must be bonded together and to the grounding electrode used to ground the wiring system in the building.

Grounding to neutral

Connection between a grounded neutral (or grounded phase leg) and equipment enclosures is permitted by the **NEC** for the purpose of grounding the enclosures to the grounded circuit conductor; however, such general permission is limited to the supply side of service equipment and the supply side of a main disconnect for a separate building. Figure 400 shows such applications. At *A*, the grounded service neutral is bonded to the meter housing by means of the bonded neutral terminal lug in the socket—and the housing is grounded by this connection to the grounded neutral, which itself is grounded at the service equipment as well as at the utility-transformer secondary supplying the service. At *B*, the service equipment enclosure is grounded by connection (bonding) to the grounded neutral—which itself is grounded at the meter socket and at the supply transformer. These same types of grounding connections may be made for current-transformer cabinets, auxiliary gutters, and other enclosures on the line side of the service entrance disconnect means, including the enclosure for the service disconnect. (In some

FIG. 400 Grounded circuit conductor may be used to ground equipment housings on the line (supply) side of the service disconnect.

areas, utilities and inspection departments will not permit the arrangement shown in Fig. 400 because the connecting lug in the meter housing is not always accessible for inspection and testing purposes.)

Aside from the above-described connections, the **NEC** generally prohibits connection between a grounded neutral and equipment enclosures on the load side of the service. So aside from the few specific exceptions, bonding between any system grounded conductor, neutral or phase leg, and equipment enclosures is prohibited on the load side of the service, as indicated in Fig. 401. The use of a neutral to ground a panelboard, or other equipment on the load side of service equipment would be extremely hazardous if the neutral became loosened or disconnected. In such cases any line-to-neutral load would energize all metal components connected to the neutral, creating a dangerous potential above ground. Hence, the prohibition of such practice.

Although the **NEC** prohibits neutral bonding on the load side of the service, it clearly requires such bonding at the service entrance. And the exceptions to the prohibition of load-side neutral bonding to enclosures are few and very specific:

- In a system, when voltage is stepped down by a transformer, a grounding connection must be made to the secondary neutral of a 208/120- or 240/120-V system, even though it is on the load side of the service.

FIG. 401 Neutral must *not* generally be bonded to equipment enclosures on the load side of the service.

- When a circuit is run from one building to another, it may be necessary or simply permissible to connect the system "grounded" conductor to a grounding electrode at the other building, as covered under "Grounding at separate buildings" in Section 4 of this handbook.

- The **NEC** permits grounding of meter enclosures to the grounded circuit conductor (generally, the grounded neutral) on the load side of the service disconnect if the meter enclosures are located near the service disconnect and the service is not equipped with ground-fault protection, as shown in Fig. 402. There is no definition for the word "near," but it can be taken to mean in the same room or general area. This rule applies, of course, to multioccupancy buildings (apartments, office buildings, etc.) with individual tenant metering.

If a meter bank is on the upper floor of a building, as in a high-rise apartment house, or otherwise away from the service disconnect, the meter enclosures would not satisfy the requirement that they be "near" the service disconnect. In such cases, the enclosures must not be grounded to the neutral. And, if the service has ground-fault protection, meter enclosures on the load side must not be connected to the neutral, even if they are "near" the service disconnect.

Main service disconnect without ground-fault protection

Disconnect and protection for each tenant feeder

Service

Meter sockets on load side may be grounded to grounded neutral

FIG. 402   Grounding of meter enclosures to the grounded conductor (neutral) on the load side of the service is limited.

Unused-neutral service

Whenever a service is derived from a grounded-neutral system, the grounded neutral conductor must be brought into the service entrance equipment, even if the grounded conductor is not needed for the load supplied by the service. This is required to provide a low-impedance ground-fault-current return path to the neutral to ensure operation of the overcurrent device, for the safety of personnel and property (see Fig. 403).

In such cases, the neutral functions strictly as an equipment grounding conductor, to provide a closed-circuit path back to the transformer for automatic circuit opening in the event of a phase-to-ground fault anywhere on the load side of the service equipment. If the neutral were not brought in as shown in Fig. 403, fault current would be forced to take the high-impedance path through the earth in flowing from the service grounding electrode to the transformer ground. Not enough current would flow back to the transformer neutral to cause the protective device to operate. (Only one phase leg is shown in these diagrams to simplify the concept. The other two phase legs have the same relation to the neutral.)

The same requirements apply to separate power and light services derived from a common 3-phase, 4-wire, grounded "red-leg" delta system. The neutral from the outdoor center-tapped

FIG. 403 Grounded neutral must be brought into service entrance equipment even if it is not used for any circuits to loads.

transformer winding must be brought into the 3-phase power-service equipment as well as into the lighting service, even though the neutral will not be used for power loads. (This is shown in Fig. 404.) Such an unused neutral must be at least equal in size to the required minimum size of grounding-electrode conductor specified in **NEC** Table 250-94 for the size of the phase conductors. In addition, if the phase legs associated with that neutral are larger than 1100 MCM, the grounded neutral must not be smaller than 12½ percent of the area of the largest phase conductor. An example of this requirement is shown in Fig. 405.

The minimum required size for the grounded neutral conductor from the supply transformer to the service is based on the size of the service phase conductors. In this case, the overall size of the service phase conductors is 4 × 500 MCM per phase leg, or 2000 MCM. Because that is larger than 1100 MCM, Table 250-94 may not be used in sizing the neutral. Instead, 2000 MCM must be multiplied by 12½ percent (0.125) to obtain 2000 MCM × 0.125 = 250 MCM, which is the minimum allowable size for the neutral conductor run from the transformer to the service equipment. But, because the service is made up in parallel—with four conduits, each containing three 500-MCMs (one for each of phase A, phase B, and phase C)—**NEC** Section 250-95 requires that a 250-MCM neutral be run in each of the four conduits. Because the

FIG. 404 Grounded neutral must be brought into both sets of service equipment.

480/277V grounded
secondary of utility
transformer

Service-entrance phase legs
made up of four 500MCM
conductors in parallel per leg

3-phase service
brought into bldg, but
system neutral is
not needed.

Service
equipment

Min. 250 MCM
neutral brought in and bonded to
service equipment enclosure (the
service ground bus or block)

3-phase, 3-wire distribution to power loads, with
local dry-type transformers stepping down
voltage for lighting

FIG. 405 Unused service neutral must be sized as an equipment grounding conductor.

neutral here is an equipment grounding conductor, the second paragraph of Section 250-95 calls for a full-size equipment grounding conductor in each conduit of this parallel makeup.

Pole-type service

As described at the beginning of this section, the basic approach to service grounding is to bond the neutral of a grounded service to the service equipment enclosure (as in a service panel or a service switchboard) and then run a grounding-electrode conductor from the service neutral bus or block to the common grounding electrode (or electrode system). But **NEC** Section 250-23(a) states a very important requirement concerning connection of the grounding-electrode conductor when a building is fed by service conductors from a meter on a yard pole.

The basic rule of Section 250-23 is that the grounding-electrode conductor required for the grounding of both the grounded service conductor (usually a grounded neutral) and the metal enclosure of the service equipment must be connected to the grounded service conductor within or on the supply side of the service disconnect. But the rule further requires that the grounding-electrode conductor be connected at some point between the load end of the service drop or lateral and the service equipment.

As a result of that requirement, if a service is fed to a building from a meter enclosure on a pole or other structure some distance away (as is commonly done on farm properties), and an overhead or underground run of service conductors is made to the service disconnect in the building, the grounding-electrode conductor will not satisfy the **NEC** if it is connected to the neutral in the meter enclosure at the pole; it must be connected at the load end of the underground or overhead service conductors. And, the connection should preferably be made within the service-disconnect enclosure.

This rule on grounding connection is shown in Fig. 406. If, instead of an underground lateral, an overhead run were made to the building from the pole, the overhead line would be a "service drop." The rule would likewise require the grounding connection to be at the load end

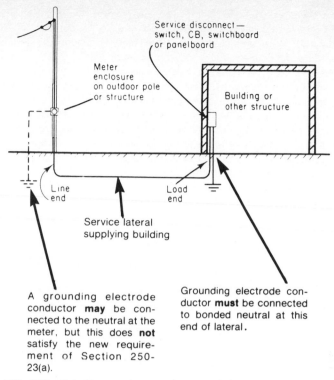

Service disconnect—
switch, CB, switchboard
or panelboard

Meter
enclosure
on outdoor pole
or structure

Building or
other structure

Line
end

Load
end

Service lateral
supplying building

A grounding electrode
conductor **may** be con-
nected to the neutral at the
meter, but this does **not**
satisfy the new require-
ment of Section 250-
23(a).

Grounding electrode con-
ductor **must** be connected
to bonded neutral at this
end of lateral.

FIG. 406 Pole-type service must be grounded at the load end of a
service drop or lateral to a building.

of the service drop. If a fused switch or CB were installed as service disconnect and protection
at the load side of the meter on the pole, then that would establish the service at that point, and
the grounding-electrode connection to the bonded neutral terminal would be required at that
point. The circuit from that point to the building would be a feeder and not service conductors.
But, electrical safety and effective operation would require that an equipment grounding conductor be run with the feeder circuit conductors to ground the interconnected system of conduits
and metal equipment enclosures along with metal piping systems and building steel within the
building. Or, if an equipment grounding conductor were not in the circuit from the pole to the
building, the neutral could be bonded to the main-disconnect enclosure in the building, and a
grounding-electrode connection could be made at that point also.

Grounding multiple service entrance disconnects

When the service disconnect means (one to six circuit breakers of fused switches) is contained
in a single enclosure—as with an individual main service switch or CB or a service panelboard
or switchboard—a single common grounding-electrode conductor is run from the bonded neutral
bus or block (of a grounded system) or from the equipment ground terminal (of an ungrounded
system) to the grounding electrode or electrode system. But, a different approach has to be followed where the service disconnect means consists of two to six switches or circuit breakers, each
in a separate enclosure. Two to six service disconnects might be fed by taps from an auxiliary
gutter that is supplied by a single set of service entrance conductors, or a separate set of service

entrance conductors may be run to each of the two to six switches or CBs, with the sets of service entrance conductors run down from a single service drop to the building (or fed by a single underground lateral). With such layouts of multiple service disconnects, the question arises, "Where should the grounding-electrode conductor be connected?"

Where a single service drop supplies a building using two to six service disconnecting means, each in a separate enclosure and each feeding a separate load, service grounding may be provided by a separate grounding-electrode conductor run from each disconnect to the water pipe or other grounding electrode, or a single grounding-electrode conductor may be used with a tap connected from each separate disconnect to the single grounding-electrode conductor (Fig. 407). Previous editions of the **NEC** required either that a separate grounding-electrode conductor be run from each enclosure to the grounding electrode or that a single, unspliced conductor be looped from enclosure to enclosure. A single grounding-electrode conductor used to ground all the service disconnects had to be without splice—run from one enclosure to the other and then to the water pipe or other grounding electrode. Exception No. 2 of Section 250-91(a) now says that splices may be made in the grounding-electrode conductor to tap into each of the service disconnect enclosures.

Exception No. 2 also covers the sizing of a main grounding-electrode conductor and taps from it to provide system grounding in each separate service enclosure where two to six service disconnects are used in separate enclosures. The wording of the exception requires the main grounding-electrode conductor to be sized from Table 250-94 but does not state the basis on which that table is to be used. The main grounding-electrode conductor might be sized for the sum of the cross-sectional areas of all the conductors connected to one hot leg of the service drop. Or, the main grounding-electrode conductor could be sized from Table 250-94 for the largest phase-leg conductor of one of the two to six sets of service entrance conductors. Or, the main grounding-electrode conductor could be sized from Table 250-94 on the basis of the size of one hot leg of a service of sufficient size to handle the demand load fed by the two to six service disconnects.

The wording of the rule does, however, make clear that the size of the grounding-electrode tap to each separate enclosure may be determined from Table 250-94 on the basis of the largest service hot leg serving each enclosure, as shown in Fig. 407. Although that illustration shows an overhead service to the layout, the two to six service disconnects could be fed by individual sets of underground conductors, making up a "single" service lateral as permitted by Exception No. 7 of Section 230-2.

Service "bonding"

To ensure the electrical continuity of the grounding circuit, "bonding" (special precautions to ensure a permanent, low-resistance connection) is required at all conduit connections in the service equipment and where any nonconductive coating exists which might impair such continuity. This includes bonding at connections between service raceways, service cable armor, and all service-equipment enclosures containing service entrance conductors, including meter fittings, boxes, and the like.

The need for effective grounding and bonding of service equipment arises from the electrical characteristics of utility-supply circuits. In the common arrangement, service conductors are run to a building, and the service overcurrent protection is placed near the point of entry of the conductors into the building, at the *load end* of the conductors. With such a layout, the service conductors are not protected against ground faults or shorts occurring on the supply side of the service overcurrent protection. Generally, the only protection for the service conductors is on the primary side of the utility's distribution transformer. By providing "bonded" connections (con-

WHERE THE LAYOUT IS LIKE THIS . . .

Metering from point of drop
connection or individual metering

Utility pole

Commercial or industrial
building with one or more
tenants or occupancies

Two to six service disconnects,
each fed by a separate set
of SE conductors (Sec. 230-45)

GROUNDING MAY BE DONE LIKE THIS . . .

Separate sets of
SE conductors

**Each tap to common
grounding electrode
conductor sized from
Table 250-94 based on
size of each set of SE
conductors to each
enclosure**

Splices

**Single, common grounding electrode conductor,
sized according to the size of a single set of SE
conductors that would handle the demand load on
the four disconnects**

. . . OR LIKE THIS

Service disconnects

**Separate grounding electrode conductor run from
each service disconnect enclosure to common
grounding electrode, with each conductor sized on
basis of service entrance conductors supplying
each enclosure**

FIG. 407   A choice of method is available for grounding
multiple service disconnects in separate enclosures.

necting with special attention to reliable conductivity), any ground fault in a metal service raceway or other metal enclosure for service entrance conductors is given the greatest chance of
burning itself clear—since there is not effective overcurrent protection ahead of those conductors
to provide opening of the circuit on such fault currents. For any contact between an energized
service conductor and grounded service raceway, fittings, or enclosures, bonding provides discharge of the fault current to the system grounding electrode—burning the fault clear. This
condition is shown in Fig. 408. Any fault to an enclosure of a hot service conductor of a grounded
electrical system must find a firm, continuous, low-impedance path to ground if there is to be

FIG. 408 Service bonding must ensure burn-clear on ground faults in service enclosures.

sufficient current flow either to operate the primary protective device or to burn the fault clear quickly. This means that all enclosures containing the service conductors—service raceway, cable armor, boxes, fittings, cabinets—must be effectively bonded together; that is, they must have low impedance through themselves and must be securely connected to each other to provide a continuous path of sufficient conductivity to the conductor that makes the connection to ground, as shown in Fig. 409.

Good engineering practice has long recognized that the conductivity of any equipment path should be at least equivalent to 25 percent of the conductivity of any phase conductor with which the ground path will act as a circuit conductor on a ground fault. Or, put another way, without

FIG. 409 Bonding is intended to ensure a low-impedance path through all service enclosures.

reference to insulation or temperature rise, the impedance of the ground path must not be greater than four times the impedance of any phase conductor with which it is associated. This figure should always be used as a specific guide in evaluating the suitability of raceways or conductors as a grounding path.

In ungrounded electrical systems, the same careful attention should be paid to the matter of bonding together the non-current-carrying metal parts of all enclosures containing service conductors. Such a low-impedance ground path will quickly and surely ground any hot conductor which might accidentally become common with the enclosure system.

NEC Section 230-63 requires that service raceways, metal sheaths of service cables, metering enclosures, and cabinets for service disconnect and protection be grounded. At one time, the **NEC** required that flexible metal conduit used as part of a run of service raceway be bonded around. But, the **NEC** no longer permits the use of flexible metal conduit or liquid-tight flex (Fig. 410).

As indicated in Fig. 411, service equipment that must have bonded connections includes (1) service raceway, cable trays, cable sheath, and cable armor; (2) all service-equipment enclosures containing service entrance conductors, including meter fittings, boxes, etc., interposed in the service raceway or armor; and (3) conduit or armor that encloses a grounding-electrode conductor that runs to and is connected to the grounding electrode or system of electrodes.

Bonded terminations must be used at the ends of conduit (rigid metal conduit, IMC, or EMT) or cable armor that encloses a grounding-electrode conductor. That means that conduit or cable armor must be connected with a bonding locknut or bonding bushing (with a bonding jumper around unpunched concentric or eccentric rings left in any sheet-metal knockout) or must be connected to a threaded hub or boss. A metal enclosure for a grounding-electrode conductor must be electrically in parallel with the grounding-electrode conductor. As a result, a metal conduit or EMT enclosing a grounding-electrode conductor forms part of that grounding-electrode conductor. In fact, such conduit or EMT has a much lower impedance than its enclosed grounding-electrode conductor (owing to the relation of magnetic fields) and is, therefore, even more important than the enclosed conductor in providing an effective path for current to the grounding electrode.

Figure 412 shows the "skin effect" of current flow over a conduit in parallel with a conductor run through it. This clearly shows the conduit itself to be a more important conductor than the

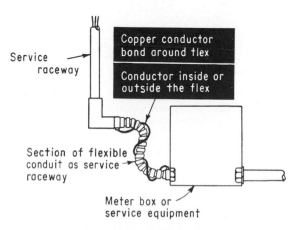

THIS IS NOW A VIOLATION !

FIG. 410 Flex may not be used as service raceway, even if bonded.

FIG. 411 "Bonding" makes use of prescribed fittings and methods to connect enclosures containing service entrance conductors.

Ammeter A₁ (indicates amount of current in conduit) = 97 amps

Ammeter A₂ (indicates amount of current in conductor) = 3 amps

FIG. 412 Grounding-electrode conductor must be bonded to both ends of any protective steel raceway sleeve for this reason.

actual conductor. In Fig. 412, the steel conduit acts as an iron core to greatly increase the inductive reactance of the conductor. This choke action raises the impedance of the conductor to such a level that only 3 A flows through the conductor while the balance of 97 A flows through the conduit. This division of current between the conduit and the conductor points up to the importance of ensuring that couplings and connectors are tight throughout every conduit system and in every metal raceway system and all metal cable jacketing. In particular, it emphasizes the need to bond both ends of any raceway used to protect a grounding conductor run to a water pipe or other grounding electrode. Such conduit protection must be securely connected to the ground electrode and to the equipment enclosure in which the grounding conductor originates.

If such a conduit is left open, lightning and other electric discharges to earth through the grounding conductor will find a high-impedance path. High conductivity in the conduit system is also important for effective equipment grounding, even when a specific equipment grounding conductor is used in the conduit. Of course, nonmagnetic conduit—such as aluminum—would have a different effect, but it should also be in parallel with the enclosed conductor.

Ensuring the continuity of raceway or armor for a grounding conductor reduces the impedance of the ground path, compared with what the impedance would be if the raceway or armor had poor connections or opens. Effective bonding of the raceway or armor minimizes the dc resistance of the ground path and reduces the overall impedance, which includes the choke action due to the presence of magnetic material (steel conduit or armor), that is, the increased inductive reactance of the circuit. Because of that, the "bonding" connection of such raceway to service enclosure must be made as shown in Fig. 413.

FIG. 413 Every grounding-conductor conduit sleeve must be "bonded" at both ends.

Bonding techniques

A variety of techniques may be used to provide the very important (and **NEC** required) bonded connections of service-conductor enclosures:

Where rigid metal conduit or IMC is the service raceway, threaded or threadless couplings can be used to couple sections of conduit together. Conduit may be connected to a meter socket by connecting a threaded conduit end to a threaded hub or boss on the socket housing, where the housing is so constructed; with a locknut and bonding bushing; with a locknut outside and a bonding wedge or bonding locknut and a standard metal or completely insulating bushing inside; or sometimes with a locknut and standard bushing where the socket enclosure is bonded to the grounded service conductor. Conduit can be connected to knockouts in sheet-metal enclosures with a bonding locknut. A bonding locknut is a recognized device for bonding a service-conduit nipple to a meter socket when the knockout is clean (no rings left in the enclosure wall) or is cut on the job. With plastic bushing permitted, this is the most economical of the several methods for making a bonded conduit termination.

The bonded termination of rigid metal conduit or IMC may also be accomplished with a bonding wedge or a bonding bushing, where no knockout rings remain around the opening through which the conduit enters. Where a knockout ring does remain around the conduit entry hole, a bonding bushing or wedge with a jumper wire must be used to ensure a path of continuity from the conduit to the enclosure.

Figure 414 summarizes the various acceptable bonding techniques. It should be noted that the use of the common locknut-and-bushing type of connection is not allowed. Neither is the use of double locknuts—one inside, one outside—and a bushing, although this technique is permitted on the load side of the service equipment. The special methods set forth for "bonded" connections are designed to prevent poor connections and the loosening of connections due to vibration. This minimizes the possibility of arcing and the damage which might result when a service conductor faults to the grounded equipment.

Similar provisions are used to ensure the continuity of the ground path when EMT is the service raceway. EMT is coupled with threadless devices—of the compression, indenter, or set-screw type, using rain-tight couplings outdoors. Standard connectors are used to connect EMT to enclosures. Bonding between the EMT fittings and the enclosures is accomplished in the same

FIG. 414   Recognized ways to "bond" wiring methods to sheet-metal enclosures.

way as with rigid conduit. And the fittings used with service-cable armor must ensure the same degree of continuity for the ground path.

Common sense and experience have dictated modern field practice in making raceway and armored service-cable connections to service cabinets. The top of Fig. 415 shows how a bonding wedge is used on existing connections at services or for raceway connections on the load side of the service—such as are required by **NEC** Section 501-16(b) for Class I hazardous locations. A bonding bushing, with provision for connecting a bonding jumper, is the common bonding device for new service installations where some of the concentric or eccentric "doughnuts" (knockout rings) are left in the wall of the enclosure, so that a bonding jumper is required. Great care must be exercised to ensure that each and every bushing, locknut, or other fitting is used in the way in which it is intended to best perform the bonding function.

FOR EXISTING INSTALLATION

Attach bond wire here and to separate screw in box if punched rings are left around the KO

FOR NEW WIRING

Bonded wire to enclosure

INSULATED THROAT

NONINSULATED THROAT

ALWAYS
NEEDS JUMPER

NO JUMPER
ON CLEAN KO

Screw here bonds to wall on clean KO

Bonding bushing with lug for jumper wire – may be used with jumper for clean KO or with rings left in wall of enclosure

Bonding bushing with screw that "bites" into enclosure wall may be used without a jumper on a clean KO or with a jumper when KO rings are left in wall

FIG. 415 Bonding bushings and similar fittings must be used in the intended manner.

Bonding jumpers

The sizing of a bonding jumper within the service-equipment enclosure or on the line or supply side of that enclosure is based on **NEC** Table 250-94, from which the jumper size is selected as if it were a grounding-electrode conductor for the particular size of service entrance conductors involved. Figure 416 shows examples of the sizing of bonding jumpers. At *A*, the bonding bushing and jumper are used to comply with **NEC** Section 250-72. In Table 250-94, with 500-MCM copper as the "largest service-entrance conductor," the minimum permitted size of grounding-electrode conductor is No. 1/0 copper (or No. 3/0 aluminum). That, therefore, is the minimum permitted size of the required bonding jumper.

At *B*, since each service phase leg is made up of two 500-MCM copper conductors in parallel, the left-hand column heading in Table 250-94 requires computation of the "equivalent for parallel conductors." The phase leg is thus taken at 2× 500 MCM, or 1000 MCM, which is the physical equivalent of the makeup. Then Table 250-94 requires a minimum bonding-jumper size of No. 2/0 copper (or No. 4/0 aluminum).

In the hookup of service bonding jumpers shown in Fig. 417, the jumper between the neutral bus and the equipment ground bus is a "main bonding jumper," and the minimum required size of this jumper for this installation is determined by calculating the size of one service phase leg. With three 500-MCMs per phase, that works out to 1500 MCM copper per phase. Because

(a)

(b)

NOTE: Bushing with jumper is acceptable bonding for a
clean KO or one with punched rings still in place.

FIG. 416   Bonding jumpers at services must be sized carefully.

Switchboard
enclosure

Equipment ground bus
bonded to switchboard
enclosure and to
neutral bus

Busbars

A   B   C   N

Main
bonding
jumper

Single bonding
jumper for three
conduits

Bonding bushing
with terminal lug
on each conduit
for attaching
bonding jumper

3 4-in. conduits stubbed
up under switchboard,
each carrying 4 500MCM copper
THW conductors of a
parallel service of 3
500MCM per phase

FIG. 417 Service bonding jumpers must be sized and installed in an acceptable manner.

that value is in excess of 1100 MCM copper, as noted in **NEC** Section 250-79(c), the minimum size of the main bonding jumper is 12½ percent of the phase-leg cross-sectional area. Then,

$$0.125 \times 1500 \text{ MCM} = 187.5 \text{ MCM}$$

Table 8 in **NEC** Chapter 9 shows that the smallest conductor with at least that cross-sectional area is No. 4/0, with an area of 211,600 cmils or 211.6 MCM. Note that a No. 3/0 has a cross-sectional area of only 167.8 MCM. Thus, No. 4/0 copper with any type of insulation would serve as the bonding jumper.

In Fig. 417, the jumper shown running from one conduit bushing to the other and then to the equipment ground bus is an "equipment bonding jumper." It is sized in the same way as a main bonding jumper (as above). In this case, therefore, the equipment bonding jumper would have to be at least No. 4/0 copper.

If each of the three parallel 4-in conduits in Fig. 417 had a separate bonding jumper connecting it individually to the equipment ground bus, the size of the separate bonding jumpers could not be less than the size of the grounding-electrode conductor for a service the size of the phase conductor used in each conduit. According to Table 250-94, a 500-MCM copper service calls for at least a No. 1/0 grounding-electrode conductor. Therefore, a bonding jumper run from the bushing lug on each conduit to the ground bus must be at least a No. 1/0 copper (or 3/0 aluminum), as shown in Fig. 418.

N

Equipment
ground bus

Separate bonding jumper
for each conduit

Each conduit contains four 500 MCM copper conductors — three phases and a neutral

FIG. 418 Separate bonding jumper may be used for each of parallel conduits.

Bonding of piping

Within a building or other structure, any available "interior metal water piping system" must be bonded to the service-equipment enclosure, to the grounded conductor (usually a neutral) at the service, to the grounding-electrode conductor, or to the one or more grounding electrodes as used. This rule applies where the metal water-pipe system does not have 10 ft of metal pipe buried in the earth and is, therefore, not a grounding electrode. In such cases, although connection is not made to the water piping as an electrode, the water-pipe system must be bonded to the service grounding arrangement. And the bonding jumper used to connect the interior water piping to, say, the grounded neutral bus or terminal (or to the ground bus or terminal) must be sized from Table 250-94, based on the size of the service conductors. The jumper is sized from that table because that is the table that would have been used if the water piping had had 10 ft buried under the ground, making it suitable as a grounding electrode. Note that the "bonding jumper" is sized from **NEC** Table 250-94 [and not from Section 250-79(c)], which means it never has to be larger than No. 3/0 copper or 250-MCM aluminum.

A bonding connection must also be made from "other" (than water) metal piping systems that "may become energized"—such as those for process liquids or fluids—to the grounded neutral, the service ground terminal, the grounding-electrode conductor, or the grounding electrodes. But, for these other piping systems, the bonding jumper is sized from **NEC** Table 250-95, using the rating of the overcurrent device of the circuit that may energize the piping.

Understanding and application of the foregoing rule of **NEC** Section 250-80 hinge on the reference to metal piping "which may become energized." What does that phrase mean? Is it not true that any metal piping "may" become energized? Or does the rule apply only to metal piping that conductively connects to metal enclosures of electrical equipment (such as pump motors, solenoid valves, pressure switches, etc.) in which failure of electrical insulation would put a potential on the metal piping system? It does seem that the latter is what the rule means. Where a particular circuit poses the threat of energizing a piping system, the equipment grounding conductor for that circuit (which could be the conduit or other raceway enclosing the circuit) may be used as the means of bonding the piping back to the service ground point. That has the effect of saying, for instance, that the equipment grounding conductor for a circuit to a solenoid valve in a pipe may also ground the piping, as shown in Fig. 419.

FIG. 419   Metal process piping may be grounded by the grounding conductor of the circuit that would energize the piping under fault conditions.

EMERGENCY SUPPLY

Where the interruption of electric power supply to a building would result in panic, hazard to life or property, or major loss of production, provision should be made for an emergency supply of power, to be used in the event of failure of the normal supply. Such provision may be made by designing an emergency system supplied by

1. Storage batteries,
2. An emergency generator set,
3. Separate emergency service, or
4. A subservice tapped ahead of the disconnect means for the normal service conductors

Detailed requirements and recommendations for emergency electrical systems are given in **NEC** Article 700. But, all the regulations of that article apply to the designated "circuits, systems, and equipment" only when the systems or circuits are required by law and classified as emergency provisions by federal, state, municipal, or other code or by a governmental authority. The **NEC** itself does not require emergency light, power, or exit signs. The **NEC** does exclude from all the rules any emergency circuits, systems, or equipment that are installed on a premises but are not legally mandated for the premises. Of course, any emergency systems that are provided at the option of the designer (or the client) must necessarily conform to all the other **NEC** regulations that apply to the work.

The placement or location of exit lights is not regulated by the **NEC** but is covered in the Life Safety Code, NFPA No. 101 (formerly Building Exits Code). But, where exit lights are required by law, the **NEC** considers them to be part of the emergency system. The **NEC** indicates how the installation will be made, but not where the emergency lighting is required, except as specified in Part **C** of Article 517 for essential electrical systems in health-care facilities.

OSHA regulations on exit signs are presented in Subpart E of the Occupational Safety and Health Standards (Part 1910). These requirements on the location and lighting of exit signs apply to all places of employment in all new and existing buildings.

Prior to OSHA, there was no universal requirement that all buildings or all places of employment have exit signs. The **NEC** does not require them. And, although the NFPA Life Safety Code does include rules on emergency lighting and exit signs, that code was enforced where state or local government bodies required it—that is, in some areas and for some specific types of occupancies. As a result, there are existing occupancies which do not have the exit signs now required by federal law.

OSHA makes clear that every building must have a means of egress—a continuous, unobstructed way for occupants to get out of a building in case of fire or other emergency, consisting of horizontal and vertical ways as required. Egress from all parts of a building or structure must be provided at all times during which the building is occupied. The law says:

Every exit shall be clearly visible or the route to reach it shall be conspicuously indicated in such a manner that every occupant of every building or structure who is physically and mentally capable will readily know the direction of escape from any point, and each path of escape, in its entirety, shall be so arranged or marked that the way to a place of safety outside is unmistakable.

Then it says:

Exit marking. (1) Exits shall be marked by a readily visible sign. Access to exits shall be marked by readily visible signs in all cases where the exit or way to reach it is not immediately visible to the occupants.

(2) Any door, passage, or stairway which is neither an exit nor a way of exit access, and which is so located or arranged as to be likely to be mistaken for an exit, shall be identified by a sign reading "Not An Exit" or similar designation, or shall be identified by a sign indicating its actual character, such as "To Basement," "Storeroom," "Linen Closet," or the like.

(3) Every required sign designating an exit or way of exit access shall be so located and of such size, color, and design as to be readily visible.

Note that marking must be supplied for the way to the exit as well as for the exit itself. Further,

(5) A sign reading "Exit," or similar designation, with an arrow indicating the direction, shall be placed in every location where the direction of travel to reach the nearest exit is not immediately apparent.

(6) Every exit sign shall be suitably illuminated by a reliable light source giving a value of not less than 5 footcandles on the illuminated surface.

A lot of discussion has been generated by that last rule. Note that an exit sign does not have to be internally illuminated, although it may be. And this lighting is required on "every exit sign," which means signs over exit doors and exit signs indicating the direction of travel.

The phrase "reliable light source" raises questions as to its meaning. Just what is reliable? Does this mean that the light source must operate if the utility supply to a building fails? And is there a difference between required applications in new buildings and those in existing buildings?

Because the OSHA regulations on exit signs do not require emergency power for lighting such signs, the light units that illuminate exit signs may be supplied from normal (nonemergency) circuits. OSHA does not make it mandatory to supply such circuits from a tap ahead of the service main or from batteries or an emergency generator. And this applies to new buildings as well as existing buildings. As far as OSHA is concerned, **NEC** Article 700, on emergency systems, does not apply to circuits for exit-sign lighting. Of course, OSHA does not object to the extra reliability such arrangements give. But the **NEC** does classify exit lights as emergency equipment.

Emergency Equipment

Where emergency lighting and/or power is mandated by law, the **NEC** requires the use of emergency equipment that is third-party certified—that is, listed for such application by UL or another qualified testing laboratory. Under the heading "Emergency Lighting and Power Equipment," the UL *Electrical Construction Materials Directory* states:

This listing covers battery-powered emergency lighting and power equipment, for use in ordinary indoor locations in accordance with Article 700 of the **National Electrical Code**. The lighting circuit ratings do not exceed 250 volts for tungsten lamps or 227 volts ac for electric discharge lamps. Other ratings may be included (motor loads, inductive loads, resistance loads, etc.) to 600 volts. This listing covers unit equipment, automatic battery charging and control equipment, inverters, central station battery systems, distribution panels, exit lights, and remote lamp assemblies, but not lighting fixtures. The investigation of emergency equipment includes the determination of their suitability of transferring operation from normal supply circuit to an immediately available emergency supply circuit.

Wherever emergency equipment is used, good design practice, as well as the **NEC**, requires that it be carefully selected to have adequate capacity for its intended performance. It is important that power be available for the necessary supply to exit lights and emergency and egress lighting, as well as to operate such equipment as elevators and other equipment connected to the emergency system. In hospitals, there may be a need for an emergency power supply for lighting in operating rooms and for such equipment as inhalators, iron lungs, and incubators. An emergency lighting system in a theater or other place of public assembly must have adequate capacity for exit signs, the chief purpose of which is to indicate the location of the exits, and lighting equipment commonly called "emergency lights," the purpose of which is to provide sufficient illumination in the auditorium, corridors, lobbies, passageways, stairways, and fire escapes to enable persons to leave the building safely.

Any switch or other control device that transfers emergency loads from the normal power source of a system to the emergency power supply must operate automatically on loss of the normal supply, and the transfer must be made in not more than 10 s. Transfer equipment must also be automatic for legally required standby systems, as covered in **NEC** Section 701-7, but a manual transfer switch may be used to switch loads from the normal to an optional standby power source, as detailed in Section 702-6.

To ensure proper control of emergency light and power, audible and visual signal devices must be included in the system to monitor and indicate the operating condition of the system and all its components. To be effective, the signal devices should be located in some room where an attendant is on duty. Lamps may readily be used as signals to indicate the position of an automatic switching device. An audible signal in any place of public assembly should not be so located or of such a character that it will cause a general alarm.

The standard signal equipment furnished by one battery manufacturer with its 60-cell battery for emergency lighting includes an indicating lamp which is lighted when the charger is operating at the high rate; in addition, a voltmeter marked in three colored sections indicates (1) that the battery is not being charged or is discharging into the emergency system, or (2) that the battery is being trickle-charged, or (3) that the battery is being charged at the high rate. This last indication duplicates the indication given by the lamp.

Engine-driven generators (diesel, gasoline, or gas) are commonly used to provide an alternative source of emergency or standby power when normal utility power fails. Gas-turbine generators also are used. The first step in selecting an on-site generator is to consider applicable requirements of the **NEC**, which differ depending on whether the generating set is to function as an emergency system, a standby power system, or a power source in a health-care facility such as a hospital. For example, an internal-combustion-type engine-generator set selected for use under Article 700 must be provided with automatic starting and automatic load transfer, with enough on-site fuel to power the full demand load for at least 2 hours. The engine driving an emergency generator must not be dependent on a public water supply for its cooling. That means a roof tank or other on-site water supply, with its pumps connected to the emergency source, must be used (Fig. 420).

A utility gas supply may be used as the sole supply of fuel to an emergency generator—at the discretion of the local inspector—where a simultaneous outage of both electric power and gas supply is highly unlikely.

Where dual utility supply is used to minimize the chance of a total electrical outage in a building, two separate services brought to different locations in the building are always preferable; these services should at least receive their supply from separate transformers where this is practicable. In some localities, municipal ordinances require either two services from indepen-

FIG. 420 Engine cooling must be ensured for a water-cooled emergency generator set to provide continuous operation.

dent sources of supply, or auxiliary supply for emergency lighting from a storage battery, or a generator driven by a steam turbine, internal-combusion engine, or other prime mover. Figure 421 shows the separate-service type of emergency supply.

As shown in Fig. 422, an emergency energy supply derived from a tap ahead of the normal main service disconnect provides protection only against failure of the main service equipment and is of no help on utility outage. In that diagram, the tap ahead of the main could feed directly to the emergency panel, without the automatic transfer switch.

Typical wall-hanging battery-pack emergency lighting units are shown in Fig. 423. The 1971 **NEC** accepted only fixed wiring as the means of connection of emergency light units. The present **NEC** recognizes permanent wiring connection or cord-and-plug connection to a receptacle. But even though the unit equipment is allowed to be hooked up with flexible cord-and-plug connections, it is necessary that the unit equipment be permanently fixed in place.

Individual battery-pack unit equipment provides emergency illumination only for the area in which it is installed; therefore, it is not necessary to carry a circuit back to the service equipment to feed the unit. The branch circuit feeding the normal lighting in the area to be served is the same circuit that should supply the unit equipment. A battery-pack emergency unit must

Notes:-
1. Switches A and C transfer to emergency service upon failure of normal supply.
2. Switch B transfers to emergency generator upon failure of emergency service from vault #1.
3. Elevator controls permit operation of only one unit at a time during emergency conditions.
4. Operation coil of starter connects through oil contacts direct to 120-v control circuit.

FIG. 421 Two separate services constitute a form of power supply that reduces the chance of an outage.

FIG. 422 Tap ahead of the service main disconnect protects against only internal failures.

FIG. 423 Battery-pack emergency light units provide simple, direct application to local areas.

provide lighting for at least 1½ hours and must not be connected on the load side of a local wall switch that controls the supply to the unit or the receptacle into which the emergency unit is plugged. Such an arrangement exposes the emergency unit to accidental energizing of the lamps and draining of the battery supply.

For the transfer of emergency lighting from the normal source to the emergency source, if a single emergency system is installed, a transfer switch must be provided which will automatically transfer the emergency system to the other source in case of failure of the source of supply on which the system is operating (Fig. 424). Where the two sources of supply are two services, the single emergency system may normally operate on either source (Fig. 425). Where the two sources of supply are one service and a storage battery, or one service and a generator set, the

FIG. 424   Automatic transfer must be provided to switch emergency loads from the normal supply to a generator.

FIG. 425   Emergency lighting may be supplied from a separate emergency service.

single emergency system would normally be operated on the service, with the battery or generator used only as a reserve in case of failure of the service.

The use of ground-fault protection on an emergency-service disconnect means that is rated at 1000 A or more on a 480/277-V grounded-wye supply is made optional by the **NEC**. To afford the highest reliability and continuity for an emergency power supply, GFP is not required on any 1000-A or larger disconnect for an emergency generator, although it may be used if desired.

Figure 426 shows two other emergency power-supply arrangements that were used in industrial buildings to satisfy the particular requirements of the building's operations.

Emergency wiring

The **NEC** requires that the wiring for emergency systems be kept entirely independent of the wiring used for normal lighting and thus that it be in separate raceways, cables, and boxes. This requirement ensures that faults which may occur on the normal wiring will not affect the emergency-system wiring.

For transfer switches, an exception to this rule is intended to permit normal supply conductors to be brought into the transfer-switch enclosure; these conductors would then be the only ones within the transfer-switch enclosure which were not part of the emergency system. Where two separate sources supply an emergency or exit lighting fixture, both supplies are, of course, permitted to enter the fixture and its junction box.

Figure 427 shows two applications that violate **NEC** rules on the hookup and wiring of emer-

FOR
EXTERNAL
FAILURES
ONLY

FOR
INTERNAL
AND
EXTERNAL
FAILURES

FIG. 426 Emergency-supply hookups can be made to provide different types of protection.

gency circuits. The top application is a violation because the wiring to the emergency light (or to an exit light, which is classed as an emergency light) is run in the same raceway and boxes as the wiring to the decorative floodlight. Emergency wiring may not be run in enclosures with wiring for nonemergency circuits. The bottom application is a violation because the water cooler is not a piece of emergency equipment, and the **NEC** prohibits emergency circuits from supplying appliances or loads other than those prescribed for truly emergency power and/or lighting.

Standby Systems

NEC Article 701 covers standby power systems that are required by law. Legally required standby power systems are systems that are required and so classified by municipal, state, federal, or other codes or by any government

FIG 427 Emergency wiring must not be located in enclosures for nonemergency circuits, and emergency circuits must not supply nonemergency equipment.

agency having jurisdiction. These systems are intended to supply power automatically to selected loads (other than those classed as emergency systems) in the event of failure of the normal power source.

Legally required standby power systems are typically installed to serve such loads as heating and refrigeration systems, communication systems, ventilation and smoke-removal systems, sewage-disposal systems, lighting, and industrial processes that, when stopped during a power outage, could create hazards or hamper rescue or firefighting operations.

NEC Article 701 covers the circuits and equipment for such systems that are permanently installed in their entirety, including power source.

Article 702 covers "optional" standby systems. Life safety is not the purpose of optional standby systems. Instead, optional standby systems are intended to protect private business or property where life safety does not depend on the performance of the system. Optional standby systems (other than those classed as emergency or legally required standby systems) are intended to supply on-site generated power to selected loads, either automatically or manually.

Optional standby systems are typically installed to provide an alternative source of electric power for such facilities as industrial and commercial buildings, farms, and residences, to serve such loads as heating and refrigeration systems, data-processing and communications systems, and industrial processes that, when stopped during a power outage, could cause discomfort, serious interruption of the process, or damage to the product or process.

Because of the constant expansion in electrical applications in all kinds of buildings, the use of standby power sources is growing at an accelerating rate. Continuity of service has become increasingly important with the widespread development of computers and intricate automatic production processes. More thought is being given to—and more money is being spent on—the provision of on-site power sources to back up or supplement purchased utility power to ensure the needed continuity as well as provide for public safety in the event of utility failure.

The most fundamental application of standby power is the portable alternator, used for standby power where the electric-utility supply is not sufficiently reliable or is subject to frequent outages (Fig. 428). An optional "standby" power load may be manually or automatically transferred from the normal supply to the standby generator. Any generator used as an "emergency" source must always have provision for automatic transfer of the load from normal supply to the emergency generator. And automatic transfer must always be used if a standby power system is required by law.

Uninterruptible power supply

Any standby power source and system must be capable of fully serving its demand load. This requirement can generally be satisfied relatively easily by generators, but the task can be complex for UPS systems. The uninterruptible power supply (UPS) is either a motor-generator assembly or an all-solid-state power-conversion system designed to protect computers and other critical loads from blackouts, brownouts, and transients. It is usually connected in the feeder supplying the load, with bypass provisions to permit the load to be fed directly. Figure 429 shows a typical basic layout for a solid-state UPS system. Such a system utilizes a variety of power sources to ensure continuous power. When the utility supply fails, the normal ac input to the UPS rectifiers ceases. The battery is the power source while the back-up engine-generator is being started. When the generator is running properly, it is brought onto the line through its transfer switch, and the 2000 A busway is then fed from the genrator and again feeds ac power to the UPS rectifiers.

The required capacity of a UPS system must be carefully calculated. Data-processing installations normally require medium-to-large 3-phase UPS systems ranging from 37.5 to over 2000

FIG. 428 Optional standby generator with a manual transfer switch is commonly used.

UPS SYSTEM (lower right) receives power from either normal source (upper right) or gas-turbine generators (upper left). From UPS, regulated power is delivered to 3000-amp bus feeding essential computer loads. Key components of UPS system are rectifiers, inverters, batteries and transfer CBs.

FIG. 429 Uninterruptible power supply (so-called "UPS system") is a common form of optional standby supply for data processing.

kVA. Some typical ratings available from UPS manufacturers are 37.5 kVA/30 kW, 75 kVA/ 67.5 kW, 125 kVA/112.5 kW, 200 kVA/180 kW, 300 kVA/270 kW, 400 kVA/360 kW, and 500 kVA/450 kW. Larger systems are configured by paralleling two or more of these standard sizes.

The necessary rating is chosen based on the size of the critical load. If a power profile itemizing the power requirements is not available from the computer manufacturer, the load may be measured using a kilowattmeter and a power-factor meter. Since UPS modules are both apparent-power (kilovoltampere) and real-power (kilowatt) limited, a system should be specified with both kilovoltampere and kilowatt ratings. The required kilovoltampere rating is obtained by dividing the actual-load kilowatts by the actual-load power factor. For example, an actual 170-kW load with an actual 0.85 power factor would require a 200-kVA UPS. The system should be specified as 200 kVA/170 kW, and the standard 200 kVA/180 kW UPS module would be selected for the application.

Equipment
Selection
and Layout

Once the voltage and current ratings and other electrical characteristics of feeder and service entrance circuits have been established, the design is completed by determining the actual physical
makeup of the circuits and overall system. This means selecting
the physical characteristics of all switching, control, protection,
and transformation equipment—along with their housings and
interconnecting components. And for current flow between system
equipment units, either a raceway system or a cable system must
be selected.

Because circuit control and overcurrent protection are required
for every electrical installation, small or large and simple or complex, such equipment is generally centralized in concentrated load
areas in the form of switchboards and/or panelboards. For smaller
systems, one or more panelboards will usually be sufficient. For
larger installations, such as industrial plants, commercial buildings, and office and apartment buildings, main switchboards
receive the total incoming power and then distribute smaller blocks
of power to a number of panelboards conveniently located near
connected electrical loads.

Modern distribution centers for lighting and appliance feeders
include dead-front panelboards and dead-front switchboards.
Feeders are tapped through fused switches or circuit breakers
from the buses in such distribution centers.

Distribution centers frequently are supplied by the service conductors to buildings and contain the main service switches. This is
the most common arrangement in commercial and institutional
buildings, although subfeeder distribution centers are sometimes
supplied by feeders from the main distribution center (the service
panel or switchboard). In many industrial buildings, feeder distribution centers may be supplied from transformers which step
high distribution voltages to utilization levels.

Switchboards

Switchboards are assemblies of switches, circuit breakers, fuses, metering equipment, relays and/or other forms of equipment which control or record the electric current distributed throughout a building or area. Widths may range from 3 ft to more than 100 ft. Currently, most large electrical installations, particularly where loads exceed 1200 A, use factory-assembled switchboards for electrical control and distribution purposes. The major reasons for selecting this type of equipment over an installation of separate fused-switch, breaker, and gutter enclosures are overall reduced labor costs and a better quality design. Where the width (sometimes referred to as the length) of a switchboard is more than 40 in, the unit usually is constructed in sections to simplify shipping and installation.

Switchgear are selected according to voltage rating, ampere rating, number of phases, and short-circuit rating. Circuit-breaker switchgear may be of the stationary type, in which the breakers are bolted to the bus and frame, or of the draw-out type, in which the breakers are mounted on a slide-out mechanism for easy removal and maintenance and for disconnecting the breaker from the bus. Switchboards may have provisions for instrument transformers and metering. Feeders from the switchboard may be carried directly to lighting or power branch-circuit panelboards, to subfeeder distribution switchboards or panelboards, to motor control centers, or to individual motor loads.

Good switchboard design provides extra capacity and spare overcurrent devices (or space for them) so that future loads can be added or shifted with minimum cost and inconvenience.

Most switchboards contain no provision for utility meters. Generally an incoming service board contains space for current transformers; from these, metering circuits are run to a separate wall location provided for metering equipment. It is prudent to check the serving utility's metering requirements before switchboard designs are finalized, to determine how much board space is needed for current transformers and/or meters, or how much adjacent wall space will be required.

Enclosed, metal-clad switchboards are almost universally used for low- and medium-voltage systems, with all electrical components receiving maximum protection within grounded metal enclosures. Only the operating handles, levers, or pushbuttons are accessible from the outside.

Boards of this type commonly are freestanding units rated from 800 to 6000 A at 600 V or less. For most installations, they contain the service disconnecting means and overcurrent protective devices, together with any current transformers for metering or system ground-fault protection.

While the section in the **NEC** outlining minimum standards for switchboards is relatively short, there are a great many options involving maintenance and operating convenience, extra safety features, overload capability, etc. The NEMA standards publication *Switchboards* is useful in understanding construction and specification details along with test and performance specs.

Operating convenience, ease of maintenance, and proximity to areas with concentrated loads are major considerations in locating switchboards. Proximity to loads may become a problem when the switchboard includes, or constitutes, the service equipment. The location of the utility supply determines where the main switchboard must be placed. However, service raceways can be run within a building in a 2-in concrete or brick envelope and still be considered as being "outside" the building up to the point where the concrete-encased raceway emerges from the concrete at the switchboard or panelboard.

If a switchboard is to be installed in other than dry areas, or outdoors, it must be of weatherproof construction. In any hazardous location, the equipment must be of a type approved for use in the relevant Class and Division of hazardous area as indicated in the **NEC**. It is usually safer and much less costly to locate switchboards outside such areas.

The place of installation of a switchboard must provide the minimum safe working clearances shown in Fig. 430. In addition, there must be work space at least 30 in wide in front of electrical equipment, and sufficient "elbow room" in front of such equipment as column-type panelboards and single enclosed switches (e.g., 12 in wide) to permit the equipment to be operated or maintained safely.

The clearances shown in Fig. 430 must be measured from the live parts if exposed or from the enclosure containing live parts. The width measurement is to be made parallel with the face of the equipment.

To prevent situations in which personnel might be trapped in the working space around burning or arcing electrical equipment, two "entrances" or directions of access must be provided to the working space around switchboards and control panels rated 1200 A or more and over 6 ft wide. At each end of the working space of such equipment, an entranceway or access route at least 24 in wide must be provided. Because personnel have been and can be trapped in work spaces when fire erupts between them and the only exit route from the space, design engineers should routinely provide two exit paths in their drawings and specs. Although two doors are not required into an electrical equipment room, it may be necessary to provide two doors to obtain the required two entrances to the work space—especially where the switchboard or control panel is located in tight quarters that do not afford a 24-in-wide exit path at each end of the work space.

At the top, Fig. 431 shows one way to provide two paths for entering or leaving the defined work space. In that sketch, placing the switchboard with its front to the larger area of the room and/or other layouts would be acceptable. It is only necessary to have a sure means of exit from the defined work space. If the space in front of the equipment is deeper than the required depth of work space, then a person could simply move back out of the work space at any point along the length of the equipment.

FIG. 430 These minimum clearances must be provided for work spaces around switchboards.

THIS IS REQUIRED . . .

24-in. min. width of each exit route

Required work space

Front

Switchboard or control panel

COMPLIES WITH CODE

TWO EXIT ROUTES FROM WORK SPACE

. . . TO PREVENT THIS

Fire here . . .

. . . could trap personnel here

Required work space

Front

Only one exit route

Switchboard or control panel

Over 6 ft

VIOLATION!

Electrical equipment room

FIG. 431 Positioning of a switchboard or motor control center must assure safe, ready exit from the work space.

Figure 431 shows a layout that must be avoided. With sufficient space available in the room, the layout of any switchboard or control center rated 1200 A or more with only one exit route from the required work space would most likely be a violation of the **NEC** rule. A door at the right end of the work space would eliminate the violation.

Figure 432 shows minimum design requirements for headroom and lighting in work spaces around switchboards and other electrical equipment. Although lighting is required for the safety of personnel in work spaces, the kind of lighting (incandescent, fluorescent, mercury vapor), the minimum footcandle level, and such details as the position and mounting of lighting equipment are left to the designer and/or installer, with the inspector the final judge of acceptability.

A minimum headroom of 6¼ ft is required in working spaces around electrical equipment rated up to 600 V. That requirement applies to "service equipment, switchboards, panelboards, or motor control centers."

The design drawings and specifications for an electrical system must include the requirement that all disconnect devices (that is, all switches and circuit breakers) for load devices and for circuits be clearly and permanently marked to show the purposes of the disconnects. This is a "must" that, under OSHA, applies to all existing electrical systems, no matter how old, and to all new, modernized, expanded, or altered electrical systems. This marking requirement has been widely neglected in the past. Panelboard circuit directories must be fully and clearly filled out. And all such marking on equipment must be in painted lettering or other substantial identification (Fig. 433).

Some type of lighting <u>must</u> be provided for any defined work space around the equipment

Ceiling

Headroom <u>must</u> be at least 6 ft 3 in.

Floor

Any service equipment, switchboard, panelboard, or motor control center installed indoors <u>in other than a</u> "dwelling unit"

Elevation

If the space on this side of a switchboard is a work space required by Section 110-16 (a)...

Switchboard (plan view)

...as well as the space on this side, then a lighting unit here, over only one of the work spaces, would not be adequate. Another light would be required for the work space on the other side of the equipment.

Plan

Illumination not required

Ceiling

Note the exception given:

For service equipment or panelboard rated not over 200 amps...

No code minimum on headroom

Floor

... BUT only in a one-family house, an apartment unit of an apartment house or other "dwelling unit."

FIG. 432 Adequate headroom and illumination must be provided at work spaces around electrical equipment.

Labels to identify circuits

Circuit schedule filled out to show loads

Lettering or label to identify function

A/C
ROOF
MOTOR
DISCON.

Floor

SAFETY SWITCH
OR CIRCUIT BKR.   PANELBOARD   SWITCHBOARD

FIG. 433 Clear identification of the purpose of each disconnect is a vital safety matter.

Effective identification of all disconnect devices is a critically important safety matter. When a switch or CB has to be opened to deenergize a circuit quickly—as when dictated by a threat of injury to personnel—it is absolutely necessary to be able to identify quickly and positively the disconnect for the circuit or equipment that constitutes the hazard. Painted labeling or embossed identification plates affixed to enclosures would comply with the requirement that disconnects be "legibly marked" and that the "marking shall be of sufficient durability." Ideally, the marking should tell exactly what piece of equipment is controlled by each disconnect (switch or CB) and where the controlled equipment is located.

FIG. 434 Switchboard design and layout should strive for simple, direct circuit makeup.

Figure 434 shows the single-line arrangement of an effectively designed switchboard for a large modern office building. The design of the complete distribution system provided for separation of the lighting and power loads, and skillful loading of the lighting feeders produced the uniform condition in which only 400-A CBs are used in the lighting section. The single size of CB assures ready and effective maintenance of the assembly. In all phases of electrical design, the kind of simplicity shown in this switchboard layout is a mark of excellence—and especially in large, high-capacity systems. Note the emergency lighting supply tapped from the line side of one of the power CBs. Both voltage levels are available for connection to essential power and lighting loads, which can be kept operating when the lighting switchboard fails. Of course, this actually provides power in only limited emergencies; it would not cover the loss of utility power.

Panelboards

Panelboards comprise the major item of distribution equipment between main or subdistribution points and individual connected loads. Almost every electrical installation involves one or more panelboards for lighting and receptacle circuits. A single panel permits the distribution of electric power from a feeder to as many as 42 lighting and appliance branch circuits, each with overcurrent protection and disconnect.

The **NEC** makes it mandatory that panelboard assemblies be installed in metal cabinets. A metal trim covers the gutter space and circuit connections, leaving only the switch or circuit-breaker handles accessible to personnel. All electrical panelboard parts can be replaced or serviced from the front, after trim removal—an important feature when panels are either surface-mounted or installed flush in walls or partitions. Special colors and metallic finishes are available for the exposed trim on flush-mounted panelboards that are located in conspicuous positions in esthetically designed areas.

A wide variety of panelboards is available to meet specific load and control requirements and installation conditions. Though somewhat similar electrically, the panels fall into five categories:

- *Service-equipment panelboards* are available for loads up to 1600 A. They contain six or fewer fused switches, fused pullouts, or circuit breakers connected to the incoming mains. They may also have split buses, one set supplying branch-circuit overcurrent devices installed in the same enclosure.

- *Feeder distribution panels* generally contain circuit overcurrent devices rated at more than 50 A to protect subfeeders that run from this panel to smaller branch-circuit panelboards throughout the area it serves.

  An interesting application of a CB distribution panelboard for individually metered subfeeds to unit occupancies in an apartment house is shown in Fig. 435. Five meters were mounted along each of the three horizontal gutters connecting the panel to the vertical trough. The three neutrals were installed as shown to provide the potential taps. Hot legs for the subfeeds were then carried through the gutters to supply the current coils in the meters. All subfeed runs were made from the pullbox at the top.

- *Lighting and appliance branch-circuit panelboards* have more than 10 percent of their overcurrent devices rated at 30 A or less, and neutral connections are provided for these. Generally, these panels have bus ratings up to 225 A at 600 V or less. However, larger bus ampacities must sometimes be selected to meet load requirements.

- *Power distribution panels,* similar to the feeder distribution type, have buses normally rated up to 1200 A at 600 V or less. They contain control and overcurrent devices sized to match connected motor or other power-circuit loads. Generally, the devices are 3-phase.

- *Special panelboards* containing relays and contactors can be obtained and installed where remote control of specific equipment is specified. Some panelboards contain branch-circuit

FIG. 435 Meter-bank distribution circuiting demands skillful design techniques.

snap switches rated 30 A or less. These boards must have overcurrent protection rated no higher than 200 A on the line side of the switches.

Any panel installed in a wet or damp location, in a dust-laden area, outdoors, or in any hazardous area must be of the type approved for use in such a location. NEMA standards list three panelboard enclosures: Type 1, indoor—general purpose; Type 3R, outdoor—rainproof and sleet-resistant; and Type 12, for indoor industrial use (dust-, drip-, and oil-tight).

If only qualified persons are to have access to the installed panelboards, then panel trims with locking-type doors should be specified.

A few precautions, if observed by designers in developing plans and specifications, can simplify panelboard installation, improve workmanship, and produce a neater job:

- Coordinate panelboard cabinet installation with the raceway system, and maintain proper alignment.
- Align floor-slab conduit stubups with openings in the bottoms of cabinets, particularly with surface-mounted panels. Maintain alignment.
- Ensure that surface-mounted panelboard enclosures are securely fastened to walls or other structural surfaces. The addition of supplementary pedestals or "legs" (structural steel or pipe sections with floor flanges) will increase rigidity and help distribute the weight of large, heavy panelboards.
- Make sure the gutter space around an overcurrent-device panel assembly is sufficiently wide to permit conductor installation in accordance with Table 373-6(a). If more space is necessary, use an oversized enclosure, supplementary wiring troughs, or auxiliary gutters.
- Where a number of panelboard enclosures are to be mounted side by side, the addition of a suitably sized auxiliary gutter, wiring trough, or pullbox above and/or below the grouped cabinets will facilitate cable pulling and circuit installation. In many cases, it will eliminate the need for oversized cabinets to accommodate the required conductors.

Switching Devices

Many types of switches are used for on-off control of loads and for no-load circuit opening. The selection of the switch best suited to a specific task involves a number of considerations:

- The electrical rating and characteristics of the switch must be proper for the circuit function to be performed. The switch must have the right number of poles and the proper voltage and current ratings to carry the load impressed on it. UL-listed enclosed switches are rated up to 4000 A, 500 hp, and 600 V, and nonfusible switches are tested to provide continuous operation (periods over 3 hours) at marked rated loads.
- Enclosed switches rated 800 or 1200 A at more than 250 V are available in two classes. One is for general use and may be used as a disconnect up to its rating; the other is for isolating use only and must be so marked. Any enclosed switch rated over 1200 A must be marked "For isolating use only—do not open under load." (Bolted-pressure switches and CB-switches are rated for load-break above 1200 A.)
- For motor loads, horsepower rating must be considered. Enclosed motor circuit switches are listed by UL up to 500 hp and 600 V. A switch that is marked "Motor circuit switch" is intended for use only in motor circuits. Enclosed switches with horsepower ratings in addition to current ratings may be used in motor circuits as well as for general-purpose circuits.

In sizing feeder disconnect means in switchboards or panelboards, care must be exercised to ensure that the ampere rating of each switch or circuit breaker is adequate for the load, for the switching duty to which it will be subjected, and for any load growth to be designed into the system. A first step in such sizing of a switch or circuit breaker is as follows (see Fig. 436):

Motor load only in power panel: The switch must have an ampere rating at least equal to the required current rating of the feeder conductors. This current rating is equal to 125 percent of the full-load current of the largest motor plus the sum of the full-load currents of the other motors fed from the panel. But a switch must accommodate the fuses which protect the feeder, and these may have a maximum rating equal to the largest rating or setting of branch-circuit protective device in the power panel plus the sum of the currents of the other motors fed from the panel. Or, a CB may have a higher trip setting. This may require a larger device. And if two or more motors fed from the panel must be started simultaneously, the starting-current drain may be so heavy as to require larger feeder conductors and disconnect.

Motor and lighting load in power panel: The switch must have an ampere rating at least equal to that required for the motor load (as described above) plus the current for the lighting and appliance load on the panel.

When fused switches are used, the heat load on the fuses should be considered in sizing the assembly. Because fuses are rated according to their ability to carry current continuously when placed in open air, on a bench, in a horizontal position, the effect on fuses of heat accumulation within enclosed mountings in safety switches must be considered. To ensure the continuous operation of loads—minimizing the chance of nuisance fuse blowing—fusible switches must be rated to accommodate fuses which are at least 25 percent higher in rating than the load currents which will be flowing for several hours or longer. Another way of stating this application requirement is that the load current must not exceed 80 percent of the fuse rating.

In selecting feeder disconnect and protection equipment, consideration should be given to the provision of spare capacity in this equipment. Depending upon anticipated future requirements and the manner in which extra capacity was included in the feeders, space should be provided in switchboards and/or feeder panelboards for additional switches or circuit breakers or for future installation of larger switching and protection units. And the design of the distribution system must integrate all provisions for future expansion of feeder capacities. This includes the routing of spare feeder conduits, accessibility of feeder raceway in which capacity has been allowed for additional feeder conductors, and ease of connection in switching and protection assemblies.

FIG. 436 Fused switches for feeders and subfeeders must be carefully selected.

Modern switchboards and feeder panelboards are constructed to accommodate all these design provisions for load growth. Care in selecting the proper types of switching and protection assemblies—with buses of substantial capacity for future expansion—is essential to economy of distribution design.

An important consideration in the selection of fusible switches is coordination between the electrical characteristics of the switch and the particular type of fuses used in the switch. With very high short circuits available in the large feeders of modern distribution systems, the fused switches used in panels, switchboards, or individual enclosures must be carefully rated for the following two conditions:

1. Fault-interruption service: In addition to being fully rated to make and break load currents up to its full current rating, every switch must have the ability to interrupt ground-fault or short-circuit currents which it might be called upon to break. This is shown in Fig. 437 and works as follows:

The contact mechanism has a repetitive load-break capability of up to 12 times the continuous-current rating of the contact assembly. This means that the contact mechanism of the switch can safely break any overload up to that value. But the opening time–current line for the fuses shows that at 12 times the contact current rating, the fuse will open in 0.1 s—which is so short a time that the operator could not possibly open the breaker switch while the fault current is flowing. For values of fault current lower than 12 times the switch rating, it would be possible to open the switch before the fuse operated. But, in such cases, the switch contacts would be operating within their rated load-brake range. For values of fault current higher than 12 times the switch rating, the fuse will always operate to open the circuit before anyone could possibly open the switch contact mechanism. This is effective coordination between switch interrupting capacity and speed of fuse operation.

An example of a dangerous application would be the use of the fuses in a switch which had an interrupting capacity of, say, two times the contact continuous rating. As shown, a fault current of better than five times the contact rating would take 10 s to operate the fuse. If someone opened the switch during that 10 s (which is possible), the switch contacts would be breaking a current higher than that for which they are safely rated—with hazard to life and property as well as to the switch itself.

Switches are coordinated with their current-limiting fuses to assure safe operation. Closing and quickly reopening a switch (without intentional delay) before the fuse clears the fault is called "fuse racing." Coordination between a switch and fuses depends upon the construction

FIG. 437   Switch should be rated to interrupt fault currents that fuses may not quickly clear.

and operating characteristics of the switch and the time-current characteristics of the particular fuses used in it. When a switch and its fuses are properly coordinated, the fuses will always operate fast enough to break any current which the switch could not safely open, before any person can operate the switch.

2. Short-circuit withstandability: Every switch must be capable of safely withstanding the amount of energy which passes through it from the time a short-circuit fault develops until the fuses in the switch open the fault. Every switch—fusible or nonfusible—must be able to take the thermal and magnetic forces produced by let-through fault current. This ability of a switch is a function of the construction of the switch and the operating speed of the fuses used in it. To be properly applied, a switch must remain operable after any fault that is cleared by the fuses in the switch.

Fig. 438 shows the coordination between a typical time-delay fuse and a heavy-duty interrupter switch. The melting time versus current curve is compared with two different switch interruption ratings—with ¼ to ½ s taken as the fastest possible time for closing and reopening the switch. Note that an adequate switch interrupting capacity could pose a hazard.

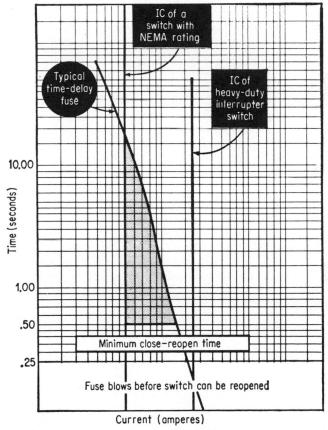

FIG. 438 Heavy-duty switches offer greater safety and reliability under fault conditions.

The above discussion leads to the following factors as considerations in selecting a fused switch:

1. The required fuse interrupting capacity is determined by the short-circuit current available from the system at the point of application.
2. The minimum switch withstandability is determined by the I^2t of fuse let-through current (from the fuse manufacturer).
3. The minimum switch interrupting capacity is based on fuse speed and switch operating time.

It can be seen that switches with high withstandability and high interrupting capacity can be used with slower-operating fuses which have longer melting times and higher let-through currents. But switches with lower withstandability and interrupting capacity must be protected by fuses with lower let-through and faster opening characteristics. The heavy-duty switches, for instance, could be used with Class K-5, RK-5, and K-9 fuses in high-capacity circuits; lighter-duty switches would require the use of Class RK-1, K-1, or J fuses in the same circuits.

Transformer inrush

Another important consideration in selecting a fusible switch is the effect of inrush current on the operation of the fuse. In selecting an overcurrent protective device (fuse or CB) to protect a transformer, consideration must be given to the time-current inverse-trip characteristics of the protective device to ensure that the inrush current (which flows when the circuit to the transformer primary is closed) will not open the protective device. This is particularly critical where the fuse or CB is set at 125 percent of the transformer primary current [as required in **NEC** Section 450-3(b)(1)] but is not so much of a concern where the primary protective device is set at 250 percent of the primary current [as permitted in **NEC** Section 450-3(a)(1) or 450-3(b)(2)]. To cover the most extreme condition of inrush current—when the circuit is closed at that point in the voltage wave where maximum asymmetry occurs and causes the highest current—tests have indicated that a fuse or CB will operate properly and will not open on inrush if its time-current trip curve will permit a current value equal to 12.1 times the transformer rated primary current to flow for at least 0.1 s. Effective operation therefore depends on the shape of the time-current curve for the particular type of fuse or CB being used.

EXAMPLE

A 300-kVA, dry-type transformer is used to step a 480-V, 3-phase, 3-wire supply down to a 208/120-V, 3-phase, 4-wire wye secondary. A fuse switch at the supply end of the primary feeder circuit to the transformer will be equipped with fuses rated not over 125 percent of the transformer primary rated full-load current. That level of protection represents the maximum value permitted for primary protection by **NEC** Section 450-3(b)(1). And because the primary-circuit conductors must be protected at their ampacity by the primary protective device (**NEC** Section 240-3), using a device rated at 125 percent provides the economy of using the smallest acceptable size of circuit conductors and the smallest size of switch. The protection is provided as follows:

Step 1

The transformer primary current is equal to 300,000 VA divided by 480 × 1.732, or 361 A.

Step 2

The maximum permitted rating of fuses in switch to protect the transformer is 1.25 × 361 A, or 451 A.

Step 3

In evaluating the possible nuisance tripping of the fuses due to transformer inrush magnetizing current, consideration must be given to the basic rule that the fuse will operate properly, without tripping or inrush, if its time-current characteristic will permit a current equal to 12.1 times the rated primary current to flow for at least 0.1 s. In this case, that value of inrush current works out to be 12.1 × 361 A, or 4368 A.

Step 4

The largest rating of standard fuse that could be used in this application is 450-A fuses. (Or, 500-A fuses might be used as the next larger standard size above 451 A. But use of the larger fuse would call for circuit conductors of such ampacity that the 500-A fuses protect them in accordance with their ampacity, as per **NEC** Section 240-3).

Step 5

In Fig. 439, the time-current plot of the 450-A fuse in the graph at the left shows that the intersection of 0.1 s and 4368 A falls to the right of the fuse melting curve, indicating that the fuse can blow on transformer inrush current.

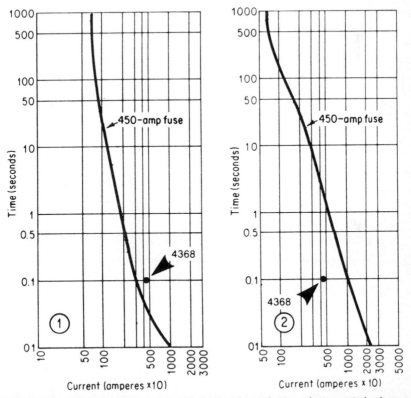

FIG. 439 If the intersection of 0.1 and 12.1 times the transformer rated primary current is to the right of the protective-device curve, inrush current can open the circuit.

Step 6

But, in the graph at the right in Fig. 439, which is for a different type of 450-A fuse, with a greater time delay in its operation (a left-to-right slope that is not as steep as that in the curve at left), the intersection of 0.1 s and 4368 A falls to the left of the fuse melting curve—assuring that, on closing of the switch to energize the transformer, the maximum possible inrush current will not blow the fuse.

In evaluating the possibility of nuisance tripping on transformer inrush current as described above, the designer should use the minimum melting curve of a fuse (not the average melting curve or the total clearing-time curve) to ensure that the inrush current I^2t will not operate even those fuses that happen to be at the low-current side of the manufacturer's calibration tolerance band. And the same evaluation should be performed for circuit-breaker tolerance bands.

Figure 440 shows how a dual-element Class RK-1 fuse would be exposed to possible blowing on inrush current to a 150-kVA transformer. The 150-kVA, 480-V, 3-phase transformer primary has a rated full-load current of 150,000 ÷ (480 × 1.732), or 180 A. The maximum rating of the primary fuses that would be used to protect such a transformer is 180 A × 1.25, or 225 A. To ensure that transformer inrush current will not blow the fuses, the fuse time-current curve must be evaluated at the intersection of 0.1 s and 180 A × 12.1, or 2178 A. In Fig. 438, that intersection point falls far to the right of the curve for the 200-A Class RK-1 dual-element fuse—far enough to indicate that a 225-A fuse of the same type would be exposed to blowing on inrush.

FIG. 440   Inrush current of 2178 A would blow a 200-A fuse and probably a 225-A fuse.

Other selection considerations

- Enclosed switches for service use must be UL-marked "Suitable for use as service equipment."
- The construction of the switch assembly must be suited to the anticipated switch usage. Switches used frequently must have sturdy makeup to assure long, troublefree life.
- Auxiliary provisions—such as fuseholders and electrical operation of the contact mechanism—must be considered.
- Enclosure requirements, including the type of housing required for the place of installation, an interlock between the enclosure cover and operating handle, and provisions for locking the switch in the OFF position, must be carefully weighed.

- For switches with dual horsepower ratings, the higher rating is based on the use of time-delay fuses in the switch fuseholders to hold in on the inrush current of the higher-horsepower-rated motor.

Safety switches are air-break switches whose current-carrying parts are enclosed in metal cases and which are manually operable by means of external handles. The word "safety" derives from the enclosed form, which reduces the possibility of external contact with energized parts of the switch assembly. Safety switches are referred to as "enclosed switches" in NEMA standards, UL regulations, and the **NEC**.

In practice, general-use and motor-circuit switches are referred to as "safety" switches. However, specific industry standards classify general-use, motor-circuit, and isolating switches as "enclosed switches." NEMA standards list two categories of enclosed switches: general duty and heavy duty.

General-duty devices include door-operated fused switches, fusible pullouts, and enclosed switches designated "general duty." None of these is meant for frequent opening and closing. They do not have quick-make, quick-break contacts and should not be used where more than 10,000 rms symmetrical amperes of short-circuit current is available.

Heavy-duty switches include fused bolted-pressure contact switches and NEMA fused heavy-duty switches. They are designed for use where duty is severe or frequent, or where the fault current may exceed 10,000 rms amperes. NEMA fused heavy-duty switches carry ratings of 30 to 1200 A; bolted-pressure switches conforming to UL standards are rated 800 through 4000 A. Both classes of switches are also rated in horsepower.

High-capacity switches

A wide range of types and sizes of switches is available for use in modern high-capacity feeder and service entrance applications. In the range from 200 to 1200 A, up to 600 V, service entrance and feeder applications can make use of heavy-duty quick-make, quick-break fusible safety switches. In the range from 600 to 6000 A, either a "fused power circuit device" or a power circuit-breaker–type load interrupter is used.

Manual and electrically operated switches designed to be used with Class L current-limiting fuses rated 601 to 4000 A and 600 V ac are listed by UL as "fused power circuit devices." This category covers bolted-pressure contact switches and high-pressure, butt-type contact switches. Such devices "have been investigated for continuous-duty use at 100 percent of their rating on circuits having available fault currents of 100,000, 150,000 or 200,000 rms symmetrical amperes" as marked. Units suitable for use as service switches are marked "Suitable for use as service equipment." Electrically tripped and/or operated switches of this type that are suitable for ground-fault protection are so marked.

High-capacity secondary-voltage service entrances commonly make use of bolted-pressure contact or butt-contact switches. Typical switches of this type are made in sizes from 600 A up. They are generally manually operated with external handles. Their most common application is in general-purpose, ventilated enclosures, either individually enclosed in an incoming line section or combined within a main secondary switchboard. Such switches are generally used with high-capacity fuses of the current-limiting type.

Modern fused power-circuit devices are rated for breaking load, for withstanding let-through thermal and magnetic stresses during fuse operation on short circuit, and for closing against a short circuit. Typical units are rated for load breaks up to 12 times their nameplate current rating.

With proper fusing, the assembly is capable of fault clearance through the fuses. With the switch closed and fitted with high-interrupting-capacity fuses, faults up to 200,000 rms amperes at 600 V can be cleared promptly and safely without damage to equipment or personnel.

Another type of high-capacity switching assembly is the fusible-CB load interrupter. One form of this assembly consists of a fully coordinated combination of a circuit-breaker mechanism that provides switching and a set of high-capacity current-limiting fuses that provide overload and short-circuit protection. Another form consists of a molded-case or air CB to provide overload protection up to the interrupting duty of the CB contact mechanism, with fuses connected on the load terminals to provide short-circuit protection at current levels above the safe interrupting capacity of the breaker, and up to 200,000 A.

IMPORTANT: Switches, motor starters, panelboards, and other equipment that has been investigated and found suitable for use with specific fuses are marked with the class of fuse intended to be used in the equipment and with an available current rating applicable to that piece of equipment. With the designated fuses installed in equipment that is so marked, the switch or panel or other equipment is suitable for use on circuits that can deliver short-circuit currents up to the current rating of the equipment or the interrupting rating of the fuse—whichever is lower.

NOTE: An interrupting rating on a fuse included in a piece of equipment does not automatically qualify that equipment for use on circuits with available current higher than the current rating of the equipment itself.

Automatic transfer switches

Automatic transfer switches are used to automatically transfer load from normal to emergency power upon failure of the normal power supply. An automatic transfer switch (ATS) also automatically returns the load to normal power when the normal source returns.

An ATS can be designed as a single device with a double-throw switch arrangement. Basic types are the mechanically held and the magnetically held forms, with either single-coil or dual-coil operators and interlocking protection. They can also consist of two circuit breakers or magnetic contactors furnished with individual operators and mechanical and electrical interlocking to provide two definite positions: normal closed/emergency open, and normal open/emergency closed. Ratings of automatic transfer switches range up to about 4000 A at 600 V.

Figure 441 shows two types of automatic-transfer operating mechanisms. In (a) is shown the single-coil, double-throw contactor with magnetic operation for either transfer, which is mechanically held in either operating position. Relays 1V, 2V, and 3V provide protection against phase failure or voltage drop on one line. Relay SE is the circuit selector, and relay LO supervises emergency. The ATS in Fig. 441(b) makes use of two separate contactors, mechanically and electrically interlocked to prevent simultaneous opening or closing. Undervoltage relay UV senses loss of normal power and initiates opening of the NORMAL contactor. The availability of emergency power then closes the EMERGENCY contactor. Each contactor has a normally closed auxiliary contact in the coil circuit of the other contactor.

Figure 442 shows an ATS making use of two separate circuit breakers that are mechanically and electrically interlocked with a common operator or individual motor operators. Overcurrent (short-circuit) protection is provided for both normal and emergency supply. The control panel may incorporate individual-phase supervisory relays and automatic starting and connection of the emergency generator or other source.

Circuit-breaker transfer switches in the larger ampere sizes are made with motor-driven breaker handles. Again, these are electrically and mechanically interlocked assemblies, with double-source design of the transfer control circuit, and with no neutral position in the mechanism. Some use a single motor for transfer; others use two motors.

FIG. 441 Automatic transfer switches are made with either one or two operating coils. (*a*) Circuits shown with normal source energized. (*b*) Circuits shown without any energy.

FIG. 442 Two circuit breakers combine to provide an automatic transfer operation.

Depending upon the electrical control circuit, circuit-breaker–type transfer switches may incorporate protection against voltage drop or failure on any one phase or may simply respond to failure of all phases. Full-phase protection is provided by adjustable voltage relays connected line to neutral or line to line.

Although different control-circuit elements are indicated above for each of the three types of automatic transfer switching, any one of them may use any control circuits to provide a desired response of the transfer assembly to load, normal-source, or emergency-supply conditions. And a wide variety of control arrangements is possible for controlling the source of emergency power.

NEC Article 700, "Emergency Systems," Article 750, "Standby Power Generation Systems," and Article 517, "Health Care Facilities," incorporate requirements for the automatic transfer of normal and alternative power sources. However, a specific type is not defined, so any of the above-described ATS assemblies may meet application requirements.

In selecting an automatic transfer switch, the designer should refer to the UL *Electrical Construction Materials List (Green Book)* and Standard UL-1008, which covers automatic transfer switches. In the *Green Book,* the information given under "Transfer Switches" also applies and is supplementary to the data given under "Automatic Transfer Switches." This information provides requirements on an approved switch. For example, automatic transfer switches with or without integral overcurrent protection are suitable for continuous use at 100 percent of rated current. Some designs may be marked "Continuous load current not to exceed 80 percent of switch rating," and they should be so applied. Also, listed switches are generally considered to be suitable for all types of loads—motors, tungsten lamps, electric heaters, etc.

Underwriters Laboratories Standard 1008 requires that automatic transfer switches be marked with (1) load classification, (2) voltage sensing on normal supply, (3) voltage and frequency on alternative supply, (4) time-delay features, and (5) closing rating and withstandability rating. The entire standard should be studied and used as a guide in selecting ATSs. Other important selection considerations, most of which are covered in UL Standard 1008, include:

1. The continuous-current rating must be sufficient for the load, in both the normal and emergency positions.

2. The interrupting capacity of the switching units must be sufficient to handle the currents encountered during closing and opening, on either the normal or the emergency supply.

3. The short-circuit rating must be sufficient to withstand let-through fault currents. The UL minimum requirement is 20 times the switch rating but not less than 10,000 A.

4. The voltage, enclosure type, number of poles, whether ac or dc, and related factors, such as mounting space and location and installation requirements, must be considered.

5. Electrical and mechanical interlocking must ensure that both switching mechanisms cannot possibly be closed at the same time. Single-throw units have inherent mechanical interlocking. On dual-coil contactor types, each contactor must have normally closed auxiliary contacts in the coil circuit of the other contactor to provide electrical interlocking. Circuit-breaker types should have mechanical as well as electrical interlocking.

6. The protection may be full-phase protection (protection against single-phasing) or feeder-failure protection. Full-phase protection ensures that transfer to the emergency source will occur when the voltage on any phase drops to approximately 70 percent or less. Feeder-failure protection provides for load transfer only on complete feeder failure; it does not protect against partial voltage loss or phase failure.

7. Proper time-delay characteristics should be incorporated in the control circuit to isolate the response of the assembly from transient voltage fluctuations in the normal system. Sensitivity to such fluctuations causes false operation of the switch and may damage the starting circuit of the generator.

Other important selection considerations include: (1) adjustable time delay (on transfer to emergency, or retransfer, and on engine starting); (2) adjustable-voltage sensing relay (for full protection on 3-phase systems, three units are required; on single-phase systems only one relay is used; for accurate voltage sensing, some of these relays utilize circuits incorporating transistors and zener diodes); (3) manual operator; (4) test switch; (5) frequency relays; (6) speed of transfer; (7) pilot lights to indicate the position of the switch.

Motor Centers

A power panelboard is a fused-switch, circuit-breaker, or fuse panelboard from which motor branch circuits originate. It may provide only protection for the branch-circuit conductors, but in most cases the power panel also provides disconnect means for each motor.

A motor control center is a dead-front assembly of cubicles, each of which contains branch-circuit overcurrent protection, motor disconnect means, motor controller, and motor running overcurrent protection. It is a type of switchboard which contains all the protective and control means for the motors supplied from it. The selection and application of power panels and motor control centers should be related to future requirements. Such units must have necessary spare capacity in their buses and must be adaptable to anticipated changes or expansion. The layout of power panels and motor control centers should be related to voltage drop and the lengths of feeders.

Miscellaneous motor loads in the majority of commercial and institutional buildings and in many industrial buildings are usually circuited from power panels to which feeders deliver power. This type of distribution is a standard method of motor circuiting, generally limited to handling a number of small integral-horsepower or fractional-horsepower motors located in a relatively small area such as a fan room or pump room.

The feeder to a power panel may be a riser in a multistory commercial or institutional building, run from a basement switchboard. In a one-level industrial area, the feeder to a power panel may be run from a main or loadcenter switchboard or from a loadcenter substation in a high-voltage distribution system. In commercial and industrial buildings, motor feeders may be run from a main switchboard or a loadcenter substation to a motor control center serving a large group of motors in a machine room or in any compact area where the motors are relatively close together and close to the control-center assembly.

In industrial buildings, where a large number of motors are used over a large area, power is generally distributed to the individual motor loads by tapping motor branch circuits from the feeder. As was indicated in the discussion of branch circuits, a feeder may be tapped to motor branch circuits in several ways, by providing various combinations of branch-circuit overcurrent protection and motor control disconnect means. The feeder-tap method of supplying motor branch circuits is, of course, the basis for plug-in busway distribution.

Wiring Methods

The selection and layout of suitable wiring methods to provide for circuiting of electrical systems is a vital design task.

Although busway and interlocked armor cable have been constantly gaining popularity for use in distribution systems, wire in rigid conduit (metal or nonmetallic), intermediate metal conduit, and electrical metallic tubing are still the most used methods. For underground applications, a number of metal-jacketed cables and non-metal-jacketed cables have proved extremely reliable for direct earth burial. And for in-building feeders, Type AC and MC cables offer advantages in many locations. Cable assemblies—such as aluminum-sheathed cable, corrugated-metal-jacketed cables, and prewired plastic conduit—also offer advantages in many types of installations.

For outdoor distribution, a choice has to be made between overhead and underground circuits. Primary and secondary distribution requirements between buildings can usually be served best by underground installations for a number of reasons:

1. Accident hazards are minimized and isolated.
2. The ambient temperature is lowest, and cable capacities are highest.
3. Cables do not get in the way, obstructing the expansion of buildings or interfering with work.
4. The appearance of installations is neater, and architectural and landscape design is not marred by unsightly poles.

The cost of underground installation is not always justified under the following conditions:

1. Where chemical ingredients in the soil constitute a serious corrosion threat to directly buried metal conduits (galvanized steel or aluminum).
2. Where extensive changes to or expansion of buildings or plants would require the removal or alteration of underground feeders.
3. Where proper installation of underground circuits is complicated by such ground conditions as excessive rock.

Conductors

Feeder conductors for primary and secondary circuits should be selected carefully, on the basis of the many factors discussed in previous sections of this book.

Conductor size is determined by:

1. Load current to be supplied
2. Voltage drop and regulation
3. Temperature rise, within the limitations of the insulation
4. Reasonable energy losses
5. Ability to withstand short-circuit heating
6. Spare capacity for load growth

Although copper has been the long-time standard electric conductor material, there has been steady and sometimes rapid growth in the use of aluminum conductors. Often, the material cost savings effected with aluminum conductors can be very substantial. And the light weight of such conductors provides a reduction in the cost of installation labor. Lugs and connectors suitable for aluminum conductors are widely available.

Figure 443 presents data on the use of aluminum conductors for large feeders in commercial, institutional, and industrial buildings. The top chart shows feeder makeups using copper and aluminum conductors, giving equivalents (reading across) on the basis of ampacity and voltage drop. The bottom chart shows how compact-strand aluminum conductors (second column from left) of the same ampacity as copper conductors (first column at left) permit the same number of conductors in a given size of conduit as their copper equivalents—except for the few cases as shown there. [See Note 4 in Part **A** (Tables) of **NEC** Chapter 9.] Allowable current-carrying capacities of both copper and aluminum conductors in raceway or cable, for direct burial and in free air, are given in **NEC** Tables 310-16 through 310-19. Although copper and aluminum conductors are usually compared on the basis of current rating alone, voltage drop is a better basis for comparison because practical equivalents are thereby indicated. Of course, aluminum

Equivalent aluminum and copper conductors based on voltage drop on three single conductors enclosed in a magnetic conduit, 80% power factor, 75°C insulation. Wire temperature and corresponding current-carrying capacities at 30°C ambient.

Copper conductors			Aluminum conductors		
Size	Voltage drop phase to phase per amp per 1000 ft	Amps	Size	Voltage drop phase to phase per amp per 100 ft	Amps
1/0	.232	150	3/0	.228	155
2/0	.193	175	4/0	.190	180
3/0	.163	200	250MCM	.169	205
4/0	.138	230	350MCM	.134	250
250MCM	.126	255	400MCM	.124	270
350MCM	.104	310	500MCM	.108	310
400MCM	.097	335	600MCM	.098	340
500MCM	.088	380	750MCM	.089	385

Copper size AWG/kcmil	Compact size AWG/kcmil	CONDUIT TRADE SIZE IN INCHES															
		1		1¼		1½		2		2½		3		3½		4	
		Cu THW	Al XHHW	Cu THW	Al XHHW	Cu THW	Al XHHW	Cu THW	Al XHHW	Cu THW	Al XHHW	Cu THW	Al XHHW	Cu THW	Al XHHW	Cu THW	Al XHHW
4	2	3	3	5	5	7	7										
2	1/0	2	1	4	3	5	5										
1	2/0					3	3	4	4	6	7	9	9				
1/0	3/0					2	2	3	3	5	6	8	8				
2/0	4/0							3	3	5	4	7	6	10	10		
3/0	250							2	2	4	4	6	5	9	8		
4/0	300									3	3	5	5	7	7	10	10
250	350									2	3	4	4	6	6	8	9
350	500											3	3	4	5	6	6
400	600									2	2	4	4	5	5	7	7
500	700											3	3	4	4	6	6

FIG. 443 Feeder makeups of equivalent ratings can be developed for copper or aluminum (bottom table shows number of conductors in each size of conduit).

conductors of the same current rating or voltage-drop rating as copper conductors will be larger and may require a larger conduit. This must be considered in an economic analysis of aluminum conductor applications.

Single insulated conductors—such as TW or THHN or any of the other so-called "building wires"—must be combined with some type of raceway to make up an **NEC**-recognized "wiring method" or may be used in cable tray or in a messenger-supported assembly (as covered in **NEC** Article 321).

Insulation must be selected according to:

1. Cost
2. Temperature rise, including both ambient temperature and temperature due to the heating effect of the load
3. Ease of installation
4. Environment of application, such as moisture, fumes, chemicals, petroleum products, acids, alkalies, etc.

Protective coverings must be selected according to:

1. Excellence of protection of the insulation against environmental factors (moisture, chemicals, acids, etc.)
2. Excellence of protection against physical damage, as by abrasion, impact, and cutting
3. Ability of metallic coverings to withstand corrosion and electrolysis

A very important consideration, indicated above, is the provision of a conductor size sufficient to handle the potential heating loads of short-circuit currents. With the expanded use of circuit-breaker overcurrent protection, coordination of protection from loads back to the source has introduced time delays in the operation of overcurrent devices. Cables in such systems must be able to withstand any impressed short-circuit currents for the duration of the overcurrent delay. For example, a motor circuit to a 100-hp motor might be required to carry as much as 15,000 A for a number of seconds. To limit damage to the cable due to heating, a much larger size conductor than is necessary for the load current alone may be required.

That concern for the ability of conductors to withstand the heating of short circuits has been expressed often by design engineers with respect to the smaller sizes of wires, as shown in the following example.

EXAMPLE

A panelboard has 20-A breakers rated at 10,000 A interrupting capacity and No. 12 copper branch-circuit wiring. The available fault current at the point of breaker application is 8000 A. The short-circuit withstandability of a No. 12 copper conductor with plastic or polyethylene insulation rated 60°C is approximately 3000 A of fault or short-circuit current for one cycle.

QUESTION

Assuming that the CB will take at least one cycle to operate, would the use of the No. 12 conductor where it is exposed to 8000 A violate **NEC** Section 240-2? This section states that overcurrent protection for conductors and equipment is provided for the purpose of opening the electric circuit if the current reaches a value which will cause an excessive or dangerous temperature in the conductor or conductor insulation. The 8000 A of available fault current would seem to call for conductors with that rating of short-circuit withstandability. This could mean that the branch-circuit wiring from all 20-A CBs in this panelboard must be at least No. 6 copper (the next larger size suitable for an 8000-A fault current).

ANSWER

As noted in UL Standard 489, a CB is required to operate safely in a circuit where the available fault current is up to the short-circuit current value for which the breaker is rated. The CB must clear the fault without damage to the insulation of conductors of proper size for the rating of the CB. A UL-listed 20-A breaker is, therefore, tested and rated to be used with wire rated for 20 A (say, No. 12 THW) and will protect the wire when applied at a point in a circuit where the available short-circuit current does not exceed the value for which the breaker is rated. This is also true of a 15-A breaker on No. 14 (15-A) wire, for a 30-A breaker on No. 10 (30-A) wire, and for all wire sizes.

Underwriters Laboratories Standard 489 states:

A circuit breaker shall be capable of performing successfully when operated under short-circuit conditions as described in paragraphs 193–209. There shall be no electrical or mechanical failure of the device; cotton or cloth placed in contact with the enclosure at any hole, slot, or other opening as described in the following paragraphs shall not be ignited; there shall be no damage to the insulation of conductors, of proper size for the rating of the breaker, used to wire the device; and after the final operation the breaker shall be capable of making and breaking 200 percent of its rated current when manually operated.

Any conductor used in a "wet location" (refer to the definition under *location* in **NEC** Article 100) must be one of the designated types, each of which has the letter "W" in its marking to indicate suitability for wet locations. Any conduit run underground is assumed to be subject to water infiltration and is, therefore, a wet location, requiring the use of only the listed conductor types within the raceway.

Although Type RHW conductor is suitable for wet locations, it is not approved for direct burial. If, however, the conductor is marked "RHW-USE"—that is, it is listed and recognized as both a single-conductor RHW and a single-conductor Type USE (underground service entrance) cable—then it is suited for direct-burial application as Type USE. "Conductors" for direct burial must be "listed"—which means certified by some kind of testing lab as suitable for direct burial in the earth.

Type UF cable is acceptable for direct earth burial (Fig. 444). The UL *Green Book* notes that listed USE cable is recognized for "burial directly in the earth."

All cables that are directly buried must be installed at a depth of at least 24 in below grade—as specified in **NEC** Section 300-5, but subject to reduced burial depths as covered there.

Raceways

For branch circuits, feeders, and services, conduit may be rigid metal conduit, intermediate metal conduit, rigid nonmetallic conduit, or electrical metallic tubing (for 4-in or smaller conduit). Rigid metal conduit and intermediate metal conduit of steel or aluminum provide excellent mechanical protection for conductors, enclose possible faults, and provide a low-resistance path to ground to assure quick operation of circuit protective devices in the event of fault currents. EMT offers light weight with ease of handling, cutting, and bending.

A big advantage in the use of conduit for feeders instead of a single high-capacity busway feeder is that a conduit installation reduces the distance to which short-circuit problems extend into a building from the service. Breaking the load into the smaller feeders inherent in a conduit-and-cable distribution system introduces more impedance between distribution point and final branch-circuit overcurrent protective devices, thereby reducing the size of the short circuits which these final overcurrent protective devices must handle, while total impedance and consequent distribution-system losses introduced by breakup into small feeders are not appreciably

Overcurrent protective devices

Service entrance cables

Feeder cable or branch circuit

Direct burial in earth cables without over-current protective devices, such as service drop and service entrance, shall be type USE.

Direct burial in earth cables protected by over-current devices may be
a) type USE, or
b) type UF.

Cables of one or more conductors for direct burial in the earth shall be type USE or type UF.

All conductors for single cables, including the neutral, for each service, feeder, subfeeder or branch circuit shall be run continuously in the same trench or raceway.

Supplementary mechanical protection, such as a covering board, concrete pad, raceway, etc, may be required by the authority enforcing the Code.

FIG. 444 Type UF and Type USE cables ("U" for "underground") are recognized for direct burial in the earth.

increased for the overall project. And, in a conduit installation, electrical faults are localized to relatively small portions of the distribution system: Repairs can be made quickly, with little interruption of service.

The lower reactance of circuits in aluminum conduit can produce an important reduction in voltage drop. For example, in the initial design of an office building, the sizing of feeders for the given loads was based on the voltage-drop characteristics of circuits in steel conduit. In the original plan, the available kilovoltamperes at each lighting panel conformed to anticipated initial and future demands of tenants. When, however, a decision was made to install the feeders in aluminum conduit, the available tenant kilovoltampere capacity had to be recalculated on the basis of the lower reactance—and, consequently, voltage drop—of circuits in nonmagnetic enclosures. As a result, it was found that the load could be increased by 10 to 15 percent without exceeding the conductor current rating or the limit of 2 percent voltage drop in each feeder.

Galvanized rigid steel conduit and galvanized intermediate steel conduit do not generally require supplementary corrosion protection when they are directly buried in soil. The use of the word "generally" in the UL instructions indicates that it is the responsibility of the designer and/or installer to use supplementary protection where specific soils are known to corrode such conduits. Where the corrosion of underground galvanized conduit is known to be a problem, a protective jacketing or a field-applied coating of asphalt paint or equivalent material must be used on the conduit. But, UL notes on "supplementary nonmetallic coatings" must be observed with regard to resistance to corrosion.

Aluminum conduit that is directly buried in soil is said by UL to require supplementary corrosion protection. But, again, it is completely the task and responsibility of the designer and/or installer to select an effective protective coating for the aluminum conduit, because UL says "supplementary nonmetallic coatings presently used have not been recognized for resistance to corrosion." That could also be interpreted as prohibiting the use of directly buried aluminum conduit.

Galvanized rigid steel conduit and galvanized intermediate steel conduit installed in concrete do not require supplementary corrosion protection. Aluminum conduit installed in concrete definitely requires supplementary corrosion protection, but the supplementary protective coatings presently used "have not been recognized for resistance to corrosion." Watch out for this! The UL warns, "Wherever ferrous metal conduit runs directly from concrete encasement to soil burial, severe corrosive effects are likely to occur on the metal in contact with the soil." Supplementary protective coating on conduit at the crossing line can eliminate this condition.

Intermediate metal conduit (IMC) is a conduit whose wall thickness is smaller than that of rigid metal conduit but greater than that of EMT. IMC has the same threading and standard fittings, and the same general application rules, as rigid metal conduit. It actually is a lightweight rigid steel conduit which requires about 25 percent less steel than heavy-wall rigid conduit.

IMC may be used in any application for which rigid metal conduit is recognized, including use in all classes and divisions of hazardous locations as covered in **NEC** Sections 501-4, 502-4, and 503-3. Its thinner wall makes it lighter and less expensive than standard rigid metal conduit, but it has outstanding strength. Its lighter weight facilitates handling and installation at lower labor costs than for rigid metal conduit. Because it has the same outside diameter as rigid metal conduit of the same trade size, it has greater interior cross-sectional area. Although this extra space is not recognized by the **NEC** as permitting the use of more conductors than can be used in the same size of rigid metal conduit, it does make wire pulling easier.

All approved rigid nonmetallic conduits are suitable for underground installations. Some types are approved for direct burial in the earth, while other types must be encased in concrete for underground applications. The nonmetallic conduits include fiber conduit, asbestos-cement conduit, soapstone, rigid polyvinyl chloride (PVC) conduit, fiberglass epoxy, polyethylene conduit, and styrene conduit. Of these, medium-density polyethylene conduit and styrene conduit are not UL-listed. High-density polyethylene conduit and the others are UL-listed. The listed and labeled conduits differ widely in weight, cost, and physical characteristics, but each has certain application advantages.

The only nonmetallic conduit approved for use above ground at the present time is rigid polyvinyl chloride (PVC Schedule 40 or Schedule 80). Since not all PVC conduits are suitable for use above ground, the UL label on each conduit length will indicate if it is suitable for such use. UL applications data are detailed and divide "rigid nonmetallic conduit" into three categories, with specific instructions concerning the use of each category.

Rigid nonmetallic conduit is a general-purpose raceway for interior and exterior wiring, concealed or exposed in wood or masonry construction *under the stated conditions*. Only PVC is acceptable as a rigid nonmetallic conduit for in-building use (above ground).

For above-ground applications, rigid nonmetallic conduit must be Schedule 40 or Schedule 80 PVC conduit. Rigid nonmetallic conduit may be used above ground to carry high-voltage circuits without the need for encasing the conduit in concrete. Above-ground use is permitted indoors and outdoors.

Other rules on rigid nonmetallic conduit are:

1. Rigid nonmetallic conduit may not be used to support boxes or other equipment (Fig. 445).

2. When equipment grounding is required for metal enclosures of equipment used with rigid nonmetallic conduit, an equipment grounding conductor must be provided. Such a conductor must be installed in the circuit along with the circuit conductors.

3. In using nonmetallic conduit, care must be taken to ensure temperature compatibility between the conduit and the conductors in it. For instance, a conduit that has a 75°C temperature rating (the temperature at which it might melt and/or deform) must not be used with conductors which have a 90°C temperature rating and which will be loaded so they are operating at or near their temperature limit. Some available PVC rigid conduit is listed by UL and marked to indicate that it is suitable for use with conductors rated at 90°C. The UL data give the acceptable ambient temperatures and conductor temperature ratings. Conductors with 90°C insulation may be used at the higher ampacities of that temperature rating only when the conduit is encased in concrete.

Box supported by two threaded conduits stubbed up out of concrete or ground

18-in. max.

FIG. 445 With threaded connections (not locknut and bushing), rigid metal conduit or IMC may support boxes, but rigid nonmetallic conduit may not support equipment or boxes.

EMT is a general-purpose raceway of the same nature as rigid metal conduit and IMC. Although rigid metal conduit and IMC afford maximum protection for conductors under all installation conditions, in many instances it is permissible, feasible, and more economical to use EMT to enclose circuit wiring rated 600 V or less. Because EMT is lighter than conduit, however, and is less rugged in construction and connection details, its use is restricted to locations (either exposed or concealed) where it will not be subjected to severe physical damage or (unless suitably protected) to corrosive agents. EMT distribution systems are constructed by combining wide assortments of related fittings and boxes. Connection is simplified by threadless components that include compression, indentation, and setscrew types.

Some questions have been raised about the acceptability of EMT directly buried in soil. The **NEC** gives EMT exactly the same recognition for direct burial that it gives to rigid steel conduit. It certainly seems clear that if galvanizing is enough corrosion protection for rigid steel conduit, it must provide equivalent protection for EMT in direct burial. In the UL listing "Electrical Metallic Tubing," a note says that "galvanized steel electrical metallic tubing in a concrete slab below grade level may require supplementary corrosion protection." (The word "may" leaves the decision up to the designer and/or installer.) But note that the rule carefully refers to "galvanized steel" EMT and not just to steel EMT.

The next note says, "In general, steel electrical metallic tubing in contact with soil requires supplementary corrosion protection." That sentence certainly admits that there are locations where soil conditions are such that supplementary corrosion protection is not required. Past experience and local soil conditions should be considered in determining the acceptability of direct burial of EMT as well as IMC and rigid conduit, with appropriate attention given to additional protection against corrosion if necessary. Of course, the ruling of the local electrical inspector should be sought and followed.

EMT up to 2 in in size has the same interior cross-sectional area as corresponding sizes of rigid metal conduit. EMT sizes of 2½ in and larger have the same outside diameter as rigid metal conduits of corresponding sizes. Accordingly, the interior cross-sectional areas (in square inches) are proportionally larger, and this provides more wiring space for greater ease of installation of conductors in these larger sizes of EMT. It is significant that the greater internal diameters of EMT provide cross-sectional areas that are from 16 to 22 percent greater than those of corresponding rigid metal conduit. But it should be stressed that these greater internal cross-sectional areas do not offer the capability of filling EMT runs with more conductors than could be used in corresponding sizes of rigid conduit. The number of conductors permitted in "conduit or tubing" is the same for rigid, IMC, and EMT, according to Tables 1 through 8 of **NEC** Chapter 9.

NEC Table 300-5 sets the basic rules on burial depths for underground applications of raceway. Rigid metal conduit and intermediate metal conduit must be buried at least 6 in below grade. EMT and plastic raceways have to be buried at least 18 in. Figure 446 shows some of the exceptions to the basic requirements as given in Section 300-5.

Conductors in raceway

Circuits of different voltages up to 600 V may occupy the same wiring enclosure (cabinet, box, housing), cable, or raceway, provided all the conductors are insulated for the maximum voltage of any circuit in the enclosure, cable, or raceway. For instance, motor power conductors and motor control conductors are permitted in the same conduit. In the past, there has been a long-standing controversy about the use of control-circuit conductors in the same conduit with power leads to motors.

As shown in Fig. 447, a common raceway may be used for both motor power conductors and control-circuit conductors only where two or more motors are required to be operated together to serve their load function. Many industrial and commercial installations contain machines, manufacturing operations, or processes in which a number of motors perform various parts or stages of a task. In such cases, either all the motors operate or none do; the placement of all control wires and power wires in the same raceway cannot produce a situation in which a fault in one motor circuit disables another circuit to a motor that might otherwise be kept operating.

But when a common raceway is used for power and control wires to supply separate, independent motors, a fault in one circuit could knock out others that do not have to be shut down when one goes out. With motor circuits so closely associated with vital equipment such as elevators, fans, and pumps in modern buildings, it is a matter of safety to separate such circuits and minimize the outage due to a fault in a single circuit. For safety's sake, "do not put all your eggs in one basket." But the objectionable loss of more than one motor on a single fault does not apply where all motors must be shut down when any one is stopped—as in multimotor machines and processes.

For those cases where each of several motors serves a separate, independent load with no

Burial-depth requirements of Table 300-5 do not apply to raceways installed like this.

FIG. 446 Exceptions to the basic rules on burial depths of raceways and cables.

interconnection of control circuits and no mechanical interlocking of driven loads, the use of a separate raceway for each motor is required—but only when control wires are carried in the raceways, as shown in Fig. 448. For the three motors shown, it would be acceptable to run the power conductors for all the motors in a single raceway and all the control circuit wires in another raceway. Of course, deratings would have to be made, and there is the definite chance of losing more than one motor on a fault in only one of the circuits in either the power raceway or the control raceway.

1. This common raceway (conduit, EMT, wireway, or etc.) may contain **all** power conductors and **all** control conductors for two or more motors . . .

Motors operate together and are functionally associated as integral parts of a machine or process

Individual conduit runs for power and control wires to each motor   Pushbuttons   MOTOR CONTROL CENTER

2 . . . **but**, the intent of this **Code** rule, along with that of Section 725-15, permits a common raceway **only** where the power and control conductors are for a number of motors that operate integrally—such as a number of motors powering different stages or sections of a multi-motor process or production machine. Such usage complies with Section 725-15 (last sentence), which permits power and control wires in the same raceway, cable, or other enclosure when the equipment powered is "functionally associated"—that is, the motors have to run together to perform their task.

FIG. 447   Mixing of power and control wires in a common raceway is limited.

Separate conduit or raceway is required for power and control wires to each motor . . .

Pushbutton stations

3 power wires plus 3 control wires in each conduit

MOTOR CONTROL CENTER

. . . When these are individual motors that do not operate together as parts of a machine or process—that is, each motor has a separate, independent load that may operate by itself.

FIG. 448   Separate conduit must be used to carry power and control wires to independent motors.

Conduit fill: For the makeup of all circuits, the **NEC** regulates the maximum number of conductors that may be pulled into rigid metal conduit, rigid nonmetallic conduit, intermediate metal conduit, electrical metallic tubing, flexible metal conduit, and liquid-tight flexible metal conduit. The number of conductors permitted in a particular size of conduit or tubing is covered in **NEC** Chapter 9: Tables 1 and 3 cover conductors all of the same size used for either new work or rewiring; Tables 4 to 8 cover combinations of conductors of different sizes used for new work or rewiring. For non-lead-covered conductors, three or more to a conduit, the sum of the cross-sectional areas of the individual conductors must not exceed 40 percent of the interior cross-sectional area of the conduit or tubing, both for new work and for rewiring in existing conduit or tubing, as shown in Fig. 449. Note 3, which precedes the tables of **NEC** Chapter 9, permits a 60 percent fill of conduit nipples not over 24 in long, without the need for derating of ampacities.

FOR NEW WORK OR REWIRING:

Total cross-section area of conductors must not be over 40% of the internal cross-section area of the conduit

Example:
From Table 3B,
NEC Chapter 9
with the 90C
conductors used at
the ampacity of 75 C
500 MCM conductors,
unless equipment is
marked to permit
connection of
90C conductors

4 No. 500MCM THHN or
XHHW in 3-in. conduit

FIG. 449   For three or more conductors in a conduit, the sum of their cross-sectional areas must not exceed 40 percent of the conduit cross-sectional area.

Tables 3A, 3B, and 3C of Chapter 9 give the maximum allowable fill for conduit or tubing in which all conductors are the same size up to 750 MCM, and for conduit sizes from ½ to 6 in.

QUESTION

What is the minimum size of conduit required for six No. 10 THHN wires?

ANSWER

Table 3B of Chapter 9 shows that six No. 10 THHN wires may be pulled into a ½-in conduit.

QUESTION

What is the minimum conduit size for use with four No. 6 RHH conductors with outer covering?

ANSWER

Table 3C of Chapter 9 shows that a 1¼-in minimum conduit size must be used for three to five No. 6 RHH conductors.

QUESTION

What is the minimum conduit size required for four 500-MCM XHHW conductors?

ANSWER

Table 3B shows that 3-in conduit may contain four 500-MCM XHHW (or THHN) conductors.

When all the conductors in a conduit or tubing are not the same size, the minimum required size of conduit or tubing must be calculated. Table 1 of Chapter 9 says that conduit containing three or more conductors of any type except lead-covered, for new work or rewiring, may be filled to 40 percent of the conduit cross-sectional area. Note 2 to this table refers to Tables 4 through 8 of Chapter 9 for the dimensions of conductors, conduit, and tubing to be used in calculating conduit fill for combinations of conductors of different sizes.

EXAMPLE

What is the minimum size of conduit required to enclose six No. 10 THHN, three No. 4 RHH (without outer covering), and two No. 12 TW conductors (Fig. 450)?

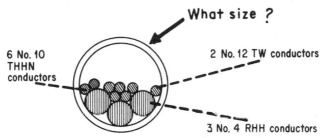

FIG. 450 Minimum permitted conduit size must be calculated when conductors of different sizes are used.

SOLUTION

The cross-sectional areas of the conductors are found in Table 5 of Chapter 9:

No. 10 THHN: 0.0184 in^2

No. 4 RHH: 0.1087 in^2

No. 12 TW: 0.0172 in^2

(Note that RHH without outer covering has the same dimensions as THW.) The total area occupied by conductors is found as follows:

Six No. 10 THHN: 6 × 0.0184	0.1104 in^2
Three No. 4 RHH: 3 × 0.1087	0.3261 in^2
Two No. 12 TW: 2 × 0.0172	0.0344 in^2
Total	0.4709 in^2

In Table 4 of Chapter 9, the fifth column from the left gives 40 percent of the cross-sectional area of each of the conduit sizes listed in the first column at the left. The 40 percent column shows that 0.34 in^2 is 40 percent fill for a 1-in conduit, and 0.60 in^2 is 40 percent fill for a 1¼-in conduit. Therefore, a 1-in conduit would be too small, and a 1¼-in conduit is the smallest that may be used for the 11 conductors.

EXAMPLE

What is the minimum size of conduit for four No. 4/0 TW and four No. 4/0 XHHW conductors?

SOLUTION

From Table 5, a No. 4/0 TW has a cross-sectional area of 0.3904 in^2. Four of these come to 4 × 0.3904 or 1.5616 in^2. From column 11 of Table 5, four No. 4/0 XHHW have a cross-sectional area of 4 × 0.3278, or 1.3112 in^2. Then the total cross-sectional area is

$$1.5616 + 1.3112 = 2.8728 \text{ in}^2$$

The fifth column of Table 4 shows that 2½-in conduit would be too small. A 3-in conduit, with a 40 percent fill of 2.9500 in^2, must be used.

Filling conduit to the **NEC** maximum is a minimal design practice; it is also frequently difficult or impossible from the practical standpoint of pulling the conductors into the conduit, owing to the usual twisting and bending of the conductors within the conduit. Bigger-than-minimum conduit should generally be used to provide some measure of spare capacity for load growth, and in many cases the conduit should be upsized considerably to allow for future installation of anticipated larger sizes of conductors.

In the sizing of conduit, neutral conductors are included in the total number of conductors because they occupy space. A completely separate consideration, however, is the relation of the neutral conductors to the number of "current-carrying" conductors in a conduit, which determines whether a derating factor must be applied:

Neutral conductors which carry only unbalanced current from phase conductors (as in the case of normally balanced 3-wire, single-phase or 4-wire, 3-phase circuits) are not counted in determining the current derating of conductors on the basis of the number in a conduit. Of course, a neutral conductor used with two phase legs of a 4-wire, 3-phase system to make up a 3-wire feeder is not a true neutral in the sense of carrying only current unbalance. Since the neutral carries the same current as the other two conductors under balanced load conditions, it must be counted as a phase conductor in derating for more than three conductors in a conduit.

Figure 451 shows four basic conditions of neutral loading relative to the need for counting the neutral conductor in derating a circuit to fluorescent or mercury ballasts:

Case 1: With balanced loads of equal power factor, there is no neutral current, and consequently no heating due to the neutral. For purposes of heat derating according to the **NEC**, this circuit produces the heating effect of three conductors.

Case 2: With two phases loaded and the third unloaded, the neutral carries the same current as the phases, but there is still the heating effect of only three conductors.

Case 3: With two phases fully loaded and the third phase partially loaded, the neutral carries the difference in current between the full phase value and the partial phase value, so again there is the heating effect of only three full-load phases.

Case 4: With a balanced load of fluorescent ballasts, third-harmonic generation causes a neutral current approximating the phase current, and there is the heating effect of four conductors. The neutral conductor should be counted with the phase conductors in determining conductor derating due to conduit occupancy.

Because the neutral of a 3-phase, 4-wire wye feeder to a load of fluorescent or mercury ballasts will carry current even under balanced loading on the phases, such a neutral is not a

FIG. 451 All neutrals must be counted for conduit fill, but only "true" neutrals do not count for load-current derating.

true non-current-carrying conductor and should be counted as a phase wire in determining the number of conductors to arrive at a derating factor for more than three conductors in a conduit. As a result, the conductors of a 3-phase, 4-wire feeder to a fluorescent load should have a load-current rating of only 80 percent of their normal ampacity from **NEC** Table 310-16.

Ampacity derating

IMPORTANT: Refer to the discussion "Note 8: ampacity derating" in Section 2 of this handbook. Note 8 to Table 310-16 on ampacity requires that the ampacity of a conductor be reduced to a value lower than the ampacity value given in the table when a raceway contains more than three current-carrying conductors. As a result, the ampacity of the conductors is changed and the conductors must be protected in accordance with the derated ampacity value and not in accordance with the tabulated value.

Because conductor "ampacity" is reduced when more than three conductors are used in a conduit, the overcurrent protection for each phase leg of a parallel makeup in a single conduit may be equal to the sum of the derated ampacities of the number of conductors used per phase leg. That would satisfy Section 240-3, which requires conductors to be protected at their ampacities. Because ampacity is reduced in accordance with the percentage factors given in Note 8 for more than three conductors in a single conduit, that derating dictates the use of multiple conduits for parallel-makeup circuits to avoid the penalty of loss of ampacity.

Figures 452 and 453 show examples of circuit makeups based on the unsafe concept of load limitation instead of ampacity derating, as applied to overcurrent rating and conductor ampacity for continuous loading [Section 220-10(b)].

Bunched or bundled cables must also have their ampacities reduced from the ampacity values shown in Table 310-16.

Note 8 does not, however, apply to conductors in wireways and auxiliary gutters. Wireways or auxiliary gutters may contain up to 30 conductors at any cross section (excluding signal circuits and control conductors used for starting duty only between a motor and its starter). The total cross-sectional area of the group of conductors must not be greater than 20 percent of the interior cross-sectional area of the wireway or gutter. And ampacity derating factors for more than three conductors do not apply to wireway the way they do to wires in conduit. However,

THIS IS DANGEROUS
AND MUST NOT BE DONE...

Min. 2½-in. c

400 A

CB or
fused
switch

Six 3/0 THW,
two per phase.
Ampacity of
each = 200 A.
No derating.

Max. permitted
load on overcurrent
protection:
Noncontinuous, 400 A
Continuous, 320 A
[Sec. 220-10(b)]

Note 8 max. conductor
load rating per phase
= 0.8 × 400 = 320 A
continuous or
noncontinuous

Possible current in excess
of conductor thermal limit
= 400 − 320 = 80 amps

... BUT THIS IS O.K.

Min. 3-in. c

400 A

CB or
fused switch

Three 500 MCM THW, one per phase.
Ampacity of each = 380 A

Max. conductor load
rating per phase:
Noncontinuous, 380 A
Continuous, 304 A
[Sec. 220-10 (b)]

Possible current in excess
of conductor thermal limit
= 400 − 380 = 20 amps

... OR THIS

Min. 3-in. c

400 A

CB or
fused switch

Three 500 MCM THHN, one per phase
Ampacity of each = 430 A

Max. conductor load
rating per phase:
Noncontinuous, 430 A
Continuous, 344 A

Possible current in excess
of conductor thermal limit
= 400 − 430 = 0 amps

FIG. 452 Ampacity reduction is required for more than three current-carrying conductors in a conduit.

if the ampacity derating factors from Note 8 of Table 310-16 are used, there is no limit to the number of wires permitted in a wireway or an auxiliary gutter. But, the sum of the cross-sectional areas of all conductors contained at any cross section of the wireway must not exceed 20 percent of the cross-sectional area of the wireway or auxiliary gutter. More than 30 conductors may be used under those conditions.

Min. 3-in. c

600 A

Nine 3/0 THW, 3 per phase
Ampacity of each = 200 A
No derating

CB rated for
continuous operation
at 100% load

DANGEROUS !

Note 8, max. conductor load rating per phase
= 0.7 × 600 = 420 A, continuous or noncontinuous

Possible current in excess of conductor thermal limit
= 600 − 420 = 180 amps

Min 4-in. c

800 A

Six 500 MCM THW, 2 per phase
Ampacity of each = 380A
No derating

CB or fused switch

DANGEROUS!

Note 8, max. conductor load rating per phase
= 0.8 × 760 = 608 A, continuous or noncontinuous

Possible current in excess of conductor thermal limit
= 800 − 608 = 192 amps

FIG. 453 Parallel-conductor feeder makeup must not be used in a single conduit without ampacity reduction.

Conductor temperature ratings

IMPORTANT: Refer to the discussion of "Conductor temperature rating" in Section 2 of this handbook. As discussed there, a general UL rule says that any electrical equipment rated over 100 A must be used with either 60°C conductors or 75°C conductors operating at the respective ampacities of those conductor insulation-temperature ratings unless the equipment is marked for use with conductors rated at higher temperatures. Equipment rated up to 100 A is generally limited to use with 60°C conductors or higher-temperature conductors operating at the 60°C ampacities for the particular conductor sizes, unless equipment is marked otherwise.

The UL temperature limitation is applied to the selection of feeder conductors in the steps that follow. The procedure is generally similar to the procedure described in Section 2 for selecting branch-circuit conductors. It can also become very involved and complex.

For the feeder in Fig. 454:

Step 1

Because the load on the feeder is continuous, the 100-A, 3-pole CB must have its load current limited to 80 A [80 percent of its rating, per **NEC** Section 220-10(b)]. This is sufficient for the 76-A load. (A CB or fused switch may be loaded continuously to 100 percent of the CB or fuse ampere rating only when the assembly is UL-listed for such use.)

Step 2

The CB load terminals are recognized by UL for use with 60°C conductors or higher-temperature conductors loaded not over the 60°C ampacity of the given size of conductor.

100A, 3∮ CB

100A, 3∮, 4-wire feeder
to a panelboard for continuous
mercury-vapor lighting, with
a 76A load on each phase
busbar

N

Distribution
panelboard

FIG. 454 Feeder conductors for equipment rated up to 100
A must not exceed their 60°C ampacity.

Step 3

The feeder phase conductors must have an ampacity of not less than 125 percent of the continuous load of 76 A, which means not less than 95 A, because that is also required by **NEC** Section 220-10(b). The feeder neutral conductor is not subject to this limitation because it is not connected to the terminals of a heat-producing device (such as a CB or fusible switch). **NEC** Section 220-22 is the basic rule that covers the sizing of a feeder neutral. Section 220-10(b) covers the sizing of feeder phase conductors.

Step 4

If 60°C copper conductors are used for this feeder, reference must be made to the second column of Table 310-16. This feeder supplies electric-discharge lighting; therefore, Note 10(c) to the table requires that the feeder neutral be counted as a current-carrying conductor because of the harmonic currents present in the neutral. Then, because there are four current-carrying conductors in the conduit, the ampacity of each conductor must be derated to 80 percent of its value in the second column of Table 310-16. After the conductor is derated to 80 percent, it must have an ampacity of at least 95 A, as required by Section 220-10(b). From Table 310-16, a No. 1/0 TW conductor is rated at 125 A when only three conductors are used in a conduit. With four conductors in a conduit, the 125-A rating is reduced to 80 percent of that value (0.8 × 125 A), or 100 A, which is properly protected by the 100-A CB.

Note that this derating of ampacity to 80 percent, based on Note 8 of Tables 310-16 through 310-19, is in addition to the 80 percent load limitation (not a conductor derating) of Section 220-10(b). This is, in effect, a "double derating." **NEC** Sections 210-22(c) and 220-2(a) used to contain exceptions that seemed to make such a double derating unnecessary for branch circuits. *But,* Section 220-10(b) does not contain an exception that eliminates the need for double derating of feeder conductors. It is certainly true that such a double derating provides a valuable and important amount of reserve capacity that is needed in feeder circuits to effectively minimize damage due to careless and/or "temporary" overloading during the operating life of the system.

In Note 8 to **NEC** Tables 310-16 through 310-19, Exception No. 2 states that it is not necessary for a feeder protective device to be rated at 125 percent of a continuous load current [as normally required by Section 220-10(b)] *if* the circuit conductors have had their ampacity derated in accordance with Note 8. But that is illogical and completely inconsistent with UL rules that require a continuous load to be limited to not more than 80 percent of the rating of any circuit breaker or fused switch that is *not* UL-listed and marked for continuous loading to 100 percent of its rating—without any relationship to conductor derating. When conductors are used in a raceway, the conductors must be protected at the derated ampacity. And then—*in addition*—if the circuit breaker or fused switch that provides the overcurrent protection is *not* UL-listed and marked for 100 percent continuous load, the continuous circuit loading must not exceed 80 percent of the rating of the circuit breaker or fuses.

The feeder circuit of four No. 1/0 TW conductors, rated at 100 A, would require a minimum of 2-in conduit.

NOTE: A reduced size of neutral could be used, because Section 220-10(b) does not apply to the neutral.

Step 5

If 75°C conductors were used for this feeder circuit instead of 60°C conductors, the calculations would be different. For THW copper conductors, Table 310-16 shows that No. 1 conductors, rated at 130 A for not more than three current-carrying conductors in a conduit, would have an ampacity of 0.8×130 A or 104 A when four are used in the conduit and derated. Because 125 percent of 76 A is 95 A, the 104-A conductor ampacity would satisfy Section 220-10(b) as to the ampacity of the feeder conductors for a continuous load.

Although UL listing and testing of the CB is based on the use of 60°C conductors, the use of No. 1 75°C THW conductors is acceptable because the terminals of the breaker in this case would not be loaded to more than the ampere rating of a 60°C conductor of the same size. A No. 1 60°C TW conductor is rated at 110 A when not more than three current-carrying conductors are used in a conduit. When four conductors are in one conduit, the 60°C No. 1 wires are derated to 80 percent of 110 A, or 88 A. Because that value is greater than the load of 76 A on each CB terminal, the CB terminals are not loaded in excess of the 88-A allowable ampacity of 60°C No. 1 conductors, and the UL limitation is satisfied.

Four No. 1 THW conductors, rated at 104 A, would require a minimum of 1½-in conduit. Or, four No. 1 XHHW conductors could be used in 1½-in conduit.

Step 6

If 90°C THHN conductors were to be used for this feeder (in a dry location), No. 2 copper conductors could be used. From Table 310-16, No. 2 THHN with a basic ampacity of 130 A would be derated to 80 percent of 130 A, or 104 A—which satisfies Section 220-10(b) as in item 3 above. The ampacity of a 60°C No. 2 conductor is 95 A normally; derated to 80 percent, it is 0.8×95 or 76 A, which gives the conductors the same rating as the load. Under such a condition the load current is not in excess of the 76-A allowable ampacity of a 60°C No. 2 conductor, and the UL limitation is satisfied.

Four No. 2 THHN conductors, rated exactly at the required minimum rating of 96 A, would require a minimum of 1¼-in conduit; or, four No. 2 XHHW could be used in 1¼-in conduit (in dry locations only).

In all the cases of conductor makeup described above, the 100-A CB would constitute proper protection.

NOTE: Of course, the voltage drop in the feeder will vary with the size of the conductors and must be accounted for.

Figure 455 shows an example in which feeder conductors could be used at up to their 75°C ampacity, as follows:

Step 1

Because the load on this feeder is continuous, Section 220-10(b) limits the load current to not more than 80 percent of the rating of the fuses and not more than 80 percent of the ampacity of

FIG. 455 Equipment rated over 100 A may make use of conductors up to their 75°C ampacity.

the conductors. It is, then, immediately obvious that the feeder load may not exceed 80 percent of the 400-A fuse rating, or 0.8 × 400 A = 320 A.

Step 2

If 75°C conductors are to be used, because they are permitted by UL test conditions, Table 310-16 shows that 700-MCM THW aluminum conductors rated at 375 A could be used and suitably protected by the 400-A fuses in accordance with Exception No. 1 of Section 240-3, because 400 A is the next higher standard fuse rating (from Section 240-6) above the 375-A rating of the conductors. In that case the maximum continuous feeder load could be no more than 375 A × 0.8, or 300 A, as required by Section 220-10(b).

Step 3

If the maximum capacity of the fusible switch (0.8 × 400 A = 320 A) is desired from 75°C conductors, 800-MCM THW aluminum conductors rated at 395 A could be used and loaded up to 80 percent of 385 A, or 316 A. If 900-MCM THW aluminum is used, 80 percent of its 425-A ampacity is 340 A; but the load may not exceed 80 percent of the 400-A fuse rating—that is, 320 A.

Step 4

If 90°C conductors are to be used, they must be used at no more than the ampacity of a 75°C conductor of the same size as the 90°C conductor. Table 310-16 shows that 600-MCM XHHW aluminum conductors (in a dry location) have a 385-A rating, and the 400-A fuses constitute acceptable protection for those conductors in accordance with Exception No. 1 of Section 240-3. The continuous load on the feeder phase legs would have to be limited to 80 percent of 385 A, or 308 A [Section 220-10(b)]. To check the suitability of the 90°C conductors, note that the ampere rating of a 75°C 600-MCM aluminum conductor is given in Table 310-16 as 340 A. Because the load of 308 A is not in excess of the 340-A rating of a 75°C conductor, the UL limitation on the maximum rating of conductor termination is satisfied.

Step 5

The smaller conduit size required for the reduced size of higher-temperature conductors is a labor and material advantage.

Conductor terminations

An extremely important element of the task of selecting switches, breakers, panelboards, switch-boards, and other electrical-system equipment is the determination of the suitability of the con-ductor terminals for use with the circuit conductors. When copper conductors are used, there are no special problems with terminations—assuming normal care and skillful workmanship are applied. But the increasing use of aluminum conductors in feeder sizes (No. 4 and larger) demands diligent attention to the terminal arrangements to avoid undesirable effects due to the creep and cold-flow characteristics of aluminum. Effective hookup of switches, breakers, panels, motor starters, and other distribution equipment can be assured if aluminum terminations are made in prescribed ways.

The vast majority of distribution equipment has always come from the manufacturer with mechanical setscrew-type lugs for connecting circuit conductors to the equipment terminals. And with copper conductors, such terminals have been highly effective. Lugs on such equipment are commonly marked "AL-CU" or "CU-AL," indicating that the setscrew terminals are suitable for use with either copper or aluminum conductors. But, such marking on the lug itself is not sufficient evidence of suitability for use with aluminum conductors. UL requires that equipment whose terminals are suitable for use with either copper or aluminum conductors must be marked to indicate such suitability on the *label* or *wiring diagram* of the equipment—completely inde-pendently of any marking on the lugs themselves. A typical safety switch, for instance, would have lugs marked "AL-CU," but it also must have a notation on its label or nameplate that reads "Lugs suitable for copper or aluminum conductors."

Although equipment of the type described above may satisfy UL regulations and **NEC** rules [Section 110-14(a)] when aluminum conductors are terminated directly in mechanical setscrew lugs, some design engineers specify the use of only "compression-type" lugs for the terminations of aluminum conductors at all equipment terminals. Work quality and application of the correct torque are critical in eliminating the creep and cold flow of aluminum conductors in setscrew terminations.

When aluminum conductors are used with distribution equipment that comes with mechan-ical setscrew terminals:

- A termination device or copper pigtail may be put on the end of each aluminum conductor to provide an "end" that is suitable, tested, and proved for effective use in a setscrew-type lug. A number of manufacturers make "adapters" which are readily crimped onto the end of an aluminum conductor to "convert" the end to copper or an alloy that will not exhibit the creep and cold-flow disadvantages of aluminum in a setscrew termination.

- Or, the setscrew lug may be removed from the switch or breaker or panel and replaced with a crimp-type lug designed to accept an aluminum conductor. But, because unauthorized alter-ation or modification of equipment in the field voids the UL listing and can lead to very dangerous conditions, the arbitrary or unspecified changing of terminal lugs on equipment is not acceptable unless such field modification is recognized by UL and spelled out very care-fully in the manufacturer's literature and on the label of the equipment itself.

Busways

One of the most popular of modern methods of electrical distribution is the busway system. Busways consist of metal enclosures containing insulator-supported busbars, constituting a com-plete prefabricated conductor wiring method with their own raceway-like enclosure. Although the use of busway for industrial feeders, plug-in subfeeders, and branch-circuit systems has been developing for many years, the use of feeder and plug-in busway in large commercial and institutional buildings has gained widespread acceptance over recent years, stimulated by developments in busway construction.

Busways are available for either indoor or outdoor use as point-to-point feeders or as plug-in takeoff routes for power. Progressive improvements in busway designs have enhanced their electrical and mechanical characteristics, reduced their physical size, and simplified the methods used to connect and support them. These developments have, in turn, reduced installation labor to the extent that busways are most favorably considered when large blocks of power must be moved to loadcenters (via low-impedance feeder busway), current must be distributed to closely spaced power-utilization points (via plug-in busway), or rows of lighting fixtures or power tools must be energized and supported (via lighting busway or trolley busway).

Figure 456 shows how busway applications are advantageous in many types of structures. For example, the upper sketch shows low-impedance feeder busways (*a*) used to bring power from an outside source to the main interior switchgear of a tall commercial building and then carrying power to upper-floor control centers (*b* and *c*). The lower sketch shows weatherproof feeder busway (*d*) connecting an outdoor transformer with the interior switchgear of an industrial plant, feeder busway (*e*) then connecting with the center tap of a plug-in system, and plug-in busways (*f* and *g*) carrying power to various equipment.

Busway conductors may be of copper or aluminum, hollow or solid, rectangular, oval, or beam-shaped. They may be insulated or bare, with phase bars grouped or interleaved, and with silver-plating applied either for entire lengths or only at connection joints and power-tap points. Busways are factory assembled and supplied in standard 10-ft sections.

Busway has proved to provide a versatile, flexible, and economical method of distribution. Offering safety, reliability, ease of layout, and efficient operation, busway systems require only basic engineering in their design. Layout possibilities are extensive; elbows, tees, offsets, and

FIG. 456   Busway has a wide variety of design applications for electrical distribution.

crosses permit either edgewise or flatwise orientation. Flanges permit busways to connect with switchboards or transformer tap boxes or to pass through walls, partitions, floors, and roofs in accordance with **NEC**-prescribed requisites. Power feeds or takeoffs may be located at busway ends or at center points. And they are capable of carrying large and small blocks of power from main switchboards to loadcenters to loads. For point-to-point power distribution, the overall economy of busway—in both labor and material costs—makes it particularly effective for heavy loads.

Either rigid or flexible connection may be made between adjacent straight busway sections or to power-supply sources or load takeoff devices. Utilizing metallic braid or lengths of armored cable and suitable busway attachments for flexible connections not only isolates vibration and accommodates expansion and movement; it also permits physical obstructions to be bypassed without requiring the use of special offsets or angle sections.

Busways are available specifically designed to minimize voltage drop and reactance, to serve high-amperage intermittent loads such as those created by resistance welders, or to carry high-cycle current for special lighting, laboratory, or induction-heating purposes. They also are available as slotted assemblies to accommodate moving trolleys.

Low-reactance feeder busway provides low-voltage-drop characteristics through close spacing and special construction of the busbars, which minimize reactance. This type of busway is used for all types of high-capacity feeders and risers. Because of its low reactance, it has also found application in high-frequency distribution systems.

The ampere rating of busway must be taken as the nameplate current value and is a continuous load rating. Busway is also given a short-circuit rating in rms symmetrical amperes and must be applied only where the available short-circuit current at its supply end does not exceed that rating.

Typical ratings on feeder busway range from 600 to 6000 A, for single-phase (two or three poles) or 3-phase (three or four poles) systems, for 120/240- or 480/277-V 3-phase, 4-wire systems, or for 3-phase, 3-wire systems—all up to 600 V. Weatherproof feeder busways are also available for exposed outdoor use. And busways are available with an extra busbar to provide equipment grounding throughout a system.

Plug-in distribution busway, with easily accessible plug-in openings for tapping to loads directly or through switches and protective devices, is available in ratings from 225 to 1500 A.

Plug-in and clamp-on devices include fused and nonfused switches and plug-in circuit breakers rated up to 800 A. Other plug-in devices include ground detectors, temperature indicators, capacitors, and transformers designed to mount directly on the busway.

Figure 457 is an isometric and floor plan of the complete riser and subfeed distribution system, which uses a variety of busway, in a modern commercial-industrial building housing research and development laboratories for telephone equipment. The low-impedance 1000-A busway risers feed 225-A plug-in runs on the floors, with taps made to 100-A plug-in busway along workbenches for tapping single-phase or 3-phase power to load devices as required by lab work.

Although busways have broad application potential, actual applications are limited. According to **NEC** rules, busway may be used only for exposed work. Some exception has, however, been made in cases where concealment is not permanent—such as above lift-out types of suspended ceilings. As long as it can be reached without removing any permanent part of a building, the busway is considered to be exposed. But, use above a suspended ceiling is limited to totally enclosed, nonventilated busway, without plug-in switches or CBs on the busway and only in ceiling space that is not used for air handling.

Busways carry various markings to indicate the use for which the busway was investigated and listed. Busway intended to supply and support industrial and commercial lighting fixtures is marked "Lighting busway." Busway with sliding or other continuously movable means for

FIG. 457 In typical commercial and industrial systems, busway may serve as feeders, subfeeders, and branch circuits.

tapping off current to load circuits is marked "Trolley busway." And, if the same busway is also acceptable for supporting and feeding lighting fixtures, it is also marked "Lighting busway." If busway is designed to accept plug-in devices at any point along its length and is intended for general use, it is marked "Continuous plug-in busway." Busway marked "Lighting busway" and protected by overcurrent devices rated in excess of 20 A is intended for use only with fixtures having heavy-duty lampholders—unless each fixture is equipped with additional overcurrent protection to protect the lampholders.

Busway risers (vertical runs) must be supported at least every 16 ft and may be supported by a variety of spring-loaded hangers, wall brackets, or channel arrangements where they pierce floor slabs or are supported on masonry walls or columns (Fig. 458). Spring mounts for vertical busways may be located at successive floor-slab levels or supported by wall brackets located at intermediate elevations. Springs provide floating cradles for absorbing transient vibrations or

FIG. 458 Specific design details should be provided to ensure safe, effective support for busway risers and horizontal runs.

physical shocks. Fire-resistant material must be packed into spaces between busway and the edges of slab-piercing throat. Where it passes through a floor slab, busway must be totally enclosed within the floor slab and for 6 ft above it. The **NEC** prohibits the use of ventilated busway in such places.

Overcurrent protection—either a fused switch or a CB—is usually required in each busway subfeeder tapping power from a busway feeder of higher ampacity, where the larger busway is protected at the higher ampacity; this is necessary to protect the lower current-carrying capacity of the subfeeder. The protective device should be placed at the point at which the subfeeder connects into the feeder. However, the **NEC** permits this overcurrent protection to be omitted where busways are reduced in size if the smaller busway does not extend more than 50 ft and

Reduced size of plug-in busway subfeeders over 50 feet long must have overcurrent protection

Less than 50 ft

Feeder bus

Protection not required here for subfeeder less than 50 feet, if it has an amp rating at least equal to one-third the rating of the CB protecting the main feeder bus

Example:

Overcurrent protection — 1000-amp duct — Reducer — 400-amp duct

1200-amp overcurrent device

1200-amp busway   800-amp busway — 25 ft —   600-amp busway — 25 ft —

Tap box

3 sets of 500 MCM THW (CU) in multiple 3 - 3 1/2" conduits

— 50 ft —

or *400 amp*

THIS IS O.K.

1200 amp X 1/3 = 400 amps

FIG. 459 Busway subfeeder may sometimes be used without overcurrent protection at its point of supply.

has a current rating at least equal to one-third the rating or setting of the overcurrent device protecting the main busway feeder, as shown in Fig. 459.

Branch circuits or subfeeders tapped from busway must have overcurrent protection on the busway at the point of tap. And if they are out of reach from the floor, all fused switches and CBs must be provided with some means for a person to operate the handle of the device from the floor (hookstick, chain operator, rope-pull operator, etc.) A switch or CB is "out of reach" if the center of its operating handle, when in its highest position, is more than 6½ ft above the floor or platform on which personnel would be standing.

Figure 460 shows an extensive application of many busway runs as the main feeders from a loadcenter unit substation in a large industrial plant. The runs of plug-in busway supply motors and power panels directly at 480/277 V, and lighting panels at both 480/277 and 208/120 V, through local dry-type transformers.

FIG. 460   Plug-in feeder busway is the basic distribution method in this system.

Armored cable

Armored cable has been used in electrical work for many years and still finds widespread application. **NEC** Article 334, entitled "Metalclad Cable," identifies so-called "interlocked armor" cable as "Type MC" and distinguishes its construction, application, and installation from standard BX-type armored cable, which is designated "Type AC."

Type AC cable is the cable assembly long used and known as BX cable. All regulations on the use of Type AC cable are given in **NEC** Article 333. Type AC armored cable is distinguished from Type MC metal-clad cables in the **NEC**, and the latter is covered in Article 334.

Type AC cable (BX) is listed and labeled by UL as "armored cable" in the *Electrical Construction Materials Directory*. The cable assembly consists of the conductors and a bonding strip (an equipment grounding conductor) within a jacket made of a spiral wrap of steel strip, with the edges of the strip interlocked, as shown in Fig. 461.

Steel metal
covering

Copper or aluminum bonding
strip in contact with armor

FIG. 461 Basic assembly of Type AC armored cable, known as "BX."

Type AC armored-cable assemblies of two, three, or four conductors in sizes No. 14 AWG to No. 1 AWG conform to UL standards. These standards cover multiple-conductor armored cables for use in wiring systems of 600 V or less, at temperatures of 60°C, 75°C, or 90°C, depending upon the conductor insulation. Armored cables of other types which do not come under these UL standards are listed by UL as "metal-clad cable, Type MC" and covered by **NEC** Article 334. One type of MC cable is commonly called "interlocked armor cable."

Type AC cable is recognized for branch circuits and feeders, but not for service entrance conductors, which must be as specified in **NEC** Section 230-43. Type MC (metal-clad), cable, such as interlocked armor cable and the other cables covered in **NEC** Article 334, is recognized for use as service entrance conductors.

Because the armor of Type AC cable is recognized as an equipment grounding conductor, its effectiveness must be ensured with an "internal bonding strip," or conductor, placed within the armor and shorting the turns of the steel jacket. The ohmic resistance of finished armor, including the bonding conductor that is required to be furnished as a part of all except lead-covered armored cable, must be within values specified by UL and checked during manufacturing. The bonding conductor that is run within the armor of the cable assembly is required by the UL standard. The function of the bonding conductor in Type AC cable is simply to short adjacent turns of the spiral-wrapped armor.

Typical assemblies of Type MC (metal-clad) cable are shown in Fig. 462. The top sketch shows the most common type of Type MC cable, used for many years under the name "interlocked armor cable." It is the heavy-duty industrial-feeder type of armored cable that is similar in appearance to but really different from standard BX armored cable. MC cable is a heavier-duty assembly than BX (Type AC), and great care must be taken to carefully distinguish between the two.

Type MC cable includes Type ALS and Type CS cables, in addition to the interlocked-armor and corrugated-tubing types of armor. Aluminum-sheathed (ALS) cable has insulated conductors with color-coded coverings, cable fillers, and an overall wrap of Mylar tape—all in an impervious, continuous, close-fitting, seamless tube of aluminum. It may be used for both exposed and concealed work in dry or wet locations, with approved fittings. CS cable is very similar but has a copper exterior sheath, instead of aluminum.

Type MC cable is rated by UL for use up to 5000 V, although cable for use up to 15,000 V has been available and used for many years. Type MC cable is recognized in three basic armor designs: (1) interlocked metal tape, (2) corrugated tube, and (3) smooth metallic sheath.

Because there are different forms of Type MC cable, care must be exercised to distinguish between the different constructions, owing to their varying suitability for design layouts and applications. For a long time, the interlocked-armor Type MC and the corrugated-sheath Type MC have been designated by UL as "intended for above-ground use." But **NEC** Section 334-3 recognizes Type MC cable as suitable for direct burial in the earth. However, such use must be checked with local inspection authorities.

Interlocked armor (IA) cable is available for use at 600, 5000, and 15,000 V. The high-voltage types of IA cable are very effective as primary feeders in loadcenter distribution systems. The 600-V class of IA cable is rapidly gaining in popularity for feeder applications in commercial and industrial systems.

600-VOLT CABLE

- Varnished cambric or varnished glass cloth
- Tape wrap or thermoplastic jacket
- Bare stranded grounding conductors

5000-VOLT CABLE

- Thermoplastic jacketing or resistant braid
- Steel, aluminum or bronze single strip armor — positively interlocked
- Rubber filled tape
- Jute fillers
- Rubber insulation
- Stranded copper or aluminum conductors

ALS CABLE

- Impervious, continuous, closely fitting, seamless tube of aluminum
- Binder tape
- One or more insulated conductors with filler

FIG. 462 Typical cable assemblies that are UL-listed as Type MC cable.

- Interlocked armor cable feeders to 400-cycle receptacles (on cable rack)
- 600-v, 400-cycle main dist. switchboards
- 2 No. 6-1"c
- 6 No. 4/0 in interlocked armor cable on aluminum cable rack
- Cable-on-rack to 400-cycle receptacles
- M–G set
- 6-500 mcm in two 3-in. conduits from main 480-v switchboard
- 6-500 mcm-2-3"c
- 400 hp syn. motor starters and exciter panel
- 2 No. 6-1"c

FIG. 463 Cables with interlocked aluminum armor provide minimum inductive reactance in high-frequency circuits for airplane servicing.

FIG. 464 Armored-cable feeders were readily routed around many obstructions to supply busway runs.

Interlocked armor cable for secondary feeders is a completely flexible and protected cable assembly, available in sizes from No. 1/0 to 750-MCM, with three or four conductors, and available in assemblies of conductors from No. 14 up for control purposes. The IA cable consists of a galvanized steel, interlocked, spiraled armor wrap around an insulated assembly of individually insulated conductors. The conductors are stranded and may be copper or aluminum. The armor may be of steel, aluminum, or bronze. Figure 463 shows how aluminum-armored cables were run in aluminum cable tray to minimize inductive reactance by eliminating magnetic material (steel) from 400-H circuits in a large airplane hangar.

There are some general criteria for choosing between IA cable and busway. Where the routing is devious and there are many obstructions, IA cable often can be used to greater advantage. For straight runs, particularly long runs of high current capacity, busway is usually better. Above 1000-A capacity, busway for a feeder generally costs less. Below 500 A, a feeder may more economically consist of interlocked armor cable. However, where the advantages of either armored cable or busway are required, the cost of materials and installation is not of primary importance. As shown in Fig. 464, the electrical distribution system in one industrial plant made extremely effective use of both armored cable and busway. The IA-cable feeders were used to provide flexible runs around various overhead obstructions, from the 480-V switchgear to cable junction boxes on the busway runs.

High-Voltage Systems (over 600 volts)

The term *high-voltage* is commonly used to refer to a circuit operating at a voltage above 600 V, phase to phase. It should be noted, however, that circuits operating at from 601 to 35,000 V are frequently called *medium-voltage* circuits; the American National Standards Institute (ANSI) uses the latter term for electrical circuits operating between 1001 and 72,500 V, reserving the term *high-voltage* for circuits operating above 72.5 kV.

High-voltage circuits and systems used for commercial and industrial distribution, both outdoors and indoors, most commonly operate at voltages up to 15,000 V (15 kV). But there is a trend to even higher voltages (26 kV, 35 kV) for extremely large installations. Typical circuits today operate at 4160/2400 and 13,200/7,600 V—both in 3-phase, 4-wire wye hookups.

There are many reasons for the great growth in the use of high-voltage electrical systems. The most important is the increasing need to distribute large amounts of electric power as efficiently as possible. For a given amount of electric power, the higher the distribution voltage, the lower the current and, therefore, the smaller the size and number of circuit conductors and raceways and the greater the economy in materials and labor for transformers and control and protection equipment. But the growing popularity of high-voltage systems has been stimulated by other factors. Spurred by continuing advances in chemical technology, new and better insulations are making possible more effective, safer, and more economical electrical equipment—switchgear, transformers, cables, terminators, splices, etc.

For distributing electric power in buildings, economic factors continue to motivate a shift toward high voltages (5 to 35 kV) and away from low voltages (up to 600 V). And as the average density of electric-power usage increases, high-voltage distribution is given further stimulus. In addition, the present concern for energy conservation—with its emphasis on efficiency—definitely points to high-voltage distribution.

Designs of high-voltage electrical systems for modern buildings vary widely with the nature of the occupancy. However, such distribution has the same general objective within all occupancies—to get required power to load devices as economically and efficiently as possible. Among the factors that affect the design of a high-voltage distribution system are: the importance of power continuity and the possible effects of power outages on production or processing operations; economy of installation and operating costs; flexibility requirements of the particular type of activity; and safety and the protection of personnel and equipment.

Depending upon the conditions and requirements in any particular case, the design of a high-voltage distribution system may incorporate modern methods such as loadcenter distribution from unit substations, high-voltage distribution to substations, and alternative supply lines to each substation. The loadcenter substation is the key to this modern distribution.

In the layout of loadcenter distribution, each loadcenter unit substation is located in the approximate center of the area supplied from that substation. Of course, the exact location of a substation with respect to the equipment it supplies depends upon the number and relative sizes of the load devices and the construction of the building. The number of unit substations required in any installation depends upon the total load to be served and the capacities of the individual unit substations.

Each unit substation is fed by high-voltage (2400 to 35,000 V) incoming power lines. The substation transforms the power down to an intermediate high-voltage level (such as 35 kV down to 4160 V) to supply other loadcenter layouts, or directly to low voltage (under 600 V) to supply feeders to motors, to other power loads, and to lighting loads. The latter are fed either directly—for 277-V lighting—or indirectly through step-down transformers for 120-V circuits.

The use of high-voltage loadcenter substations allows efficient and economical distribution of large amounts of power throughout a building. Power distribution at high voltages requires much lower feeder current-carrying capacities than would be required to distribute the same amount of power at low voltage. As a result, the primary feeders are relatively small in size and are not subject to the severe voltage drops which result from the use of long, low-voltage but high-current feeders to load concentrations. And the locations of the unit substations, in the center of each load concentration, permit the use of very short secondary feeders or branch circuits to loads, again minimizing voltage drop and copper loss.

Many advantages may be realized through the use of loadcenter distribution in industrial plants. It has been estimated that the saving in feeder copper alone may run as high as 30 percent. And other substantial savings are effected through the elimination of much heavy-capacity secondary switchgear. In addition, loadcenter distribution provides unusual flexibility for, and adaptability to, expansion of facilities at a later date, offering long range economy.

Design procedure

The design of a high-voltage loadcenter distribution system for any building begins with analysis and breakdown of the total load within the building. From this study, the number and sizes of loadcenter subs are determined.

Generally, the total load in the building is broken into blocks and apportioned among a number of unit substations, or "subs." The number of unit subs selected for an installation depends on the economics of equipment application, along with other factors. Typical considerations include the following:

1. Generally, the smaller the unit sub, the more expensive it is per kilovoltampere.
2. The greater the number of unit subs provided to handle a given total load, the smaller (and more expensive per kilovoltampere) each unit sub will be, and the greater the number of primary cables will be.

3. The use of a large number of unit subs, each covering only a small area of load accumulation, gives the many advantages of reduced lengths of secondary feeders.

4. The use of a small number of very large subs requires fewer primary conductors but more secondary conductors.

From investigations of standard kilovoltampere ratings of unit subs and system costs per kilovoltampere, it has been found that economy usually dictates the use of unit subs rated from 500 to 1500 kVA. Maximum economy can be realized from the use of substations rated from 750 to 1000 kVA, as clearly indicated by the graphs in Fig. 465.

A typical loadcenter substation contains either high-voltage switchgear or only terminals in an incoming section, a transformer section in which the power is stepped to low voltage, and a section containing switching and protection for the outgoing secondary feeders. High-voltage switchgear may utilize circuit breakers, air-, vacuum- or oil-break interrupting switches (with or without power fuses), or oil-filled cutouts.

The most common arrangement for the incoming line sections of modern loadcenter substations consists of a high-voltage load interrupter switch, with suitable power fuses for the protection of the substation. In a typical case, the primary feeder to the substation might originate in a metal-clad CB switchgear center. In some cases, where the primary-feeder switchgear is located near the unit sub and no other sub is fed by the feeder, additional protection and switching are not required in the incoming line section of the substation, and an air-filled terminal chamber can satisfy incoming line requirements. Where fuses cannot be coordinated for primary transformer protection, a power circuit breaker is required.

FIG. 465 Relative economy of the loadcenter unit substation varies widely with kilovoltampere rating and secondary voltage.

Liquid-filled or dry-type transformers, offering advantages for indoor and outdoor applications, are used in unit subs. Secondary switching and protection usually consists of completely enclosed assemblies of drawout-type circuit breakers or molded-case circuit breakers. Figure 466 shows the various components that may be used to make up the complete, packaged assembly of a modern loadcenter unit substation.

Another type of loadcenter unit substation that finds extensive use is the double-ended unit sub. A double-ended substation is really two single loadcenter unit subs with a tie breaker between them (Fig. 467). Three breakers (one tie and two main) provide additional reliability by providing for separate primary feeder circuits to each end of the double-ended substation. If one transformer is down, its load is easily transferred to the other, with the single primary feeding both transformers.

Primary voltage: The primary voltage used with loadcenter distribution depends upon the available utility lines, the total load to be supplied, and the number of unit subs to be used. Consultation with the local utility company is essential to the selection of the best primary voltage.

FIG. 466 Basic loadcenter unit substation is a factory-made assembly of a transformer with primary and secondary control and protection.

Primary distribution at 13.8 kV usually offers the best loadcenter design at lowest cost for very large industrial plants with demand loads above 20,000 kVA. Between 10,000 and 20,000 kVA, plant load may be economically served by either 13.8-kV or 4160-V primaries. For plants with demands below 10,000 kVA, primary distribution at 4160 V is generally recommended.

System Layout

One of the most important considerations in the use of loadcenter substations is the layout or circuiting arrangement. Typical systems are radial, secondary selective, primary selective, and network. Figure 468 shows basic layout diagrams for nine different distribution systems, with a tabulated evaluation of the initial costs, operating costs, and other important characteristics or advantages of the systems.

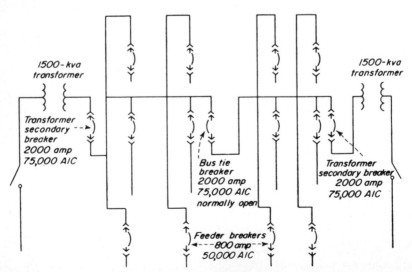

FIG. 467 Double-ended unit sub may be fed by both separate primary circuits or by only one of the primaries with the tie CB closed.

1 Low Voltage Simple Radial, A

2 Modern Simple Radial, B

3 Modern Simple Radial, C

4 Loop Primary Radial

5 Banked Secondary Radial, Looped Primary

6 Primary Selective Radial

7 Secondary Selective Radial

Typical Lighting Transformer and Panel Board

8 Primary Selective Network

9 Spot Network

☐ Primary Breaker   ⎍ L.V. Breaker   Limiter

SYSTEM CHARACTERISTICS

System Type	Flexibility	Service Dependability	Regulation	Efficiency	Operation & Maintenance Cost	Initial Cost
			Grading			
1	3	5	6	5	1	4
2	6	6	4	3	1	1
3	4	5	4	3	1	3
4	4	5	4	2	1	1
5	1	3	2	1	2	4
6	4	4	4	2	1	2
7	4	2	2	4	4	4
8	1	1	2	1	3	4
9	3	1	1	4	4	5

Key to Grading Figures: 1, 91-100%; 2, 81-90%; 3, 71-80%; 4, 61-70%; 5, 51-60%; 6, 41-50%

FIG. 468 Distribution systems differ in performance capabilities.

Of the loadcenter distribution systems in use today, the majority follow a simple radial pattern. This system offers high reliability and service continuity and is safe, flexible, stable, and easy to operate and maintain. Figure 469 shows the basic radial system, using a single primary feeder. This layout is generally used where the total system load exceeds several hundred kilowatts and where the use of a single substation or secondary voltage service would result in long,

SINGLE PRIMARY

Basic radial system for high-voltage distribution
with load-center substations. Used where the
total system load to be served exceeds several
hundred kilowatts and where the use of a single
substation or secondary voltage service would
result in long heavy feeders with characteristic
high cost, poor voltage regulation and low efficiency.
Offers advantages of high voltage distribution at
lowest cost.

FIG. 469 Basic radial distribution system has a single
feeder to all unit substations.

heavy feeders with characteristic high cost, poor voltage regulation, and low efficiency. It offers
the advantages of high-voltage distribution at lowest cost.

A variation on the basic system (Fig. 470) has individual primary feeders to unit subs for
more flexibility. This reduces the effect of outages due to faults on the primary feeder in the
basic system. The use of automatic protective devices in each primary feeder increases the cost
of this system relative to the basic system. But its adaptability to load growth and plant changes
is much better than that of the basic radial system with a single primary feeder.

Where more flexibility is required, a secondary selective system—two unit subs with a normally open secondary tie breaker between them—may be used. A double-ended substation, with
primary supply to each end and a normally open secondary tie within the assembly, may also
be used to form a secondary selective system, as shown in Fig. 471. In operation, each substation
or each end of a double-ended sub operates as a straight radial system. If one primary feeder or
transformer should fail or be taken out of service, the load normally carried on that substation
or that end of the double-ended substation is automatically transferred to the still-operating
substation.

The secondary selective system represents an increase of only 5 to 15 percent over the cost
of a simple radial system.

Similar to the basic radial system, the system in Fig. 472 has the primary feeder looped back
on itself at the point of power supply, to increase system reliability at a slight increase in cost
over that of the basic system. The use of main primary feeder breakers at the source point and

MULTI – PRIMARY

A variation on the basic system; has individual
primary feeders to unit subs. Provides more
flexible system. Reduces the effect of outages
due to fault on primary feeder in basic system.
Use of automatic protective devices in each
primary feeder increases cost of system over
basic system. Adaptability to load growth and
plant changes is improved over that of the
basic system.

FIG. 470 Radial distribution may utilize individual feeders.

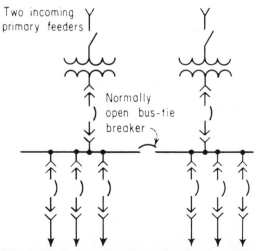

FIG. 471 Secondary selective distribution is readily
provided by a double-ended substation.

FIG. 472 Loop primary distribution system adds a measure of reliability over the basic radial layout.

Similar to the basic system, with the primary feeder looped back on itself at the point of power supply. Increases system reliability at slight cost increase over basic system. By the use of main primary feeder breakers at the source point and automatic sectionalizing breakers at the points of tap to the load-center subs, the occurrence of primary feeder faults will not interrupt service to any loads. By opening one of the sectionalizing breakers in the loop, the system may be operated as two simple radial systems with independent primary feeder to each group of substations. The system is then a form of primary selective system.

automatic sectionalizing breakers at the points of tap to the loadcenter subs prevents primary feeder faults from interrupting service to any loads. By opening one of the sectionalizing breakers in the loop, the system may be operated as two simple radial systems.

A primary selective radial system provides a choice of primary feeder for improved service reliability in the event of a primary feeder fault. Figure 473 shows a comparison between a secondary selective system and a primary selective system. In the primary selective system, a transfer means is provided to ensure rapid reconnection from one primary to the other. Figure 474 shows a system layout that provides very high reliability of supply to each loadcenter sub— a radial system and a loop primary system are combined into a primary selective system. Normally, each sub is fed by its own radial feeder. On a fault in any of the radial feeders, its transformer can be switched to the loop system, which has a tie into each substation.

Two types of network layout of unit subs are shown in Fig. 475. In the spot network, two or three transformers in one location or "spot" are connected to a common secondary bus and divide the load. Upon a primary or transformer fault, the secondary is isolated from the faulted section by automatic operation of the network protector; this provides a high order of supply continuity in the event of faults. In general form, the network system is similar to the spot network, except that widely separated individual substations are used, and associated network protectors and tie circuits are run between the secondary bus sections. This system provides for the interchange of power to accommodate unequal loadings on the transformers. Limiters (fusible-element cable-terminating lugs) provide short-circuit protection for the secondary tie conductors that tie secondary load buses together in a general network hookup.

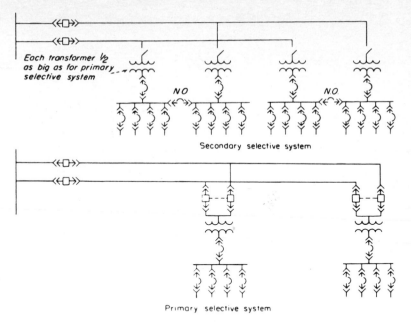

Each transformer ½ as big as for primary selective system

N.O.   N.O.

Secondary selective system

Primary selective system

The secondary selective system permits disconnecting of transformers for maintenance while maintaining service to loads through secondary ties The primary selective system allows choice of primary feeders for improved service reliability.

FIG. 473 Primary-selective and secondary-selective systems differ in operating advantages.

Figure 476 shows a general network distribution layout in an industrial plant. The single-line diagram represents a 3-phase system. The primary feeders may operate at any standard voltage up to 13,800 V, and the secondary voltage is commonly 480 V. The rating of the transformers used in such a system would usually be within the range of 300 to 1000 kVA. The diagram shows two primary feeders, both of which are carried to each transformer so that, by means of a double-throw switch, each transformer can be connected to each feeder. Each feeder is large enough to carry the entire load. The feeders are protected so that no primary overcurrent devices are required at the transformers. The secondary ties consist of two conductors in multiple per phase, and these conductors form a closed loop. Switches are provided so that any section of the loop, including the limiters protecting that section, can be isolated in case repairs or replacements should be necessary.

As shown, both types of network systems contain a network protector that consists of a CB and a reverse-power relay. The protector is necessary because, without this device, if a fault develops in the transformer (or, in some cases, in the primary feeder), power will be

Main 4160-v switchgear bus

Loop to all unit subs

Radial feeder to each unit sub

Selector switch in each sub

FIG. 474 This hybrid system combines the advantages of the radial and loop primary systems.

FIG. 475 Network distribution systems provide high-capacity input to the electrical system supplied.

fed back to the fault from the other transformers through the secondary ties. The relay is set to trip the breaker on a reverse-power current not greater than the rated secondary current of the transformer. The breaker is not arranged to be tripped by an overload on the secondary of the transformer.

NEC Section 450-5(a)(3) requires that:

1. Where two or more conductors are installed in parallel, an individual protective device is provided at each end of each conductor.
2. The protective device (fusible link or CB) does not provide overload protection, but provides short-circuit protection only.

In case of short circuit, the protective devices must open the circuit before the conductors reach a temperature that would injure their insulation. The principles involved are (1) that the entire system is so designed that the tie conductors will never be continuously overloaded in normal operation (and, hence, protection against overloads of less severity than short circuits is not necessary) and (2) that the protective devices should not open the circuit and thus cause an interruption of service on load peaks of such short duration that the conductors do not become overheated.

FIG. 476   General network distribution system in an industrial plant.

A limiter is a special type of fuse having a very high interrupting capacity. A typical limiter consists of a cable lug, the fusible section, and the extension for connection to the bus—all made in one piece from a length of copper tubing; the enclosing case is also copper. A typical device of this type is rated to interrupt a current of 50,000 A without perceptible noise and without the escape of flame or gases from the case.

In practice, high-voltage distribution systems may utilize one or more of the hookup configurations described above. Figure 477 shows an industrial system that includes a loop-primary radial at 12 kV with selector switches that permit the four double-ended substations (power centers) to be fed from both or either of the two feeds (at right) into the loop. And the tie CB in each substation provides for either of the two transformers in each sub to carry the entire load on both sections of secondary bus.

High-Voltage Service

For any building or premises that is supplied with electric power at higher than 600 V, an important design task is to determine what part of the supplied system constitutes the service entrance. This is necessary to establish which conductors are the service entrance conductors and which disconnect and overcurrent devices are the service equipment. Without accurate identification of the service components, it is impossible to ensure compliance with the rules of **NEC** Article 230 on services. And where it is determined that the service conductors and service equipment are operating at over 600 V, the rules of **NEC** Sections 230-200 to 230-211 must be satisfied.

The **NEC** itself can be used as a guide in determining where the service entrance is in a system involving one or more high voltages (over 600 V). Section 230-201 attempts to identify the service entrance conductors of such systems. Once these conductors are identified, the disconnect and protection devices associated with them are considered service equipment and are subject to the applicable rules of Article 230.

NEC Section 230-201 says that secondary conductors, not the primary conductors, are

FIG. 477 Overall industrial distribution system uses a variety of layout and switching techniques.

regarded as constituting the service conductors to a building proper where: (1) step-down transformers are located outdoors; (2) step-down transformers are installed in a mechanical-equipment building that is separate from the building or structure being supplied; (3) step-down transformers are located in a transformer vault in the building or structure served; (4) step-down transformers are located in the building or structure served, in a locked room or locked enclosure meeting **NEC** rules on the construction of such places and accessible only to qualified persons; or (5) step-down transformers are located in the building or other structure served and the transformers are within metal-enclosed gear. Figure 478 shows these concepts. In all other cases, the primary circuit conductors to the transformer must be considered to be the service conductors. But that rule must be applied very carefully, as follows:

Section 230-201 establishes whether the primary or secondary circuit of a service transformer must be considered as the "service conductors" for an installation. This whole matter has long been a difficult area of **NEC** application.

Under the wording of condition 4 above, if a loadcenter unit substation that handles the electrical load for a building is in a locked room in the building and is fed, say, by an underground high-voltage (over 600 V) utility line from outdoors, then the secondary conductors from the transformer of the unit sub are the "service conductors" to the building. And the switching and control devices (up to six CBs or fused switches) in the secondary section of the unit sub would constitute the "service equipment" for the building. Under such a condition, if any of the secondary section "service disconnects" is rated 1000 A or more, at 480/277 V grounded wye, it must comply with Section 230-95, requiring ground-fault protection for service disconnects.

TRANSFORMER NOT IN BLDG. SERVED

Primary service to premises

Step-down transformer outside or in a separate building

Building wall

Service main

Inside building

The voltage of these secondary conductors is the service voltage. It may be above or below 600 volts.

① Where transformers are located in a transformer vault (constructed per S 450-41 through 450-48)

All types of transformers permitted (oil, dry, etc.)

Locked door

② Where transformers are located in locked rooms

Locked door

Dry-type or askarel-filled trans. only

③ Where transformers are in locked enclosures or in "metal-enclosed gear"

Totally enclosed and/or locked transformer enclosure

TRANSFORMER IN BLDG. SERVED

Conditions where SECONDARY conductors are considered service conductor to a building when transformers are located within the building

Secondary conductors are most commonly designated as the service entrance conductors to a building or premises.

FIG. 478 Secondary conductors are most commonly designated as the service entrance conductors to a building or premises.

And the same determination would apply to an arrangement where the transformer is not part of a unit substation but is an independent transformer unit located in a vault or locked room and fed by a primary disconnect with overcurrent protection in the vault or locked room, with the transformer secondary feeding one or more low-voltage (under 600 V) disconnect and protection devices—either inside or outside the vault or locked room.

However, under the wording of condition 4, if the same unit sub is inside the building but not in a locked room, it might seem that the primary conductors would be the service conductors; and the primary switch or CB would be the "service disconnect." In that case, no ground-fault protection would be needed on the "service disconnect" because Section 230-95 applies only up to 600 V, and there is no requirement for GFP on high-voltage services. Also, in that case, there would be no need for GFP on the secondary-section disconnects, because they would not be "service disconnects."

But, condition 5 of **NEC** Section 230-201, as described above, adapts the concept of condition 4 to "metal-enclosed gear," which covers loadcenter subs and any other assembly in which the transformer is actually in metal-enclosed gear. Under condition 5, for any loadcenter sub located inside a building and used for service to the building, the secondary conductors from the transformer section would be considered as the "service conductors," and the secondary control and protection section would have to satisfy all **NEC** rules on service equipment—including service ground-fault protection as required by Section 230-95 (Fig. 479).

In applying Section 230-201, care must be taken to distinguish between "secondary conductors" that are not over 600 V (such as a 480/277-V unit-sub secondary) and those that are over 600 V (such as 34 kV stepped down to 4160/2400 V). For the latter, all of Part **K** of Article 230 applies to the secondary.

CAUTION: A persistent and perplexing controversy revolves around **NEC** Section 230-201. This is where the **NEC** attempts to identify the "service entrance" for systems fed at more

FIG. 479 Secondary of an indoor loadcenter unit substation constitutes the service conductors.

FIG. 480 The "service point" does not determine where "service entrance conductors" originate.

than 600 V. After noting in Section 230-200 that "service conductors and equipment used on circuits exceeding 600 volts" must comply with all the rules in Article 230 (including any "applicable provisions" that cover services up to 600 V), the **NEC** proceeds to say that in many cases where the utility feeder is operating above 600 V, the "service" must be taken to be located on the secondary sides of transformers, where the voltage may be either above or below 600 V. Design engineers have difficulty in determining where the service entrance exists on premises served by high-voltage (over 600 V) utility lines—especially on multibuilding properties with both outdoor and indoor transformers and/or with two outdoor voltage transformations.

Because there has been so much controversy over what is and what is not "service" equipment in the many complicated layouts of outdoor high-voltage circuits and transformers, a definition in Section 230-200 defines "service-point" as the "point of connection between the facilities of the serving utility and the premises' wiring" as shown in Fig. 480. All equipment on the load side of that point is subject to **NEC** rules. Any equipment on the line side is the concern of the power company and is not regulated by the **NEC**. In any particular installation, identification of that point can be made by the utility company and design personnel. The definition simply makes clear that the property line is not the determinant as to where **NEC** rules must first be applied. This is particularly important in multibuilding industrial complexes where the utility company has distribution circuits on the property. But this definition of "service-point" must not be construed as establishing that "service conductors" originate at that point. The whole matter of precisely identifying the "service conductors" is covered by Section 230-201(a).

The circuit that crosses the property line of the premises, connecting the utility system off the premises to the customer's system on the premises, is the "service" for premises fed at voltages up to 600 V. But for high-voltage utility supplies, the "service" starts sliding around on the plot plan, and pinning it down is often like trying to squeeze quicksilver. The definition of "service-point" indicates where the **NEC** becomes applicable but does not pinpoint the actual service conductors. And that is a critical task because a corollary of that determination is identification of the equipment which is, technically, "service equipment" and hence is subject to all applicable **NEC** rules on such equipment. Any conductors between the "service-point" of a particular installation and the point at which "service conductors" are identified would have to be considered as "feeder" conductors and subject to **NEC** rules on feeders.

NOTE: The above layout poses a conflict between condition (1) and condition (5) of Section 230-201(a).

FIG. 481 "Service" is best identified as the circuit that must be opened to isolate a building from its electric supply.

The wording of Section 230-201 bases the identification of "service conductors" on the location of "step-down transformers." Assuming the rule refers to any "step-down transformers" on the premises (whether the transformers belong to the power company or the property owner), the application of Section 230-201(a) to modern high-voltage systems on multibuilding properties is, at best, extremely difficult in many cases. Exactly which "step-down transformers" is the rule referring to? Today's commercial and industrial systems commonly have step-down transformers both outdoors and indoors. Figure 481 shows a typical case involving both indoor and outdoor step-down transformers. In such a layout, there is no way to identify which transformation—the outdoor or the indoor—produces the secondary conductors that "constitute the service conductors." It is basically consistent with **NEC** concern for service disconnect to take the 4160-V circuit as the service conductors and the 4160-V gear as service equipment. And there are many systems in which two voltage transformations are made outdoors—say from 34 kV to 4160 V and then to 480 V. Which secondary circuit constitutes the service conductors: the 4160-V circuit or the 480-V circuit? The 4160-V level is the secondary for one outdoor transformer and the primary for another. However, when the voltage level is stepped down to 600 V or less outdoors, it is much easier to establish the circuit to the building as the service conductors.

From the commentary and discussions that have been involved in changes in these **NEC** rules, as reported in NFPA literature, it does seem clear that the intent of the rule in Section 230-201 is to identify as "service conductors" those conductors that supply a "building or other structure." And it is probably the intent of the **NEC** that "other structure" should mean any constructed place where people are normally present, as in a building. From the standpoint of safety, it seems that the concern is for a building or structure that somehow encloses personnel and from which they must exit in an emergency situation such as a fire. And service rules are aimed at the ability to cut the building or structure off from incoming electric power. That concept would correspond to the basic purpose of the **NEC** as stated in Section 90-1: to safeguard "persons and [not *or*] property." If that is a reasonable analysis, then outdoor switchgear enclosures (even of the walk-in type) and substations—both open-type and enclosed unit substations—are not "other structures" in the meaning of that phrase as used in Section 230-201. All this is consistent with the **NEC** concern for safety in being able to disconnect a building from all electric supply conductors in case of a fire or other abnormal condition, in which the presence of energized circuits might constitute a hazard. Fast, ready service disconnect is a major **NEC** concept that accounts for the limiting of disconnect means to not more than six switches or CBs—not more than six operations of the hand.

Application of the foregoing analysis to an equipment layout will usually reveal where the **NEC** would designate service conductors in accordance with Section 230-201. But many layouts require the application of **NEC** rules on multibuilding properties [Sections 230-84, 230-90(c), and 230-72]. In Fig. 482, each of the three outdoor loadcenter unit subs supplies two buildings. In such a layout, the transformer supplying each building is outdoors. From Section 230-201(a)(1), the secondary conductors from the transformer clearly are service conductors, and the main secondary fused switch in each case must be the service disconnect. Then the 480-V circuit feeding each building is simply a feeder, and not service conductors. However, for the layout shown, Section 230-84 would require a disconnect means in each building to consist of one to six switches or CBs installed "either inside or outside" the building "nearest the point of entrance" of the feeder to the building. Section 230-90(c) would recognize overcurrent protection in the 480-V substation secondary as suitable for the feeder, so that a nonfusible switch could be used as the disconnect in each building. Note that although those conductors to a building are technically not service conductors, they must be treated as such; and the disconnect means must be "suitable for use as service equipment." That is consistent with **NEC** concern for fast and ready disconnect, regardless of where the supply to the building originates. Even grounding

FIG. 482   The actual "location" of the "service" must be carefully established.

at individual buildings is generally treated the same as at a service entrance—as required by Section 250-24.

As indicated in Fig. 482, ground-fault protection may be required for the secondary main fused switch in each unit sub. But there is no requirement for GFP on the feeder disconnect at each building.

Figure 483 specifies the wiring methods that are acceptable for use as service entrance conductors where it has been established that primary conductors (over 600 V) are the service con-

(a) HIGH–VOLTAGE SERVICE CONDUCTORS FOR LOCATIONS ACCESSIBLE TO OTHER THAN QUALIFIED PERSONS

Conductors rated for the service voltage, installed in rigid metal conduit or IMC   Min. No. 6 conductors

Rigid nonmetallic conduit with at least 2 in concrete envelope

Multiple conductor cable approved for purpose   Min. No. 8

Note: Underground runs may be in conduit or duct or approved cable assemblies and must conform to Sec. 710-3(b)

In cablebus—5 kv to 35 kv
[Article 365]

(h) SERVICE CONDUCTORS OPERATING AT MORE THAN 15 KV

Service ⟶

Voltage over 15 kv between conductors

Conductors must either metal-enclosed switchgear or a code constructed transformer vault

Building wall

(f) POTHEAD ON SERVICE CONDUCTORS

Load conductors to transformer of switchgear from capnut terminals

Conductor insulation protected where conductors emerge from assembly

Supply

Service entrance cable conductors in lead sheath wiped to sleeve on pothead (or this could be conduit with a fitting on pothead)

(i) CONDUCTORS ENCLOSED IN MASONRY ARE CONSIDERED AS INSTALLED OUTSIDE THE BUILDING

Concrete or brick at least 2 in. thick all around conduit or duct

2"

2"

Encased conduit or duct

2"

FIG. 483 High-voltage service conductors may be used in these wiring methods.

ductors or where the secondary conductors are the service conductors and operate at more than 600 V. Cable-tray systems are also acceptable for high-voltage services, provided that the cables used in the tray are identified for use as service entrance conductors. Section 318-2(a) recognizes "multiconductor service-entrance cable" for use in tray, for cables rated up to 600 V. High-voltage (over 600 V) service entrance cables may be used if the cables are "identified"—which, in today's strict usage, virtually means that such cable must be listed by a nationally recognized testing lab (UL, etc.).

High-Voltage Feeders and Branch Circuits

The basic rules on high-voltage feeders and branch circuits require that conductors be suitable for the voltage and conditions under which they are installed. They must be installed as follows:

1. Conductors above ground must be in rigid metal conduit, in intermediate metal conduit (IMC), in rigid nonmetallic conduit (concrete encasement is *not* needed), in cable trays, in cable bus, in other suitable raceways (inspectors should be checked as to the suitability of EMT), or installed as open runs of metal-armored cable suitable for the specific use and purpose. IMC is fully equivalent to rigid metal conduit for indoor and outdoor (including underground) applications for high-voltage circuits. In locations accessible to qualified persons only, open runs of nonmetallic sheathed cable, bare conductors, and busbars may be used.

2. Underground conductors may be installed in raceways identified for the use or in approved direct-burial cable assemblies.

NEC Table 710-3(b) makes clear that underground circuits may be installed in rigid metal conduit, in intermediate metal conduit, or in rigid nonmetallic conduit. Rigid metal conduit or IMC must be buried at least 6 in and does not have to be concrete-encased—but it may be, of course. Direct-burial nonmetallic conduit must be approved (UL-listed and labeled) for use without concrete encasement. If rigid nonmetallic conduit is approved for use only with concrete encasement, at least 2 in of concrete must enclose the conduit. Full data on the use of rigid nonmetallic conduits underground are made available by UL in the *Electrical Construction Materials Directory*—the basic book on product acceptability.

Direct-burial cables must be of a type identified (as by UL) for use directly buried in the ground. Direct-burial cables rated above 2000 V must be of the shielded type, although non-shielded multiconductor cables rated 2001 to 5000 V may be used if they have an outer metallic sheath or armor. But any unshielded cable (i.e., cable without electrostatic shielding on the insulation) must be installed in rigid metal conduit, in IMC, or in rigid nonmetallic conduit encased in not less than 3 in of concrete.

Bare neutrals

As with ungrounded conductors operating at 600 V or less, high-voltage conductors must be insulated for the rated circuit voltage. But the **NEC** does permit a neutral conductor of a solidly grounded wye system to have insulation rated at only 600 V. It also notes that a bare copper neutral may be used for service entrance conductors or for directly buried feeders, and bare copper or copper-clad aluminum may be used for overhead sections of outdoor circuits. But high-voltage feeders that are not service conductors and are installed in raceway indoors or outdoors— above ground or underground—must have at least 600-V insulation on a solidly grounded neutral.

For such high-voltage systems, the phase legs—the ungrounded conductors—must be insulated for the circuit phase voltage. It is interesting, however, that there is no specific **NEC** rule that requires insulation of any circuit to be rated for phase-to-phase voltage. **NEC** rules do not distinguish between phase-to-phase voltage and phase-to-neutral voltage on grounded systems with respect to insulation. Thus, the use of such circuit conductors with insulation rated only for phase-to-neutral voltage would not constitute a violation of any specific rule, and such practice is used on high-voltage systems.

Circuit conductors

Modern high-voltage circuits for buildings include:

- Overhead bare or covered conductors. The conductors are installed with space between them, and they are supported by insulators at the tops of wood poles or metal tower structures.
- Overhead cable assemblies of insulated conductors entwined together, supported on building walls or on poles or metal structures.
- Insulated conductors installed in metal or nonmetallic conduits or ducts run underground, either directly buried in the earth or encased in a concrete envelope under the ground.
- Insulated conductors in conduit run within buildings.
- Multiple-conductor cable assemblies (such as nonmetallic jacketed cables, lead-sheathed cable, or interlocked armor cable), installed in conduit or on cable racks, trays, or other types of supports within buildings.

The metal conduit used with these wiring methods is invariably intermediate metal conduit (IMC) or rigid metal conduit—steel, wrought iron, or aluminum. Nonmetallic conduits are commonly used for high-voltage circuits installed underground. Although direct-burial cables are available and are widely used for economy, the fact that high-voltage circuits are essential arteries in electrical systems recommends the location of key underground circuits in raceways to assure access to the conductors (ready removal from raceway) for elimination of fault conditions.

An other wiring method gaining wide acceptance consists of plastic conduit containing factory-installed conductors, affording a readily used direct-earth-burial cable assembly for underground circuits but still permitting removal of the cable for repair or replacement.

An important consideration in the selection and application of conductors for systems operating at more than 600 V is careful correlation of **NEC** regulations and the data made available by third-party testing laboratories. Underwriters Laboratories and Electrical Testing Laboratories have been deeply involved in listing cables for use in high-voltage systems. The expanded testing and listing of high-voltage conductors and cable assemblies have increased the need for the electrical designer to be particularly thorough and careful in establishing full and effective compliance with all applicable codes and standards.

NEC Article 326, "Medium Voltage Cable," defines Type MV cable in a limited way, and the article itself is relatively brief. However, because there is rapidly growing concern for the greater potential hazards of high-voltage systems as compared with systems up to 600 V, the full significance of the sketchy **NEC** rules must be evaluated.

Although the **NEC** description of Type MV cable is brief, the UL *Green Book* amplifies the explanation of Type MV cable as follows:

> Medium voltage cables are rated 2001 to 35,000 volts.
>
> They are single or multi-conductor, aluminum or copper, with solid extruded dielectric insulation and may have an extruded jacket, metallic covering or combination of both over the single conductors or over the assembled conductors in a multi-conductor power cable.
>
> All insulated conductors 8001 volts and higher have electrostatic shielding. Cables rated 2001 to 8000 volts may be shielded or nonshielded.

Nonshielded cables are intended for use where conditions or maintenance and supervision ensure that only competent individuals service and have access to the installation.

Cables marked MV-75, MV-85 or MV-90 are suitable for use in wet or dry locations at 75°C, 85°C or 90°C, respectively.

Cables which are suitable for use in dry locations only are so marked. Cables marked "oil resistant I" or "oil resistant II" are suitable for exposure to mineral oil at 60°C or 75°C, respectively.

Cables marked "sunlight resistant" may be exposed to the direct rays of the sun.

Cables intended for installation in cable trays in accordance with Article 318 of the National Electrical Code are marked "for CT use" or "for use in cable trays."

Cables with aluminum conductors are marked with the word "aluminum" or the letters "AL."

Cables are marked with their conductor size, voltage rating and insulation level (100 percent or 133 percent).

NOTE: Although the UL data refer to use in cable trays on the condition that the cable is marked for tray use, **NEC** Section 318-2(b) permits both single-conductor and multiconductor Type MV cables to be used generally in trays in industrial occupancies.

Designers should refer to **NEC** Table 310-31 regarding Type MV conductors, and to Section 318-2(b) for information on the use of Type MV cables in trays. Section 310-13 clearly and specifically requires that any insulated conductors for use over 600 V "shall comply" with Tables 310-31 through 310-37. Table 310-31, "Conductor Application and Insulation," makes reference to only one "type" of cable—"medium voltage solid dielectric," which is listed as Type MV-75, Type MV-85, or Type MV-90, depending upon its maximum rated operating temperature. Therefore, the net effect of the "shall comply" phrase in Section 310-13 is to require that any and every circuit conductor operating over 600 V be a listed Type MV cable.

As noted above, the construction of insulated high-voltage conductors must comply with Tables 310-31 through 310-37, which show minimum insulation thicknesses for the various types and voltage ratings of conductors. These tables should be followed carefully in selecting and specifying such conductors.

The UL data quoted above indicate that Type MV cable is marked with a percentage insulation level—either 100 percent or 133 percent. That marking refers to the thickness of the insulation (in mils) on Type MV cables rates from 2001 to 35,000 V. The two insulation levels are defined as follows:

100 Percent Insulation Level. Cables in this category shall be permitted to be applied where the system is provided with relay protection such that ground faults will be cleared as rapidly as possible, but in any case within 1 minute. While these cables are applicable to the great majority of cable installations which are on grounded systems, they shall be permitted to be used also on other systems for which the application of cables is acceptable provided the above clearing requirements are met in completely de-energizing the faulted section.

133 Percent Insulation Level. This insulation level corresponds to that formerly designated for ungrounded systems. Cables in this category shall be permitted to be applied in situations where the clearing time requirements of the 100 percent level category cannot be met, and yet there is adequate assurance that the faulted section will be de-energized in a time not exceeding 1 hour. Also they shall be permitted to be used when additional insulation strength over the 100 percent level category is desirable.

In recent years, cable-insulation failures on circuits up to 35 kV have focused attention on the need for adequate insulation thickness; they have emphasized the desirability of the 133 percent level to ensure the high reliability so essential in modern high-voltage circuits. Value rather than price should be the goal in selecting high-voltage cables.

Insulation and jacketing: The UL data quoted above note that Type MV cable must have a "solid extruded dielectric insulation." Insulated copper and aluminum conductors, for use individually or in cable assemblies, have been made in a wide variety of configurations, with a number of different solid dielectric insulations and jacket materials. Advanced chemical technology is producing a wide range of conductor insulations based on the use of natural rubber, butyl rubber, silicone rubber, ethylene-propylene polymer, polyvinyl chloride (PVC), polyethylene, cross-linked polyethylene, varnished cambric (and other varnished fabrics), and paper. These insulations have varying characteristics with respect to maximum temperature rating, resistance to heat, resistance to moisture, resistance to corona discharge (which can destroy a cable from within), dielectric strength, flexibility at low temperatures, and thickness of insulation required.

The voltage ratings of conductor insulations are usually given as maximum values.

Over recent years, cross-linked polyethylene and ethylene-propylene rubber have gained wide use as insulations. But the opinions and preferences of design engineers vary, and other insulations are often preferred. Certainly, the booming application of high-voltage circuits for industrial and commercial distribution and field experience with various insulations—particularly cable failures—have created wide discussion and controversy about the relative quality and merits of the different insulations. Among designers, cable insulation is a dynamic and turbulent topic. Experience reveals many changes in attitudes and evaluations as new developments alter the characteristics and economics of each type of insulation. Some of the insulations that have been used are described in the next few paragraphs.

Butyl rubber has been widely used as an insulation on high-voltage conductors up to 35 kV. It is inherently resistant to ozone, heat, moisture, and aging. Butyl-rubber–insulated conductors are suitable for use aerially, in conduit, in ducts, or directly in the earth, when properly jacketed. They can operate as high as 90°C, depending upon voltage (for example, at 90°C up to 5 kV, and 85°C up to 15 kV).

Oil-base insulation used on conductors rated up to 35 kV offers 90°C operation and heat, moisture, and ozone resistance. This insulation has higher dielectric strength and better heat-aging properties than butyl insulation. Figure 484 shows the basic makeup of a shielded conductor with oil-base rubber insulation used in conjunction with inner and outer stress-control layers. Copper tape is used as the major electrostatic shielding, and the two layers of nonconductive material provide electrical and physical stress control throughout the lengths of cable runs. The outer stress-control layer is not stripped back at the cable ends, and it permits terminations of this cable assembly without the need for stress-relief cones or any other additional

FIG. 484   Oil-base, rubber-insulated cable has built-in (manufactured) stress control and does not require stress cones.

stress relief. The simplified termination speeds installation and improves reliability at voltages from 9 to 35 kV.

The numbered components in Fig. 484 perform as follows:

1. The *outer jacket* protects inner components against physical and environmental damage.
2. The *fabric tape* reinforces the jacket and enhances slip between the jacket and metal tape when the cable is flexed.
3. The *metal shielding tape* is 5-mil tinned copper.
4. The *semiconducting fabric cloth tape* serves as bedding for the metal shielding tape.
5. The *insulation system* consists of inner and outer layers of nonconducting stress-control material, separated by oil-base rubber insulation.

Ethylene-propylene polymer is another 35-kV, 90°C insulation that can handle 130°C overloads. Like butyl, it provides extremely high corona resistance, and can be three or more times as effective as cross-linked polyethylene or other plastic insulation. It has superior heat resistance (over 20 times the heat life of butyl at 80 to 90°C) and superior low-temperature flexibility and requires less insulation material, making for smaller diameters and lighter weight. Ethylene-propylene insulation is more flexible than butyl and much more flexible than cross-linked polyethylene, offering easy handling and installation, particularly in cold weather. It has better resistance to moisture than butyl and higher dielectric strength. Ethylene-propylene insulation combines the advantages of rubber (high corona resistance and flexibility) and cross-linked polyethylene (high tensile strength and long heat life).

Silicone-rubber–insulated conductors offer continuous operation at 125°C conductor temperature, and up to 200°C for intermittent duty; they may be used where the conductors will be exposed to higher temperatures continuously.

Polyethylene-insulated conductors and *cross-linked polyethylene conductors* have found wide application in underground and above-ground circuits up to 15 kV, and voltage-stabilized polyethylene-insulated conductors are available for use up to 69 kV. Vulcanized polyethylene insulation has been used in primary underground circuits rated from 15 to 115 kV; it exhibits high heat resistance in withstanding 30-s overloads at 250°C. Conventional polyethylene-insulated conductors are rated at 75°C, with lower current ratings than higher-temperature conductors, but they offer effective low-cost application in certain cases.

Cross-linked polyethylene is a compound that combines the physical advantages of rubber insulation (ozone resistance and thermal stability) with the superior mechanical strength and electrical characteristics of polyethylene. It offers 90°C operation, small diameter, low cost, and high resistance to chemicals, moisture, heat, crushing, sunlight, electrical tracking, and ozone. Two typical constructions of cross-linked polyethylene–insulated conductors are shown in Fig. 485.

Varnished-cambric and *varnished-glass* fabric–insulated conductors are used in high-voltage armored-cable assemblies and in single- and three-conductor cable in conduit in dry locations. A lead jacket is installed on cable for use in wet locations.

Paper-insulated, lead-jacketed, high-voltage cable is used for high-capacity distribution systems operating at up to 69 kV, where high performance reliability and economy are vital factors. However, the newer rubber and plastic insulations are more popular for industrial and commercial feeders.

Although some individual conductors for use in raceway or for direct burial in the ground are made without an outer jacketing, most high-voltage conductors do have some type of outside covering—neoprene, asbestos braid, polyethylene, or polyvinyl chloride—to protect the insulation against damage.

Metal cable jacketing for high-voltage multiconductor cables is available in two basic types:

Copper shielding tape

| Conductor of copper or aluminum | Conductive strand shielding | Cross−linked polyethylene insulation | Semi−conducting shielding of nonmetallic tape wrap | Jacket |

Typical high−voltage cable with metal−tape shielding

| Conductor | Extruded strand shielding | Cross−linked polyethylene insulation | Extruded non−metallic conductive shielding | Concentrically wrapped wires-may be a bare grounded neutral or grounded drain wires as part of the electrostatic shielding |

Typical cable with conductive shielding and concentric neutral or drain wires (URD and UD type)

FIG. 485   Two types of cross-linked polyethylene cable construction, using different shielding methods.

a wrapped interlocking metal tape and an impervious, close-fitting corrugated tube. Either jacket may be of steel, aluminum, copper, or bronze. Metal-sheathed cables rated 5 and 15 kV have gained wide popularity for primary circuiting on cable ladder racks (**NEC** Article 318, "Cable Trays") in industrial installations. And in certain applications, plastic jackets are applied over the metal jackets of cable assemblies to provide protection against rust or other corrosion. Lead jacketing can be used on single-conductor and multiconductor assemblies with any type of insulation, such as paper, varnish-cambric, or cross-linked polyethylene, for application to wet or dry locations.

"Type MC metal-clad cable" is covered in **NEC** Article 334 and listed in the UL *Green Book* under the same designation. This is the heavy-duty, industrial-feeder type of armored cable that is similar to but really different from standard BX armored cable. Type MC is rated by UL for use up to 5000 V, although cable for use up to 15,000 V has been available and used for many years. And because no testing lab lists the 15-kV Type MC cable, this is no violation of **NEC** Section 110-2 or OSHA rules—both of which require listing only when a product is of a type that is available in a listed version. Type MC is recognized in two basic armor designs: (1) interlocked metal tape and (2) corrugated tube. Both these assemblies are recognized by UL for above-ground use, although Section 334-3(5) erroneously permits all MC cables to be "direct buried." Section 334-4 prohibits "direct burial in the earth."

It should be noted that the UL description of Type MV cable includes individual single-conductor cable and "multiconductor" cable that may have a "metallic covering." That latter wording suggests that a Type MV cable could conform to the construction of Type MC. Again Section 310-31 literally requires that conductors in a Type MC cable be individual Type MV conductors.

For effective, trouble-free operation and long life, the selection of high-voltage cable insulation must be based on a number of considerations. The operating voltage of the circuit is, of course, the first determinant. Then insulation thickness and voltage rating must be related to either grounded-neutral or ungrounded circuit operation. Manufacturers' specs and applications literature must be thoroughly studied, and all questions or potential problems should be fully resolved by discussion with qualified cable engineers representing the manufacturers. All this discussion and evaluation must take place against the background of sound understanding of **NEC** rules and the recommendations of the IPCEA (Insulated Power Cable Engineers Association). The quality-control capabilities and practices of manufacturers should always be fully considered, along with the manufacturers' reputations and warranties.

Conductor shielding: A basic decision in the selection of high-voltage conductors is whether or not electrostatic insulation shielding is required on the cable. **NEC** Section 310-6 contains rules on shielding that must be observed. Solid-dielectric–insulated conductors operated above 2000 V in permanent installations must have ozone-resistant insulation and must be shielded. The shielding is for the purpose of confining the voltage stresses to the insulation. However, non-shielded insulated conductors listed by a nationally recognized testing laboratory are permitted for use up to 8000 V under the conditions given in Section 310-6.

The net effect of that rule is to require all conductors operating over 2 kV to be shielded, unless the conductor is testing-lab–listed for unshielded operation at voltages above 2 kV and up to 8 kV. Because the 2400-V delta (which is over 2 kV) is the lowest general-purpose, high-voltage circuit in use today, listed conductors must be shielded for such circuits and for circuits with voltages above that—such as 4160/2400-V, 3-phase, 4-wire wye (grounded or ungrounded neutral) circuits—unless the listing specifically recognizes nonshielded use of the cable. It is critically important, therefore, to check not only that a cable is listed by a testing lab but also that its listing data conform to the application.

Both UL and ETL (Electrical Testing Laboratories) list high-voltage (over 600 V) cables. ETL lists an ethylene propylene–insulated cable for use at 5 to 35 kV, with copper drain wires embedded in a semiconductive jacket over the insulation. UL lists shielded polyethylene-insulated conductors up to 35 kV. And, in accordance with **NEC** Table 310-34, UL lists Type RHH insulated conductors (rubber or cross-linked polyethylene insulation) with electrostatic shielding for operation up to 5 kV.

In addition to applicable **NEC**, IPCEA, and UL data on the use of cable shielding, manufacturers' data should be consulted to determine the need for shielding on the various types and constructions of available cables.

Shielding on high-voltage cables protects the conductor assembly against surface discharge or burning (due to corona discharge in ionized air), which can be destructive to the insulation and jacketing. It does this by distributing stress in the insulation and eliminating charging-current drain to intermittent grounds. It also prevents ionization of any tiny air spaces at the surface of the insulation by confining electrical stress to the insulation. When effectively grounded, shielding increases safety by eliminating the shock hazard presented by the external surfaces of unshielded cables. By preventing electrical discharges from cable surfaces to ground, shielding also reduces fire and explosion hazards and minimizes radio interference caused by high-voltage circuits.

Both nonmetallic and metallic materials are used for the electrostatic shielding of cables. As shown in the sketches of typical cable assemblies, semiconductive tapes or extruded coverings of semiconductive materials are combined with metal shielding to perform the shielding function.

For years, high-voltage shielded power cables for indoor distribution circuits rated from 5 to 15 kV were of the type using copper-tape shielding and an outer overall jacket. But in recent years, cables shielded by concentrically wrapped bare wires have also come into widespread use—particularly for underground outdoor systems up to 15 kV. These latter cables are the

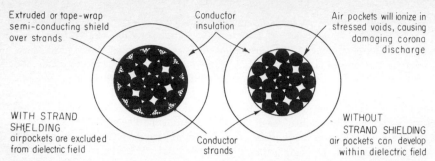

FIG. 486   Strand shielding prevents cable damage due to internal corona discharge.

ones commonly used for underground residential distribution (called "URD"). Such a conductor is shown at the bottom of Fig. 485.

In addition to their use for URD (directly buried with the concentric wire shield serving as the neutral or second conductor of the circuit), concentric-wire-shielded cables are also available for use in indoor power circuits, such as in conduit, with a nonmetallic outer jacket over the concentric wires. Such cable assemblies are commonly called "drain-wire-shielded" cable rather than "concentric-neutral" cable because the bare wires are used only as part of the electrostatic shielding and not also as a neutral. Smaller-gauge wires are used where they serve only for shielding and not as a neutral.

Another form of drain-wire-shielded high-voltage cable incorporates electrostatic shielding by means of drain wires embedded in a semiconducting jacket over the conductor insulation. This type of drain-wire-shielded conductor is designed to be used for high-voltage circuits in conduit or duct for commercial and industrial distribution, as an alternative to tape-shielded cables. For the same conductor size, this embedded drain-wire-shielded cable has a smaller outside diameter and lighter weight than a conventional tape-shielded cable. The drain-wire cable reduces the cost of installation labor, permits reduced bending radii in tight locations and easier pulling, and affords faster terminations (with stress cones) and splices. An extremely important result of the smaller overall cross-sectional area of the drain-wire-shielded cable is the chance to use smaller conduits—with lower materials and labor costs—in filling conduit to 40 percent of its cross-sectional area (based on the actual cross-sectional area of the cable, as covered by Note 4 to the tables in **NEC** Chapter 9).

Another consideration in conductor assemblies is strand shielding. As shown in Fig. 486, a semiconducting material is tape-wrapped or extruded onto the conductor strands to prevent voids between the insulation and the strands, thereby reducing the possibility of corona cutting on the inside of the insulation.

Conductor ampacities

NEC Section 310-15 says that maximum continuous ampacities for copper and aluminum solid-dielectric–insulated conductors rated 2001 to 35,000 V must be as given in Tables 310-39 through 310-54, with accompanying notes. These tables cover single conductors and multiconductor cables in air, in conduit, in underground ducts, and under various specific conditions of use. The tables are based on the "power cable ampacities," with notes, made available by IPCEA. Very careful reference should be made to these **NEC** rules and the accompanying notes, along with the data on ampacities made available by cable manufacturers.

Cable terminations

Terminating and splicing methods for high-voltage cables vary widely because of the many different cable insulations—rubber, plastic, varnished cloth, paper, etc. For each type of insulation and jacketing material, there are certain tapes and other materials which are best suited to the construction of splice joints and terminals.

The metallic shielding (or any other conducting or semiconducting static shielding components on shielded cable) must be stripped back to a safe distance, depending on the circuit voltage, at all terminations of the shielding. At such points, stress reduction must be provided by a method such as the use of potheads, terminators, stress cones, or similar devices. However, if a cable contains built-in stress-relief components, separate stress cones need not be installed at the ends of the cable. Or, heat-shrinkable-tubing terminations may be used with stress-control material that provides the needed relief of electrostatic stress. And metallic shielding components such as tapes, wires, braids, or combinations of such elements and their associated conducting or semiconducting materials must be grounded.

Stress cones: A stress cone (or stress-relief cone) is a field-installed device or a field-assembled buildup of insulating tape and shielding braid which must be located at a terminal of high-voltage shielded cable, whether a pothead is used or not. A stress cone is required to relieve the electrical stress concentration in cable insulation directly under the end of cable shielding. The cable assembly shown in Fig. 484 contains stress-control components that afford the cable sufficient stress relief without the need for stress cones.

At a cable terminal, the shielding must be cut back some distance from the end of the conductor to prevent any arcing over from the hot conductor to the grounded shield. When the shield is cut back, a stress is produced in the insulation. Providing a flare-out of the shield (i.e., extending the shield a short distance in the shape of a cone) relieves the stress, as shown in Fig. 487.

Stress-relief cones provide protection against insulation failure at the terminals of shielded

FIG. 487 How a stress cone functions to protect against damage to insulation at cable ends.

APPLICATION:
Cable shield is cut back about 12 in. Then, using silicone lubricant, the cable insulation surface and the inner bore of the stress cone are lubricated. The stress cone is simply pushed down over the cable end until it bottoms on the cable shield. After cable is prepared, termination takes about 30 seconds.

    REFER TO DIAGRAM:
    1. Cable insulation with shielding cut back
    2. Tight fit between insulation and bore of stress cone
    3. Insulating rubber
    4. Stress relief provided by conductive rubber flaring away from insulation
       along bond between insulating rubber and conductive rubber
    5. Conductive rubber of cone tightly fit to conductive cable shield

FIG. 488   Typical preformed stress cone is readily applied to indoor cable terminations up to 35 kV indoors.

high-voltage cables. Manufacturers of cables and cable terminators provide special preformed stress cones, and kits for preparing cable terminals with stress cones, for cables operating at specified levels of high voltage.

    A stress-cone cable termination can be made by using a pennant-type, preshaped insulation wrap to form the stress cone. Kits are available for this purpose, in various forms, for use with polyethylene, butyl, ethylene-propylene, cross-linked polyethylene, and other insulations. They can be used for indoor terminations on shielded conductors and for outdoor terminations either with or without rainshields. The quick, easy pennant method of stress-cone termination is also suitable for use on concentric-wire-shielded and URD-type cable constructions with conductive nonmetallic jacketing and concentric neutral strands.

A completely preformed stress-cone assembly may be used for simple slip-on connection to cable. A type of slip-on stress-relief cone for use on a concentric neutral cable with a conductive, nonmetallic shield extruded over the cable insulation is shown in Fig. 488. The adapter shown converts the stress cone for use with metallic-tape-shielded cables. The adapter is fitted on the cable, and the stress cone is slipped down over it.

These simple devices are made in many different types for indoor or weather-protected terminations of shielded cables up to 35 kV. The one-piece molded stress cone greatly reduces the time and skill required, as compared to the taped buildup and assembly of a job-fabricated stress cone.

Metallic shielding tape on shielded cable must be grounded, according to **NEC** Sections 710-6 and 300-5(b), which refer to "metallic shielding." The shield on shielded cables must be grounded at one end at least, but it is better to ground the shield at two or more points. Grounding of the shield at all terminals and splices keeps the entire length of the shield at about ground potential for the safest, most effective operation of the cable. Cable with improperly or ineffectively grounded shielding can present more hazards than unshielded cable.

Pothead terminators: When high-voltage cables are terminated outdoors, where they are exposed to the weather, a pothead is commonly used to protect the conductor insulation against moisture or mechanical injury where such conductors emerge from a metal sheath. Figure 489 shows a typical pothead termination on a single-conductor cable.

A pothead is one specific form of stress-reduction means referred to in **NEC** Section 710-6. And, as noted, it is also a means of protecting insulation against moisture and mechanical injury where conductors emerge from a metal sheath. Such protection for metal-sheathed cables (such as lead-covered, paper-insulated cables) is required by Section 710-8. A pothead is a cable terminal which provides sealing to the sheath of the cable for a moistureproof connection between the wires within the cable and those outside:

1. Paper-insulated cables must be terminated in potheads. This requirement extends to such cables operated at under 600 V.

2. Varnished-cambric-insulated cables should be terminated in potheads but may be terminated with taped connections in dry locations.

3. Rubber-insulated cables are commonly terminated in potheads in locations where moisture protection is critical, but they may be terminated without potheads in accordance with manufacturers' instructions.

4. Although many modern high-voltage cables can be terminated without potheads, many engineers consider potheads the best terminations for high-voltage cable.

5. The use of potheads offers a number of advantages:
   a. Seals cable ends against moisture that would damage the insulation.

FIG. 489   Typical single-conductor pothead.

b. Provides a compartment for surrounding the termination with insulating compound to increase the strength of electrical insulation.

c. Seals cable ends against loss of insulating oils.

d. Provides engineered support for connections.

Cable terminators: In addition to the conventional, long-time standard type of pothead, which requires considerable skill and time to install and is filled on the job with insulating compound, there are newer types of high-voltage cable terminators for use outdoors or in wet locations; the latter provide simple slip-on connection to shielded and nonshielded cables, and they are particularly suited for use with the modern plastic and rubber-type insulations that have become so popular for URD and other outdoor and indoor applications up to 35 kV. Requiring no taping, no filling, and no soldering, these so-called "instant terminators" provide sealing, terminating, and grounding of the shield on shielded conductors.

Available in ratings from 15 to 69 kV, these devices are made in a wide range of sizes for single-conductor terminations. They require no special tools, no fillers or other accessories, and no special or critical dimensions for installation, and they greatly simplify high-voltage termination. Figure 490 shows one slip-on terminator. These terminators are made for use on concentric-neutral cables and, with adapters, for use on jacketed metal-tape-shielded cables. They are provided with a full line of terminal fittings, mounting brackets, and other accessories for installation on poles, on other outdoor structures, and in pad-mount and other enclosures.

Another indoor-outdoor high-voltage termination technique is based on the use of heat-shrinkable tubing applied to prepared cable ends. This type of termination is available for use from 5 to 69 kV and is a fast, simple, effective, and time-tested method. The heat-shrinkable material is low in cost and nontracking and can be applied as termination for all types of solid-dielectric or nonmigrating oil-filled paper/lead high-voltage power cables.

The "heat-shrink" method may be used for outdoor terminations as well as for indoor and water- and weather-protected terminations. Without special skills or tools, one person can make the basic termination in minutes on concentric-neutral (URD), drain-wire-shielded, or tape-shielded cable.

Heat-shrink stress-control material and heat-shrink insulation material are also made available in kits for cable-splice assemblies and for insulating bare busbars. The heat-shrink technique has the advantage that each tubing size can be used on a range of cable sizes, and there is no problem in matching cable outside diameter to tubing inside diameter.

Molded elastomer housing is solid lightweight one-piece insulator

Terminal connector

Grounding conductor for bracket grounds shield and neutral strands

Stress cone is part of terminator

Adapter must be added for use with tape-shielded cables

High-voltage (15 kv) cable with concentric wires

FIG. 490 Various outdoor slip-on high-voltage cable terminators are available. This lightweight slip-on terminator contacts the conductive shield, provides a stress cone, and seals the end, all with a one-piece assembly.

Cable splices: Cable and insulation manufacturers make available the materials necessary for splicing and provide these materials in unit kits. Each kit contains the amount of tape, solder, and other material necessary to make one splice of a given cable, at a given voltage.

Various kits are available for splicing unshielded cables, jacketed tape-shielded cables, and concentric wire-shielded cables. Each cable manufacturer provides detailed, illustrated instructions on the most effective technique for splicing and terminating its various cables, of both the shielded and nonshielded types.

The use of splice and terminal kits ensures that the proper materials are being used in a proven method, simplifies specifications for the engineer, and clarifies estimating for the contractor. Typical kits are made for in-line splicing of a shielded cable using copper shield braid and ground leads. Other splice kits provide for T and Y splices.

In addition to the wide variety of field-constructed splices, modern high-voltage distribution systems are making great use of many types of molded-rubber in-line, T, and Y splices, of both permanent and disconnectable types. These devices can be used for concentric-neutral UD cables and, with adapters, for jacketed metal-tape-shielded cables. A permanent straight splice can be obtained for joining one end of cable off a reel to the start of cable off another reel, or for repairing cable that is cut through accidentally by a backhoe or other tool digging into the ground. The T and Y splices are made with similar techniques. Disconnectable splice devices provide watertight plug-and-receptacle assembly for all types of shielded cables and are fully submersible.

Based on molded-rubber techniques, various load-break and non-load-break connectors are made for terminating UD cables at submersible transformers and other apparatus.

Raceways

For high-voltage circuits, the following are requirements and recommendations on the use of the several available raceways:

Rigid-metal (steel, wrought iron, or aluminum) conduit has been the most common raceway for in-building distribution of high-voltage circuits. It will, of course, continue to be used. But the 1978 **NEC** gave equal acceptability to intermediate metal conduit (IMC) for use with high-voltage circuits. Both rigid conduit and IMC are recognized for above-ground and underground use, with or without concrete encasement.

For outdoor runs or feeders between buildings, conductors are generally carried underground in rigid metal or nonmetallic conduit (either directly buried or concrete-encased), in IMC, or as direct-burial cables. The advantages of underground installation are:

1. Accident hazards are reduced to a minimum.

2. Ambient temperatures are lowest, and cable capacities are highest.

3. Cables are not in the way if it is necessary to make changes above ground.

The cost of underground construction is not always justified in cases where there are:

1. Unsuitable soil conditions

2. Destructive chemical ingredients in the soil

3. Imminent major changes in building layout

When high-voltage conductors are installed underground in rigid metal conduit or in IMC, the conduit must be at least 6 in below grade if directly buried, or less than 6 in if beneath 2 in of concrete. When concrete-encased nonmetallic conduit is used "in the ground," it must be installed at the burial depths given in **NEC** Table 710-3(b), with at least 2 in of concrete above the conduit.

All conduit runs should be laid so they will drain throughout the run (cables are always safer in dry ducts). Loops should never be used, and vertical U turns should be avoided as much as possible. Straight runs are always preferable. Where a bend is necessary, its radius should be made as large as possible (and the cable manufacturer should be consulted for the minimum bending radius)—especially when heavy cable is involved. Standard factory ells and field bends of steel conduit are often not of sufficiently large radius for heavy, high-voltage cable because they were designed for smaller wires.

Of the metal conduits used for high-voltage circuits, rigid steel conduit, wrought iron conduit, and rigid aluminum conduit have proven durability, strength, and particular advantages when installed properly. IMC now has **NEC** status comparable to that of rigid metal conduit and is a standard alternative to rigid metal conduit.

Galvanizing on steel conduit has suited it to indoor and outdoor installations, both above and below ground. Aluminum conduit has proved itself highly resistant to many corrosive atmospheres (such as sewage-disposal atmospheres) and can be used both indoors and outdoors. Where aluminum conduit is used in concrete, care should be exercised to avoid corrosion due to soluble chlorides in the concrete by painting the conduit or otherwise coating it with asphalt or other protection.

Wrought-iron conduit has a long record of effective resistance to corrosion, particularly in wet locations and salt atmospheres. This conduit has been used in highly corrosive environments, where extremely high reliability is required. Available as black or galvanized wrought iron, such conduit has been used for bridges, tunnels, piers, highways, and such difficult applications as direct submersion in salt water and the brackish water of rivers.

The nonmetallic conduits include fiber conduit, asbestos-cement conduit, soapstone, rigid polyvinyl chloride conduit, polyethylene conduit, and styrene conduit. Of these, medium-density polyethylene conduit and styrene conduit are not UL-listed. High-density polyethylene conduit and the others are UL-listed. The listed and labeled conduits differ widely in weight, cost, and physical characteristics, but each has certain application advantages.

Of the nonmetallic conduits, rigid polyvinyl chloride (PVC) is the only one recognized by the **NEC** for use above ground. But PVC and all the other UL-listed nonmetallic conduits are suitable for underground use with primary-voltage circuits. Underground PVC conduit may be of the type that is suitable for direct burial or, if marked and so indicated by UL data, may be of the type that requires 2-in-thick concrete encasement. As indicated in UL listings of "Conduit—Rigid Nonmetallic—Underground," Type I fiber conduit may be used for underground circuits only when its entire length is encased in concrete. Type II fiber conduit is recognized for direct earth burial, without concrete.

In using nonmetallic conduits, care must be exercised to ensure temperature compatibility between the conduit and the conductors used in it. For instance, a conduit that has a 75°C temperature rating (at which it might melt and/or deform) must not be used with conductors which have a 90°C temperature rating and which will be loaded so they are operating at their temperature limit. This is covered in Section 347-3(e), which says that rigid nonmetallic conduit may not be used "for conductors whose insulation temperature limitations would exceed those for which the conduit is approved." There is available rigid PVC conduit that is listed by UL and marked to indicate its suitability for use with all 90°C conductors, and that conduit may be used with conductors rated at 90°C.

In many cases where nonmetallic conduit is used to enclose conductors suitable for direct burial in the earth, inspectors and engineering authorities have accepted the use of any type of conduit—PVC, polyethylene, styrene, etc.—without concrete encasement and without considering application of **NEC** rules to the conduit. The reasoning is that because the cables are suitable for direct burial in the earth, the conduit itself is not required at all, and its use is above and beyond **NEC** rules. But temperature considerations are real and are related to effective, long-time operation of an installation.

Cable tray

As a support method for armored-cable high-voltage circuits, cable trays and racks, covered by **NEC** Article 318, have gained wide popularity. Such racks and troughs provide economy and speed in installing high-voltage metal-armor cables. Both aluminum and steel types are available.

The many advantages that cable tray offers to high-voltage distribution have dramatically expanded its application. And **NEC** Section 710-3(a) includes cable tray as a suitable raceway for high-voltage conductors above ground for both service conductors and feeders or other circuits.

For high-voltage applications, **NEC** Section 318-2 recognizes cable tray as a support for wiring methods that may be used without a tray (metal-clad cable, conductors in rigid conduit or in intermediate conduit). **NEC** Section 318-2(b) permits the use of single-conductor Type MV (medium voltage) cables in tray in sizes 250-MCM and larger. The basic rule states that the use of such single-conductor cables in tray is permitted in industrial establishments only, where conditions of maintenance and supervision ensure that only competent individuals will service the installed cable-tray system. And Section 501-4(b), in its last sentence, permits the use of metallically shielded cables over 600 V in cable tray in Class I, Division 2 hazardous locations.

NEC Section 318-5 covers the use of both high-voltage and low-voltage cables in the same cable tray. Although, in general, cables rated over 600 V must not be installed in the same cable tray with cables rated 600 V or less, there are two exceptions to that rule. High-voltage cables and low-voltage cables may be used in the same tray if a solid, noncombustible, fixed barrier is installed in the tray to separate the high-voltage cables from the low-voltage cables. Where the cables are of the metal-clad Type MC, it is not necessary to have a barrier in the cable tray, and MC cables operating above 600 V may be installed in the same tray with MC cables operating below 600 V (Fig. 491). But for cables other than Type MC, a barrier must be used.

Cable tray is both a raceway and a wiring method. Steel or aluminum cable tray may serve

Multiconductor cables rated up to 600 volts may be used in the same tray, even when voltage ratings differ.

Cable rated 600V

Cable rated 300V

Cable tray

Cable rated 5000V

Cable rated 600V

A solid, noncombustible, fixed barrier must be used in tray to separate high-voltage and low-voltage cables with nonmetallic jackets—**but** barrier is not needed in tray if **all** of the cables are Type MC.

FIG. 491 Cable tray may require a barrier between high-voltage cables and 600-V cables.

as an equipment grounding conductor for the circuits in the tray, in much the same way that conduit and EMT may serve as the equipment grounding conductor—a return path for fault current—for the circuit conductors they contain.

Free-air ampacities may be used for single-conductor cables in tray under certain conditions. Table 310-39 shows the free-air ampacity of a 250-MCM cable to be 445 A at 5 kV. In conduit, the ampacity of the same cable is 315 A, from Table 310-43. Figure 492 shows cable tray used to carry single-conductor Type MV cables in an industrial plant [Section 318-2(b)], or single-conductor metal-shielded cables in a Class I, Division 2 hazardous area. Conductors that are 250 MCM and larger may use ampacities of conductors in free air from **NEC** Table 310-39 for copper conductors, or Table 310-40 for aluminum conductors.

Control and Protection

High-voltage distribution systems in the range 2400 to 15,000 V utilize four principal devices for indoor load switching and the short-circuit protection of circuits. These are:

1. Circuit-breaker switchgear
2. Enclosed, fused, load-interrupter switchgear
3. Vacuum switches
4. Oil-filled, fused cutouts

High-voltage power CBs provide load switching, short-circuit protection, electrical operation, adjustable time delays for trip characteristics for selectively coordinated protection schemes, quick reclosing after tripping, and various protective hookups such as differential-relay protection of transformers. There are oil-type, oilless (or air-magnetic), and vacuum-break CBs. The air-magnetic CB is the common type for indoor applications in systems up to and beyond 15 kV. Oil CBs are generally used for indoor and outdoor high-voltage service equipment, where they provide economical disconnect and protection on the primary of a transformer.

Modern high-voltage circuit-breaker equipment meets all the needs of control and protection for electrical systems from the simplest to the most complex and sophisticated. In particular, its capability of use for selectively coordinated protection of services, feeders, and branch circuits is unique. In current ratings up to 3000 A, CB gear has the very high interrupting ratings required for today's high-capacity systems. It is available in "metal-clad" assemblies, in which all live parts are completely enclosed within grounded metal enclosures for maximum safety. For applications exposed to lightning strikes or other transient overcurrents, CB equipment offers quick reclosing after operation. The draw-out construction of the circuit-breaker units provides ease of maintenance and ready testing of breakers. Such gear offers unlimited arrangements of source and load circuits and is suited to a variety of ac and dc control power sources. Accessory devices are available for special functions.

Indoor installations of circuit breakers must consist of metal-enclosed units or fire-resistant, cell-mounted units, except that open-mounted CBs may be used in places accessible to qualified persons only. All CBs must be rated for short-circuit duty at the point of application.

The basic rules on high-voltage circuit-breaker installations are shown in Fig. 493.

Switching for modern high-voltage electri-

Cable installed in a single layer. Space between adjacent conductors is equal to at least one cable diameter.

Uncovered tray

FIG. 492   When installed like this, cables in tray may operate at their free-air ampacity.

Circuit breakers controlling oil-filled transformers *must* either be located outside the transformer vault *or* be capable of being operated from outside the vault.

Oil CBs must be arranged or located so that adjacent readily combustible structures or materials are safeguarded in an approved manner. Adequate space separation, fire-resistant barriers or enclosures, trenches containing sufficient coarse crushed stone, and properly drained oil enclosures such as dikes or basins are recognized as suitable safeguards.

FIG. 493 High-voltage CB must be applied according to these rules.

cal systems can be provided in a number of different equipment arrangements. For any particular installation, the best arrangement depends on several factors: the point of application (either for outside or inside distribution or as service equipment); the voltage; the type of distribution system (radial, loop, selective, or network); conditions such as accessibility and the actual layout of the equipment; job atmosphere; specific use; projected system expansion; and economic considerations.

Types of switches include:

1. Enclosed air-break load-interrupter switchgear with or without power fuses
2. Oil-filled cutouts (fused or unfused)
3. Oil-immersed disconnect switches

Modern load-interrupter switchgear in metal safety enclosures finds wide application in high-voltage distribution systems, in combination with modern power fuses. **NEC** Section 230-208 covers the use of air load-interrupter switches, with fuses, for disconnect and overcurrent protection of high-voltage service entrance conductors. Section 710-21(e) covers the use of fused air load-interrupter switches for high-voltage feeders in distribution systems:

Load-interrupter switches shall be permitted if suitable fuses or circuit breakers are used in conjunction with these devices to interrupt fault currents. Where these devices are used in combination, they shall be so coordinated electrically that they will safely withstand the effects of closing, carrying, or interrupting all possible currents up to the assigned maximum short-circuit rating.

Metal-enclosed fused load interrupters offer a fully effective alternative to the use of power CBs, with substantial economies, in 5- and 15-kV distribution systems for commercial, institutional, and industrial buildings. Typical applications for such switchgear parallel those of power CBs and include the following:

1. In switching centers. Switchgear is set up for the control and protection of individual primary feeders to transformer loadcenters. Here an incoming service bay is switched to bus, from which the feeders are tapped through individual fused load interrupters.
2. In substation primaries. Load-interrupter switchgear is used for transformer switching and protection on the primary sides of substations.
3. In substation secondaries. Here the switchgear is used as a switching center closely coupled to a high-voltage transformer secondary.
4. In service entrances. This is a single-unit application of a switchgear bay for service entrance disconnect and protection in a primary supply line. In many applications, more than one such assembly is used for multiple supply circuits, often in combination with emergency automatic transfer.

Fused load-interrupter switchgear, typically rated up to 1200 A, can match the ratings and performance capabilities of power circuit breakers for a large proportion of the applications in which either might be used. Load-interrupter switchgear does not provide as quick reclosing after fault opening as a CB because fuses have to be replaced. And it does not offer differential-relay protection of transformers against internal faults. For many industrial and commercial systems, these shortcomings may not be considered sufficiently important to warrant the substantially higher cost of CB equipment and its installation. As always, economics greatly influences design decisions.

Fused interrupter switches for high-voltage circuits are available with manual or power operation—including types with spring-powered, over-center mechanisms for manual operation or motor-driven, stored-energy operators. Available in indoor and outdoor housings, assemblies can be equipped with a variety of accessory devices, including key interlocks for coordinating switch operation with remote devices such as transformer secondary breakers, padlocks for securing the switch in the open or closed position, and door interlocks to prevent the door from being opened when the switch is closed.

Vacuum switches

Vacuum switchgear, with its contacts operating in a vacuum "bottle" that is enclosed in a compact cylindrical assembly, has gained wide acceptance for all kinds of high-voltage switching and sectionalizing. Available in 200- and 600-A ratings for use at 15.5, 27, and 38 kV, this switching equipment is suited to full-load interruptions and is rated for 15,000- or 20,000-A short-circuit current under momentary and make-and-latch operations. Basic impulse level (BIL) ratings are 95, 125, and 150 kV.

Vacuum-switch assemblies, with a variety of accessories including stored-energy operators and electric motor operators for remote control, are suited to all indoor and outdoor switching operations (including submersible operation for underground systems). Vacuum-switch assemblies takes up much less space than alternative oil- or air-break switches or breakers and are economical. The units offer fireproof and explosion-proof operation, with virtually maintenance-free life for their rated 5000 load interruptions. Units are available in standard two-way, three-way, and four-way configurations, along with automatic-transfer options. Accessory cur-

FIG. 494 Warning signs and layout techniques can minimize potential dangers.

rent transformers and relays can be used with stored-energy operators to apply vacuum switches for fault-interrupting duty.

Careful layout and application of switching components of all types is important for high-voltage electrical systems because of their involvement with transformers, loop systems, and network hookups—which can produce extremely undesirable feedback conditions. Figure 494 shows the kind of condition that can be extremely hazardous in high-voltage layouts where there is the chance of a secondary-to-primary feedback—such as the intentional one shown to provide standby power to essential circuits in Building 2. Under emergency conditions, the main fused interrupter is opened and the secondary CB for the generator is closed, feeding power to the 480-V switchboard in Building 1 and then feeding through two transformers to supply power to the 480-V circuits in Building 2. This hookup makes the load side of the main interrupter switch live, presenting the hazard of electrocution to any personnel who might go into the switch thinking that it is dead because it is open. A second switch can eliminate this difficulty, if it is applied with interlocks.

NEC Section 710-24 is extremely detailed and includes many specific rules on metal-enclosed power switchgear and industrial control assemblies. Part (o) of that section, on fused interrupter switches, requires a warning sign where the type of backfeed shown above could occur.

Power fuses

Power fuses used in load-interrupter switches are available in current-limiting and non-current-limiting types. The current-limiting types offer reduction of thermal and magnetic stresses on

Typical solid-material (boric acid) power fuse as used in fused load-interrupter switches (fuses could be non-expulsion current-limiting type)

In operation, such fuse normally expels hot gases

Condenser, "muffler" or "snuffer" to prevent hazard to persons or property, attached to bottom of fuse for indoor switchgear applications

Load brk sw.

Supply terminals

Load terminals

Fuses

Floor

FIG. 495 Expulsion-type fuses must be arranged to prevent hazardous conditions.

fault by reducing the energy let-through. They are constructed with a silver-sand internal element, similar to that in 600-V current-limiting fuses. Such fuses generally have higher interrupting ratings at some voltages, but their continuous current-carrying ratings are limited.

Non-current-limiting power fuses are made with two types of operating characteristics: expulsion and nonexpulsion. The expulsion fuse gets its name from the fact that it expels hot gases when it operates. Such fuses should not be used indoors without a "snuffer" or other protector to contain the exhaust, because a hazard is presented by the expelled gases. **NEC** Section 710-21(b)(5) requires that fuses expelling flame in operation must be designed or arranged to prevent hazard to persons or property. As shown in Fig. 495, the boric acid fuse with a condenser for protection against arcing and gas expulsion is a typical nonexpulsion non-current-limiting fuse.

Sections 710-21(b)(6) and (7) cover some very important safeguards in the use of fuses and fuseholders:

(6) Fuseholders shall be designed or installed so that they will be de-energized while replacing a fuse.

Exception: Fuse and fuseholder designed to permit fuse replacement by qualified persons using equipment designed for the purpose without de-energizing the fuseholder.

(7) Metal-enclosed switchgear and substations that utilize high-voltage fuses shall be provided with a gang-operated disconnecting switch. Isolation of the fuses from the circuit shall be provided by either connecting a switch between the source and the fuses or providing roll-out switch and fuse type of construction. The switch shall be of the load-interrupter type, unless mechanically or electrically interlocked with a load-interrupting device arranged to reduce the load to the interrupting capability of the switch.

Extreme care must be taken in evaluating and selecting from the various types of fuses available for high-voltage circuit applications. The American National Standards Institute publishes standards for power fuses. The continuous current ratings of power fuses are given with the letter "E" following the number of continuous amperes—for instance, 65E or 200E or 400E. The E indicates that the fuse has a melting time-current characteristic in accordance with the standard for E-rated fuses:

The melting time-current characteristics of fuse units, refill units, and fuse links for power fuses shall be as follows:

(1) The current-responsive element with ratings 100 amperes or below shall melt in 300 seconds at an rms current within the range of 200 to 240 percent of the continuous current rating of the fuse unit, refill unit, or fuse link.

(2) The current-responsive element with ratings above 100 amperes shall melt in 600 seconds at an rms current within the range of 220 to 264 percent of the continuous current rating of the fuse unit, refill unit, or fuse link.

(3) The melting time-current characteristic of a power fuse at any current higher than the 200 to 240 or 264 percent specified in (1) or (2) above shall be shown by each manufacturer's published time-current curves, since the current-responsive element is a distinctive feature of each manufacturer.

(4) For any given melting time, the maximum steady-state rms current shall not exceed the minimum by more than 20 percent.

The fact that E-rated fuses are given melting times at 200 percent or more of their continuous current ratings explains why **NEC** Sections 230-208 and 240-100 set 300 percent of conductor ampacity as the maximum fuse rating but permit CBs up to 600 percent. In effect, the 300 percent for fuses times 2 (the 200 percent) becomes 600 percent—the same as for CBs.

In coordinating power fuses, care must be exercised to account for ambient-temperature adjustment factors, because time-current curves are based on an ambient of 25°C. Adjustment also must be made for preheating of fuses due to load current to ensure effective coordination of fuses with each other and/or with circuit breakers. Figure 496 shows one manufacturer's curves of adjustment factors for ambient temperature and fuse preloading.

Figure 497 shows the time-current characteristic for an R-rated, current-limiting (silver-sand) fuse designed for 2400- and 4800-V motor applications. Such fuses must be selected to coordinate with the motor-controller overload protection, with the controller clearing overloads up to 10 times the motor current and the fuse taking over for faster opening at higher currents up to the interrupting rating of the fuse. The ampere ratings of R-rated fuses are given in values such as 2R or 12R or 24R. If the number preceding the R is multiplied by 100, the value obtained is the ampere level at which the fuse will blow in 20 s. Thus, the rating designation is not continuous current but is based on the operating characteristics of the R-rated fuse. The continuous-current rating of such a fuse is given by the manufacturer at some value of ambient temperature.

Oil-filled cutouts

In addition to air circuit breakers, oil circuit breakers, and fused load-interrupter switches, another device frequently used for the control of high-voltage circuits is the oil-filled cutout. Compared to breakers and fused switchgear, oil-filled cutouts are inexpensive devices that provide economical switching and, where desired, overload and short-circuit protection for primary-voltage circuits.

The oil-filled cutout is a completely enclosed, single-pole assembly with a fusible or nonfusible element immersed in the oil-filled tank that makes up the major part of the unit, and with two terminals on the outside of the housing. Figure 498 shows the basic construction of a typical

FIG. 496 Power-fuse applications must include adjustments for different ambient temperatures and for preheating of fuses due to load current.

cutout with a variety of available entrance fittings for the terminals (to suit them to various cable and job requirements). The circuit is broken or closed safely and rapidly by the internal switching mechanism. This mechanism is made up of a rotating element that, in the closed position, bridges two internal contacts—each contact connecting to one of the outside terminals. The rotating element is completely insulated from the external case and from the external handle that operates the element. The rotating element may simply be a shorting blade when the cutout is used as an unfused switch. When the cutout is to be used as a fused switching unit, the rotating bridging element is fitted with a fuse. The operation of an oil-filled cutout is controlled at the top end of the shaft extending out through the top of the housing.

As a single-pole switching device, the oil-filled cutout is not polarized; that is, either terminal may be a line or load terminal. This is a result of the symmetrical construction of the switching element, and it suits the device to use in circuit sectionalizing or as a tie device in layouts involving two or more primary supply circuits.

With an oil-filled cutout, the switching of load current or the breaking of fault current is confined within a sturdy metal housing. Operation is made safe and quiet by confining arcs and current rupture forces within the enclosure. This operating characteristic of cutouts especially suits them to use where explosive gases or flammable dusts are present, where complete submersion is possible, where severe atmospheric conditions exist, or where exposure of live electrical parts might be hazardous.

Oil-filled (sometimes called *oil-fuse*) cutouts are made in three sizes, based on continuous current: 100, 200, and 300 A, up to 15 kV. In one line there are three basic types. Pole-type cutouts are equipped with rubber-covered leads from the terminals for use in open wiring. Pothead-type cutouts have a cable lead from one terminal for open wiring, and a sleeve on the other for connecting a lead or rubber-covered cable from an underground circuit. Subway-type cutouts are for underground vaults and manholes, particularly where submersion might occur; they are equipped with a sleeve on each terminal for rubber- or lead-sheathed cable. Figure 499 shows the various available termination techniques for oil-filled cutouts.

FIG. 497 R-rated fuses have time-current curves that coordinate with overload protection for high-voltage motors.

For multiphase circuits, two or three single-phase cutout units can be group-mounted with a gang-operating mechanism for simultaneous operation. For pole mounting, linkage and long handles are available for operating cutouts from the ground.

Because oil-filled cutouts provide load-break capability and overcurrent protection, they may be used for industrial and commercial service equipment, for switching the outdoor lighting of sports playing fields and shopping centers, for transformer loadcenters, for primary-voltage motor circuits, or in vaults and manholes of underground systems.

In the 100- and 200-A ratings, oil-filled cutouts are available in combination with current-limiting power fuses in double-compartment indoor or outdoor enclosures. These fused oil interrupter switches provide moderate load-break and high fault-current interrupting capability in an economical package.

NEC Section 710-21(c) gives detailed application data on the use of "distribution cutouts and fuse links—expulsion type," and Section 710-21(d) covers "oil-filled cutouts."

Oil switches

Oil-immersed disconnect switches are used for load control and for sectionalizing primary-voltage underground distribution systems for large commercial and industrial layouts. Designed for handling high power—such as 400 A up to 34 kV—this type of switch can be located at trans-

Expansion chambers increase short-circuit rating

Single-pole cutout

Mounting bracket

Handle inserts here to rotate top and internal switching mechanisms

Fill oil here

Fuse or solid link used in rotating assembly which bridges contact posts

This is a wiping sleeve for lead cable connection

Line or load | Load or line
Non-polarized terminals

FIG. 498 Oil-filled cutout is a fused or unfused single-pole disconnect device.

former loadcenter primaries or at other strategic points in high-voltage circuits, allowing a wide variety of sectionalizing arrangements to provide alternative feeds for essential load circuits.

Oil-immersed disconnect switches are available for as many as five switch positions and ground positions to ground the feeder or test-ground positions for grounding or testing. Ground positions are used in such switches to connect circuits to ground while they are being worked on, to ensure personnel safety.

Oil switches for load-break applications up to 15 kV are available for either manual or elec-

Application	Cable cover	No. of conduits	Entrance type	Method of sealing
Indoor	Rubber or neoprene	Single	Porcelain	Tape Always tape this connection!
		Multiple	Stud bushing	
Indoor or outdoor	Rubber, lead or neoprene	Single	Stuffing box	Compression fittings
	Polyethylene	Single		Compression fittings and tube seal
	Lead	Single	Wiping sleeve	Solder wipe

FIG. 499 Termination provisions available on oil-filled cutouts suit them for use in a variety of applications and with different cable types.

trically powered switching for all types of circuits. When electrical operation is used, the switch functions as a high-voltage magnetic contactor.

In its regulations on high-voltage load-interrupter switches of all types, the **NEC** requires that suitable fuses or circuit breakers be applied in conjunction with such switches to provide for interruption of fault currents. When these devices are used in combination, they must be coordinated electrically to safely withstand the effects of closing, carrying, or interrupting all possible currents up to the assigned maximum short-circuit rating.

Switches intended only for isolating duty must be interlocked with other devices to prevent opening of the isolating switch under load, or the isolating switch must be provided with an obvious sign warning against opening the switch under load.

Rating of overcurrent protection

NEC Section 240-100 covers the protection of high-voltage feeders, as follows:

1. A high-voltage feeder must have a "short-circuit protective device in each ungrounded conductor," which means a fuse in each conductor or a circuit breaker with three overcurrent relays operated from three current transformers, *or*
2. It may be protected by a circuit breaker equipped with overcurrent relays and current transformers in only two phases and arranged as described in Section 230-208(d)(2) or (d)(3); Section 230-208 covers overcurrent protection for high-voltage service conductors.

The requirements regarding the maximum value of high-voltage overcurrent protection are as follows:

- A *fuse* must be rated in continuous amperes at not more than *three times* the ampacity of the circuit conductor.
- A *circuit breaker* must have a long-time trip element rated not more than *six times* the ampacity of the feeder conductor.

Section 240-101 covers overcurrent protection for high-voltage branch circuits. A high-voltage branch circuit must be protected by a short-circuit protective device in each ungrounded conductor, *or* it may be protected by a circuit breaker with relays in only two phase legs, as described in Section 230-208(d)(2) or (d)(3). The protection must be rated to clear the maximum short-circuit capability of the circuit, but no maximum value of rating or trip setting is given.

Section 230-208 gives very extensive data on overcurrent protection for high-voltage service entrance conductors. Section 710-20 presents general rules on high-voltage overcurrent protection.

Service equipment

Whenever it is established that a circuit operating above 600 V constitutes the service entrance conductors to a premises, the following design requirements relative to the service equipment should be observed to ensure compliance with **NEC** rules:

1. *Warning signs.* High-voltage warning signs must be posted where unauthorized persons might contact live parts.
2. *Disconnecting means.* The CB, fused load-interrupter switch, or alternative overcurrent devices selected in accordance with Sections 230-206 and 230-208 will provide a disconnecting means. The disconnect must simultaneously open all ungrounded conductors and must be rated to close on a fault within the rating of the overcurrent protection. Section 230-205

FIG. 500   Rules on control and protection for high-voltage service must be carefully observed.

requires that the service disconnect be capable of closing, safely and effectively, on a fault equal to or greater than the maximum short-circuit current that is available at the line terminals of the disconnect. Where fuses are used within the disconnect or in conjunction with it, the fuse characteristics may contribute to the fault-closing rating of the disconnect. The idea behind this rule is to ensure that the disconnect switch may be safely closed on a level of fault that can be safely interrupted by the fuse.

3. *Isolating switches*

   a. Where an oil switch, air circuit breaker, or oil circuit breaker is used as the disconnect means for a high-voltage service, an air-break isolating switch must be used between the oil switch or CB and the supply conductors (Fig. 500). An exception to that rule is as follows:

   > Where such equipment is mounted on removable truck panels or metal-enclosed switchgear units, which cannot be opened unless the circuit is disconnected, and which, when removed from the normal operating position, automatically disconnects the circuit breaker or switch from all live parts.

   b. Where the fuses are of the type that can be operated as a disconnecting switch, a set of such fuses may be used as the isolating switch required by Section 230-204(a), provided that the oil switch is a nonautomatic switch and that the set of fuses disconnects the oil switch and all associated service equipment from the service entrance conductors.

   c. Air-break isolating switches must be accessible to qualified attendants only. They must be arranged for grounding connection on the load side.

4. *Overcurrent protection*

   a. Where service equipment is installed in a code vault or consists of metal-enclosed switchgear, overcurrent protection and disconnect means must be provided by one of the following (as shown in Fig. 500):

(1) A nonautomatic oil switch, an oil-filled cutout, an air load interrupter, or other approved switch may be used with suitable fuses—provided that the switching contact assembly has an interrupting rating at least equal to the continuous current rating of the fuses used in the switch unit.

(2) An automatic-trip CB of suitable current-carrying and interrupting capacity may be used.

(3) A switch capable of breaking the no-load current of the transformer may be used with suitable fuses if the switch is interlocked with a single switch or circuit breaker in the transformer secondary, arranged so the primary switch cannot be opened when the secondary device is closed.

b. Where the service equipment is not in a vault or metal enclosure, overcurrent protection and disconnect must be *either* of the following:

(1) An air load-interrupter switch or other switch capable of breaking the rated load, with suitable fuses, on a pole or elevated structure outside the building, provided it can be operated by persons using the building.

(2) An automatic-trip CB of suitable current-carrying and interrupting capacity, with an overcurrent unit in each ungrounded conductor. This CB must be located outside the building, as near as possible to the entrance of the service conductors into the building, and may be on a roof, a pole, a foundation, or other structure.

General design and layout

Figure 501 illustrates the **NEC** rules that cover the installation of high-voltage equipment indoors, in places accessible to unqualified persons.

Installation must be in a locked vault or locked area, or equipment must be enclosed in metal:

Electrical installation in a vault, room, or closet or in an area surrounded by a wall, screen, or fence, access to which is controlled by lock and key or other approved means, shall be considered to be

FIG. 501   High-voltage equipment must be guarded to protect unqualified persons.

accessible to qualified persons only. The type of enclosure used in a given case shall be designed and constructed according to the nature and degree of the hazard(s) associated with the installation.

A wall, screen, or fence less than 8 ft in height shall not be considered as preventing access unless it has other features that provide a degree of isolation equivalent to an 8-ft fence.

Metal-enclosed unit substations, switchgear, transformers, pull boxes, connection boxes, and other associated equipment must be marked with appropriate caution signs. Openings in ventilated, dry-type transformers and other equipment must be designed to prevent inserted objects from contacting energized parts.

Outdoor installations with exposed live parts must not provide access to unqualified persons. Elevation may be used to prevent access [Section 110-34(e)], or equipment may be enclosed by a wall, screen, or fence under lock and key, as described in Section 110-31, quoted above. Outdoor installations that are open to unqualified persons must comply with **NEC** Article 225, "Outside Branch Circuits and Feeders."

Where high-voltage equipment is installed in places accessible only to qualified persons, the rules of **NEC** Sections 110-34, 710-32, and 710-33 apply. In such areas, circuit conductors may be installed in conduit, in duct systems, in metal-clad cable, as bare wire, cable, and busbars, or as nonmetal-sheathed cables or conductors as permitted by Sections 710-3 through 710-6.

The entrances to all buildings, rooms, or enclosures containing live parts or exposed conductors operating in excess of 600 V must be kept locked, except where such entrances are under the observation of a qualified attendant at all times.

Where the voltage exceeds 600 V, permanent and conspicuous warning signs reading "Warning—high voltage, keep out" should be provided. This is required by **NEC** Section 110-34(c). It is a safety measure that alerts unfamiliar or unqualified persons who may for some reason gain access to a locked high-voltage area.

Sufficient space for ready and safe operation and maintenance must be provided around all high-voltage electrical equipment—switchgear, transformers, motor controllers, switches, etc. Where energized parts are exposed at any time:

1. Minimum headroom of 6½ ft must be maintained about any piece of high-voltage equipment (Section 110-32).
2. A space that is a minimum of 3 ft wide must be provided in front of equipment (Section 110-32).
3. The minimum depth of clear working space in front of electric equipment is given in **NEC** Table 110-34(a).
4. The work space must permit at least the 90° opening of doors or hinged panels.

All high-voltage equipment must be installed in conformance with the rules of **NEC** Sections 110-32 and 110-34. A minimum work-space depth is needed for access to deenergized parts of high-voltage switchgear or control enclosures. The Exception after Section 110-34(a)(3) notes that working space is not required behind dead-front equipment when no fuses, switches, other parts, or connections require rear access. But, the rule adds that if rear access is necessary to permit work on "de-energized" parts of the enclosed assembly, the work space must be at least 30 in deep. This prohibits switchgear requiring rear access to be installed too close to a wall behind it, with personnel having to work in cramped quarters to reach taps, splices, and terminations. It applies where only "de-energized" parts are accessible from the back of the equipment. If energized parts are accessible, then Condition 2 of Section 110-34(a) holds, and the depth of the working space must be anywhere from 4 to 10 ft, depending upon the voltage [see **NEC** Table 110-34(a)].

High-voltage switches and circuit breakers must be marked to indicate the·circuit or equipment controlled. Each disconnecting means required for motors and appliances, and each service, feeder, and branch circuit at the point where it originates, must be legibly marked to indicate its purpose unless it is located and arranged so that the purpose is evident. The marking must be sufficiently durable to withstand the environment.

Section 110-30 requires all high-voltage equipment to comply with "all applicable provisions of the preceding sections of this article," which means those **NEC** rules covering equipment rated up to 600 V. Because these rules are particularly aimed at personnel safety, great care must be taken to ensure that any safety provisions made for 600-V equipment are also made for high-voltage equipment. Figure 431 in Section 8 of this handbook shows how to ensure clear exit from work spaces around high-voltage equipment.

High-Voltage Transformers

Transformers for high-voltage electrical systems (above 600 V) include both dry-type and liquid-filled units, selected as individual units for connection to control and protection gear or combined in unit substations. The selection of the best transformer for a modern circuit operating at 4160 V, 13.2 kV, or some other voltage level depends upon a variety of conditions and requirements as related to transformer characteristics.

Dry-type transformers have gained increased acceptance over recent years for services and systems operating up to 15 kV. A wide range of open-ventilated, sealed, and cast-coil transformers are available today. Improvements in insulation systems on modern dry-type transformers have given them BIL ratings of 95 kV and higher. These higher BIL levels give such units vastly improved resistance to lightning-strike damage, and they satisfy electric-utility requirements for transformers with high BIL ratings for service applications in indoor and outdoor installations. Modern insulations also have led to more favorable prices for these dry-type units; and their low maintenance requirements—with no need to test or change insulating oils or askarels—have provided a significant advantage over liquid-filled types. The elimination of the space and expense required to construct an **NEC** transformer vault for indoor use of oil transformers, and elimination of the need for pressure-relief venting and gas absorbers for askarel units, also have promoted dry-type units.

Of the liquid-filled high-voltage transformers recognized by the **NEC**, the oil-filled type—with its high BIL ratings, moderate price, and long, effective record of use—finds continued use outdoors for pad-mount and mat installations. But because these transformers require a vault when used indoors, they are commonly limited to outdoor applications.

Although askarel-filled transformers up to 35 kV have been used for many years for indoor applications without vaults, there has been a steady decline in their use over recent years. Growth in the ratings, characteristics, and availability of dry-type high-voltage transformers—especially the "cast-coil" dry-type transformers, with their excellent capabilities—has accounted for a major part of the reduction in askarel-unit usage. But another factor that has weighed against the use of askarel transformers in very recent years is the environmental objection to the askarel liquid itself.

A major component of any askarel fluid is polychlorinated biphenyl (PCB), a chemical compound designated as a harmful environmental pollutant because it is nonbiodegradable and cannot be readily disposed of. Thus, although the askarels are excellent coolants where freedom from flammability is important, increased pressures for restrictions on the sale, use, and disposal of PCBs have reduced the use of askarel transformers and stimulated a search for a nontoxic, environmentally acceptable substitute. The production of askarel has been stopped because suitable alternative fluids have been accepted by the electrical industry.

Proper handling and disposal of askarel is important if such units are selected. An EPA regulation (No. 311) requires that all PCB spills of 1 lb or more be reported. Failure to report a spill is a criminal offense punishable by a $10,000 fine and/or 1-year imprisonment. A manufacturer of askarel has established a program for disposing of spent or contaminated PCB fluid, using an incinerator that completely destroys the fluid by burning it at over 2000°F.

Non-PCB dielectric coolant fluids are available for use in small and medium-sized power transformers, as a safe alternative to askarels. Transformers using these new "high-fire-point" dielectric coolants have found wide and effective application.

Extensive data from tests on available substitutes show that they provide a high degree of safety. Such fluids do have **NEC** and OSHA recognition. Responsibility for proper clearances with insurance underwriters, government regulating agencies, and local code authorities rests with the user or purchaser of the fluid in new or refilled transformers. Underwriters Laboratories does not test or list liquid-filled equipment. Both UL and Factory Mutual Research Laboratory have been involved in providing a classification system for flammability. The EPA has commented favorably on such high-fire-point fluids.

Use of the new fluid dielectrics requires few physical changes to transformers. However, load ratings on existing units may be reduced about 10 percent because of the difference in fluid viscosity and heat conductivity, compared with askarel. The high-fire-point fluids cost about twice as much as askarel. The purchase price of a new transformer filled with one of these fluids (say, a typical 1000-kVA loadcenter unit) is about 10 to 15 percent more than that of an askarel-filled unit. But the economic aspects vary for different fluids and must be carefully evaluated.

High-fire-point liquid-insulated transformers require no special maintenance procedures. The liquids exhibit good dielectric properties over a wide range of temperatures and voltage stress levels, and they have acceptable arc-quenching capabilities. They have a high degree of thermal stability and a high resistance to thermal oxidation that enable them to maintain their insulating and other functional properties for extended periods of time at high temperatures.

Since silicone liquids will ignite at 750°F (350°C), they and other liquid dielectrics with some degree of flammability are not classed as fire-resistant. However, if the heat source is removed or the fluid temperature drops below 750°F, burning will stop. The silicones and other high-fire-point dielectric fluids are thus self-extinguishing.

Indoor and outdoor use is permitted for transformers that utilize a noncombustible fluid dielectric, which is one that does not have a flash point or fire point and is not flammable in air. As an alternative to askarel-insulated transformers, these transformers offer high BIL ratings and other features of operation similar to high-fire-point dielectric-insulated transformers— without concern for flammability. Such transformers do not, therefore, have the restrictions that are set down in Section 450-23 for the high-fire-point-liquid transformers.

There are no installation restrictions on indoor applications of nonflammable fluid-insulated transformers, except that units rated over 35 kV must be installed in a vault.

Modern high-voltage transformers should, in general, be selected in accordance with the criteria used for 600-V transformers. After a unit is selected on the basis of its electrical rating and suitability for a particular circuit design, mechanical considerations such as housing, mounting, connecting, and ventilating must be fully and carefully factored into the selection task.

Basic rules

Important **NEC** rules on transformers are as follows:

1. Liquid-filled or dry-type transformers used for research, development, or testing are exempt from the normal **NEC** rules that apply to general-purpose transformers used for distribution within buildings and for energy supply to utilization equipment, controls, signals, communication equipment, and the like.

2. Any high-voltage transformer (one operating at more than 600 V) must be readily accessible to qualified personnel for inspection and maintenance. This means that any high-voltage

transformer—liquid-filled or dry-type—must be capable of being reached quickly for maintenance, repair, or inspection without requiring the use of a portable ladder, and personnel must not have to climb over or remove obstacles to reach it. As shown in Fig. 502, a high-voltage transformer may be mounted on a platform or balcony, but there must be fixed stairs or a fixed ladder for access to the transformer. [The **NEC** does, however, permit dry-type transformers rated 600 V or less to be located "in the open on walls, columns, or struc-

NOTE: Installation at far right is O.K. for dry-type transformer rated up to 600 V.

FIG. 502 Every high-voltage transformer must be located where it will always be readily accessible.

tures"—without the need to be readily accessible. A transformer suspended from the ceiling or hung on a wall (in which cases a ladder would be required to reach it) would be acceptable. And a dry-type transformer rated up to 600 V and 50 kVA may be installed in a fire-resistant hollow space of a building, not permanently closed in by structure, provided the transformer is designed so that such installation provides adequate ventilation].

3. Rules for the overcurrent protection of transformers apply to any transformer (dry-type or liquid-filled) rated over 600 V. Protection may be provided either by a protective device of specified rating on the transformer primary or by a combination of protective devices of specified ratings on both the primary and secondary. Figure 503 shows the basic rules for such overcurrent protection.

4. Transformers must be guarded against physical damage, and exposure to shock hazard must be prevented, as shown in Fig. 504.

5. Oil-filled transformers used indoors require transformer vaults, and the high-fire-point liquid-insulated transformer and nonflammable-fluid-insulated transformer offer alternatives to oil-filled transformers without the need for vaults. High-fire-point liquid-insulated transformers and the nonflammable-fluid type may be used indoors or outdoors. However, such transformers, if over 35 kV, must be installed in vaults.

Design layout

Figure 505 shows the basic **NEC** rules regarding the installation of dry-type transformers indoors. A weatherproof enclosure must be used for a dry-type transformer installed outdoors. Essential design considerations are as follows:

Accessibility: Accessibility is an important location feature. As stated above, the **NEC** requires a transformer to be readily accessible to qualified personnel for hookup and subsequent maintenance. Also, the unit should not be dangerously near combustible materials, and it should be kept away from very quiet areas where its audible noise would override ambient noise levels.

On the primary side, either at the transformer or at the supply end of the primary circuit, by fuses rated at not more than 250% of rated primary current

NOTE: Where 250% of primary current does not correspond to a standard fuse rating, the next higher size is permitted.

On the primary side, either at the transformer or at the supply end of the primary circuit, by a circuit breaker rated at not more than 300% of rated primary current

NOTE: Where 300% of primary current does not correspond to a standard CB trip setting, the next lower trip must be used.

On the primary side by a feeder overcurrent device sized from Table 450-3(a)(2), provided that the transformer is equipped with a coordinated thermal overload protection or has a secondary overcurrent device sized from Table 450-3(a)(2)

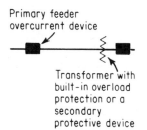

Primary feeder overcurrent device

Transformer with built-in overload protection or a secondary protective device

FIG. 503   Every high-voltage transformer must be protected in one of these ways.

Safety: Safety to personnel is always important, particularly where a transformer is to operate with live components. To protect against accidental contact with such components, units should be isolated in a room or place accessible only to qualified personnel, and live parts should be guarded, as with a railing. When elevation is to be used to safeguard live parts, **NEC** Sections 110-34(e) and 110-17 should be consulted.

Atmosphere: A place that has the driest and cleanest air possible should be selected for the location of open-ventilated units. Exposure to dripping or splashing water or other wet conditions should be avoided. If this is not possible, proper protection should be provided for the units. Outdoor application requires suitable housings.

The ambient temperature in the installation area must be normal, or the transformer may have to be derated. Modern standard, ventilated, dry-type transformers are designed to provide the rated kilovoltampere output at the rated voltage when the maximum ambient temperature of the cooling air is 40°C and the average ambient temperature of the cooling air over any 24-hour period does not exceed 30°C. At higher or lower ambients, the maximum transformer loading can be adjusted as follows:

1. For each degree Celsius by which the average ambient temperature exceeds 30°C, the maximum load on the transformer must be reduced by 1 percent of the kilovoltampere rating.

Exposed live parts must be protected against accidental contact by putting the transformer in a room or place accessible only to qualified personnel **or** by keeping live parts above the floor in accordance with Table 110-34(e)

ISOLATION

ELEVATION – OUTDOORS

Signs or other visible markings must be used on equipment or structure to indicate the operating voltage of exposed live parts.

WARNING FOR HIGH VOLTAGE

FIG. 504   Transformer installation must minimize the chances of exposure to shock.

2. For each degree Celsius by which the average ambient temperature is below 30°C, the maximum load on the transformer may be increased by 0.67 percent of the kilovoltampere rating.

Ventilation: Ventilation at the place of installation is another important requirement. For proper cooling, dry-type transformers depend upon circulation of clean air—free from dust, dirt, and corrosive elements. Filtered air is preferable and may be mandatory in some cases of extreme air pollution. In any case, it can reduce required maintenance.

In small mechanical rooms and similar restricted spaces, ventilation must always be checked. A typical requirement calls for 100 ft³/min of air movement for each kilowatt of transformer loss. The height of the vault, locations of openings, and transformer loading affect the ventilation requirements. It is best to provide ventilation openings in a vault based on full-load operation of the transformer.

Free-standing units mounted on the floor with metal grilles at the bottom must be set up off the floor a sufficient distance to provide the intended ventilation draft up through their housings.

The location must not expose the transformer housing to damage by normal movement of personnel, trucks, or equipment. Ventilation openings should not be exposed to accidental or mischievous poking of sticks or rods into the windings. Adequate protection against the possible entry of small animals or other foreign objects must also be considered.

Handling: Transformers must be handled with care to prevent damage to enclosures and core-coil assemblies. Lifting eyes or openings provided by the manufacturer should be used to lift the units off trucks or storage docks and to raise them into mounting position.

Transformers should always be moved in the upright (top end up) position. Hoisting eyes, installed in the concrete slab above transformer locations in cramped rooms or installation areas, facilitate installation and removal for maintenance or repair.

TRANSFORMERS RATED 112½ KVA OR LESS

Wood wall or any other combustible material

Separation of at least 12 in.

Dry-type transformer rated not over 112½ kva and over 600 volts

Fire resistant, heat-insulating barrier between wall and transformer—no spacing required

Wall made of wood or other combustible material

Dry-type transformer rated not over 112½ kva and not over 600 volts

Transformer comletely enclosed

No separation and no barrier required

TRANSFORMERS RATED OVER 112½ KVA

Completely enclosed and ventilated unit with 80°C rise or higher insulation...

...may be installed in any room or area (need not be fire-resistant)

Clearances from combustible materials in any room or area (not fire-resistant)

12 ft

6 ft   6 ft

Dry-type transformer with 80°C rise or higher insulation but not enclosed and ventilated

Room of fire-resistant construction to house transformer

Dry transformer with less than 80°C rise insulation

Dry transformer rated over 35 KV...

...must be in **NE Code** constructed transformer vault (Part C, Art. 450)

FIG. 505 Construction of a particular dry-type transformer dictates its basic installation requirements.

Liquid-filled transformers

The various high-fire-point liquid dielectrics are less flammable than the mineral oil used in oil-filled transformers but not as fire-resistant as askarel. Because these askarel substitutes will burn to some degree, design and layout must aim at minimizing fire hazards:

1. The high-fire-point dielectrics used in transformers must, first of all, be listed—that is, tested and certified by a testing laboratory or organization and shown in a published listing as suitable for the particular application.

**IF THE TRANSFORMER IS INSTALLED WHERE COMBUSTIBLE
MATERIALS ARE ANYWHERE AROUND IT, AN . . .**

High-fire-point-liquid
transformer

**REQUIRED LIQUID
CONTAINMENT AREA**

Concrete-curbed mat
forms enclosure to
contain liquid in case
of leak

**. . . AUTOMATIC FIRE-
EXTINGUISHING SYSTEM
MUST BE PROVIDED IF
THE TRANSFORMER IS
NOT IN A VAULT.**

FIG. 506   High-fire-point liquid-insulated transformers must satisfy
installation rules.

2. Transformers containing high-fire-point dielectrics may be used without a vault but only
within noncombustible buildings (brick, concrete, etc.) and then only in rooms or areas that
do not contain combustible materials.

3. A liquid-confinement area must be provided around all high-fire-point liquid-insulated
transformers, because tests indicate that these liquids are not completely nonpropagating; that
is, if such a liquid is ignited, flame will be propagated along the liquid. A propagating liquid
must be confined to a given area to confine the flames if it burns. The details are shown in
Fig. 506.

In addition, the location of a transformer with a high-fire-point liquid-insulated dielectric
must be equipped with an automatic fire-extinguishing system, or the transformer must be
installed in a transformer vault constructed according to **NEC** rules, under either of the following
conditions:

1. The transformer is installed in a combustible building, *or*

2. The transformer is installed in a noncombustible building but in an area containing combustible materials.

Figure 507 presents the basic rules for oil-filled transformers. Indoor oil-insulated transformers must be installed in a vault constructed according to **NEC** specs, except as follows:

1. Units rated not over 112.5 kVA may be installed in vaults constructed of reinforced concrete
not less than 4 in thick.

2. Units installed in a detached building used only for providing electric service do not require
an **NEC**-specified vault if no fire hazard is created and the interior is accessible only to qualified persons.

FIG. 507   Installations of oil-filled transformers must minimize hazards due to oil fires.

For oil-filled transformers installed outdoors, the following layout provisions must be observed:

1. Combustible material, combustible buildings and parts of buildings, fire escapes, and door and window openings must be safeguarded from fires originating in oil-insulated transformers that are installed on roofs or attached to or adjacent to a building or combustible material.

2. Space separators, fire-resistant barriers, automatic water-spray systems, and enclosures that confine the oil of a ruptured transformer tank are recognized safeguards. One or more of these safeguards must be applied, according to the degree of hazard involved, in cases where the transformer installation presents a fire hazard.

3. Oil enclosures may consist of fire-resistant dikes, curbed areas or basins, or trenches filled with coarse crushed stone. Oil enclosures must be provided with trapped drains where the exposure and the quantity of oil involved are such that removal of oil is important. Transformers installed on poles or structures or underground must conform to the National Electrical Safety Code.

Transformer vault

The purpose of a transformer vault is to isolate transformers and related apparatus and to confine any fire that might be caused by the failure of any of this apparatus. It is important that

Readily vented to outside without flues or ducts (where practicable)

Common recommendations on thickness of walls and roof: (Not **Code** requirements)

1. 6-in. reinforced concrete, or
2. 8-in. brick, or
3. 12-in. hollow, load-bearing tile or blocks- with inside coating of ¾-in. thick cement or gypsum plaster

Ground

Min. floor thickness: 4-in. concrete when in contact with earth

Approved minimum 3-hr fire-rated door with lock to exclude unqualified persons

Door sill or curb to contain transformer oil-min of 4 in.

FIG. 508 Transformer vaults must satisfy minimum construction requirements.

the door as well as the enclosure be of proper construction and that a substantial lock be provided. Vault requirements are shown in Fig. 508 and include the following details:

1. The walls and roofs of a vault must be constructed of reinforced concrete, brick, load-bearing tile, concrete block, or other fire-resistant materials with adequate strength and a fire resistance of 3 hours according to ASTM Standard E119-75.

2. A vault must have a concrete floor not less than 4 in thick if it is in contact with the earth. When the vault is constructed with space below it, the floor must have adequate structural strength and a minimum fire resistance of 3 hours. A typical 3-hour-rated construction is 6-in-thick reinforced concrete.

3. Building walls and floors that meet the above requirements may serve for the floor, roof, and/or walls of the vault.

An exception to the basic regulations establishing construction standards for fireproof transformer vaults notes that the transformer-vault rating may be reduced where transformers are protected with automatic sprinklers, water spray, or carbon dioxide. The usual construction standards for transformer vaults (such as 6-in-thick reinforced concrete) provide a minimum fire-resistance rating of 3 hours. Where automatic sprinklers, water spray, or carbon dioxide is used, a construction rating of only 1 hour is permitted.

Each doorway (and door) to a vault must have a 3-hour fire rating as defined in the Standard for the Installation of Fire Doors and Windows (NFPA No. 80-1977). Such construction may be desired for doorways leading from the vault to the outdoors, in addition to any doorways leading into adjoining space in the building.

Vent openings must be installed in a vault where ventilation is required by ANSI C57-12.00-1973, "General Requirements for Distribution, Power, and Regulating Transformers," as noted in Section 450-8. Figure 509 shows the openings as regulated by **NEC** Section 450-45. One or

Net area of vent openings, after deducting for screens and grates —

min. 3 sq. in. per kva of transformer capacity or 1 sq. ft. for capacity less than 50 kva

Vault ventilated without flues or ducts

FIG. 509 Adequate ventilation is specified for transformer vaults.

Vault layout to break up direct reflections and
standing waves. Door air lock on outside of room.

FIG. 510   Angle installation of a transformer can significantly reduce its noise
level.

more openings may be used; however, if only a single vent opening is used, it must be in or near
the roof of the vault—and not near the floor.

A transformer vault should have direct ventilating openings (grilles or louvers through the
walls) to an outdoor area. The use of ducts or flues for ventilating is not as effective as direct
outside openings and should be avoided.

Figure 510 shows a transformer vault whose designer sought to reduce the noise level to an
absolute minimum, to prevent noise intrusion into very quiet adjacent areas. That example of
refined vault construction includes the following:

1. Air-conditioning equipment located in a separate room
2. No building ventilating ducts passing through the transformer vaults
3. Conduits sealed to walls with sound-absorbing material rather than "cemented in"
4. Doors of "tight" construction (no air passage around doors)
5. Ventilating ducts lined or baffled where they open onto a quiet street or areaway

section

10

Design
Reference Data

DC CIRCUIT CHARACTERISTICS

Ohm's Law:

$$E = IR \qquad I = \frac{E}{R} \qquad R = \frac{E}{I}$$

E = voltage impressed on circuit (volts)
I = current flowing in circuit (amperes)
R = circuit resistance (ohms)

In direct current circuits, electrical power is equal to the product of the voltage and current:

$$P = EI = I^2 R = \frac{E^2}{R}$$

P = power (watts)
E = voltage (volts)
I = current (amperes)
R = resistance (ohms)

AC CIRCUIT CHARACTERISTICS

The instantaneous values of an alternating current or voltage vary from zero to a maximum value each half cycle. In the practical formulae which follow, the "effective value" of current and voltage is used, defined as follows:

Effective value $= 0.707 \times$ maximum instantaneous value

Impedance:

Impedance is the total opposition to the flow of alternating current. It is a function of resistance, capacitive reactance and inductive reactance. The following formulae relate these circuit properties:

$$X_L = 2\pi f L \qquad X_C = \frac{1}{2\pi f C} \qquad Z = \sqrt{R^2 + (X_L - X_C)^2}$$

X_L = inductive reactance (ohms)
X_C = capacitive reactance (ohms)
Z = impedance (ohms)
f = frequency (cyles per second)
C = capacitance (farads)
L = inductance (henrys)
R = resistance (ohms)
π = 3.14

Ohm's Law for AC Circuits:

$$E = I \times Z \qquad I = \frac{E}{Z} \qquad Z = \frac{E}{I}$$

POWER FACTOR

Power factor of a circuit or system is the ratio of actual power (watts) to apparent power (volt-amperes), and is equal to the cosine of the phase angle of the circuit:

$$PF = \frac{actual\ power}{apparent\ power} = \frac{watts}{volts \times amperes} = \frac{KW}{KVA} = \frac{R}{Z}$$

KW = kilowatts
KVA = kilovolt-amperes = volt-amperes $\times$ 1,000
PF = power factor (expressed as decimal)

SINGLE-PHASE CIRCUITS

$$KVA = \frac{EI}{1,000} = \frac{KW}{PF} \qquad KW = KVA \times PF$$

$$I = \frac{P}{E \times PF} \qquad E = \frac{P}{I \times PF} \qquad PF = \frac{P}{E \times I}$$

$P = E \times I \times PF$
P = power (watts)

THREE-PHASE CIRCUITS, BALANCED STAR OR WYE

$$I_N = O \qquad I = I_p \qquad E = \sqrt{3}\,E_p = 1.73\,E_p$$

$$E_p = \frac{E}{\sqrt{3}} = \frac{E}{1.73} = 0.577 E$$

I_N = current in neutral (amperes)
I = line current per phase (amperes)
I_p = current in each phase winding (amperes)
E = voltage, phase to phase (volts)
E_p = voltage, phase to neutral (volts)

THREE-PHASE CIRCUITS, BALANCED DELTA

$$I = 1.732 \times I_p \qquad I_p = \frac{I}{\sqrt{3}} = 0.577 \times I$$

$E = E_p$

POWER: BALANCED 3-WIRE, 3-PHASE CIRCUIT, DELTA OR WYE

For unity power factor (PF = 1.0):

$$P = 1.732 \times E \times I$$

$$I = \frac{P}{\sqrt{3}\,E} = \frac{0.577P}{E} \qquad E = \frac{P}{\sqrt{3} \times I} = \frac{0.577P}{I}$$

P = total power (watts)

For any load:

$$P = 1.732 \times E \times I \times PF \qquad VA = 1.732 \times E \times I$$

$$E = \frac{P}{PF \times 1.73 \times I} = \frac{0.577 \times P}{PF \times I}$$

$$I = \frac{P}{PF \times 1.73 \times E} = \frac{0.577 \times P}{PF \times E}$$

$$PF = \frac{P}{1.73 \times I \times E} = \frac{0.577 \times P}{I \times E}$$

VA = apparent power (volt-amperes)
P = actual power (watts)
E = line voltage (volts) phase to phase
I = line current (amperes)

POWER LOSS: ANY AC OR DC CIRCUIT

$$P = I^2 R \qquad I = \sqrt{\frac{P}{R}} \qquad R = \frac{P}{I^2}$$

P = power heat loss in circuit (watts)
I = effective current in conductor (amperes)
R = conductor resistance (ohms)

STANDARD LOADS FOR LIGHTING IN COMMERCIAL BUILDINGS

Watts/Sq Ft

Armories
Drill Sheds and Exhibition Halls   5
This does not include lighting circuits for demonstration
booths, special exhibit spaces,
etc.

Art Galleries
a. General   3
b. On Paintings—50 watts per
running foot of usable wall
area.

Auditoriums   4

Automobile Show Rooms   6

Banks
a. Lobby   4
b. Counters—50 watts per running foot including service
for signs and small motor
applications, etc.
c. Offices and Cages   5

Barber Shop and Beauty Parlors   5
This does not include circuits
for special equipment.

Billiards
a. General   3
b. Tables—450 watts per table.

Bowling
a. Alley Runway and Seats   5
b. Pins—300 watts per set of
pins.

Churches
a. Auditoriums   2
b. Sunday School Rooms   5
c. Pulpit or Rostrum   5

Club Rooms
a. Lounge   2
b. Reading Rooms   5
The above two uses are so
often combined that the
higher figure is advisable.
It includes provision for
convenience outlets.

Court Rooms   5

Dance Halls   2
No allowance has been in-
cluded for spectacular lighting, spots, etc.

Drafting Rooms   7

Fire Engine Houses   2

Gymnasiums
a. Main Floor   5
b. Shower Rooms   2
c. Locker Rooms   2
d. Fencing, Boxing, etc.   5
e. Handball, Squash, etc.   5

Halls and Interior Passageways
—20 watts per running foot.

Hospitals
a. Lobby, Reception Room   3
b. Corridors—20 watts per running foot.
c. Wards   3
Including allowance for convenience outlets for local
illumination.

d. Private Rooms   5
Including allowance for convenience outlets for local
illumination.
e. Operating Room   5
f. Operating Tables or Chairs
Major Surgeries—3000 watts
per area.
Minor Surgeries—1500 watts
per area.
This and the above figure
include allowance for direc-
tional control. Special wiring for emergency systems
must also be considered.
g. Laboratories   5

Hotels
a. Lobby   5
Not including provision for
conventions, exhibits.
b. Dining Room   4
c. Kitchen   5
d. Bed Rooms   3
Including allowance for convenience outlets.
e. Corridors—20 watts per running foot.
f. Writing Room   5
Including allowance for convenience outlets.

Library
a. Reading Rooms   6
This includes allowance for
convenience outlets.
b. Stack Room—12 watts per
running foot of facing stacks.

**Motion Picture Houses and
Theatres**
a. Auditoriums   2
b. Foyer   3
c. Lobby   5

Museums
a. General   3
b. Special exhibits—supplementary lighting   5

Office Buildings
a. Private Offices, no close
work   4
b. Private Offices, with close
work   5
c. General Offices, no close
work   4
d. General Offices, with close
work   5
e. File Room, Vault, etc.   3
f. Reception Room   2

Post Office
a. Lobby   3
b. Sorting, Mailing, etc.   5
c. Storage, File Room, etc.   3

Professional Offices
a. Waiting Rooms   3
b. Consultation Rooms   5
c. Operating Offices   7
d. Dental Chairs—600 watts per
chair.

Railway
a. Depot—Waiting Room   3
b. Ticket Offices—General   5
On Counters-50 watts per
running foot.
c. Rest Room, Smoking Room   3
d. Baggage, Checking Office   3
e. Baggage Storage   2
f. Concourse   2
g. Train Platform   2

**Restaurants, Lunch Rooms and
Cafeterias**
a. Dining Areas   3
b. Food Displays—50 watts per
running foot of counter (including service aisle).

Schools
a. Auditoriums   3
If to be used as a study hall
—5 watts per sq. ft.
b. Class and Study Rooms   5
c. Drawing Room   7
d. Laboratories   5
e. Manual Training   5
f. Sewing Room   7
g. Sight Saving Classes   7

Show Cases—25 watts per running foot.

Show Windows
a. *Large Cities
Brightly Lighted District—
700 watts per running foot
of glass.
Secondary Business Loca-
tions—500 watts per running
foot of glass.
Neighborhood Stores—250
watts per running foot of
glass.
b. *Medium Cities
Brightly Lighted District—
500 watts per running foot
of glass.
Neighborhood Stores—250
watts per running foot of
glass frontage.
c. *Small Cities and Towns—
300 watts per running foot
of glass frontage.
d. Lighting to Reduce Daylight
Window Reflections—1000
watts per running foot of
glass.

*Wattages shown are for white light with
incandescent filament lamps. Where color
is to be used, wattages should be doubled.

**Stores, Large Department and
Specialty**
a. Main Floor   6
b. Other Floors   6

Stores in Outlying Districts   5

Wall Cases—25 watts per running foot.

NOTE: Figures based on use of fluorescent equipment for large-area application, incandescent
for local or supplementary lighting.

Watts/Sq Ft

AISLES, STAIRWAYS, PASSAGEWAYS
10 watts per running foot.

ASSEMBLY
a. Rough — 3.0
b. Medium — 4.5
c. Fine — *4.5
d. Extra Fine — *4.5

AUTOMOBILE MANUFACTURING
a. Assembly Line — *4.5
b. Frame Assembly — 3.0
c. Body Assembly — 4.5
d. Body Finishing and Inspecting — *4.5

BAKERIES — 4.0

BOOK BINDING
a. Folding, Assembling, Pasting — 3.0
b. Cutting, Punching, Stitching, Embossing — 4.0

BREWERIES
a. Brew House — 3.0
b. Boiling, Keg Washing, etc. — 3.0
c. Bottling — 4.0

CANDY MAKING — 4.0

CANNING AND PRESERVING — 4.0

CHEMICAL WORKS
a. Hand Furnaces, Stationary Driers and Crystallizers — 2.0
b. Mechanical Driers and Crystallizers, Filtrations, Evaporators, Bleaching — 2.0
c. Tanks for Cooking, Extractors, Percolators, Nitrators, Electrolytic Cells — 3.0

CLAY PRODUCTS AND CEMENTS
a. Grinding, Filter Presses, Kiln Rooms — 2.0
b. Moldings, Pressing, Cleaning, Trimming — 2.0
c. Enameling — 3.0
d. Glazing — 4.0

CLOTH PRODUCTS
a. Cutting, Inspecting, Sewing
(1) Light Goods — 4.5
(2) Dark Goods — *4.5
b. Pressing, Cloth Treating (Oil Cloth, etc.)
(1) Light Goods — 3.0
(2) Dark Goods — 6.0

COAL BREAKING, WASHING, SCREENING — 2.0

DAIRY PRODUCTS — 4.0

ENGRAVING — *4.5

FORGE SHOPS
a. Welding — 2.0

FOUNDRIES
a. Charging Floor, Tumbling, Cleaning, Pouring, Shaking Out * — 2.0
b. Rough Molding and Core Making — 2.0
c. Fine Molding and Core Making — 4.0

GARAGES
a. Storage — 2.0
b. Repair and Washing — *3.0

GLASS WORKS
a. Mixing and Furnace Rooms, Pressing and Lehr Glass Blowing Machines — 3.0
b. Grinding, Cutting Glass to Size, Silvering — 4.5
c. Fine Grinding, Polishing, Beveling, Etching, Inspecting, etc. — *4.5

GLOVE MANUFACTURING
a. Light Goods
(1) Cutting, Pressing, Knitting, Sorting — 4.5
(2) Stitching, Trimming, Inspecting — 4.5
b. Dark Goods
(1) Cutting, Pressing, etc. — *4.5
(2) Stitching, Trimming, etc. — *4.5

HANGARS— AEROPLANE
a. Storage—Live — 2.0
b. Repair Department — *3.0

HAT MANUFACTURING
a. Dyeing, Stiffening, Braiding, Cleaning and Refining
(1) Light — 2.0
(2) Dark — 4.5
b. Forming, Sizing, Pouncing, Flanging, Finishing and Ironing
(1) Light — 3.0
(2) Dark — 6.0
c. Sewing
(1) Light — 4.5
(2) Dark — *4.5

ICE MAKING
a. Engine and Compressor Room — 2.0

INSPECTION
a. Rough — 3.0
b. Medium — 4.5
c. Fine — *4.5
d. Extra Fine — *4.5

JEWELRY AND WATCH MANUFACTURING

LAUNDRIES AND DRY CLEANING

LEATHER MANUFACTURING
a. Vats — 2.0
b. Cleaning, Tanning and Stretching — 2.0
c. Cutting, Fleshing and Stuffing — 3.0
d. Finishing and Scarfing — 4.5

LEATHER WORKING
a. Pressing, Winding and Glazing
(1) Light — 2.0
(2) Dark — 4.5
b. Grading, Matching, Cutting, Scarfing, Sewing
(1) Light — 4.5
(2) Dark — *4.5

LOCKER ROOMS — 2.0

MACHINE SHOPS
a. Rough Bench and Machine Work — 3.0
b. Medium Bench and Machine Work, Ordinary Automatic Machines, Rough Grinding, Medium Buffing and Polishing — 4.5
c. Fine Bench and Machine Work, Fine Automatic Machines, Medium Grinding, Fine Buffing and Polishing — *4.5
d. Extra Fine Bench and Machine Work, Grinding
(1) Fine Work — *4.5

MEAT PACKING
a. Slaughtering — 2.0
b. Cleaning, Cutting, Cooking, Grinding, Canning, Packing — 4.5

MILLING—GRAIN FOODS
a. Cleaning, Grinding and Rolling — 2.0
b. Baking or Roasting — 4.5
c. Flour Grading — 4.5

OFFICES
a. Private and General
(1) No close work — 3.0
(2) Close work — 4.5
b. Drafting Rooms — 7.0

The figures given in this table are average design loads for general lighting. In those cases marked with an asterisk (*), the load values provide only for large area lighting applications. Local lighting must then be provided as an additional load.

The figures given are based on the use of fluorescent and mercury vapor equipment of standard design. Adjustments may be made for use of higher or lower efficiency equipment. For equal lighting intensities from incandescent units, the figures must be at least doubled.

Watts/Sq Ft

PACKING AND BOXING 3.0

PAINT MANUFACTURING 3.0

PAINT SHOPS
a. Dipping, Spraying, Firing, Rubbing, Ordinary Hand Painting and Finishing 3.0
b. Fine Hand Painting and Finishing *3.0
c. Extra Fine Hand Painting and Finishing (Automobile Bodies, Piano Cases, etc.) *3.0

PAPER BOX MANUFACTURING
a. Light 3.0
b. Dark 4.0
c. Storage of Stock 2.0

PAPER MANUFACTURING
a. Beaters, Grinding, Calendering 2.0
b. Finishing, Cutting, Trimming 4.5

PLATING 2.0

POLISHING AND BURNISHING 3.0

POWER PLANTS, ENGINE ROOMS, BOILERS
a. Boilers, Coal and Ash Handling, Storage Battery Rooms 2.0
b. Auxiliary Equipment, Oil Switches and Transformers 2.0
c. Switchboards, Engines, Generators, Blowers, Compressors 3.0

PRINTING INDUSTRIES
a. Matrixing and Casting 2.0
b. Miscellaneous Machines 3.0
c. Presses and Electrotyping 4.5
d. Lithographing *4.5
e. Linotype, Monotype, Typesetting, Imposing Stone, Engraving *4.5
f. Proof Reading *4.5

RECEIVING AND SHIPPING 2.0

RUBBER MANUFACTURING AND PRODUCTS
a. Calendars, Compounding Mills, Fabric Preparation, Stock Cutting, Tubing Machines, Solid Tire Operations, Mechanical Goods Building, Vulcanizing 3.0

b. Bead Building, Pneumatic Tire Building and Finishing, Inner Tube Operation, Mechanical Goods Trimming, Treading 4.5

SHEET METAL WORKS
a. Miscellaneous Machines, Ordinary Bench Work 3.0
b. Punches, Presses, Shears, Stamps, Welders, Spinning, Medium Bench Work 4.5
c. Tin Plate Inspection *4.5

SHOE MANUFACTURING
a. Hand Turning, Miscellaneous Bench and Machine Work 2.0
b. Inspecting and Sorting Raw Material, Cutting and Stitching
(1) Light 4.5
(2) Dark *4.5
c. Lasting and Welting 4.5

SOAP MANUFACTURING
a. Kettle Houses, Cutting, Soap Chip and Powder 3.0
b. Stamping, Wrapping and Packing, Filling and Packing Soap Powder 4.5

STEEL AND IRON MILLS, BAR, SHEET AND WIRE PRODUCTS
a. Soaking Pits and Reheating Furnaces. 2.0
b. Charging and Casting Floors 2.0
c. Muck and Heavy Rolling, Shearing (Rough by Gauge), Pickling and Cleaning 2.0
d. Plate Inspection, Chipping *4.5
e. Automatic Machines, Light and Cold Rolling, Wire Drawing, Shearing (fine by line) 4.5

STONE CRUSHING AND SCREENING
a. Belt Conveyor Tubes, Main Line Shafting Spaces, Chute Rooms, Inside of Bins 2.0
b. Primary Breaker Room, Auxiliary Breakers under Bins 2.0
c. Screens 3.0

STORAGE BATTERY MANUFACTURING
a. Molding of Grids 3.0

STORE AND STOCK ROOMS
a. Rough Bulky Material 2.0

b. Medium or Fine Material requiring care 3.0

STRUCTURAL STEEL FABRICATION 3.0

SUGAR GRADING 5.0

TESTING
a. Rough 3.0
b. Fine 4.5
c. Extra Fine Instruments, Scales, etc. *4.5

TEXTILE MILLS
a. Cotton
(1) Opening and Lapping, Carding, Drawing, Roving, Dyeing 3.0
(2) Spooling, Spinning, Drawing, Warping, Weaving, Quilling, Inspecting, Knitting, Slashing (over beam end) 4.5
b. Silk
(1) Winding, Throwing, Dyeing 4.5
(2) Quilling, Warping, Weaving, Finishing
Light Goods 4.5
Dark Goods 6.0
c. Woolen
(1) Carding, Picking, Washing, Combing 3.0
(2) Twisting, Dyeing 3.0
(3) Drawing-in, Warping
Light Goods 4.5
Dark Goods 6.0
(4) Weaving
Light Goods 4.5
Dark Goods 6.0
(5) Knitting Machines 4.5

TOBACCO PRODUCTS
a. Drying, Stripping, General 3.0
b. Grading and Sorting *4.5

TOILETS AND WASH ROOMS 2.0

UPHOLSTERING
a. Automobile, Coach, Furniture 4.5

WAREHOUSE 2.0

WOODWORKING
a. Rough Sawing and Bench Work 2.0
b. Sizing, Planing, Rough Sanding, Medium Machine and Bench Work, Gluing, Veneering, Cooperage 4.5
c. Fine Bench and Machine Work, Fine Sanding and Finishing 6.0

Use of these figures in computing number and loading of circuits and feeders should always be checked against the conditions and requirements of the particular area. The figures are not substitutes for lighting design.

In any case, need for particular color quality of light or special intensities or control of light must be determined as part of the lighting design, and wattages and circuit requirements must be provided accordingly.

POWER LOAD DATA

Appliance, Device or Machine	Domestic		Commercial — Industrial	
	Watts	Horsepower	Watts	Horsepower
	From-To	From-To	From-To	From-To
LIGHTING EQUIPMENT				
Airport Floods			240-3000	
Airport Landing Lights.........................			to 1 Kw.	
Blue Printing			3-10 Kw.	
Borderlights, Prof. Stage, per ft...............			200-2000	
Borderlights, Schools, per ft..................			100-500	
Cove, Strips, per ft..........................	10-200		20-500	
Exit Signs			40-150	
Floodlights, Outdoor	75-750		200-2000	
Floodlights, Window			100-1000	
Footlights, Prof. Stage, per ft................			100-1000	
Footlights, Schools, per ft...................			100-300	
Infrared lamps, per lamp.....................	100-250		250-1000	
Luminaires (Commercial Lighting Fixtures).....			100 up	
Luminous Tubing (Cold Cathode) per ft.......			5-10	
Operating Rooms (Hospital)...................			1-10 Kw.	
Photostat Machines			1-5 Kw.	
Projectors, Amateur Movie	300-750			
Projectors, Amateur Movie and Sound.........	650-1250			
Projectors, Prof. Movie			1500-3500	¼-1
Projectors, Visual Lecturing			400-1000	
Reflectors, Show Case, per ft................			10-150	
Reflectors, Show Window, per ft.............			100-750	
Spotlights, Ball Room	100-500		100-2000	
Spotlights, Projection Booth			2750-3300	
Spotlights, Stage or Balcony Rail			200-1500	
Spotlights, Show Windows			100-1000	
Spotlights, Statuary (Residence)	25-300			
Sterilamps, per ft............................	10		10	
Vapor, Mercury, High Intensity...............			250-3000	
ELECTRICALLY HEATED EQUIPMENT				
Blankets	50-100			
Broilers	1000-1500			
Casseroles	100-1000			
Cookers, Food	1200			
Dishes, Chafing	160-660			
Driers, Clothes	1500-5000	¼	1-10 Kw.	½
Driers, Hair	200-550		300-1200	
Friers, Deep Fat	1300		1-2 Kw.	
Frying Pans	1200			
Heaters, Air	.4-9 Kw.		.4-9 Kw.	
Heaters, Immersion Type	150-1000		200-2500	
Heaters, Organ Chamber	1-3 Kw.		2-10 Kw.	
Heaters, Permanent Wave Mach			2-4 Kw.	
Heaters, panel per sq. ft.....................	17-50		17-50	
Heaters, Soil per 60-ft. & 120-ft. Lengths......	400-800			
Heaters, Space Elements	1-3 Kw.		1-3 Kw.	
Heat Pumps		1-5		1 up
Heaters, Tank Type Water	1-5 Kw.		1-5 Kw.	
Ironers, Clothes	1200-3300	1/20-¼		

Appliance, Device or Machine	Domestic		Commercial — Industrial	
	Watts	Horsepower	Watts	Horsepower
	From-To	From-To	From-To	From-To
Irons, Flat	500-1200		500-2500	
Irons, Soldering	60-500		200-400	
Irons, Waffle	300-1320		1-3 Kw.	
Lamps, Health and Sun	30-1500		250-1500	
Machines, Vending			100-1000	
Machines, Popcorn			3-6 Kw.	
Makers, Coffee	450-1200		1-6 Kw.	
Ovens, Baking & Roasting	660-4000		5-15 Kw.	
Ovens, Bread & Pie			12-55 Kw.	
Ovens, Industrial Annealing			5-30 Kw.	
Ovens, Industrial Enamelling			10-100 Kw.	
Plates, Hot, Grills, Griddle Tops	480-2000		1-6 Kw.	
Pots, Glue			100-1500	
Ranges	5-17 Kw.		8-25 Kw.	
Roaster	1-2 Kw.			
Sterilizers, Dental & Doctor			1000-3000	
Toasters, Bread & Sandwich	420-1400		2-5.5 Kw.	
Waffle Iron	1000-1500		2-6 Kw.	
Warmers, Bottle	300-600			
Warmers, Cafeteria Food			4-6 Kw.	
Warmers, Plate	110-500		1-2 Kw.	
Warmers, Soup & Seafood			450-1000	
MOTOR-OPERATED EQUIPMENT				
Air Conditioning Systems		1-5		1 up
Automatic Heating Equipment		1/8-1/2		1/4-10
Blowers, Organ		1-3		2-7 1/2
Blowers, Pneumatic Tube Systems				10-30
Blowers, Portable Cleaning				1/6-3
Cash Registers				1/20-1/8
Cleaners, Vacuum Built-in		1/2-5		1-7 1/2
Cleaners, Vacuum Portable		1/30-3/4		
Clippers, Hedge		1/30-1/4		
Compressors, Air (Gasoline Station)				1-5
Compressors, Air (Temp. Regul. System)		1/4-2		1/2-3
Compressors, Refrigeration				5-5000
Conditioners, Air (Room Type)		1/3-1 1/2		1/3-5
Coolers, Water		1/8-1/3		1/8-1/3
Cranes, Travelling Lift				5-30
Cranes, Travelling Bridge				1
Dehumidifiers				1/8-1/3
Dental Chair Units			500-1500	
Disposal Units (Garbage)		1/3-		
Door Openers, Private Garage		1/8-1		
Door Openers, Commercial				1/4-5
Drills, Portable 1/8 to 1/2 in.		1/8-1/4		1/6-1/4
Drills, Portable 5/8 & Larger				1/4-1
Dumbwaiters				1/2-5
Elevators, 1-Ton Freight				3-10
Elevators, 5-Ton Freight				7 1/2-20
Elevators, 10 Pass.				7 1/2-20
Elevators, 25 Pass.				10-50

Appliance, Device or Machine	Domestic		Commercial — Industrial	
	Watts	Horsepower	Watts	Horsepower
	From-To	From-To	From-To	From-To
Escalators				10-40
Extractors Steam Laundry				5-20
Fans, Bracket & Desk	30-100		30-100	
Fans, Ceiling	80-125		80-125	
Fans, Pedestal	125-300		125-300	
Fans, Ventilating 10-in.	35-45		35-45	
Fans, Ventilating 12-24 in.		1/40-¼		1/40-¼
Fans, Ventilating 30-in. & up				⅜-3
Fans, Attic		¼-½		
Freezers, Food		⅛-½		⅛ up
Freezers, Ice Cream		1/20-¼		⅛ up
Grinders, Coffee	⅛-¼			1/6-¾
Grinders, Meat		1/20-¼		¼-1
Hoists, Ash & Cinder				1-5
Hoists, Tramrail 1-ton				1½-3
Hoists, Tramrail 5-ton				6-10
Hoists, Warehouse Loading				1-3
Lathes, Home Shop		⅛-1		
Machines (Floor) Sanding		⅛-1		1-3
Machines (Floor) Terrazzo		⅛-1		1-5
Machines (Floor) Waxing		⅛-1		¼-1
Machines Sewing		1/50-1/20		¼-2
Machines Office, Adding				1/20-1/10
Machines Office, Addressing				⅛-⅓
Machines Office, Billing				1/10-½
Machines Office, Bookkeeping				1/10-½
Machines Office, Computing			1-5 Kw.	
Machines Office, Dictation				-1/30
Machines Office, Record Shaving				-1/20
Machines Office, Sealing & Stamping				1/10-1/6
Machines Office, Typewriters				1/30-1/10
Mangles, Laundry				7½-20
Mixers, Beverage	30-100		30-100	
Mixers, Dough		1/10-¼		5-20
Mixers, Food		1/10-¼		1/6-2
Mowers, Lawn		⅓-½		⅓-1
Pumps, Boiler Feed				1-5
Pumps, Brine				2-20
Pumps, Drinking Water Circ.				½-5
Pumps, Fire Protection				20-150
Pumps, Fuel		⅛-1		¼-3
Pumps, Household Water		1/6-1		
Pumps, Milking Machines				½-2
Pumps, Pool & Illum. Fountain				¼-5
Pumps, Roof Storage Tank				7½-25
Pumps, Sump		⅛-½		¼-3
Pumps, Vacuum				2-5

Appliance, Device or Machine	Domestic		Commercial — Industrial	
	Watts From-To	Horsepower From-To	Watts From-To	Horsepower From-To
Refrigerators		1/8-1/3		1/3-2
Saws, Band (Home Work Shop)		1/4-1/2		
Sprayers, Paint & Insecticide		1/20-3/4		
Stage, Curtain Control Motor				1/4-1
Stage, Orchestra Lift				15-25
Stage, Organ Lift				3-7 1/2
Stokers, Coal		1/6-1/3		1/2-5
Tumblers, Laundry Drying				3-10
Washers, Clothes	300-1800	1/8-1/2		
Washers, Dish	500-1800	1/8-1/2		1/2-5
Washers, Steam Laundry				3-15
MAGNETS, RECTIFIERS, TRANSFORMERS				
Chargers, Battery	600-750		1-15 Kw.	
Diathermy, Therapeutic			250-750	
Electroplating			5-20 Kw.	
Furnaces, Induction			5-500 Kw.	
Magnets, Lifting Metal			4-15 Kw.	
Magnets, Metal Extracting			1 1/2-5 Kw.	
School Laboratory Panel			5-50 Kw.	
Transformers, Bell Ringing	25-50			
Transformers, Signal Systems, Relay	25-50		50-500	
Valves, Gas & Liquids, 1-in. & Less			200-1000	
Valves, Above 1 in.			500-3000	
Welders, Light Duty Spot & Arc			3-20 Kw.	
Welders, Heavy Duty & Arc			20-100 Kw.	
X-Ray — Dental & Doctor			2-25 Kw.	
X-Ray Hospital			10-40 Kw.	
COMMUNICATIONS AND SIGNALLING EQUIPMENT				
Airport Communications			500-5000	
Alarms, Burglar	10-60		100-1000	
Alarms, Fire	10-60		100-1000	
Amplifiers, Radio Distribution			200-1000	
Annunciators, Home 5/8- to 2 1/2-in. Lamps, Each	1.8-2.4			
Annunciators, Large Systems—(110-Volt Lamps, Each)	-10		-10	
Bells, 2 1/2 in. to 4 in.	5-10		5-10	
Bells, Larger			10-30	
Buzzers	5-6		5-6	
Chimes, Door Single and Multiple-Tone	15-25		15-25	
Chimes, Church Systems				1/2-2
Clocks, Master Impulse	1-2		1-2	
Clocks, Secondary Type	.5-1		.5-1	
Gongs, Horns, Howlers	10-30		10-30	
High Fidelity Systems	75-500		100-750	
Radio, Amateur Transmitting	100-2000			
Radio, Home Receivers	50-500			
Sirens, Small & Heavy-duty			65-250	1-7 1/2
Television receivers	300-1000		300-1000	
Whistles, Motor Compressor				1/20-1

MOTOR SELECTION CHART

Basic Motor Types	HP Ratings	Motor Operating Classifications	Application Data
Direct-current motors 115, 230 volts d.c.	Fractional to several hundred hp.	Series-wound	For high starting torque, with speed control depending upon load—cranes, hoists, elevators, electric railway cars, locomotives and trucks. In fractional hp. sizes—sewing machines, vacuum cleaners, electric fans, and hair driers (usually the universal type motor).
	Fractional to several thousand hp.	Shunt-wound	For constant-speed or adjustable-speed, with good speed regulation—drives for milling machines, centrifugal pumps, lathes, conveyors, grinders, blowers, shapers and elevators.
		Compound-wound	Offers constant-speed and variable-speed action, with a definite zero load speed and high starting torque. Used with fly wheels on punch presses, power shears, hoists, rolling mill drives and conveyors.
Universal motor 115, 230 volts d.c. or 1 phase a.c.	Fractional and several hundred hp.	Series-wound or compensated series-wound	A special adaptation of the direct-current series motor. Applications are same as for fractional hp. d.c. series type. In larger sizes, for electric railways.
Single-phase a.c. motors 110-220-440-550 volts	Fractional hp.	Shaded-pole induction motor	For small appliances and devices requiring low starting torque—small fans, motion picture projectors and similar small constant-speed devices.
	Fractional to 5 hp.	Split-phase start induction motor	For moderate starting torque and constant-speed—small machine tools, oil burner motors and small appliances.
		Reactor-start split-phase induction motor	Same as split-phase start induction motor, but requires less starting current.
	Fractional to 15 hp.	Capacitor-start split-phase induction motor	For high starting torque, constant speed and low noise—air conditioning equipment, large fans and commercial refrigeration equipment.
		Capacitor-start and run induction motor	Similar to capacitor-start split-phase induction motor, but in larger sizes, with increased power factor and smoother, quieter operation.
	Fractional to 40 hp.	Repulsion motor	For varying-speed, high starting torque and low starting current—pumps, stokers, conveyors, compressors and similar applications when polyphase current is not available.
		Compensated repulsion motor	Similar to repulsion motor, but with improved power factor and constant- or varying-speed.
		Repulsion-induction motor	Similar to repulsion motor, but with constant- or varying-speed.
		Repulsion-start induction motor	Similar to repulsion motor, with high starting torque and fairly constant speed.
3-phase a.c. motors 208-220-440-550-2200-2300-4000 volts	$\frac{1}{2}$ to 400 hp.	Squirrel-cage induction motor	For high reliability and efficiency at essentially constant-speed, requiring little maintenance. Depending upon construction, classifications are as follows: normal-torque, normal starting current; normal-torque, low starting current; high-torque, low starting current; high-slip; low starting torque, normal starting current; low starting torque, low starting current. For rotary compressors, machine tools, large fans, light conveyors, milling machines, agitators, elevators, hoists, punch presses, centrifugal pumps and blowers.
	$\frac{1}{2}$ to several thousand hp.	Wound-rotor (slip-ring) induction motor	For limited speed control and speed adjustments under fluctuating load, with low starting current—conveyors, fans, lift bridges, cranes, hoists and drives for metal-rolling mills.
	20 to several thousand hp.	Synchronous	For power factor correction and for exact slow-speed drives and maximum efficiency on continuous loads above 75 hp.

Average Sound Levels (Decibels) For Various Occupancies

Occupancy	Decibel Range
Apartments and hotels	35–45
Average factory	70–75
Classrooms and lecture rooms	35–40
Hospitals, auditoriums, churches	35–40
Private offices, conference rooms	40–45
Offices—small	53
—medium (3 to 10 desks)	58
—large	64
—factory	61
Stores—average	45–55
—large (5 or more clerks)	61
Residence—without radio	53
—with radio, conversation	60
Radio, recording, television	25–30
Theaters, music rooms	30–35
Street—average	80

Note: Manufacturers now sound rate dry-type transformers to meet or exceed NEMA Audible Sound Level standards. Select a transformer with a decibel rating lower than the ambient sound level of the area in which it is to be installed.

NEMA AUDIBLE SOUND LEVELS
For dry-type general-purpose specialty transformers
600-volt or less, single or 3-phase

TRANSFORMER RATING (KVA)	AVERAGE SOUND LEVEL (DECIBELS)
0-9	40
10-50	45
51-150	50
151-300	55
301-500	60

NEMA AUDIBLE SOUND LEVELS
Oil-immersed and dry-type self-cooled transformers
15,000-volt insulation class and below

KVA	OIL IMMERSED (DECIBELS)	DRY TYPE (DECIBELS)	
		Ventilated	Sealed
0-300	55 DB	58 DB	57 DB
301-500	56 DB	60 DB	59 DB
501-700	57 DB	62 DB	61 DB
701-1000	58 DB	64 DB	63 DB
1001-1500	60 DB	65 DB	64 DB
1501-2000	61 DB	66 DB	65 DB
2001-3000	63 DB	68 DB	66 DB

Calculation Of Short-Circuit Currents—Point-To-Point Method.

Adequate interrupting capacity and protection of electrical components are two essential aspects required by the ——— National Electrical Code in Sections 110-9, 110-10, 230-98, and 240-1. The first step to assure that system protective devices have the proper interrupting rating and provide component protection is to determine the available short-circuit currents. The application of the point-to-point method permits the determination of available short-circuit currents with a reasonable degree of accuracy at various points for either 3ϕ or 1ϕ electrical distribution systems. This method assumes unlimited primary short-circuit current (infinite bus).

Basic Short-Circuit Calculation Procedure.

Procedure		Formulae
Step 1	Determine transf. full-load amperes from either: a) Name plate b) Table SC-2 c) Formula	3ϕ transf. $\quad I_{FLA} = \dfrac{KVA \times 1000}{E_{L\text{-}L} \times 1.73}$ 1ϕ transf. $\quad I_{FLA} = \dfrac{KVA \times 1000}{E_{L\text{-}L}}$
Step 2	Find transf. multiplier.	— $\quad$ Multiplier $= \dfrac{100}{\text{Transf. }\% Z}$
Step 3	Determine transf. let-thru short-circuit current (Table SC-4 or formula).	— $\quad$ †$I_{SCA} = \text{Transf.}_{FLA} \times \text{multiplier}$
Step 4	Calculate "f" factor.	3ϕ faults $\quad f = \dfrac{1.73 \times L \times I}{C \times E_{L\text{-}L}}$ 1ϕ line-to-line (L-L) faults on 1ϕ, center-tapped transformers $\quad f = \dfrac{2 \times L \times I}{C \times E_{L\text{-}L}}$ 1ϕ line-to-neutral (L-N) faults on 1ϕ, center-tapped transformers $\quad f = \dfrac{2 \times L \times I^*}{C \times E_{L\text{-}N}}$ L = length (feet) of circuit to the fault. C = constant from Table SC-1. For parallel runs, multiply C values by the number of conductors per phase. I = available short-circuit current in amperes at beginning of circuit.
Step 5	Calculate "M" (multiplier) or take from Table SC-3.	$M = \dfrac{1}{1 + f}$
Step 6	Compute the available short-circuit current (symmetrical) at the fault.	I_{SCA} at fault $= I_{SCA}$ at beginning of crk. $\times M$

†**Note 1.** Motor short-circuit contribution, if significant, may be added to the transformer secondary short-circuit current value as determined in Step 3. Proceed with this adjusted figure through Steps 4, 5, and 6. A practical estimate of motor short-circuit contribution is to multiply the total load current in amperes by 4.

Reprinted, with permission, from the "Electrical Protection Handbook" of the Bussman Division, McGraw-Edison Company, St. Louis, Missouri.

Example Of Short-Circuit Calculation.

FAULT #1

Step 1 $\quad I_{FLA} = \dfrac{KVA \times 1000}{E_{L-L} \times 1.73} = \dfrac{300 \times 1000}{208 \times 1.73} = 834A$

Step 2 $\quad Multiplier = \dfrac{100}{Trans. \% Z} = \dfrac{100}{2} = 50$

Step 3 $\quad I_{SCA} = 834 \times 50 = 41,700A$
At Transformer Secondary

Step 4 $\quad f = \dfrac{1.73 \times L \times I}{C \times E_{L-L}} = \dfrac{1.73 \times 20 \times 41,700}{18,100 \times 208} = .383$

Step 5 $\quad M = \dfrac{1}{1 + f} = \dfrac{1}{1 + 0.383} = .723$ (See Table SC-3)

Step 6 $\quad I_{SCA} = 41,700 \times .723 = 30,150A$
Fault #1

FAULT #2

Step 4 $\quad$ Use I_{SCA} @ Fault #1 to calculate

$\qquad f = \dfrac{1.73 \times 20 \times 30,150}{4760 \times 208} = 1.05$

Step 5 $\quad M = \dfrac{1}{1 + f} = \dfrac{1}{1 + 1.05} = 0.49$ (See Table SC-3)

Step 6 $\quad I_{SCA} = 30,150 \times 0.49 = 14,770A$
Fault #2

Note: For simplicity, the motor contribution was not included.

*Note 2. The L-N fault current is higher than the L-L fault current at the secondary terminals of a single-phase center-tapped transformer. The short-circuit current available (I) for this case in Step 4 should be adjusted at the transformer terminals as follows:

At L-N center tapped transformer terminals
I = 1.5 × L-L Short-Circuit Amperes at Transformer Terminals
At some distance from the terminals, depending upon wire size, the L-N fault current is lower than the L-L fault current. The 1.5 multiplier is an approximation and will theoretically vary from 1.33 to 1.67. These figures are based on change in turns ratio between primary and secondary, infinite source available, zero feet from terminals of transformer, and 1.2 × %X and 1.5 × %R for L-N vs. L-L resistance and reactance values. Begin L-N calculations at transformer secondary terminals, then proceed point-to-point.

Calculation Of Short-Circuit Currents At Second Transformer In System.

Use the following procedure to calculate the level of fault current at the secondary of a second, downstream transformer in a system when the level of fault current at the transformer primary is known.

MAIN TRANSFORMER

KNOWN FAULT CURRENT

$I_{P(SCA)}$ $I_{S(SCA)}$

H.V. UTILITY CONNECTION

KNOWN FAULT CURRENT

$I_{P(SCA)}$ $I_{S(SCA)}$

Procedure For Second Transformer In System.

Procedure	Formulae
Step 1 Calculate "f" ($I_{P(SCA)}$, known).	3φ transformer ($I_{P(SCA)}$ and $I_{S(SCA)}$ are 3φ fault values.) $$f = \frac{I_{P(SCA)} \times V_P \times 1.73\ (\%Z)}{100{,}000 \times KVA_{TRANS.}}$$ 1φ transformer ($I_{P(SCA)}$ and $I_{S(SCA)}$ are 1φ fault values; $I_{S(SCA)}$ is L-L.) $$f = \frac{I_{P(SCA)} \times V_P \times (\%Z)}{100{,}000 \times KVA_{TRANS.}}$$
Step 2 Calculate "M" (multiplier) or take from Table SC-3.	— $$M = \frac{1}{1 + f}$$
Step 3 Calculate short-circuit current at secondary of transformer. (See Note 1 under "Basic Procedure")	— $$I_{S(SCA)} = \frac{V_P}{V_S} \times M \times I_{P(SCA)}$$

††$I_{P(SCA)}$ = Available fault current at transformer primary

$I_{S(SCA)}$ = Available fault current at transformer secondary

V_P = Primary voltage L-L.

V_S = Secondary voltage L-L.

KVA_{TRANS} = KVA rating of transformer.

%Z = Percent empedance of transformer.

Note — To calculate fault level at the end of a conductor run, follow Steps 4, 5, and 6 of Basic Procedure.

Table SC-1A. "C" Values For Conductors and *Busway.

AWG Or MCM	Copper Three Single Conductors Magnetic Duct 600V And 5KV Nonshielded	Copper Three Single Conductors Magnetic Duct 5KV Shielded And 15KV	Copper Three Single Conductors Nonmagnetic Duct 600V And 5KV Nonshielded	Copper Three Single Conductors Nonmagnetic Duct 5KV Shielded And 15KV	Copper Three Conductor Cable Magnetic Duct 600V And 5KV Nonshielded	Copper Three Conductor Cable Nonmagnetic Duct 600V And 5KV Nonshielded	Aluminum-Three Single Conductors Or Three Conductor Cables Magnetic 600V And 5KV Nonshielded	Aluminum-Three Single Conductors Or Three Conductor Cables Nonmagnetic Duct 600V And 5 Nonshielded
12	617	—	—	—	—	—	—	—
10	982	—	—	—	—	—	—	—
8	1230	1230	1230	1230	1230	1230	1180	1180
6	1940	1940	1950	1940	1950	1950	1870	1870
4	3060	3040	3080	3070	3080	3090	2360	2360
3	3860	3830	3880	3870	3880	3900	2960	2970
2	4760	4670	4830	4780	4830	4850	3720	3750
1	5880	5750	6020	5920	6020	6100	4670	4690
1/0	7190	6990	7460	7250	7410	7580	5800	5880
2/0	8700	8260	9090	8770	9090	9350	7190	7300
3/0	10400	9900	11500	10700	11100	11900	8850	9170
4/0	12300	10800	13400	12600	13400	14000	10300	10600
250	13500	12500	14900	14000	14900	15800	11900	12400
300	14800	13600	16700	15500	16700	17900	13500	14200
350	16200	14700	18700	17000	18600	20300	14800	15800
400	16500	15200	19200	17900	19500	21100	—	—
450	17300	15900	20400	18800	20700	22700	—	—
500	18100	16500	21500	19700	21900	24000	17200	18700
600	18900	17200	22700	20900	23300	25700	18900	21000
700	—	—	—	—	—	—	20500	23100
750	20200	18300	24700	22500	25600	28200	21500	24300
1000	—	—	—	—	—	—	23600	27600

*Note—See next page for Busway

Table SC-1B. "C" Values For Busway.

Ampacity	Plug-In Busway		Feeder Busway		High Imped. Busway
	Copper	Aluminum	Copper	Aluminum	Copper
225	28700	23000	18700	12000	—
400	38900	34700	23900	21300	—
600	41000	38300	36500	31300	—
800	46100	57500	49300	44100	—
1000	69400	89300	62900	56200	15600
1200	94300	97100	76900	69900	16100
1350	119000	104200	90100	84000	17500
1600	129900	120500	101000	90900	19200
2000	142900	135100	134200	125000	20400
2500	143800	156300	180500	166700	21700
3000	144900	175400	204100	188700	23800
4000	—	—	277800	256400	—

Table SC-2A. Three-Phase Transformer—Full-Load Current Rating (In Amperes).

Voltage (Line-To-Line)	Transformer KVA Rating								
	150	167	225	300	500	750	1000	1500	2000
208	417	464	625	834	1388	2080	2776	4164	5552
220	394	439	592	788	1315	1970	2630	3940	5260
240	362	402	542	722	1203	1804	2406	3609	4812
440	197	219	296	394	657	985	1315	1970	2630
460	189	209	284	378	630	945	1260	1890	2520
480	181	201	271	361	601	902	1203	1804	2406
600	144	161	216	289	481	722	962	1444	1924

Table SC-2B. Single-Phase Transformer—Full-Load Current Rating (In Amperes).

Voltage	Transformer KVA Rating									
	25	50	75	100	150	167	200	250	333	500
115/230	109	217	326	435	652	726	870	1087	1448	2174
120/240	104	208	313	416	625	696	833	1042	1388	2083
230/460	54	109	163	217	326	363	435	544	724	1087
240/480	52	104	156	208	313	348	416	521	694	1042

Table SC-3. "M" (Multiplier).*

f	M	f	M
0.01	0.99	1.50	0.40
0.02	0.98	1.75	0.36
0.03	0.97	2.00	0.33
0.04	0.96	2.50	0.29
0.05	0.95	3.00	0.25
0.06	0.94	3.50	0.22
0.07	0.93	4.00	0.20
0.08	0.93	5.00	0.17
0.09	0.92	6.00	0.14
0.10	0.91	7.00	0.13
0.15	0.87	8.00	0.11
0.20	0.83	9.00	0.10
0.25	0.80	10.00	0.09
0.30	0.77	15.00	0.06
0.35	0.74	20.00	0.05
0.40	0.71	30.00	0.03
0.50	0.67	40.00	0.02
0.60	0.63	50.00	0.02
0.70	0.59	60.00	0.02
0.80	0.55	70.00	0.01
0.90	0.53	80.00	0.01
1.00	0.50	90.00	0.01
1.20	0.45	100.00	0.01

* $M = \dfrac{1}{1 + F}$

Table SC-4. **Short-Circuit Currents Available from Various Size Transformers.**

Voltage And Phase	KVA	Full Load Amps	% Impedance	†Short Circuit Amps
	25	104	1.6	10,300
	37½	156	1.6	15,280
120/240	50	209	1.7	19,050
1 ph.	75	313	1.6	29,540
	100	417	1.6	38,540
	167	695	1.8	54,900
	150	417	2.0	20,850
	225	625	2.0	31,250
	300	834	2.0	41,700
120/208	500	1388	2.0	69,400
3 ph.	750	2080	5.0	41,600
	1000	2776	5.0	55,520
	1500	4164	5.0	83,280
	2000	5552	5.0	111,040
	2500	6950	5.0	139,000
	150	181	2.0	9,050
	225	271	2.0	13,550
	300	361	2.0	18,050
277/480	500	601	2.0	30,059
3 ph.	750	902	5.0	18,040
	1000	1203	5.0	24,060
	1500	1804	5.0	36,060
	2000	2406	5.0	48,120
	2500	3007	5.0	60,140

†Three-phase short-circuit currents based on "infinite" primary. Single-phase short-circuit currents on 100,000 KVA primary.

Estimated Short-circuit at End of Low-voltage Feeder

(See Figs. 1 through 30)

Power-system maximum estimated short-circuit currents, as functions of distance along feeder conductors fed from standard three-phase radial secondary unit substations, can be read directly in rms symmetrical amperes from a series of curves, Fig. 1 through 30). The one-line diagram shows the typical radial circuit investigated.

The conditions on which the curves are based were as follows:

1. The fault was a bolted three-phase short circuit.

2. The primary three-phase short-circuit duty was 500 MVA (60 cycles) for all curves. A typical supply-system X/R at the low-voltage bus was used in calculating the curves for each case.

3. Motor contributions through the bus to the point of short circuit were included in the calculations on the basis of 100-percent contribution for the 240-, 480-, and 600-volt systems and 50-percent contributions for the 208-volt systems.

4. The feeder-conductor impedance values used in the calculations are indicated for various conductor sizes.

These curves can also be used to select feeder conductor sizes and lengths needed to reduce short-circuit duties to desired smaller values. Note that conductors thus selected must be further checked to assure adequate load and short-circuit capabilities and acceptable voltage drop.

Coordinated ratings are based on two protective devices operating in series with all short-circuit current flowing through the upstream device. If any current bypasses the upstream device (such as motor contribution fed in on load side of upstream device) a fully rated system, not coordinated rated system, should be used.

Typical circuit investigated to show effect on short-circuit duty as point of fault is moved away from the low-voltage bus along the feeder conductors.

Feeder Impedance Values Used in Investigation

Feeder Conductor Size/Phase	Resistance (R) Ohms/Phase/1000 Ft	60-cycle Inductive Reactance (X) Ohms/Phase/1000 Ft
#4	0.3114	0.0492
#1/0	.1231	.057
250 MCM	.0534	.0428
2-500 MCM	.0144	.0201
4-750 MCM	.0053	.0099

Reprinted, with permission, from "Short-circuit Current Calculations," publication EESG II-AP-1, of the Apparatus Distribution Sales Division, General Electric Company.

Fig. 1   Transf: 150 kVA, 208 V, 2.0% Z

Fig. 2   Transf: 150 kVA, 208 V, 4.5% Z

Fig. 3 Transf: 150 kVA, 240 V, 2.0% Z

Fig. 4 Transf: 150 kVA, 240 V, 4.5% Z

Fig. 5 Transf: 225 kVA, 208 V, 2.0% Z

Fig. 6 Transf: 225 kVA, 208 V, 4.5% Z

Fig. 7 Transf: 225 kVA, 240 V, 2.0% Z

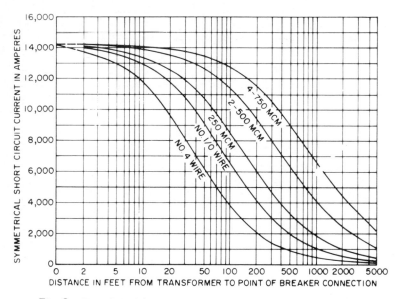

Fig. 8 Transf: 225 kVA, 240 V, 4.5% Z

Fig. 9 Transf: 300 kVA, 208 V, 4.5% Z

Fig. 10 Transf: 300 kVA, 240 V, 4.5% Z

Fig. 11   Transf: 300 kVA, 480 V, 4.5% Z

Fig. 12   Transf: 300 kVA, 600 V, 4.5% Z

Fig. 13   Transf: 500 kVA, 208 V, 4.5% Z

Fig. 14   Transf: 500 kVA, 240 V, 4.5% Z

Fig. 15 Transf: 500 kVA, 480 V, 4.5% Z

Fig. 16 Transf: 500 kVA, 600 V, 4.5% Z

Fig. 17   Transf: 750 kVA, 208 V, 5.75% Z

Fig. 18   Transf: 750 kVA, 240 V, 5.75% Z

Fig. 19   Transf: 750 kVA, 480 V, 5.75% Z

Fig. 20   Transf: 750 kVA, 600 V, 5.75% Z

Fig. 21   Transf: 1000 kVA, 208 V, 5.75% Z

Fig. 22   Transf: 1000 kVA, 240 V, 5.75% Z

10.29

Fig. 23 Transf: 1000 kVA, 480 V, 5.75% Z

Fig. 24 Transf: 1000 kVA, 600 V, 5.75% Z

Fig. 25 Transf: 1500 kVA, 208 V, 5.75% Z

Fig. 26 Transf: 1500 kVA, 240 V, 5.75% Z

Fig. 27   Transf: 1500 kVA, 480 V, 5.75% Z

Fig. 28   Transf: 1500 kVA, 600 V, 5.75% Z

Fig. 29   Transf: 2000 kVA, 480 V, 5.75% Z

Fig. 30   Transf: 2000 kVA, 600 V, 5.75% Z

Index

1